Fire Retardancy of Polymers

New Applications of Mineral Fillers

Fire Retardancy of Polymers

New Applications of Mineral Fillers

Edited by

Michel Le Bras
Ecole Nationale Supérieure de Chimie de Lille, France

Charles A. Wilkie
Department of Chemistry, Marquette University, USA

Serge Bourbigot
P.E.R.F., Ecole Nationale Supérieure de Chimie de Lille, France

Co-editors

Sophie Duquesne
PERF, Ecole Nationale Supérieure de Chimie de Lille, France

Charafeddine Jama
PERF, Ecole Nationale Supérieure de Chimie de Lille, France

RS•C
advancing the chemical sciences

Most of the papers have been presented at the 9th European Meeting on Fire Retardancy and Protection of Materials (FRPM'03) organized jointly by the Ecole Nationale Supérieure de Chimie de Lille, the Ecole Nationale Supérieure des Arts et Industries Textiles de Roubaix and the Université des Sciences et Technologies de Lille, at the University of Lille (France) on the 15-17th September 2003.

ISBN 0-85404-582-1

A catalogue record for this book is available from the British Library

Published by The Royal Society of Chemistry,
Thomas Graham House, Science Park, Milton Road,
Cambridge CB4 0WF, UK

Registered Charity Number 207890

For further information see our web site at www.rsc.org

Typeset by the Charlesworth Group, Wakefield, UK
Printed by Athenaeum Press Ltd, Gateshead, Tyne and Wear, UK

Preface

The utilization of polymeric materials continues to increase each year. Scientists and engineers are now able to develop new materials that meet specific needs. In ancient times, warriors wore metal to protect themselves in battle; now soldiers wear synthetics, and they are quite possibly safer than the early warriors were. In the home, plastic has replaced many items that were once metal. In the transportation field, the weight savings achieved by replacing a metal with a plastic is a potent driving force.

All of these changes provide opportunities for scientists, because plastic materials are inherently flammable, unlike the metal items that they have replaced. For many years, halogens led the list of fire retardant elements; it did not work in every instance but one could frequently find a halogen-containing compound that would work. Following the change in policies in Europe in the 1990s, there has been a renaissance of activity in diverse areas, including micro- and nanocomposites.

The early history of fire retardant systems begins with the painting of wood fortifications with vinegar in 360 BC. In the 1600s, a combination of clay and gypsum was used to fire retard canvas. The first patent was granted in 1735 in England for fire retardancy of textiles using alum, borax and vitriol (zinc, copper or iron sulfate). In 1820 Gay Lussac suggested the use of a mixture of ammonium phosphate, ammonium chloride and borax for textiles.

Modern fire retardants include the use of compounds of halogen, phosphorus, boron, nitrogen, aluminium, magnesium, sulfur and others. At this moment in time the halogens and phosphorus compounds appear to find the most use, but this is certain to change. The use of halogen is diminishing in Europe, while this process has begun in the United States, it is not yet complete. Fire retardants have been applied to all types of materials, ranging from fabrics to hard plastics. There is no universally accepted material that can be used with all, or even with several, polymers. One must discover the appropriate system for each polymer. As the use of halogen has declined, the need for other materials has become evident.

In 1997, at the sixth European Meeting on the Fire retardancy of Polymeric Materials, several the papers were focused on the topic of intumescence and a book on this topic was produced. At the ninth meeting, held in 2003, the focus was on mineral additives, especially those that form either micro- or nano-composites. Typical additives that may give a microcomposite are alumina trihydrate, $Al(OH)_3$, and magnesium hydroxide, $Mg(OH)_2$, both of which decompose endothermically and release water. These materials remove a good deal of the heat evolved in a degradation and thus can prevent further degradation. To be effective, they must be used at very high loadings, which lead in some instances to the loss of mechanical properties of interest.

Nanocomposites are formed when a small amount of an organically-modified aluminosilicate clay is added to a polymer. The presence of only a small amount of clay can give a significant reduction in the peak heat release rate. In addition to clays, nanocomposites have been prepared using polyhedral oligosilsesquioxanes, POSS, graphites and carbon nanotubes. Clay systems are the most well-developed, followed by POSS; little work has been performed using graphite–polymer and nanotube–polymer nanocomposites.

The difference between the microcomposite and the nanocomposite is the dispersion of the material in the polymer. In a nanocomposite, the clay, or the nano-filler/additive, is well dispersed throughout the polymer. The typical clay consists of particles with a high aspect ratio, their length is much longer than their width. Dispersion of the filler in the nanometer scale generally gives interesting insulation properties to a polymer, the fire retardancy being generally poor.

Recent work deals with the association of micro-sized additives with other additives and/or fillers to decrease needed loadings and obtain synergistic effects resulting from the association of nanosized additives with other additives to reach optimized fire retardancy performance. Typical studies are the object of chapters of this book.

It gives me great pleasure to acknowledge the contributions of the organizing committee for this, the ninth European Meeting on Fire Retardancy of Polymeric Materials, and especially Michel Le Bras, who took on the very important function of arranging this Meeting.

Charles A. Wilkie

The latest European Meeting on Fire Retardancy and Protection of materials (FRPM'03) was held in Lille, France in September 2003. A large number of the 106 presented contributions dealt with the use of halogen-free additive systems and, more precisely, with the importance of mineral additives used alone or in association with synergistic agents (33 presentations). Two types of mineral additives were clearly presented, depending on their initial size or on their distribution in the fire retarded polymeric materials.

This book presents most of the original reviews and work presented at FRPM'03, to which the Editors have added two invited reviews and four invited papers.

The various chapters have been written by experts in fire retardancy of polymers using microsized additives, nanosized additives or hybrid nanocomposite materials. In all, twenty Academic Research Teams and nine Institutes or Industrial Groups have contributed to this book and proposed different chemical or physical approaches for the modes of protection developed by mineral additives and fillers and their eventual economical applications. A comparatively short last section (4 papers) deals with the toxicity of some of these additives and of the products resulting from the degradation of the mineral additives/polymer formulations.

I should like to express my gratitude to my co-editors, every co-author and to the numerous experts who have agreed to review these chapters.

Michel Le Bras

Acknowledgements

For their invaluable help with refereeing the papers included in this volume the editors would like to thank:

Dr P. Anna (Budapest University of Technology and Economics, Hungary),
Professor G. Camino (Politecnico di Torino, Alessandria, Italy),
Professor P. Degée (University of Mons-Hainaut, Belgium),
Professor E. Devaux (ENSAIT, Roubaix, France),
Dr X. Flambard (ENSAIT, Roubaix, France),
Dr P. Georlette (Dead Sea Bromide, Beer Sheva, Israel),
Professor R. Hull (Bolton Institute, U.K.),
Dr E. Kicko-Walczak (Instytut Chemii Przemyslowej, Warsaw, Poland)
Dr S. Levchik (Akzo Nobel Functional Chemicals, New York, U.S.A.),
Dr R. Lyons, (F.A.A., Atlantic City International Airport, NJ, U.S.A.),
Professor Gy. Marosi (Budapest University of Technology and Economics, Hungary),
Professor G. Nelson (Florida Institute of Technology, Melbourne, U.S.A.),
Dr M. Nyden (BFRL, NIST, Gaithersburg, U.S.A.),
Professor E. Pearce (Polytechnic University of New York, U.S.A.),
Dr B. Schartel (B.A.M., Berlin, Germany),
Dr K.K. Shen, (Borax/Luzenac America, Denver, U.S.A.),
Professor W.H. Starnes (William and Mary College, Williamsburg (VA), U.S.A),
Dr J. Troitzsch (Fire Protection Service, Wiesbaden, Germany),
Professor E. Weil (Polytechnic University of New York, U.S.A.)

Contents

Effect of the Addition of Mineral Fillers and Additives on the Toxicity of Fire Effluents from Polymers

Abbreviations

Polymers and Products

ABS	acrylonitrile-butadiene-styrene copolymer
BPO	benzyl peroxide
BSil	polyboroxosiloxane elastomers
EP	epoxy resin
EVA	copolymer ethylene/vinyl acetate
HDI	hexamethylene diisocyanate
HIPS	high impact polystyrene
IPP	intumescent polypropylene
LLDPE, LDPE	linear, low density polyethylene
MDI	methylenediphenyl diisocyanate oligomer
PAE	poly(acrylic ester)
PAN	poly(acrylonitrile)
PA-6	polyamide-6
PA11	polyamide 11
PA-6,6	polyamide-6.6
PBT	poly(butylene terephthalate)
PC	polycarbonate
PE	polyethylene
"PEMUBEL"	EVA/SBS/PS blend
PEO	poly(ethylene oxide)
PET	poly(ethylene terephthalate)
PLGO	polymer-layered graphite oxide
PLS	polymer-layered silicates
PMMA	poly(methyl methacrylate)
PO	propylene oxide
PP	polypropylene
PP/MA, PPgMA	polypropylene graft maleic anhydride
PPO	poly(phenylene oxide)
PS	polystyrene
PS-b-PEO	poly(styrene-ethylene oxide) block copolymer
PU	polyurethane
PVA	poly(vinyl alcohol)
PVC	poly(vinyl chloride)
PVDC	poly(vinylidene chloride)
SAN	styrene-acrylonitrile copolymer
SBS	styrene-butadiene-styrene triblock polymer
TDI	toluene diisocyanate

THF	tetrahydrofuran
TPU	thermoplastic PU
VB, VB16	styryldimethylhexadecylammonium chloride
3P	system of methyldiphenyl isocyanate oligomer and water glass

Additives

APP	ammonium polyphosphate
ATH	alumina trihydrate (aluminium hydroxide)
BD	butanediol
Bentone SD-1	organophilic montmorillonite
BEO	brominated epoxy resin
Cl	clay
Cloisite	organically modified montmorillonite
C20A	modified montmorillonite (cloisite 20A)
DPDPO	decabromodiphenyl oxide
DPO	diphenyl oxide group
DP-POSS	dodecaphenyl-POSS
E. Cat	exhaust FCC catalyst
ENC	expandable nanocomposite
FBZB, FB290, FB415, FB500	zinc borates
FCC	fluid-bed catalytic cracking
FQ-POSS	poly(vinylsilsesquioxane)
FR-245	tris(tribromophenyl) cyanurate
FR-1808	brominated trimethylphenyl indan
F-3020	tribromophenol end-capped brominated epoxy
G5	glass fibre
IPDI	isophorone diisocyanate
KC8	potassium graphite
LDHs	layered double hydroxides
M	adhesion promoter in intumescent PP
MB	magnesium tetraborate
MEG	milled and sifted ECat
melabis	bis(2,6,7-trioxa-1-phosphabicyclo[2.2.2]octane-4-methanol) phosphate
MH	magnesium hydroxide
MMT	modified montmorillonite
MWNT	multi-walled carbon nanotube
NH	melamine phosphate
NW	dipentaerythritol/melamine phosphate
Oib-POSS	octaisobutyl POSS
OM	organophilic montmorillonite
PER	pentaerythritol
POL	polyol; pentaerythritol

POSS	polyhedral oligomeric silsesquioxanes
POTM	polyoxytetramethylene glycol
PPFBS	perfluorobutane sulfonic acid potassium salts
PPh	polyhydroxyphenol
PPOL	phosphorylated polyol
P_r	red phosphorus
PTFMS	trifluoromethane sulfonic acid potassium salts
Red-P	red phosphorus
T	talc
THPC-urea	tetrakis(hydroxymethyl)phosphonium chloride urea precondensate
TMDS	tetramethyldisiloxane
ZB, Zn B	zinc borate
ZHS (Flamtard H)	zinc hydroxystannate
10A	dimethylbenzylhydrogenated tallow

General Considerations on the Use of Fillers and Nanocomposites

CHAPTER 1

An Introduction to the Use of Fillers and Nanocomposites in Fire Retardancy

CHARLES A. WILKIE

Department of Chemistry, Marquette University, PO Box 1881, Milwaukee, WI 53201, U.S.A. (charles.wilkie@marquette.edu)

1.1 Introduction

This chapter is to serve as an introduction to the very broad topic of the use of fillers, both well-dispersed and less well-dispersed, in polymers. When the filler is well-dispersed, a nanocomposite results in which a layered material has been separated into its constituent layers and these can either maintain the registry between the layers, an intercalated system, or this registry may be lost, an exfoliated system. When a well-dispersed system is obtained, loadings of 3 to 5% are sufficient to cause a large increase in mechanical properties and a significant reduction in the rate of peak heat release. Conversely, if the layers are not well-separated, or if there are no layers that can be separated, the filler is not well-dispersed and a simple filled system is obtained; typical loadings of 60% or more are required to confer fire retardancy in such systems and this invariably has an adverse effect on both strength and toughness of the composite, which can be ameliorated by judicious use of surface treatments.

1.2 Characterization of Fire Retardancy of Polymers

The evaluation of fire retardancy is carried out by a variety of techniques, most of which do not correlate well with other test protocols. The three most common methods that are used are the oxygen index, the UL-94 test, and cone calorimetry. Oxygen index is an evaluation of the ease of extinction of a fire, how rapidly does the flame chemistry lead to extinction. The measurement consists of determining the minimum concentration of oxygen in a nitrogen–oxygen mixture that will sustain combustion. The more the value of the oxygen index is above the percentage of oxygen in the air, the better the system is considered to be. This does not mean that a material with a high oxygen index will not burn, the test measures the ease of extinction of the fire. The UL-94 test measures the ease of ignition; in this test a sample is ignited and the time for self extinguishment is determined. The results of this test permit a ranking of the material. The cone calorimeter measures a third parameter, the rate at which heat is released in a fire. In many cases, this is considered to be the most definitive test, but it still does not necessarily correlate with the other tests. From a cone calorimetry experiment, one can obtain the mass loss rate, the total heat released, the quantity of smoke that is produced and the amount of carbon monoxide and carbon dioxide that are evolved.

1.3 Fire Retardant Fillers for Polymers

The major materials that are used as fire retardant fillers for polymers are alumina trihydrate, ATH, ($Al_2O_3{\cdot}3H_2O$) and magnesium hydroxide, MH, ($Mg(OH)_2$).[1,2] There are various forms for both of these materials, both naturally occurring and synthetic, and the reader is referred to references 1 and 2 for information on these forms. These two materials account for more than 50% by weight of the world-wide sales of fire retardants; as much as 400 kt annum^{-1} is currently used. Most of this is low cost ATH that is used in thermosetting resins. The use of ATH is limited to those polymers processed below about 200°C while MH is stable above 300°C and thus can be used in polymers that must be processed at higher temperatures. Their effectiveness comes from the fact that they both decompose endothermically and consume a large amount of heat, while also liberating water, which can dilute any volatiles and thus decrease the possibility of fire. For ATH, decomposition begins near 300°C and consumes 1270 joules per gram of ATH; for MH, decomposition begins at somewhat higher temperature, near 400°C, and consumes 1244 joules per gram of MH. There is some tendency for MH to catalyze the degradation of some polymers; in unsaturated polyester resins it can act as a chain extender, affecting resin rheology. A major use of both ATH and MH is in low smoke, halogen-free wire and cable applications, where there is significant commercial activity.[3]

With some polymers, the resin and the additive might interact, and so one must be aware of these possibilities as these will influence the mode of action.[4] With polypropylene, 60% loading of MH gives an oxygen index of 26, while with polyamide-6, the same loading gives an oxygen index of almost 70.[5] Both the heat capacity of the filler and the endothermic decomposition may affect the fire retardancy. Analysis of the combustion gases produced just above the oxygen index value can enable one to ascertain the relative contributions of the decomposition endotherm and the heat capacity.[2] With polypropylene, polyphenylene oxide, poly(butadiene terephthalte) and acrylonitrile-butadiene-styrene terpolymer, both MH and ATH break down to give the metal oxides, which, when combined with whatever amount of carbonaceous char is formed, provide an effective thermally insulating barrier, leading to fire retardancy.

In a cone calorimetry study, compositions of polypropylene (PP) that contain the same mass of either glass beads or MH have been examined. In both cases the heat release rates were significantly reduced, but the reduction was far greater for MH, even though both materials are considered to be inert fillers.[6] This may suggest that MH is not simply an inert filler. The degradation of MH filled PA-6 and PA-6,6 has been studied and it was found that the presence of MH enhances the degradation of the polyamide.[7] This was attributed to the release of water from the decomposition of MH and its subsequent attack on the polyamide. With PA-6,6, polymer degradation occurred before MH decomposition, while with PA-6 there is better overlap between MH and PA-6 degradations, resulting in enhanced fire retardancy.

With polyethylene, both MH and ATH give the same oxygen index at an equivalent loading level. Conversely, in EVA (30% vinyl acetate content) MH gives an oxygen index of 46 while with ATH the value is 37. It was suggested that this difference is due to the loss of acetic acid from the polymer either delaying water loss (ATH) or accelerating this process (MH).[8]

Another area in which the metal hydroxides excel is smoke suppression. These hydrated fillers not only reduce the smoke release but they also can delay the time at which it is released, and thus provide additional time for escape from a fire.[5] Little work has been done on the process by which smoke suppression may occur, but the best guess is that carbon, from polymer degradation, is deposited on the oxide and this is then volatilized as carbon dioxide, resulting in no smoke.[9] This may be an area in which someone can make a very useful contribution.

As in any fire retardant system, synergy can be useful. Combinations that have been used include ATH with MH (giving an increased range of endothermic decomposition),[10] ATH with red phosphorus (enabling lower loadings),[11] MH with melamine and novolac in PP;[12] several additional examples are given in reference 2.

1.4 Nanocomposites

Nanocomposites are a new class of inorganic materials that only somewhat recently have begun to be used to achieve fire retardancy. The initial discovery is that a polyamide-6clay nanocomposite, containing 5% clay, shows an increase of 40% in tensile strength, 68% in tensile modulus, 60% in flexural strength and 126% in flexural modulus, while the heat distortion temperature increases from 65 to 152°C and the impact strength is lowered by only 10%.[12,13]

The initial work, which was not yet recognized as nanocomposites, actually took place sometime earlier when Blumstein synthesized poly(methyl methacrylate) in the presence of a clay and found that the clay had a templating effect on the formation of the polymer.[14–19] The significance of these observations was not realized for several years and this work has taken on more importance since the advent of the nano era.

Nanocomposites may be produced using several different materials for the nano-dimensional material, including clays, graphites, carbon nanotubes, and polyhedral oligosilsesquioxanes, POSS. Most work to date has been with clays, particularly with montmorillonite clay, an alumina-silicate material. A wide variety of other clays naturally occur, but, for some reason, montmorillonite has been by far the chosen material, probably because interesting results were obtained with this clay.

Surprisingly, graphite has not been more widely used; one concern may be that the *d*-spacing in most organically-modified montmorillonites is in the range of 2 or 3 nm while graphite has a *d*-spacing of about 2 or 3 Å. To form a nanocomposite, the polymer must be able to enter into the gallery space of the nanomaterial, and this may require that this space be large enough to permit the polymer to begin to enter. Graphite does form a number of intercalation compounds in which the *d*-spacing is large. For instance, potassium graphite, KC8 has a *d*-spacing of 5.5 Å and that of graphite sulfuric acid is even larger.[20,21] Possibly, if one begins with an already expanded graphite, a *d*-spacing in the range of 2 to 3 nm at least, that graphite may become more useful as a nano-dimensional material for nanocomposite formation.

Carbon nanotubes are, of course, a newer discovery and they are still quite expensive. There is still some activity in this area;[22–24] the major difficulty with the single wall nanotubes appears to be the need to organically-modify the nanotubes to make them more organophilic, this is probably also a limitation with the graphite system also. The multi-wall nanotubes do not require organic modification for nanocomposite formation. There has been little work on the fire retardancy of nanocomposites using carbon nanotubes.[25–27] The polymers that have been investigated include polypropylene and ethylene/vinyl acetate, EVA, and the reductions in PHRR are comparable to those seen with clays.

Polyhedral oligomeric silsequioxanes, POSS, are a unique class of materials that have the general formula $(RSiO_{1.5})_n$.[28] At least some of the R groups are usually unreactive, as phenyl, methyl, *etc.*, but one can also incorporate one or more reactive groups, *e.g.*, styryl, methacrylate, *etc.* The presence of a polymerizable substituent enables the formation of polymers, either by direct polymerization or co-polymerization with another monomer. The diameter of the POSS is typically on the order of 15 Å and they are, in general, easily incorporated into a polymer matrix. The generalized structure of a POSS system is shown in Figure 1. This consists of substituents R, which are unreactive and provide for compatibilization and solubility, and reactive groups X (only one of which is shown in this figure but more are possible) attached to a chemically and thermally robust hybrid framework. The composition is intermediate between that of silica and silicones; it offers a precise three-dimensional structure for reinforcement at the molecular level of polymers segments.

There has been much less work in fire retardancy with POSS than with clays, one US patent[29] and one paper.[30] The patent shows that POSS significantly reduces the PHRR for a polyether-block-polyamide system (50–70% reduction), for polypropylene (a 40% reduction) and a styrene-butadiene-styrene (SBS) triblock polymer (40–60% reduction). The decrease in the time to ignition,

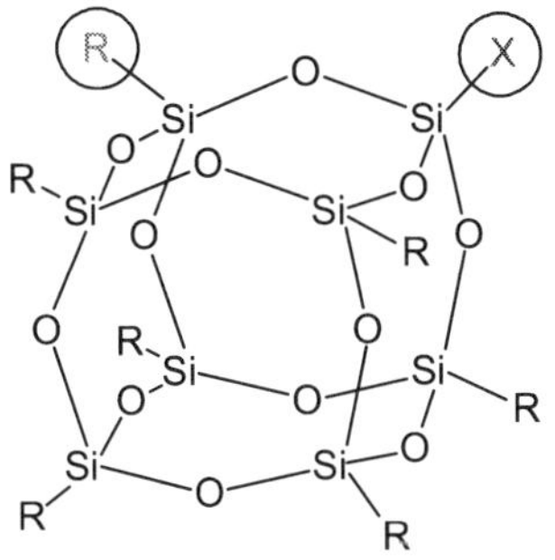

Figure 1 *Generalized structure of a POSS material*

which is common for clay-based systems, is observed for some, but not all, polymers with POSS. For POSS with polyurethane fabrics[30] the reduction in PHRR is about 55%. It appears that POSS materials should be more widely studied as fire retardant systems, since the reduction in PHRR is quite large and the time to ignition shows a more useful behavior.

1.4.1 Preparation and Modeling of Nanocomposites

A nanocomposite is formed by either a polymerization process in the presence of a clay, or similar material, or by blending of the nano-dimensional material with a polymer. At this stage of the discussion, we will speak only about clay–polymer nanocomposites. The clay begins in the form of tactoids with a high aspect ratio – for montmorillonite the length is typically in the range of 100 nm while the width is around 1 nm. Upon formation of a nanocomposite, three possible situations may arise. The clay may remain as tactoids with no penetration of the polymer between the clay layers; this is called either an immiscible nanocomposite or a microcomposite. If the clay is well-dispersed, then either an intercalated or an exfoliated (also known as delaminated) nanocomposite may be formed. Intercalation means that the clay layers maintain their registry while exfoliation indicates that this registry is lost. These situations are depicted in Figure 2.

Vaia and Giannelis have reported on a thermodynamic model for nanocomposite formation by melt blending.[31] This model indicates that the entropic

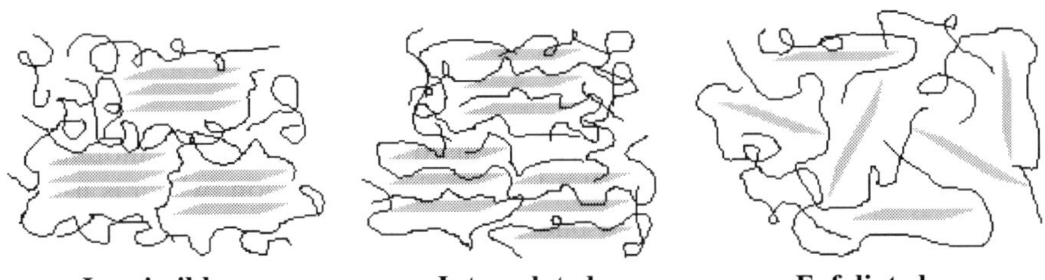

Figure 2 *Depiction of immiscible, intercalated and exfoliated nanocomposites*

penalty for polymer confinement may be compensated by the increased conformational freedom of the tethered chains as the clay layers separate. Complete layer separation depends upon the establishment of very favorable polymer–organically modified clay interactions to overcome the penalty of polymer confinement. The total entropy change is near zero, if complete layer separation is achieved, and the polymer is now not confined.

Balazs *et al.*[32] have also modeled the behavior of polymer–clay nanocomposites and they have shown that immiscibility occurs for the natural clay and polymers with a degree of polymerization of 100. When the clay is organically-modified, there can be favorable enthalpic interactions between the surfactant and the polymer, which can overcome the unfavorable entropy term and lead to efficient mixing. The formation on a intercalated or exfoliated system depends upon the length of the surfactant chain, the density of the surfactant on the clay, and the molecular weight of the polymer. It appears that if the length of the surfactant and the polymer are similar, then some of the entropic barrier is overcome and this will lead to easier nano-dispersion. When the amount of surfactant increases, the surfactant becomes denser and it becomes more difficult for the polymer chains to penetrate and good nano-dispersion will become more difficult. Finally, if one can produce attractions between the surfactants and the polymer, this highly attractive surface interaction can lead to exfoliation. Thus, one may conclude that the design of the surfactant is extremely important for success in the preparation of polymer–clay nanocomposites.

1.4.2 Organic Clay Modification

The gallery space of a typical clay is hydrophilic, based on the presence of the sodium cations and the alumino-silicate framework of the clay. To permit the insertion of a hydrophobic polymer within this gallery space, one must first render this gallery space organophilic. This is most typically accomplished by ion exchanging the sodium cation for an organophilic ammonium salt; the usual requirement is that there must be at least one long chain or twelve carbons or more on the nitrogen atom of the ammonium cation. As noted above, theoretical studies have shown that an attractive interaction between the surfactant and the polymer greatly enhances the possibility of nano-dispersion of the clay within the polymer. Thus, one should pay careful attention to the type of surfactant that is used. In addition, the thermal degradation of many surfactants begins at temperatures as low as 200°C by the Hofmann elimination, giving an olefin and a tri-substituted ammonium cation.[33,34] The loss of the long chain will frequently eliminate the possibility of nano-dispersion.

Several different counterions have been used to enhance the organophilicity of the clays; the reader will usually think of the 'onium' ion, which is usually taken to include ammonium and phosphonium ions. Brief mention should be made of the single example of a stibonium-substituted clay and its polystyrene nanocomposite.[35] The initial degradation step, which is the loss of the olefin, occurs at slightly higher temperature but the degradation stops at this stage and there is no loss of the stibine, meaning that the counterion of the clay is R_3SbH^+ and this should impart additional thermal stability to the clay and its nanocomposites. There has been some work in which oligomeric ammonium and phosphonium ions have been used to enhance the interaction between the polymer and the surfactant.[36–40] Three types of oligomers have been examined, styrene, methacrylate and butadiene. For both styrene and methacrylate, copolymers of the monomer with vinylbenzyl chloride, containing about 1 to 2 benzylic chlorides per oligomer, have been prepared and then the benzylic chloride has been used to quaternize an amine, giving a new ammonium salt. For butadiene, the authors used an oligomeric polybutadiene and graft copolymerized vinylbenzyl chloride to the butadiene. Best results were obtained with the styrenic copolymer; exfoliation was observed when this organically-modified clay was melt blended with polystyrene in a Brabender mixer. Even with unmodified polypropylene, an almost exfoliated nanocomposite is formed in the Brabender; it is assumed that complete exfoliation will be obtained if higher shear is applied. With both the methacrylate-modified and the butadiene-modified clays, immiscible materials are usually formed. Quite recently, Zhang has shown that one may use a substituted tropylium ion as the counterion for the clay and produce styrene nanocomposites.[41]

1.4.3 Determination of the Morphology of Nanocomposites

The determination of morphology is usually dominated by two techniques, X-ray diffraction, XRD, and transmission electron microscopy, TEM. XRD enables the determination of the *d*-spacing of the clay. An immiscible system is obtained if the *d*-spacing in the presence of the polymer is unchanged from that of the pristine clay. If the *d*-spacing increases, this indicates that intercalation has occurred. Since the registry between the clay layers is lost in an exfoliated system, no peak is expected. Unfortunately, this same situation will occur if the clay has extensively disordered, so XRD information alone is not enough to identify the morphology. TEM is usually used to address this question, since one can directly image the clay and polymer and identify the type of morphology. This type of measurement is frequently considered to be definitive. One must remember that to obtain one TEM requires only a miniscule piece of material and one cannot be certain that this is representative of the whole. The morphology can only be clearly determined by either sampling enough of the material by TEM so that one has statistical significance or by sampling the bulk of the sample.

A recently reported NMR technique to identify the morphology is based on proton NMR relaxation measurements.[42–44] The relaxation time depends upon the separation between nearest polymer–clay interfaces and the efficiency of paramagnetically-induced relaxation,[45] due to iron that is naturally present in the clay. An immiscible system will have the largest separation and thus the longest relaxation time while an exfoliated system has the smallest distance and the shortest time.

It appears that cone calorimetry may also be used as a method to sample the bulk. Some of the early work on nanocomposites showed that immiscible systems showed no reduction in the peak heat release rate, PHRR, while intercalated and exfoliated nanocomposites gave significant reductions.[46] In work from these laboratories, we have confirmed this observation and shown that there seems to be a correlation between the extent of nano-dispersion and the reduction in PHRR.

The classic definition of intercalation and exfoliation depends on the XRD and there is a need for new definitions based on other techniques. At this time, one can never be sure how an author is defining the nano-morphology so these terms are somewhat ambiguous.

1.4.4 Utility of Nanocomposites

There are currently believed to be four areas in which nanocomposite formation may be important: permeability, heat distortion temperature, flexural modulus and fire retardancy. A review has covered many of these enhanced properties.[47] The type of nanocomposite is important for some of these properties but unimportant for others. The permeability of a polymer is attributed to the tortuosity of the path that a gas must follow to penetrate a polymer, and the presence of exfoliated clay layers will make the path more difficult and thus lead to a decrease in permeability. It is also felt that exfoliation is an advantage for mechanical properties. However, there appears to be no difference between intercalated and exfoliated polymer–clay nanocomposites for fire properties.

1.4.5 Modeling of Fire Retardancy Due to Nanocomposite Formation

Nyden and Gilman[48] have simulated the thermal degradation of polypropylene that is nano-confined in a graphite matrix. They used graphite for convenience, since they have computational experience dealing with hydrocarbons but not with clays. The model consisted of four chains of isotatic polypropylene, each with 48 monomer units, contained within a graphite sheet of 600 carbons, end-capped with hydrogens. The degradation mechanism for the virgin polymer and the nanocomposite were unchanged in this simulation; the process involves the random scission of the CH–CH_3 bonds, followed by β-scission of the backbone to produce secondary free radicals, which then can unzip. Interactions with the graphite layer imparts a degree of stabilization when the graphite layers are separated by 2.8 to 3.2 nm. There is no reason to think that the results would be significantly different if clay were the nano-dimensional material.

1.4.6 Mechanisms by which Nanocomposites Enhance the Fire Retardancy of Polymers

Two mechanisms have been proposed to explain how nanocomposite formation can reduce the PHRR of a polymer. Gilman *et al.*[49] have proposed that the degradation of the nanocomposite produces a multi-layered carbonaceous-silicate structure that may act as an excellent insulator and also as a barrier to mass transport. Zhu *et al.*[50] have shown that the presence of iron in the clay can lead to some radical trapping reactions that will lower the heat release rate. It appears that at low amounts of clay the paramagnetic radical trapping is effective while the barrier mechanism becomes more important at higher amounts of clay. In a series of papers, Wang *et al.*[51] have shown that the alumino-silicate barrier proposed by Gilman *et al.* does form for both polystyrene and methyl methacrylate nanocomposites. For nanocomposites of poly(vinyl chloride), the surface is covered with carbon. This difference is no doubt due to the different degradation pathways of the polymers; PVC normally degrades to give char while neither PS or PMMA are char-formers.

Gilman *et al.*[49] have found that polystyrene-fluorohectorite nanocomposites do not show a reduced PHRR, even though the same polymer with montmorillonite gives a reduction of more than 50% in PHRR. They note that there is a difference in chemical composition, aspect ratio, and nano-morphology and that they cannot assign the difference to any one of these. In recent work from this laboratory, polystyrene-magadiite nanocomposites have been prepared.[52] Magadiite, like fluorohectorite, is an all silicate material. Again no reduction in PHRR is observed and the differences include the composition, aspect ratio and nano-morphology. With magadiite, the morphology, based on TEM, shows a rather large immiscible component; the improvement in mechanical properties argues that there is also a large intercalated or exfoliated component. Polystyrene hectorite nanocomposites[53] have also been examined. Here the PHRR shows a reduction, but only at greater than 3% clay. With montmorillonite, a reduction is seen even when the clay amount is as low as 0.1% organically-modified clay. Advances in fire retardancy will require an identification of what causes these various clays to behave differently.

To further complicate the situation, work has been carried out using graphite as the nano-dimensional material. The graphite that has been used is both sulfuric acid-graphite and modified graphite oxides.[54–56] In both cases, the reduction in PHRR is equivalent to the best values that have been obtained with montmorillonite. One may well expect that the nano-morphology, the aspect ratio and, certainly, the chemical composition of graphite are quite different than those of any of the clays, yet the fire retardancy, at least as measured by the reduction in PHRR, is equivalent. This is an additional area of challenge for the FR nanocomposite community to attempt to explain these observations.

The reduction in PHRR is different for each polymer and the values for both montmorillonite and graphite systems are shown in Table 1. The differences are striking, for instance with clay-PMMA, the best reduction in PHRR is 25% while polyamide-6 and polystyrene give values in the 60% range. If the mechanism is barrier formation, one would expect that the same barrier would be built in each case and this would be expected to lead to similar reductions for each case. Recent work using TGA/FTIR methods has shown that the clay appears to change the degradation pathway of polystyrene.[57] The degradation of polystyrene proceeds to give a mixture of oligomer and monomer; this is expected based upon the structure, which requires that a secondary and a primary radical

Table 1 *Reduction in PHRR for clay–polymer and graphite–polymer nanocomposites; values taken from references cited in the text. (irradiance level is 35 kW m^{-2} in every case)*

Polymer	*% reduction for clay-polymer nanocomposite*	*% reduction for graphite–polymer nanocomposite*
Polystyrene	57	48
HIPS	40	36
ABS	45	48
Polyamide-6	63	62
Poly(methyl methacrylate)	25	35
PP-g-MA	54	
EVA	47	

be produced upon bond cleavage. These unstable radicals will hydrogen abstract, giving a more stable radical with concomitant formation of oligomer. The degradation process of a PS-montmorillonite clay nanocomposite is changed so that much less monomer is seen, but oligomer is still produced.

1.4.7 Fire Retardancy Due to Nanocomposite Formation

The literature on the fire retardancy of nanocomposites has been recently reviewed[58] and the reader is referred to this review for specific information on polymers that have been studied. In this section, we will only describe the general details of fire retardancy due to nanocomposite formation. Fire retardancy is usually measured by cone calorimetry, particularly the reduction in the peak heat release rate, PHRR. Notably, a nanocomposite in which the clay is well-dispersed, whether intercalated or exfoliated, appears to give the same reduction in the PHRR. However, if one considers that all of the heat from the polymer is eventually released, the nanocomposite does not truly form a permanent barrier but rather an impermanent barrier that still permits the remaining polymer to degrade. It is the opinion of this author that nanocomposites alone will never solve the problem of fire retardancy but they can be a component of the solution. This author advocates the synergistic combination of a clay with some other fire retardant system. In such a system, the role of the clay will likely be to maintain the desirable mechanical properties of the polymer that may be lost by the presence of some other additive. In this case, the type of nano-dispersion may be very important and the formation of exfoliated systems may be required to achieve the level of fire retardancy required while maintaining the needed mechanical properties.

One advantage that nanocomposite formation may have for fire retardancy purposes is the improvement in mechanical properties that usually occurs through the formation of the nanocomposite. Many fire retardants are used at very high loadings, which can significantly impact the physical properties of the polymer. Clays may function synergistically with other fire retardants, and the presence of the clay may counteract any deleterious effects from the fire retardant and make these more useful.

The usual measure that is used to evaluate the fire retardancy of nanocomposites is the cone calorimeter, which measures the rate of heat release and mass loss rate, along with smoke and carbon monoxide and carbon dioxide, as a function of the applied radiant energy. The effects that one would like to see are that the time to ignition and the time to peak heat release are increased while the peak heat release rate (PHRR), the total heat released and the mass loss rate are lowered; if one can have every wish, the amount of smoke and CO will also be reduced. In actuality, the peak heat release rate is usually decreased upon nanocomposite formation but the time to PHRR is unchanged and the time to ignition is decreased. Significantly, the total heat released is unchanged, which means that all of the polymer does eventually burn. Nanocomposite formation appears to lengthen the time of burning but none of the polymer is retained. The mass loss rate is somewhat decreased and the smoke is not much changed.

1.5 Conclusion – the Future of Fillers and Nanocomposites in Fire Retardancy

The role of the ATH and MA type filler in fire retardancy is assured, since these are now used on a commercial scale and they are affordable. The rather high price currently charged for modified clays means that the clays must clearly outperform other systems before they will make inroads into the marketplace. It is the opinion of this author that clays alone will not be used as fire retardants but they may be a component of the solution to the problem of fire retardancy. Synergy has been demonstrated between conventional fire retardants and nanocomposite formation in a few cases.[59–62] There will need to be additional investigations of this type to confirm the observations that have been made and to evaluate the different conventional fire retardants that could be used. The advantage that the clay brings to fire retardancy is the improvement in mechanical properties and this means that one can add some other material, the fire retardant, that may cause a deterioration of the mechanical properties. This opens the door to new opportunities with combinations of materials.

1.6 References

1. W.E. Horn, Jr., in *Fire Retardancy of Polymeric Materials*, A.F. Grand and C.A. Wilkie (eds.), Marcel Dekker, New York, 2000, pp. 285–352.
2. P.R. Hornsby, *Int. Mater. Rev.*, 2001, **46**, 199.
3. J. Jow and D. Gomolka, *US Patent 5482990A*, 1996; E. Sezaki, M. Akami and H. Endo, *European Patent 0331358*; Y. Yamamoto and M. Tanmachi, *Japanese Patent 04253748*.
4. P.R. Hornsby, *Macromol. Symp.*, 1996, **108**, 203.
5. P.R. Hornsby, *Fire Mater.*, 1994, **18**, 269.
6. P.R. Hornsby and A. Mthupha, *Plast. Rubber Compos. Process. Appl.*, 1996, **25**, 347.

7. P.R. Hornsby, J. Wang, G. Jackson, R. N. Rothon, G. Wilkinson and K. Cosstick, *Polym. Degrad. Stab.*, 1996, **51**, 235.
8. J. Rychly, K. Vesely, E. Gal, M. Kummer, J. Jancar and L. Rychly, *Polym. Degrad. Stab.*, 1990, **30**, 57.
9. P.R. Hornsby and C.L. Watson, *Plast. Rubber Process. Appl.*, 1989, **11**, 45.
10. G.L. Kirshenbaum, *Kunstst. J.*, 1989, **79**, 62.
11. H. Staendeke, *FRCA meeting*, Spring 1988.
12. E.D. Weil, M. Lewin and H.S. Lin, *J. Fire Sci.*, 1998, **16**, 383.
13. Y. Kojima, A. Usuki, M. Kawasumi, A. Okada, Y. Fukushima, T. Kurauchi and O. Kamigaito, J. *Mater. Res.*, 1993, **8**, 1185.
14. A. Blumstein, *J. Polym. Sci.: Part A*, 1965, **3**, 2653.
15. A. Blumstein, *J. Polym. Sci.: Part A*, 1965, **3**, 2665.
16. A. Blumstein and F. W. Billmeyer, *J. Polym. Sci.: Part A-2*, 1966, **4**, 465.
17. A. Blumstein, R. Blumstein and T.H. Vandersppurt, *J. Colloid Interface Sci.*, 1969, **31**, 236.
18. A. Blumstein, S. L. Malhotra and A. C. Watterson, *J. Polym. Sci.: Part A-2*, 1970, **8**, 1599.
19. A. Blumstein, K.K. Parikh, S.L. Malhotra and R. Blumstein, *J. Polym. Sci.: Part A-2*, 1971, **9**, 1681.
20. W. Rudroff, *Adv. Inorg. Radiochem.*, 1959, **1**, 233.
21. G.R. Henning, *Prog. Inorg. Chem.*, 1959, **2**, 125.
22. C.A. Mitchell and R. Krishnamoorti, *Proc. Addit.* 2003, April, 2003.
23. H. Koerner, C.-S. Wang, R.A. Vaia, M.D. Alexander, N.A. Pearce and H. Bentley, *Proc. Addit.* 2003, April, 2003.
24. C.A. Mitchell, J.L. Bahr, S. Arepalli, J.M. Tour and R. Krishnamoorti, *Macromolecules,* 2002, **35**, 8825.
25. G. Beyer, *Fire Mater.*, 2002, **26**, 291.
26. T. Kashiwagi, E. Grulke, J. Hilding, R. Harris, W. Walid and J. Douglas, *Macromol. Rapid Commun.*, 2002, **23**, 761.
27. T. Kashiwagi, E. Grulke, J. Hilding, J. Shields, R. Harris, W. Awad and J. Douglas, *Abstract of 9th European Meeting on Fire Retardancy*, September, 2003.
28. G. Li, L. Wang, H. Ni and C. U. Pittman, Jr., *J. Inorg. Organmet. Chem.*, 2002, **11**, 123.
29. J.D. Lichtenhan and J.W. Gilman, "Preceramic additives as fire retardants for plastics," *US 6,362,279 B2*, issued March 26, 2002.
30. E. Devaux, M. Rochery and S. Bourbigot, *Fire Mater.*, 2002, **26**, 155.
31. R.A. Vaia and E.P. Giannelis, *Macromolecules*, 1997, **30**, 7990.
32. A.C. Balazs, C. Singh, E. Zhulina and Y. Lyatskaya, *Acc. Chem. Res.*, 1999, **32**, 651.
33. W. Xie, Z. Gao W-P. Pan, R. Vaia, D. Hunter and A. Singh, *Polym. Mater. Sci Eng.*, 2000, **83**, 284.
34. J. Zhu, A.B. Morgan, F.J. Lamelas and C.A. Wilkie, *Chem. Mater.*, 2001, **13**, 3774.
35. D. Wang and C.A. Wilkie, *Polym. Degrad. Stab.*, 2003, **82**, 309.

36. S. Su, D.D. Jiang and C.A. Wilkie, *Polym. Degrad. Stab.*, 2004, **83**, 321.
37. S. Su, D.D. Jiang and C.A. Wilkie, *Polym. Degrad. Stab.*, 2004, **83**, 333.
38. S. Su, D.D. Jiang and C.A. Wilkie, *Polym. Degrad. Stab.*, 2004, **84**, 279.
39. S. Su, D.D. Jiang and C.A. Wilkie, *Polym. Adv. Tech.*, 2004, **15**, 225.
40. S. Su, D.D. Jiang and C.A. Wilkie, *J. Vinyl Add. Tech.*, 2004, **10**, 44.
41. J. Zhang and C.A.Wilkie, *Polym. Degrad. Stab.*, 2004, **83**, 301.
42. D.L. VanderHart, A. Asano and J.W. Gilman, *Macromol.*, 2001, **34**, 2001, 3819.
43. D.L. VanderHart, A. Asano and J.W. Gilman, *Chem. Mater.*, 2001, **13**, 3781.
44. D.L. VanderHart, A. Asano and J.W. Gilman, *Chem. Mater.*, 2001, **13**, 3796.
45. S. Bourbigot, D.L. VanderHart, J.W. Gilman, W.H. Awad, R.D. Davis, A.B. Morgan and C.A. Wilkie, *J. Polym. Sci.: Part B: Polym. Phys*, 2003, **41**, 3188.
46. J.W. Gilman, T. Kashiwagi, M. Nyden, J.E.T. Brown, C.L. Jackson, S. Lomakin, E.P. Giannelis and E. Manias, in *Chemistry and Technology of Polymer Additives*, S. Al-Malaika, A. Golovoy and C.A. Wilkie (eds.), Blackwell Scientific, Oxford, 1999, pp. 249–265.
47. M. Alexandre and P. Dubois, *Mater. Sci & Eng.*, 2000, **R28**, 1.
48. M.R. Nyden and J.W. Gilman, *Comp. Theor. Polym. Sci.*, 1997, **7**, 191.
49. J.W. Gilman, C.L. Jackson, A.B. Morgan, R. Harris, Jr., E. Manias, E.P. Giannelis, M. Wuthenow, D. Hilton and S.H. Phillips, *Chem. Mater.*, 2000, **12**, 1866.
50. J. Zhu, F.M. Uhl, A.B. Morgan and C.A. Wilkie, *Chem. Mater.*, 2001, **13**, 4649
51. J. Wang, J. Hao, J. Zhu and C.A. Wilkie, *Polym. Degrad. Stab.*, 2002, **77**, 249; J. Du, J. Zhu, C.A. Wilkie and J. Wang, *Polym. Degrad. Stab.*, 2002, **77**, 377; J. Du, D. Wang, C.A. Wilkie and J. Wang, *Polym. Degrad. Stab.* 2003, **79**, 319; J. Du, J. Wang, S. Su and C.A. Wilkie, *Polym. Degrad. Stab.*, 2004, **83**, 29.
52. D. Wang, D.D. Jiang, J. Pabst, Z. Han, J. Wang and C.A. Wilkie, *Polym. Eng. Sci.*, 2004, **44**, 1122.
53. D. Wang, B.N. Jang, S. Su, J. Zhang, X. Zheng, G. Chigwada, D.D. Jiang and C.A. Wilkie, this book, chapter 5.
54. F.M. Uhl and C. A. Wilkie, *Polym. Degrad. Stab.*, 2002, **76**, 111.
55. F.M. Uhl and C.A. Wilkie, *Polym. Degrad. Stab.*, 2004, **84**, 215.
56. F.M. Uhl, Q. Yao and C.A. Wilkie, *Polym. Deg. Stab.*, submitted.
57. S. Su and C.A. Wilkie, *Polym. Degrad. Stab.*, 2004, **83**, 347.
58. D. Wang and C.A. Wilkie, in *Fire Behaviour of Composite Materials*, G. Gibson and A. Mouritz (eds.), Kluwer Press, 2005 in press.
59. G. Chigwada and C.A. Wilkie, *Polym. Degrad. Stab.*, 2003, **81**, 551.
60. X. Zheng and C.A. Wilkie, *Polym. Degrad. Stab.*, 2003, **81**, 539.
61. M. Zanetti, G. Camino, D. Canavese, A.B. Morgan, F.J. Lamelas and C.A. Wilkie, *Chem. Mater.*, 2002, **14**, 189.
62. G. Beyer, *Fire Mater.*, 2001, **25**, 193.

Micro-sized Fire Retardant Fillers

CHAPTER 2

Fire Retardant Fillers for Polymers

PETER R. HORNSBY[1] AND ROGER N. ROTHON

[1]School of Mechanical and Manufacturing Engineering, Queen's Universtity Belfast, Belfast, BT9 5AH, UK

[2]Rothon Consultants, UK, 3 Orchard Croft, Guilden Sutton, Chester, CH3 7SL, UK

The term "fire retardant fillers" usually refers to products, like the metal hydroxides, which decompose endothermically and can function as fire retardants on their own, without the addition of other additives. This terminology will be adopted in this chapter. Several other types of filler play important roles in fire retardant applications, notably ammonium polyphosphate, antimony oxides, borates, nano-layer silicates and stannates, but these are normally used in combination with other fire retardant types.

2.1 Fire Retardant Fillers Available

Fire retardant fillers make up a very significant part of the fire retardant additive market. Worldwide sales are estimated as about 500,000 tonnes or about 40% of the total market by weight. They function by endothermic decomposition, with release of inert gasses (water and/or carbon dioxide). This decomposition needs to occur above the polymer processing temperature, but near to the polymer degradation temperature. Fire retardant mechanisms will be discussed in detail later, but one of the main advantages of this type of fire retardant is that it doesn't generate the smoke and fume hazards typical of some other approaches. However, high levels of additive are needed and this can adversely affect cost, processing and mechanical properties.

Potential products have recently been summarised.[1] For successful commercial use, a candidate material must have the following properties:

(i) A significant endothermic decomposition. This should be in the temperature range 100–300°C, depending on the polymer, and by experience needs to result in the release of at least 25% by weight of water and/ or carbon dioxide.

(ii) Ready availability and low cost.
(iii) Low toxicity.
(iv) Readily processable into small particle sizes capable of giving high filler loadings.
(v) Low solubility.
(vi) Low levels of extractable salts and of potentially detrimental impurities (such as those causing premature polymer degradation).
(vii) No colour.

Few materials meet all of these requirements and are predominately hydroxides, hydroxy carbonates and hydrates of aluminium, calcium and magnesium. Despite the large volumes involved, there only about five different materials in use commercially, with one, aluminium hydroxide, making up about 90% of the market by tonnage. The main commercial products are discussed below.

2.1.1 Aluminium Hydroxides

As already mentioned, this product (also known as alumina trihydrate, ATH: $Al(OH)_3$) dominates the market. This is because it best fulfils the criteria outlined above, especially with regards to cost. Decomposition starts at about 200°C, which is very suitable for most polymers, and results in the loss of 34.5% of water. Because of its importance, its production will be described in more detail than for the other fillers.

Aluminium hydroxide is produced from the mineral bauxite (a crude form of aluminium hydroxide). The process involves dissolution with sodium hydroxide to form a solution of sodium aluminate, followed by controlled precipitation. The low cost is due to the ability to link the production of fire retardant grades to that of the same material produced on a vast scale as an intermediate in the Bayer process for the manufacture of alumina.

There are two main grades of aluminium hydroxide fire retardant. The first is produced by milling of the large (about 70 μm) aggregates produced in the Bayer process itself. These are the lowest cost, but have platy, irregular particles, not ideally suited for many applications. The second, often referred to as precipitated, grades are specially precipitated from purified sodium aluminate, using conditions that give control of shape and size and remove the need for milling.

Aluminium hydroxide is available in a wide range of sizes and shapes and with different surface treatments. Grades with specially tailored particle size distributions are available for applications requiring very high loadings. It is widely used in elastomer and thermoset applications, but is of limited use in thermoplastics, due to the decomposition temperature being too low.

2.1.2 Magnesium Hydroxide, $Mg(OH)_2$

This is the second most widely used fire retardant filler. It is more expensive than aluminium hydroxide, but has a higher decomposition temperature (about 300°C), making it more suitable for use in thermoplastics applications.

There are several origins for this product.[2] First, there is limited use of milled natural product (known as brucite). This is suitable for some applications, but currently has inadequate performance for most of the market. Secondly, there is material precipitated with lime from sea-water and brines. This is already produced in large quantities for other uses and is of relatively low cost, but again the properties are currently only suitable for a small part of the market.

Finally, there are specially made products, with optimised morphology, which are suitable for the more demanding applications. Production methods are quite complex, with raw materials ranging from serpentine (a magnesium silicate) brines and magnesium oxides and will not be covered here. A recent development has been the introduction of nickel-doped forms of magnesium hydroxide, which are claimed to have superior fire retardant properties.[3]

2.1.3 Basic Magnesium Carbonates

These products are related to the mineral hydromagnesite [$4MgCO_3 \cdot Mg(OH)_2 \cdot 4H_2O$]. This decomposes over a range of temperatures, starting about 220°C and with a total weight loss of 57%. In principle, they should be excellent fire retardants for many polymers, including some thermoplastics, but market acceptance has, so far, been limited. This is mainly due to their unsuitable morphology and the relatively high price of current products.

There are natural forms of hydromagnesite available, but these are mixed with varying amounts of other minerals, notably huntite (a calcium magnesium carbonate), which is less effective and has a platy morphology, which can affect processing.[4] The huntite content can be up to 50 wt%.

Products approximating to the hydromagnesite composition can also be precipitated from solutions of magnesium salts and this process has been used. Unfortunately, the product formed has a poor morphology for use at high loadings and is mainly sold for smoke suppressing applications, where it is effective at lower levels than those needed for fire retardancy.

2.1.4 Boehmite, AlO(OH)

Boehmite is, in effect, partly decomposed aluminium hydroxide, where two thirds of the water has been removed. It has been promoted as a fire retardant in its own right but, because of the relatively low water content, does not seem to have a high fire retardant effectiveness. However, it does seem to have some potential in mixtures with other fire retardant fillers and this is where it is now being targeted.[5]

2.1.5 Calcium Sulphate Dihydrate, (Gypsum) $CaSO_4 \cdot 2H_2O$

This is a low cost material. Its fire retardant properties are limited and it has a low onset of decomposition (under 100°C), but it is reported to find some use as a fire retardant in unsaturated polyester resins.[6]

2.2 Mechanistic Studies

Before discussing mechanistic aspects in detail, it must be pointed out that the determination of fire retardant performance is a complex and somewhat controversial topic. Most product development and mechanistic studies are carried out with very simple, small scale, laboratory tests, which may not relate well with real fire scenarios.

The main laboratory tests used with fire retardant fillers are described in reference.[7] The principal tests are Oxygen Index, Underwriters Laboratory Vertical Burn Test UL94 and the cone calorimeter.

2.2.1 Flame Retardancy

The relative performance of hydrated fire retardant fillers in polymers strongly depends on the nature and origin of the filler type and the chemical characteristics of the host polymer, in particular its decomposition mechanism. Additionally, specific interactions may exist between certain polymers and fillers, which influence their mechanism of action.[8]

Compared to alternative fire retardants, including phosphorous-based intumescent and halogen-containing formulations, hydrated fillers are relatively ineffective, requiring addition levels of up to 60% by weight to achieve acceptable combustion resistance.[9] For example, with polypropylene, 60% by weight would be required to achieve an oxygen index in excess of 26%. At the same addition level in polyamide 6, however, an oxygen index of nearly 70% is achieved (Figure 1).[10] The issue with polyamides, though, is their tendency to

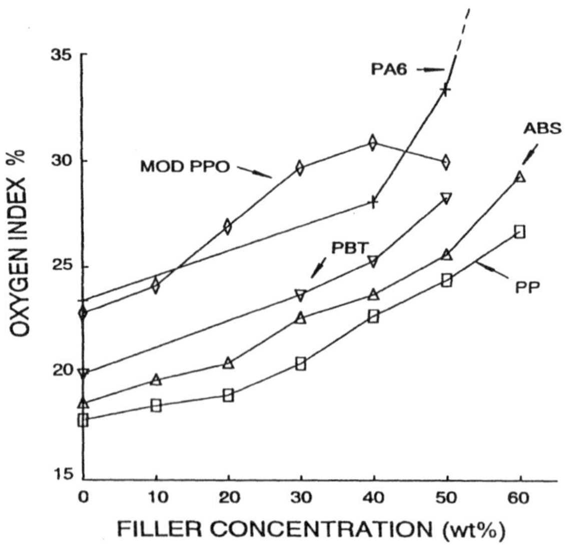

Figure 1 *Influence of magnesium hydroxide filler loading on the oxygen index of selected thermoplastics*

form ignitable drips during combustion; however, by increasing the filler level the viscosity of the decomposing polymer is increased and dripping is limited.[11] Several contributing effects that may combine to influence the mechanistic behaviour of fire retardant fillers are discussed below.

2.2.1.1 Thermal Effects from Filler

A characteristic of hydrated fire retardant fillers is that their thermal decomposition is endothermic and can adsorb significant quantities of heat. Differential scanning calorimetry (DSC) and thermogravimetric analysis (TGA) have been widely applied to study this behaviour.[12] Comparing magnesium hydroxide grades, differences have been reported in the magnitude of decomposition endotherm and the decomposition temperature.[13] Furthermore, it is significant that sample size, rate of heating, rate of inert gas flow rate and degree to which the pan is sealed can all influence the measured result.[12] Also, different grades of magnesium hydroxide may degrade at different rates, which appears to depend on the filler morphology and/or its surface area.[14]

The endothermic decomposition of aluminium hydroxide has been extensively studied. The enthalpy for complete decomposition is about 1300 kJ kg^{-1}. However, decomposition temperatures relevant to polymer applications are hard to define precisely. They depend very much on heating rates and on the ability of gaseous decomposition products to escape from the system. Isothermal studies provide the best information and show that decomposition starts about 200°C. It does appear to vary with the exact nature of the sample however, with precipitated grades being reported as significantly more stable than those produced by grinding.[15]

Although, the decomposition of aluminium hydroxide to oxide is usually written as one step, it can proceed through an intermediate mono-hydroxide [boehmite, AlO(OH)]. Boehmite is much more stabile than the hydroxide, decomposing about 550°C and its presence can reduce fire retardant effectiveness in some tests. Boehmite formation is favoured when escape of water is hindered. This can occur in the centre of large filler particles,[16] or in thick polymer composites especially with hard, impermeable polymers.[17]

The heat capacity of hydrated fillers and, in particular, their strong endotherm can greatly influence the input of heat required for polymer decomposition and release of combustible volatiles.[10] In this connection, the most thorough mechanistic work has been carried out for the oxygen index test, although this is thought to be the least realistic. A full heat balance model has been worked out for fire retardant fillers in this test.[18] With the exception of very fine fillers, this model fits experimental data very well. The fire retardant effect is found to come from three sources; the endotherm itself, the heat capacity of the oxide residue and the heat capacity of the evolved gasses. For aluminium hydroxide, the endotherm contributes about 51% of the effect, the oxide residue about 19% and the evolved gases about 30%.

Particle size of the filler also plays a role in flammability resistance. For example, below a certain particle size (about 1–2 μm), in many tests, including oxygen index, aluminium hydroxide shows enhanced fire retardant performance.[19] The reasons for this are not clear, but in the oxygen index test at least it seems that a more persistent ash is formed. This ash is raised to very high temperatures, increasing the heat capacity effect from this component. Strangely, it has been reported that the particle size effect is absent, or less pronounced, in the cone calorimeter test.[20]

At sufficiently high filler levels, hydrated fillers can also reduce the mass burning rate by inhibiting the rates of heat transfer from the flame to the underlying matrix. As a consequence, the supply of fuel necessary to continue the combustion process diminishes, causing the flame to extinguish due to fuel starvation.[21] Hence reductions in applied heat flux or increased surface heat losses will lead to a decrease in the mass burning rate of the polymer, as reported for aluminium hydroxide filled–polypropylene compositions.[22]

This thermal insulating effect has been demonstrated experimentally through forced combustion studies with a range of polymer types, including polypropylene, modified polyphenylene oxide (PPO), polybutylene terephthalate (PBT) and acrylonitrile-butadiene-styrene copolymer (ABS), designed to measure rates of heat transfer through a fire-retardant polymer composition exposed to an ignition source at its outer surface. On thermal breakdown, magnesium and aluminium hydroxides decompose to their respective oxides, which, together with any carbonaceous char produced, provide an effective thermal barrier, reducing heat transmission to the underlying substrate.[23,9]

Microscopic analysis of the oxide/char residue formed following combustion of magnesium hydroxide-filled polypropylene has revealed an oxide morphology similar in form to the parent hydroxide.[24] In this example, hexagonal platelets appear to align predominantly in the same plane and in some cases overlap, which contrasts with large aggregated oxide structures derived from hydroxide particles formed from association of small crystallites. There is some evidence of increased crystal growth and that the coherency of the oxide particles contributes to the stability of the decomposition residue observed from combustion products arising from oxygen index tests. Oxide residues may, possibly, also undergo partial sintering reactions, leading to the development of stronger ash structures, which may account for differences in performance observed between different grades of the same filler.[12] It may also be significant that the strength of agglomerates containing magnesium hydroxide pseudomorphs has been estimated as 50 MN m^{-2}, arising from physio-chemical association between magnesium oxide and water.[25]

2.2.1.2 Dilution of Combustible Polymer

The presence of up to 60% by weight of fire retardant filler results in around 35% by volume reduction of combustible polymer (in the case of magnesium hydroxide). In studies using polypropylene compositions containing different grades of magnesium hydroxide, magnesium oxide and glass beads, values of heat release rate (HRR) were determined by cone calorimetry.[24] The rates of heat release were significantly reduced, after allowing for the volume dilution of each of these fillers. However, magnesium oxide was far more effective than the glass beads, even though both are nominally considered to be inert, suggesting that even the physical and/or chemical nature of non-hydrated fillers may influence behaviour.

2.2.1.3 Filler/Polymer Interactions

TGA and DSC analytical techniques can provide useful information concerning the nature of filler–polymer interaction, together with their relative decomposition temperatures, when used in combination with evolved gas analysis (EGA) and on-line FTIR techniques. Using these methods, it was demonstrated that, on thermal breakdown, magnesium hydroxide exerts a significant pro-degradative action on polyamide 6 (PA-6) and polyamide 6.6 (PA-66), which was attributed to water release and resulting hydrolysis of the polymer chains.[26] Evolved gases released from both filled and unfilled PA compositions were shown to be water, carbon monoxide, carbon dioxide, ammonia and various hydrocarbon fragments. Importantly in PA-66, polymer degradation occurred before magnesium hydroxide breakdown, whereas there was much greater overlap in the case of decomposition of PA-6 and this filler, resulting in significantly improved flammability resistance with this composition.

Different magnesium hydroxide filler types influence the rheological behaviour of thermally decomposing polyamides in different ways and hence their resistance to dripping,[27] with plate-like filler particles being more effectively in this regard.[28]

Several comparisons exist on the relative efficiency of magnesium and aluminium hydroxides in the same polymer type. Care must be taken in their interpretation, however, due to the particle size effects described above. One study using polyethylene showed that, at the same additive loading, these fillers gave an equivalent oxygen index.[29] However, in ethylene-vinyl acetate copolymer (EVA) with 30% vinyl acetate content, magnesium hydroxide yielded an oxygen index of 46%, whereas when using aluminium hydroxide this was only 37%. Results from non-isothermal thermogravimetric analysis, suggested that water release was delayed from aluminium hydroxide, yet accelerated from magnesium hydroxide, possibly arising from reaction with acetic acid, generated during polymer decomposition.

2.2.1.4 Vapour Phase Action

Although the primary fire retardant action of hydrated fillers is in the condensed phase, the release of water and/or inert gas into the vapour phase on decomposition also contributes to the overall fire retardation mechanism. Little detailed analysis has been undertaken in this area; however, it is generally considered that water release into the vapour phase exerts a beneficial effect through dilution and cooling of flammable volatiles produced on polymer degradation.[10]

2.2.1.5 *Effects of Filler Particle Size and Morphology*

It has sometimes been observed that different grades of the same fire retardant filler can give significantly different effects, despite apparent similarities in their endothermic decomposition or release of inert gas. Whilst this may, in part, be an outcome of the flammability test procedure applied, distinct particle size and particle morphology effects have been reported. These factors also have a significant bearing on the mechanical properties and melt rheology of polymer composites containing hydrated fillers.

In relation to flammability, however, it has been shown using the UL94 vertical burn test that the effectiveness of magnesium hydroxide in polypropylene increased with decreasing particle size.[30] Similarly, in studies involving PMMA modified with ATH, fine grades (<1 μm) gave markedly higher oxygen index values than coarser (45 μm) grades, particularly at filler loadings above 50% by weight.[31,12] ATH is reported to be less thermally stable as the particle size increases.[32] Early in the decomposition process the alumina produced is very reactive, readily combining with water vapour to rehydrate to ATH. In larger particles, water escaping nearer the centre of the particle has a larger diffusion path, giving more time to react with alumina formed near the surface of the decomposing particle. During this process boehmite or psuedo-boehmite is formed and, being a partial decomposition product, is more stable than ATH, decomposing at about 450°C. In relation to the effects of particle size on thermal stability, there is a greater transition from gibbsite to pseudo-boehmite as the particle size increases.[15]

2.2.2 Smoke Suppression

An early study discussed the effects of calcium carbonate, aluminium hydroxide and magnesium hydroxide fillers on smoke production from styrene butadiene (SBR) foams.[33] All these fillers evidently reduced soot formation relative to unfilled foam, with the hydrated fillers being more effective than the calcium carbonate, which was considered to act merely as a matrix diluent. Aluminium and magnesium hydroxides gave enhanced char formation and promoted solid-state crosslinking, as opposed to pyrolytic degradation. The occurrence of afterglow, after extinction of the flame, was noted with MH and attributed to slow combustion of carbon residues. Several other reports have demonstrated the smoke suppressing tendencies of hydrated fillers in various polymers, including ethylene-propylene-diene elastomers, polypropylene, polystyrene, modified polyphenylene oxide, polybutylene terephthalate and ABS.[9,34,35] These hydrated fillers can also delay the onset of smoke evolution in the event of a fire.[10]

Although there has been extensive analysis of the composition and formation of soot from polymers undergoing combustion,[36,37] limited work has been published on the mechanism of smoke suppression using hydrated fillers. It seems likely, however, that the process is a consequence of the deposition of carbon onto the oxide surface, produced on decomposition of the hydrated filler.[9] Volatilisation of carbonaceous residue as carbon oxides subsequently occurs, which does not contribute to the obscuration effects of smoke. On hydroxide decomposition, these oxides have high surface areas and are catalytically active,[38] promoting both the carbon deposition and subsequent oxidation processes.[39] The reduced combustion rate arising from the effects of the fire retardant filler will also play a part in lowering the rate of smoke evolution and also in improving oxygen-to-fuel ratios, further limiting obscuration.

The role of evolved water from hydroxide decomposition is of interest, since water can also oxidise carbon. In this connection, the smoke yields from polypropylene compounds containing magnesium hydroxide and magnesium oxide were compared.[9] These results showed little difference in levels of smoke evolution, suggesting that water has limited effect on the smoke suppression mechanism. These data are supported by CO emission from burning ABS, which again demonstrate little distinction between oxide and hydrate forms of this magnesium compound[23] and also by the fact that the so-called water–gas oxidation reaction occurs at temperatures and pressures in excess of those normally found at the burning surface of a polymer.[40]

2.2.3 Incandescence

Other aspects of fire retardant performance include the effect of the additives on smoke, afterglow and incandescence, and these topics appear to be related. Smoke reduction is primarily due to the trapping of smoke precursors by the high surface area oxides formed on filler decomposition. These trapped species are then oxidised on the surface. This oxidation process is probably the cause of afterglow problems sometimes observed with fire retardant fillers.[41] In studies on the ignition and incandescence of filled polymers, both ATH and MH were found to increase the self-ignition temperature, but to decrease the incandescence temperature of an EVA copolymer[42] Using TGA, in these systems it was concluded that the solid state after-glow effects observed were due to oxidation of carbonaceous residues.

2.3 Synergists for Hydrated Fillers

The ability of fire retardant fillers to meet the most demanding specifications is limited. The high filler levels needed also causes problems with processing and mechanical properties. To some extent these drawbacks can be reduced by surface treatments, as will be discussed later, but there is a continuing search for synergists that allow overall filler levels to be reduced, or higher specifications to be achieved.

As the following examples demonstrate, potentially the problems resulting from high filler loadings can be reduced by introducing other fire retardant components into the composition to act in a synergistic manner, thereby increasing fire retardant action and enabling an overall reduction in filler content. However, in general, most of the literature on synergistic effects is in the patent literature and thus scientific explanation is scarce.

Although not exhaustive, Table 1 gives an overview of synergists reported or patented for use in combination with hydrated fillers, such as magnesium and aluminium hydroxides, in various polymer matrices. An indication of observed

Table 1 *Examples of reported synergists for metal hydroxides*

Co-additives	*Hydrated Filler(s)*	*Polymer(s)*	*Effect(s)*
Antimony trioxide	ATH	PVC (flexible)	Reduced overall filler level/Increased smoke
Antimony trioxide/ zinc borate	ATH	PVC (flexible)	Reduced overall filler level/lower smoke
Borate compounds (zinc borate/calcium borate)	ATH	EVA	Enhanced flammability resistance at low co-additive additionsIncreased char promotion
MH/ATH combinations	ATH MH	PVC	Reduced flammability Wider range of endotherm and water release Enhanced oxide thermal barrier(?)
Molybdenum compounds (molybdenum oxide/molybdate salts)	ATH MH	PVC	Reduced flammability and smoke emission Increased char promotion
Red phosphorus	ATH MH		Reduced overall filler levels Suppression of phosphine formation by metal hydroxide Coloured formulations Low co-additive additions
Silicon-containing compounds (organosilicones)	ATH MH	Polyolefins	Enhanced flammability resistance / reduced smoke Improved processibility and physical propertiesHandling issues
Polyacrylonitrile fibres	ATH MH	Polyolefins	Char promotion Reduced filler levels Can be pigmented
Transitionl metal oxides (nickel oxide/cobalt oxide)	ATH MH	Polyolefins	Reduced overall filler levels Colour limitations Toxicity concerns(?)
Metal nitrates (copper nitrate/iron nitrate)	ATH	EVA	Enhanced flammability resistance with low co-additive additions
Melamine	ATH MH	PP	Improved fire retardancy Reduced after-glow
Tin compounds (zinc stannate/zinc hydroxystannate)	ATH MH	PVC Cl-Rubbers EVA	Enhanced flammability resistance / reduced smoke especially with ZH/ZHS-coated filler variants
Nano-clays	ATH MH	EVA	Lower heat release rates/reduced smoke emission Used in combination with tin compounds

effects and possible modes of action is also given, where this has been proposed. Reasons for use of synergistic co-additives include enhanced fire retardancy, reduced smoke emission (particularly in halogenated formulations), and often a combination of these benefits. Examples of such systems are discussed briefly below.

Combinations of MH and ATH give improved performance when used together, due to the increased range of endothermic reaction (180–400°C) and release of water in the vapour phase.[43,44] The different metal oxides produced on dehydration may also contribute to this effect.

Mixtures of zinc borate, antimony trioxide, ATH and MH can be combined to enhance oxygen index and reduce smoke emission, the proportions varying depending on the end-use requirements.

The addition of low levels (~3%) of zinc borate to metal hydroxides can give synergistic effects.[45] For example, in an EVA/MH formulation, the oxygen index increased from 39 to 42%. In EVA, small levels of ZB (around 6% by weight) together with ATH or MH yield significant increases in oxygen index (Figure 2), presumably due to the greater char forming tendency of the co-additive and its ability to bind together oxide residue produced on decomposition of the hydrated filler.

ATH and red phosphorus (3–5%) have also been used in synergistic mixtures with metal hydroxides to increase fire retardancy and enable lower filler loadings.[46]

The addition of melamine and novolac (~1%) to PP/MH mixtures enables a UL 94 V-O rating to be reached at lower filler levels (30–50%), allowing the formulation to be mechanically more flexible. The novolac causes a structurally stabilising effect above the melting point of PP. Thermal evidence suggests that a novolac magnesia gel may be formed.[47]

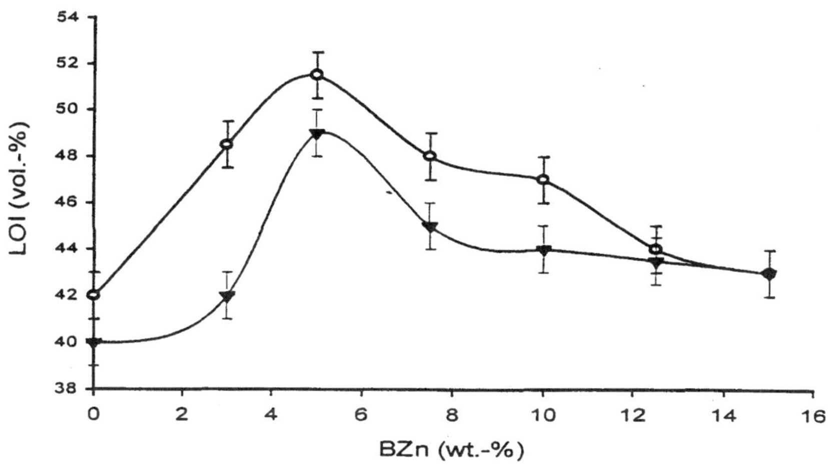

Figure 2 *Effect of zinc borate (BZn) substitution on the oxygen index of EVA24/MH and EVA24/ATH compositions (constant overall filler loading of 60% by weight; (▼) ATH, (○) MH)*

Metal hydroxides in combination with various formulations of silicon-containing compounds have been used to reduce the amount of additive required to achieve a required level of flame retardancy in various polymeric materials, including polyolefins.[48,49] There is little insight into how these formulations work; however, it seems likely that the silicon compounds bond with the metal hydroxides. Systems, which have been used, contain a combination of reactive silicone polymers, a linear silicone fluid or gum and a silicone resin, which is soluble in the fluid, plus a metal soap, in particular magnesium stearate.

The loading levels of metal hydroxides required to flame retard polyolefins can be reduced by addition of transition metal oxides as synergistic agents.[3] For example, combination of 47.6% MH modified with nickel oxide in PP gave a UL94-VO flammability rating that would require ~55% of unmodified MH.[50] These systems, however, can only be used where the colour of the product is not important.

The addition of metal nitrates to improve the flame retardancy of metal hydroxides and EVA has been reported.[51] Synergistic behaviour was observed by addition of 2% of copper nitrate to EVA containing only 33% ATH, in which the oxygen index was raised from 19.9 to 30.0%.

A natural mineral filler, containing mainly huntite and hydromagnesite, has been used, together with a blend of antimony trioxide (Sb_2O_3) and decabromodiphenyl oxide (DPDPO), to reduce the flammability of an ethylene-propylene copolymer.[52]

There are reports that additions of very small amounts (< 5% by weight) of polyacrylonitrile fibres or carbon powder to formulations, including MH or ATH hydrated fillers, can enable UL94 VO ratings to be obtained at overall filler levels well below that necessary in the absence of these co-additives.[53,54] For example, in LLDPE containing ATH, addition of only 1% of PAN fibres yielded a filler reduction of 8% for the same UL94 flammability ranking, thereby resulting in improved mechanical properties. These effects are thought to result from the charring action of the co-additives.

The flammability properties of an intumescent fire retardant PP formulation with added MH has been investigated.[55] The results show that the intumescent flame retardant ammonium polyphosphate-filled PP has superior flammability properties, but gives higher CO and smoke evolution. The addition of MH reduced smoke density and CO emissions, in addition to giving superior fire resistance.

Addition of silane cross-linkable PE copolymer to PE/metallic hydroxide systems can significantly improve the flame retardant properties of these materials, allowing lower filler levels to be used.[56]

The combination of melamine with hydrated mineral fillers can improve the fire retardancy behaviour of PP, eliminating at the same time the after-glow phenomenon associated with these fillers used in isolation.[57] Similarly, in EVA copolymer, antimony trioxide used in combination with metal hydroxides reportedly reduces incandescence.[58] Furthermore, chlorinated and brominated flame retardants are sometimes used in combination with metal hydroxides to provide enhanced fire retardant efficiency, lower smoke evolution and lower

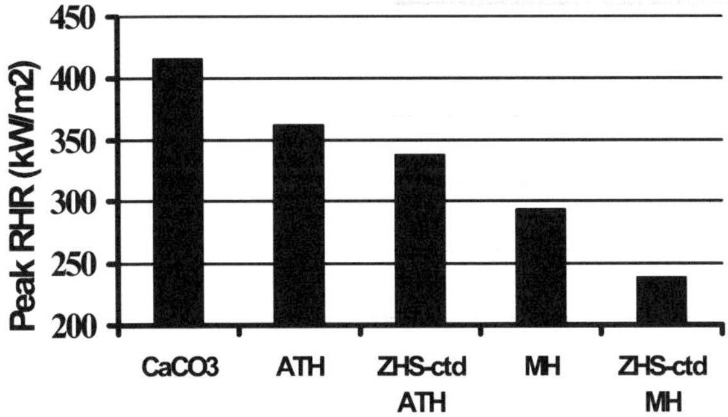

Figure 3 *Heat release rate from EVA compositions filled with ATH, MH and coated variants of these fillers*

overall filler levels. For example, in polyolefin wire and cable formulations, magnesium, hydroxide in combination with chlorinated additives was reported to show synergism and reduced smoke emission.[59]

Zinc stannates are effective fire retardants and smoke suppressants in halogenated polymers, such as PVC; however, when combined as a physical mixture with MH in ethylene-acrylate copolymer, a significant enhancement in oxygen index results compared to the hydrated filler in isolation. In addition, techniques have been developed for applying coatings of zinc stannate or zinc hydroxystannate onto hydrated fillers.[60,61] These novel systems provide a more effective means for presenting the tin-containing synergist to the polymer, yielding markedly improved flammability performance relative to physical admixtures of these fire retardant types. Results obtained using PVC offer the prospect of gaining acceptable fire retardancy behaviour at much lower overall filler levels. Similar effects have been found in EVA copolymer, where ZHS-coated hydrated fillers enable large reductions in heat release rate and smoke emission, even though halogens are absent in this system (Figure 3).

The combination of 10phr silicate layer nanoclay with hydrated fillers significantly improves fire performance (Figure 4) for EVA filled with ATH and nanoclay–ATH mixtures.[62] This apparent synergism arises from increased char promotion evident when the nanoclay is present. However, it is significant that inclusion of ZHS to this system, either as an admixture or in coated form, yields a further improvement in performance with this polymer. Using MH in place of ATH gives even better results (Figure 5).

2.4 Processing and Considerations on Mechanical Property

2.4.1 Rheological Issues

Extensive literature on the rheology of polymer suspensions containing rigid fillers demonstrates that their flow behaviour is strongly affected by the filler

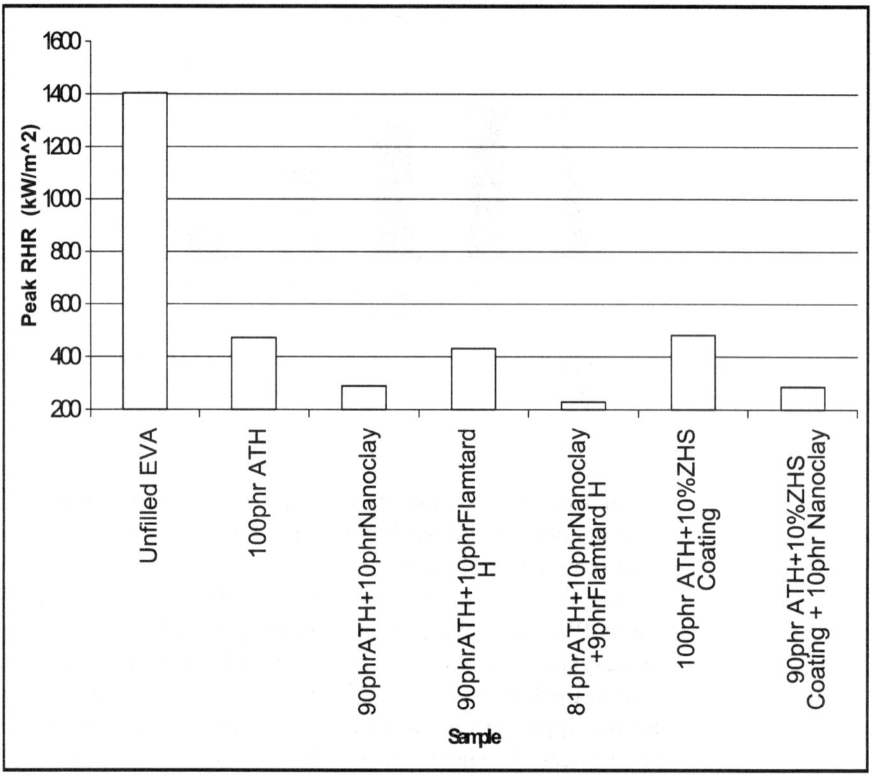

Figure 4 *Heat release rates from EVA filled with ATH, ZHS (Flamtard H), ZHS-coated ATH and nano-clay*

type, including its particle size, morphology, surface chemistry and concentration.[63] Characterisation of such materials normally involves measurement of the interdependence of shear viscosity with applied shear stress and shear rate, the effects of uniaxial extension on elongational viscosity and the consequences of applied frequency on real and imaginary components of complex dynamic modulus.[64] The inclusion of fillers into viscoelastic polymers generally reduces melt elasticity and hence the magnitude of processing related phenomena, such as die swell.[65] The occurrence of a yield stress is also a common phenomenon with highly filled polymer melts.[64] At stresses below a threshold value the material has unbounded viscosity, generally capable of only elastic deformation, but experiences flow at higher applied stress levels. The effect is strongly influenced by particle interactions within the matrix and the formation of a network structure. Many investigations have reported shear yield values for polymers through extrapolation of capillary and dynamic viscosity measurements to 'zero' shear rate or stress, using a wide range of polymer and filler types.[66–70]

Despite the relevance of this previously reported work to the rheology of filled polymer systems, little has been published on the flow characteristics of

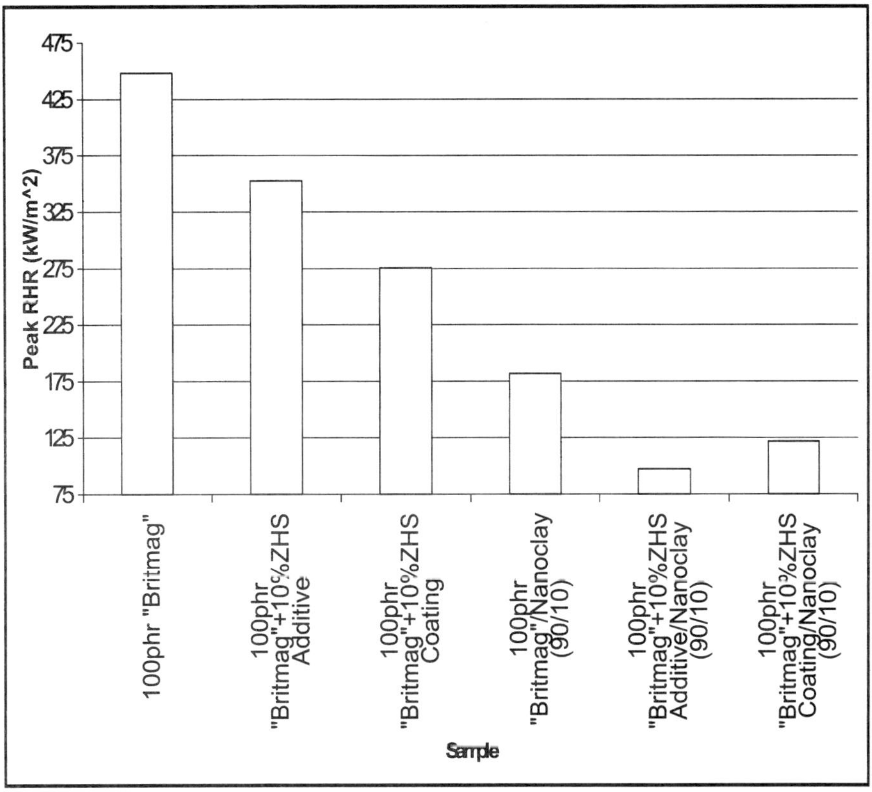

Figure 5 *Heat release rates from EVA filled with MH (Britmag), ZHS, ZHS-coated MH and nano-clay*

polymers containing hydrated fillers. However, it has been shown, using both capillary and dynamic viscosity measurements, that magnesium hydroxide particle morphology, size and applied surface treatment all influence the rheology of filled polypropylene melts.[71] At a filler concentration of 60% by weight, the presence of magnesium hydroxide caused a significant increase in the shear viscosity of polypropylene relative to unfilled polymer, although this was much less pronounced using magnesium stearate surface treated filler, especially at low shear rates. Complex viscosity and storage modulus data, obtained at very low shear rates (0.002 s^{-1}), demonstrated the presence of a critical shear yield stress for flow to occur, which was greatest for compositions containing uncoated fillers with very small particle size. These effects were attributed to the formation of network structures, arising from strong interactive forces between the particles.

Rheological effects, resulting from inclusion of magnesium hydroxide have also been investigated in relation to dripping tendencies of filled polyamide 6 and 6,6 composites during combustion.[72] Low shear dynamic viscosity measurements were undertaken on polymer compositions filled with magnesium hydroxide variants differing in particle surface area and filler morphology. Plate-like magnesium hydroxide particles gave the highest viscosity in both polymers studied, and, when using grades with less well defined morphology, surface area effects predominated. Increasing filler surface area gave rise to higher melt viscosities, reflecting greater immobilization of polymer chains at the filler interface.

Much of the foregoing discussion has focussed on thermoplastics composites modified using hydrated fillers; however ATH, as mentioned previously, is widely used in thermosets. For example, in unsaturated polyester compositions, the influence of high loadings of this filler on resin viscosity has a crucial effect on processability of dough moulding compounds (DMC) and sheet moulding compounds (SMC).[73] Selection of ATH grade is a key factor in controlling the extent of this viscosity increase.

Addition of untreated magnesium hydroxide to polyester resins causes a progressive thickening reaction, resulting in reduced workability. The observed thickening effect arises from an increase in the apparent molecular mass of the polyester resin due to polymer chain extension, through salt formation.[74]

To assess the feasibility of using magnesium hydroxide, as an alternative fire retardant to ATH in polyester resins, specially coated grades have been prepared in order to inhibit the thickening effect described above.[75] Whereas unmodified magnesium hydroxide caused substantial resin thickening after only two to three days, filler variants surface treated with maleic anhydride imparted minimal changes to resin viscosity (similar to aluminium hydroxide) for a period of up to fifty days.

2.4.2 Enhancement of Mechanical Properties

Surface modification of fillers used in plastics can be beneficial for several reasons associated with the production; storage, handleability, processability of polymer compounds and as a means for enhancing the mechanical properties of polymer composites.[7] Generally, when mineral fillers are added to plastics a large increase in elastic modulus is accompanied by a decline in strength and, in particular, toughness.[76] The magnitude of these effects may be significant, especially when filler levels are high, potentially limiting the application of the material. The degree of interfacial adhesion at the boundary between polymer matrix and filler has a critical bearing on the ability of a particulate composite to resist crack propagation and can be strongly influenced by the nature of surface treatments applied to the filler, or reactive sites within the polymer matrix.[77] Surface modification plays an especially important role in many applications of fire retardant fillers.

In thermoset systems, surfactants are frequently used to improve dispersion, thus dramatically lowering viscosity and allowing higher filler loadings to be achieved. These treatments are usually pre-dissolved in the liquid resin, rather than pre-coated onto the filler.

In addition to the surfactants, surface modifiers are frequently used to improve polymer–filler interaction and hence improve processing and mechanical properties. The types in most use are fatty acids and organo-silanes.

The fatty acids bond to suitable (basic) sites on the filler surface through the formation of carboxylate salts. The most common fatty acid used is actually a blend, with a composition close to that of stearic acid ($C_{17}H_{35}COOH$). The hydrocarbon tail renders the surface hydrophobic, and improves its compatibility with organic polymers. This improves processing, but gives a weak interface with the matrix. This weak interface promotes early filler debonding under stress that, under favourable circumstances, can improve elongation and impact strength, but often causes a decrease in tensile strength. In addition to the effects described above, fatty acid treatments can reduce water adsorption by the filler when in a composite (an important factor in electrical applications) and can also reduce carbon dioxide attack and chalking in composites filled with fillers like magnesium hydroxide.

Organo-silanes are used as coupling agents, reacting with both the filler surface and the polymer matrix and tying the two together. This delays debonding and increases strength. The effects on elongation and toughness are variable, and may be positive or negative depending on the system involved. The organo-silanes are based on a silicon atom, which carries three alkoxy groups and also has one alkyl group carrying a functionality that can react with the host polymer and is attached through a silicon–carbon bond. The polymer reactive functionality is chosen to suit the matrix, but the most common ones for use with fire retardant fillers are vinyl, methacryl and amino. Reaction with the filler is through hydrolysis of the alkoxy groups to form silanols, which then condense with metal hydroxyls at the filler surface.

Organo-silanes can be pre-coated onto the fillers, or added during the compounding operation, while fatty acids always appear to be pre-coated.

Surface treatments can both affect the thermal stability of the filler and influence the fire retardant performance (not necessarily because of the change in stability). Thus, fatty acids are reported to decrease the stability of magnesium hydroxide and hydromagnesite,[78,79] while both decreases and increases in fire retardant performance have been reported to be associated with the use of various types of modifier.[80]

2.4.3 Alternative Processing Strategies for Hydrated Fillers

As discussed earlier, the advantages to be gained from lowering filler levels has prompted a number of approaches involving co-additives for hydrated fillers, which in many cases appear to work synergistically, maintaining flammability performance, but at a reduced overall filler content. This in turn should lead improved mechanical properties.

An alternative and potentially commercially promising method to achieving the necessary balance of fire retardancy and mechanical performance in processed parts is to use multi-component polymer processing technologies widely adopted in industry, in particular co-extrusion and co-injection moulding. By this means it is possible to structure the component to control the location, type and amount of fire retardant, to achieve maximum effect without unduly compromising mechanical properties. In principle, different fire retardants may

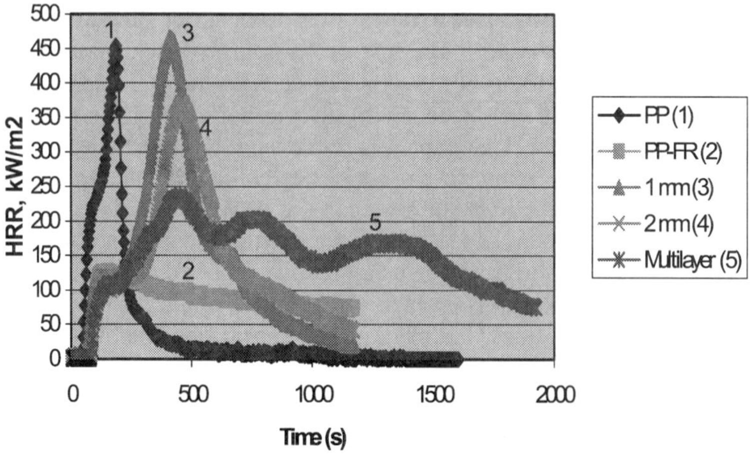

Figure 6 *Heat release rate from 6 mm thick single component and multilayer fire retardant materials*

be combined, fire retardant level may be graded and/or reinforcing additives may be judiciously introduced all within the same part.[81]

Using this concept, it has been shown that over a 3 min combustion period, 6 mm thick laminated structures, made with different skin-core combinations made by compression moulding a series of 1 mm thick injection mouldings, can give similar resistance to ignition and comparable HRR and SEA results to fully fire retardant compositions (Figure 6). Using co-injection moulding technology, similar effects were seen in 3 mm thick parts made with 0.5 and 1 mm thick fire retardant skins. Mechanical properties, in particular impact strength, were also greatly enhanced by this approach, since less fire retardant filler is present in the material (Figure 7).

2.5 Conclusions

The application of fire retardant fillers in plastics is dominated by two major types, aluminium and magnesium hydroxides, although there are many variants of these, which differ in their particle size, particle morphology, chemical purity and origin. They are widely used on account of their environmental acceptability, being free from halogen and phosphorus compounds, and their strong smoke suppressing tendencies. Their major drawback is their relative fire retarding inefficiency, necessitating use of high filler loadings to obtain acceptable performance levels, which adversely influences mechanical properties and processability. Use of filler surface treatments and melt processing aids can offset these physical limitations, however, and, in addition, the application of two-component processing technologies offers a promising alternative approach to achieve the same end.

Fire retardant fillers function principally in the condensed phase, through endothermic decomposition, water release, and oxide residue formation, which

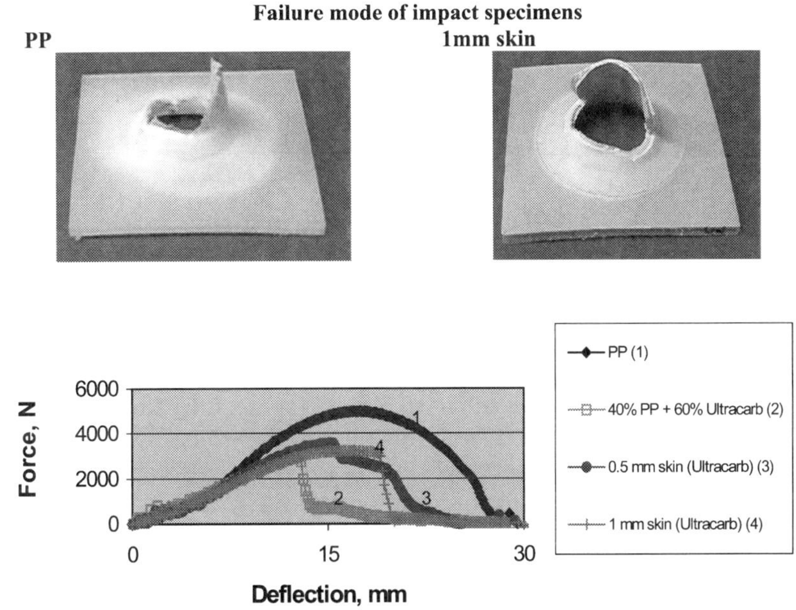

Figure 7 *Force deflection curves from falling weight impact tests on single- and two-component injection moulded polypropylene (Ultracarb filler)*

inhibits thermal feedback. Combination with fire retardant synergists can significantly reduce overall filler level to achieve an acceptable level of performance and forms an active area of current research.

2.6 References

1. R.N. Rothon, in *Particulate Filled Polymer Composites*, 2nd Edn, R.N. Rothon (ed.), RAPRA Technology, Shawbury Shrewsbury, U.K., 2003, ISBN: 1-85957-382-7, pp. 270–273.
2. M. Hancock and R.N. Rothon, in reference 1, pp. 88–90.
3. H. Kurisu, T. Kodani, A. Kawase and T. Oki, *US Patent* 5,766,568, 1998, Tateho Chemical Industries Co Ltd assignees.
4. M. Doyle, M. Clemens, G. Lees, C. Briggs and R. Day, *Proceedings of Flame Retardants '94*, Interscience, London, 1994, p. 193.
5. R. Sauerwein, *Proceedings of Fire and Materials 2001*, Interscience, San Francisco, 2001, 395.
6. D.A. Rust, *Proceedings of Functional Fillers and Reinforcements '99*, Intertech, Atlanta, Georgia, USA, Paper 22, 1999.
7. R.N. Rothon, in reference 1, pp. 264–269.
8. P.R. Hornsby, *Macromol. Symp.*, 1996, **108**, 203.
9. P.R. Hornsby and C.L. Watson, *Plastics Rubber Process. Applic.*, 1989, **11**, 45–51.

10. P.R. Hornsby, *Fire Mater.*, 1994, **18**, 269.
11. P.R. Hornsby, J. Wang, K. Cosstick, R. Rothon, G. Jackson and G. Wilkinson, *Progr. Rubber Plastics Technol.*, 1994, **10(3)**, 204–220.
12. R.N. Rothon, in *Particulate Filled Polymer Composites*, R.N. Rothon (ed.), Longman, New York, 1995, Chapter 6.
13. P.R. Hornsby and C.L. Watson, in *IOP Short Meetings* Series No 4, Institute of Physics, London, April 1997, p. 17.
14. P.R. Hornsby and A. Mthupha, "Mechanism of fire retardancy in magnesium hydroxide filled polypropylene", in *Proceedings from Society of Plastics Engineering*, Annual Technical Conference (ANTEC '93) New Orleans, May 9–13 1993, pp. 1954–1956.
15. N.R. Dando, T.R. Clever, A. Pearson, J.M. Stinson, P.L. Kolok and E.S. Martin, "Aluminium trihydroxide (ATH) as a filler for polymer composites: Improvements in thermal stability by controlled precipitation", in *Proceedings from 50th Annual Technical Conference*, Composite Institute, Society of Plastics Industry Inc, Session 1-D, p1-4, 1995.
16. I. Sobolev and E.A. Woycheshin, in *Handbook of Fillers for Plastics*, H.S. Katz and J.V. Milewski (eds.), Van Nostrand Reinhold, New York, 1987, Chapter 16.
17. G.V. Jackson, R.N. Rothon and G.A. Moorman, in *Proceedings of Fillplas'92*, Manchester, U.K., Paper No. 10, 1992.
18. R.N. Rothon, in reference 1, pp. 277–280.
19. P. Hughes, G.V. Jackson and R.N. Rothon, *Makromol. Chem. Makromol. Symp.*, 1993, **74**, 179.
20. M.J. Herbert, in *Proceedings of Flame Retardants'94*, Interscience, London, 1994, p. 59.
21. A. Tewarson, in *Flame Retardant Polymeric Materials*, M Lewin, S.M. Atlas and E.M. Pearse (eds.), Plenum Press, New York, 1982.
22. I. Spilda, M. Kosik and A. Blazej, *J. Appl. Polym. Sci.*, 1986, **31**, 589.
23. P.R. Hornsby and C.L. Watson, *Polym. Degrad. Stab.*, 1990, **30**, 73–87.
24. P.R. Hornsby and A. Mthupha, Plastics, *Rubber Composites Process. Applicat.*, 1996, **25**, 347–355.
25. K. Itatani, *J. Mater. Sci.*, 1988, **23**, 3405.
26. P.R. Hornsby, J. Wang, R. Rothon, G. Jackson, G Wilkinson and K. Cosstick, *Polym. Degrad. Stab.*, 1996, **51**, 235–249.
27. P.R. Hornsby, J. Wang, G. Jackson, R.N. Rothon, G. Wilkinson and K. Cosstick, "Analysis of fire retardancy in polyamides modified with magnesium hydroxide filler", in *Proceedings from Society of Plastics Engineers Annual Technical Conference (ANTEC '94)*, 1–5 May 1994, San Francisco, pp. 2834–2839.
28. J. Wang, *Mechanism of Flame Retardancy in Polyamides Containing Magnesium Hydroxide*, PhD thesis, Brunel University, U.K., 1994.
29. J. Rychlý, K. Vesely, E. Gal, M. Kummer, J. Jancar and L. Rychla, *Polym. Degrad. Stab.*, 1990, **30**, 57.
30. S. Miyata, T. Imahshi and H. Anabuki, *J. Appl. Polym. Sci.,* 1980, **25**, 415.

31. W.E. Horn, in *Fire Retardancy of Polymeric Materials*, A.F. Grand and C.A. Wilkie (eds.), Marcel Dekker, Basel, Swisserland, 2000, Chapter 9.
32. M. Hancock and R. Rothon, in *Particulate-Filled Polymer Composites*, R. Rothon (ed.), Longman, Harlow, 1995.
33. D.F. Lawson, E.L. Kay and D.T. Roberts, *Rubber Chem. Technol.*, 1975, **48**, 124.
34. M. Moseman and J.D. Ingham, *Chem. Technol.*, 1978, **51**, 970.
35. M.M. Hirschler and T.R. Thevaranjan, *Eur. Polym. J.*, 1985, **21(4)**, 371.
36. J. Lahaye, *Polym. Degrad. Stab.*, 1990, **30**, 111–121.
37. I. Jagoda, G. Prado and J. Lahaye, *Combus. Flame*, 1980, **37**, 261.
38. O.V. Krylov, in *Catalysis by Non-Metals*, Academic Press, London, 1970.
39. D.W. McKee in "The catalyzed gasification reactions of carbon", in *Chemistry and Physics of Carbon*, P.L. Walker and P.A. Thrower (eds.), Marcel Dekker, New York, 1981, Volume 16.
40. P.L. Walker, M. Shelef and R.A. Anderson, "Catalysis of carbon gasification" in *Chemistry and Physics of Carbon*, P.L.Walker Jr. (ed.), Edward Arnold, London, 1968, Volume 4.
41. C. Baillet, L. Delfosse, S. Antonik and M. Lucquin, *C.R. Acad. Sci., Ser. C*, 1972, **274**, 146.
42. L. Delfosse, C. Baillet, A. Brault and D. Brault, *Polym. Degrad. Stab.*, 1989, **23**, 337.
43. G.S. Kirschbaum, in *Proceedings of the 1995 Fall Conference of the Fire Retardant Chemical Association*, Rancho Mirage, CA, October 29–November 1 1995, Technomic Publishing Co., Lancaster, 1995, p. 145.
44. G.S. Kirschbaum, *Kunststoffe*, 1989, **79(11)**, 62–64.
45. S. Bourbigot, F. Carpentier, C. Fernandez, J.P. Amoureux, M. Le Bras and B. Revel, in *Proceedings from 6th European Meeting on Fire Retardancy of Polymeric Materials (F.R.P.M.'97)*, M. Le Bras *et al.* (eds.), AGIR/MITI Pub., Lille, France, 1997, pp. 120–121.
46. H. Staendeke, in *Proceedings of the Spring Conference of the Fire Retardant Chemical Association*, March 20–23, (1988), F.L. Grenelefe (ed.), Technomic Publishing Co., Lancaster, PA, 1988, p. 32.
47. E.D. Weil, M. Lewin and H.S. Lin, *J. Fire Sci.*, 1998, **16**, 383–404.
48. M.J. Chavez, D.J. Romanesco, in *Proceedings of the 1995 Fall Conference of the Fire Retardant Chemical Association*, Rancho Mirage, CA, October 29–November 1 1995, Technomic Publishing Co., Lancaster, PA, 1995, p. 169.
49. M.S. Huber, in *Proceedings of the Spring Conference of the Fire Retardant Chemical Association*, March 25–28 1990, New Orleans, FL Technomic Publishing Co., Lancaster, PA, 1990, p. 237.
50. T. Imahasi, A. Okada and T. Abe, *U.S. Patent 5,583,172*, Dec. 10 1996.
51. W. Zhu and E.D. Weil, *J. Appl. Polym Sci.*, 1995, **56**, 925–933.
52. B. Toure, J.M.L. Cuesta, P Gaudon, A. Benhassaine and A. Crespy, *Polym. Degrad. Stab.*, 1996, **53**, 371–379.
53. Y. Namike, Y. Kato, Y. Kitano, H. Kuriso and Y. Yokata, *US Patent No. 5,654,356*, August 5th 1997.

54. S.H. Chiu, and W.K. Wang, *Polymer*, 1998, **39**, 1951–1955.
55. J.T. Yeh, H.M. Yang and S.S. Huang, *Polym. Degrad. Stab.*, 1995, **50**, 229–234.
56. G. Bertelli, P. Goberti, R. Marchini, G. Camino and M.P. Luda, “Combined malamine/mineral fillers as fire retardants for polypropylene” in *Proceedings 6th European Meeting on Fire Retardancy of Polymeric Materials (F.R.P.M. ’97)*, M. Le Bras *et al.* (eds.), AGIR/MITI Pub., Lille, France, 1997, p. 34.
57. C. Baillet and L. Delfosse, *Polym. Degrad. Stab.*, 1990, **30**, 89–99.
58. R.L. Markezich and D.G. Aschbacker, *ACS Symposium Series*, 1995, **599**, 65–75.
59. S. Miyata and T. Imahasi, *US Patent No.5,094,781*, March 10th 1992.
60. R.G. Baggaley, P.R. Hornsby, R. Yahya, P.A. Cusack and A.W. Monk, *Fire Mater.*, 1997, **21**, 179.
61. P.A. Cusack and P.R. Hornsby, *J. Vinyl Additive Technol.*, 1999, **5**, 21–30.
62. M. Cross, *The Development and Application of Halogen-Free Tin-based Fire-Retardant Additives in EVA*, EngD Thesis, Brunel University, UK, 2004.
63. P.R. Hornsby, in *Advances in Polymer Science: Mineral Fillers in Thermoplastics*, J. Jancar (ed.), Springer-Verlag, Munich, Germany, 1999, Volume 139, pp. 155–217.
64. A.Y. Malkin, *Adv.* in *Polym. Sci.*, 1990, **96**, 69.
65. P.J. Carreau and P.A. Lavoie, *Macromol. Symp.*, 1996, **108**, 111.
66. Y. Suetsugu and J.L. White, *J. Appl. Polym. Sci.*, 1983, **28**, 1481.
67. C.D. Han, in *Multiphase Flow in Polymer Processing*, Academic Press, New York, 1981, Chapter 3.
68. C.D. Suh and S.L. White, *J. Non-Newtonian Fluid Mech.*, 1996, **62**, 175.
69. L. Lin and T. Masuda, *Polym. Eng. Sci.*, 1990, **30(14)**, 841.
70. G.J. Osanaiye, A.I. Leonov and J.L. White, *J. Non-Newtonian Fluid Mech.*, 1993, **49**, 87.
71. P.R. Hornsby and A. Mthupha, *J. Mater. Sci.*, 1994, **29**, 5293.
72. P.R. Hornsby, J. Wang, K. Cosstick, R. Rothon, G. Jackson and G. Wilkinson, *Progr. Rubber Plastics Technol.*, 1994, **10(3)**, 204.
73. E.A Woycheshin and I Sobolev, “Alumina trihydrate”, in *Handbook of Fillers and Reinforcements for Plastics*, H.S. Katz and J.V. Milewski (eds.), Van Nostrand Reinhold, Amsterdam, 1978, Chapter 14.
74. V.I. Szmercsanyi and A. Szilagyi, *J. Polym. Sci. Polym. Chem.*, 1974, **12**, 2155–2163.
75. W.Z.A. Wan Hanafi, *A Study of Magnesium Hydroxide as A Flame Retardant and Smoke Suppressant Filler for Unsaturated Polyesters*, MPhil Thesis, Brunel University, Uxbridge, UK, 1988.
76. P.R. Hornsby and C.L. Watson, *Plast. Rubber Proc. Appl.*, 1986, **6**, 169.
77. P.R. Hornsby and C.L. Watson, *J. Mater. Sci.*, 1995, **30**, 5347–5355.
78. M. Hancock and R. N. Rothon, in reference 1, p. 90.

79. L. Haurie, A. I. Fernandez, J. L. Velasco, J. M. Chimenos, J. R. Tico Grau and F. Espeill, in *Proceedings of Eurofillers'03*, Alicante, Spain, Paper A3, September 2003.
80. R.J. Ashley and R.N. Rothon, *Plastics, Rubber Composites Process. Applicat.*, 1991, **15**, 19.
81. P.R. Hornsby, A. Ahmadnia, G. Marosi and P. Anna, "Tailoring the fire retardant performance of polymers using multi-component processing technologies" in *Proceeding from Society of Plastics Engineers Annual Technical Conference (ANTEC '03)*, Nashville, USA, 5–8 May 2003, Soc. of Plastics Engineers, Brookfield, CT, USA, 2003.

CHAPTER 3

Lamellar Double Hydroxides/ Polymer Composites: A New Class of Fire Retardant Materials

JÉRÔME LEFEBVRE, MICHEL LE BRAS AND SERGE BOURBIGOT

PERF (Procédés d'Elaboration de Revêtements Fonctionnels), UPRES EA 1040, E.N.S.C.L.-U.S.T.L, BP108, F-59652 Villeneuve d'Ascq Cedex, France (michel.le-bras@ensc-lille.fr)

3.1 Introduction

In recent years polymeric materials have been filled with several synthetic and/or natural minerals to improve mechanical properties (such as modulus, tensile strength, hardness, *etc.*) and/or to increase the fire properties of the materials. In the field of fireproofing, the use of inorganic additives enables one to replace advantageously halogen-based fire retardants in many applications because of their better environmental impact. The most widely studied inorganic flame retardant additives are metallic hydroxides, phosphorilated compounds and clays.

Systems based on metallic hydroxides develop, during their degradation, one or more endothermic reactions. A part of the heat fed back heat to the virgin polymer is used degrade the hydroxides and so reduce the degradation of the polymer. The resulting water production acts in the gas phase by dilution.[1] Hydroxides also act as catalysts for the oxidation of the polymeric matrix and/ or the products of its degradation and lead to a decrease in the CO/CO_2 ratio.[2] The oxides, produced by the decomposition, can contribute to the formation of an insulative charred layer acting as a further protection for the polymer.[3] Currently, the processes of fireproofing using hydroxides find significant application in the field of electrical cable-making. However, to be effective, these compounds must be used in large quantities, often higher than 60% in mass, which can lead to a significant reduction in the mechanical properties of material.

Recently, polymer/clay nanocomposites[4] have attracted great interest because they often exhibit improved materials properties in comparison with the virgin polymer and micro- and macro-composites. These improvements include better mechanical properties, higher heat resistance and lower flammability.[5–7] Montmorillonite, hectorite and saponite are the most commonly used layered silicates. The "nanocomposites" flame retardant mechanism involves a high carbonaceous-silicate char, which builds up on the surface during burning. This insulates the underlying material and slows the mass loss rate of decomposition products.

A new class of nanoscale additives very close to clays is emerging, *i.e.* layered double hydroxides (LDHs) that are very tunable 2D host materials. The organic/inorganic hybrid LDHs nanocomposites are used, for instance, in many application such as catalysts,[8] electrochemical sensors[9,10] or adsorbents for organic pollutants. Recent studies on LDHs/Polymer nanocomposites describe their nanoscale arrangement as increasing the mechanical properties of the materials. Most of these articles deal with the synthesis and characterization of LDHs and LDHs/polymer nanocomposites[11] but very few deal with the modifications of the properties of the materials induced by addition of LDHs.

In this chapter, first the synthesis of LDHs and, secondly organic/inorganic LDHs hybrid materials are described. The state of the art in the properties of LDHs/polymer systems will then be presented. In conclusion LDHs/polymer composites are presented as potential candidate in the field of fire retardancy.

3.2 Description of LDHs Materials

LDHs are a kind of layered materials that consist of positively charged layers and interlayer exchangeable anions. Their chemical composition is generally represented as $[M(II)_{1-X}\,M(III)_X(OH)_2]^{X+}[A^{n-}_{X/n}\cdot mH_2O]^{X-}$ where $M(II)$ and $M(III)$ are divalent and trivalent cations respectively (Ni, Zn, Cu or Ca as divalent cations and Fe, Al, Ga, Cr as trivalent cations) and A^{n-} is an exchangeable anion (such as chloride, sulfate, nitrate or carbonate).

The structure is made of brucite-like layers constituted of edge-sharing $M(OH)_6$ octahedra between which anions are inserted (Figure 1). LDHs, like clays, are characterized by their anionic exchange capacity (AEC) and by the layer charge density. The presence of interlayer species is linked to the net layer charge value. The nature of the divalent and trivalent cations, their molar ratio and the size of the anion make it possible to modulate the properties of LDHs.

Due to the anion-exchange properties of these materials, many studies about the intercalation of organic anions into LDHs have been carried out. The low layer charge density of LDHs and so the greater area per charge is favorable for an intercalation of anions and an exfoliation process. Incorporation of various anions in the interlayer domain has been described.[11–16] An organic spacer molecule can be incorporated in the sheets to separate them and to weaken the attractive forces between layers, thus favoring exfoliation of LDHs

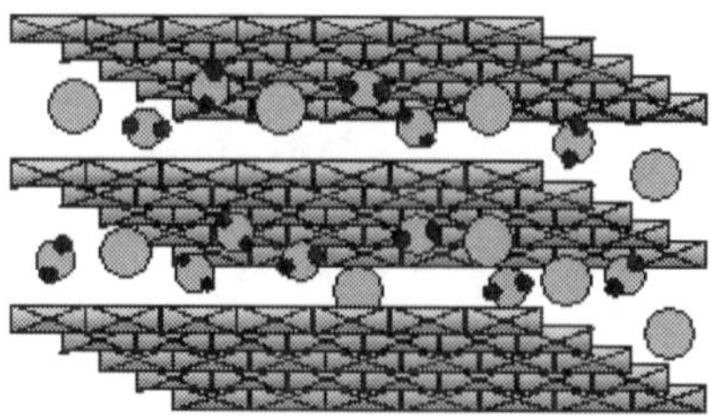

Figure 1 *Schematic structure of layered double hydroxides*

layers. The only criterion the anion has to satisfy is not to form a strong complex with the present cations. Intercalation can be obtained *via* exchange, coprecipitation or reconstruction.[17,18] A two-step method based on delamination and restacking has been also developed.[19]

3.3 Synthesis of LDHs/Polymer Nanocomposites

Polymer/LDHs nanocomposites can be referred to as organoceramics. There are many ways to synthesize LDHs/polymer nanocomposites[11] that can be listed as three principal options (Figure 2).

3.3.1 Intercalation of Monomer Molecules Followed by "*in situ*" Polymerization

Polymerization is carried out between the sheets by thermal, photoinduced or heat treatment. However, the size and the affinity of LDHs with the polymeric matrix can disturb the incorporation. This can be avoided by pre-intercalation of a spacer-molecule, a modifying layer agent or an *in situ* reactant. The "*in situ* polymerization" is limited by the degree of freedom of the monomers (affinity with the LDHs) and by some conditions of polymerization that can modify the layered structure.

3.3.2 Direct Intercalation of Extended Polymer Chains between LDHs Layers

After expansion and chemical change of the layers, the polymer can be incorporated between the layers.

3.3.3 Transformation of Host Material into a Colloid System and Precipitation in the Presence of the Polymer

Nanocomposite systems can be formed by restacking exfoliated layers. This method is only available for LDHs that can be delaminated. The layers are restacked on the monomer or on the whole polymer.

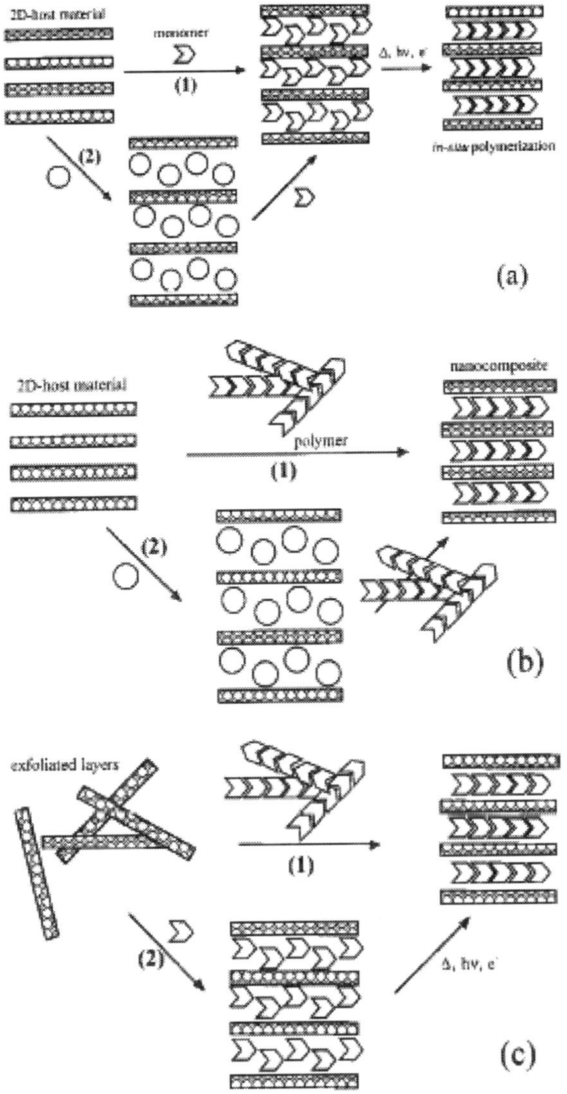

Figure 2 *Synthesis of LDHs/polymer nanocomposite:* (a) *monomer exchange and in situ polymerization,* (b) *direct polymer exchange and* (c) *restacking of the exfoliated layers*[11]

3.4 Mechanical Properties of LDHs/Polymer Composites

Generally, in inorganics/polymer composites (*e.g.* with ATH), inorganic and organic phases are separated because of their incompatibility, so reducing the mechanical properties of the materials. Organo-modification of the inorganic layer structure (by insertion between the sheets of organic anions in the case

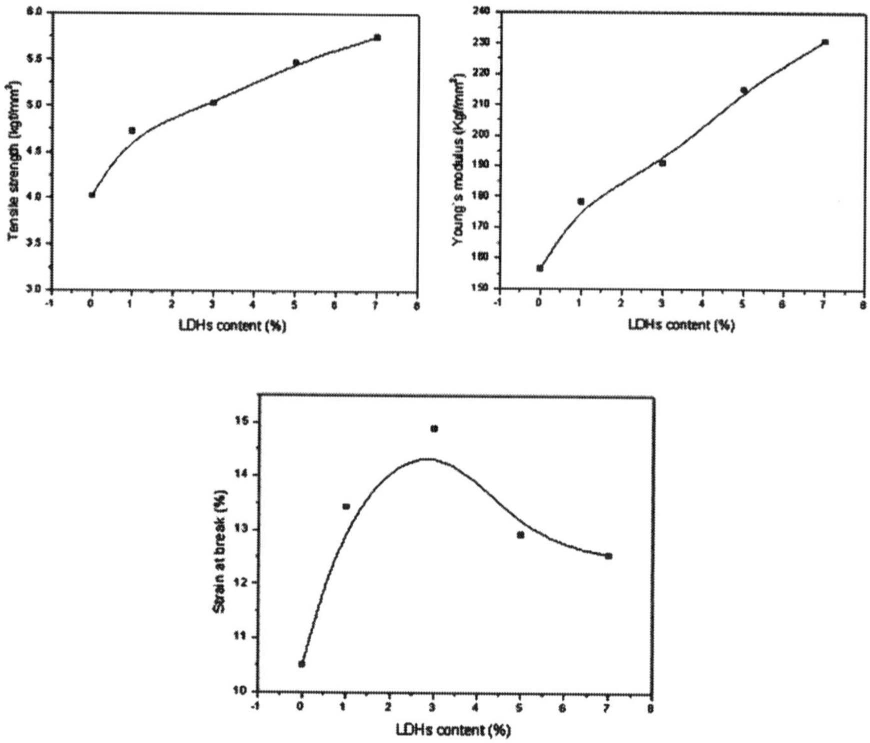

Figure 3 *Effect of LDHs on the mechanical properties of LDHs/epoxy nanocomposites*[15]

of LDHs or cations in the case of clays) allows a significant increase in the compatibility between the organic and inorganic phases, so leading to improved mechanical properties of interest. Hsueh *et al.* have discussed the modification of the tensile and viscoelastic properties of LDHs/epoxy[15] and LDHs/polyimide[14] nanocomposites systems using, respectively, amino laurate and amino benzoate as intercalated agents.

For epoxy resins, an organo-modified LDHs was prepared by intercalation of amino laurate into the LDHs layers *via* a coprecipitation method. Reaction between the intercalated amino laurate and the epoxy molecules favors the compatibility of the exfoliated nanocomposite system. A study of the mechanical properties (Figure 3) versus the LDH content clearly proves that the tensile strength and the Young's modulus are correlated with the content of LDHs: the higher the LDHs content, the higher the tensile strength and the Young's modulus. The strain at break increases and reaches a maximum at about 3 wt% LDH. With higher levels of LDHs the strain at break then decreases.

The elastics modulus of these nanocomposites increases with the LDHs content because of improved stiffness, and the loss modulus is slightly increased. This may be explained by a loss of mobility of the extensive segments of the epoxy main chain, leading to an increase in the glass transition temperature (from 48°C for 0 wt% LDH to 61°C for 7 wt% LDH; Figure 4).

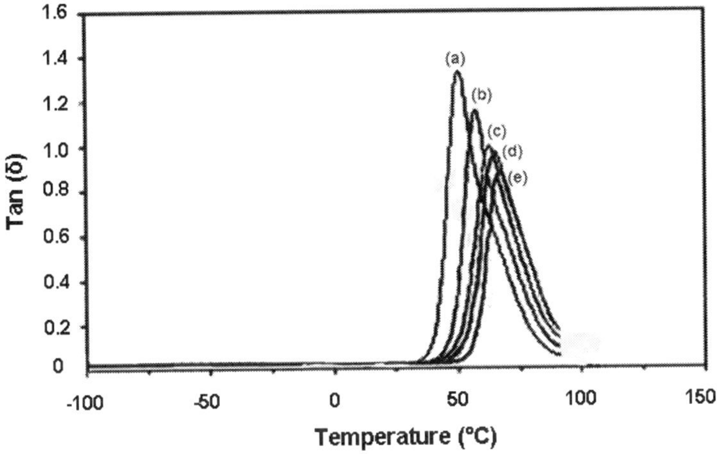

Figure 4 *Effect of LDHs on the glass transition temperature of LDHs/epoxy nanocomposites*[15]

A LDH modified by amino benzoate has also been used to synthesize LDHs/polyimide nanocomposites. The same effects on mechanical properties as observed for LDHs/epoxy systems are noted. For example, 5 wt% of organo-modified LDH leads to an increase in maximum tensile strength of the nanocomposite of about 43 relative % compared to pure polyimide. The use of only 4 wt% improves the elongation at break by about 53%. For levels of LDH higher than 5 wt% the tensile strength and the elongation at break dramatically decrease but remain higher than that of the pure polyimide.

3.5 Thermal Stability of LDHs/Polymer Nanocomposites

Thermogravimetric data under nitrogen of LDHs/polymer systems generally show that the decomposition temperatures of these nanocomposites increase with LDHs content. This temperature of decomposition depends on the dispersion behavior of the inorganic sheets and increases with very well dispersed nanolayers in the polymeric matrix. Such behavior is observed with LDHs/epoxy[15] and LDHs/polyimide[14] systems.

Studies on poly(vinyl alcohol) have shown that the incorporation of LDHs leads to a lower thermal stability at low temperature (up to 300°C) compared with pure PVA (Figure 5).[20]

Differential thermal analysis carried out on organo-modified PE-g-MA/MgAl-LDH exfoliated nanocomposite showed endothermic peaks at about 80 and 200°C that are assigned to the loss of water and to the deshydroxylation of the LDH, respectively (Figure 6). A sharp exotherm is also observed at 300°C. This event is supposed to speed up the weight loss of the nanocomposite and to

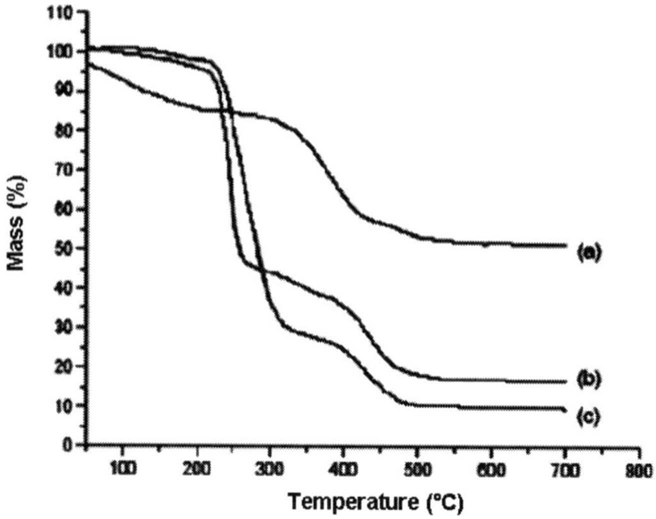

Figure 5 *Thermogravimetric analysis (under nitrogen at 10 °C min⁻¹) of the air-dried LDH* (a), *the nanocomposite LDH/PVA* (b) *and PVA* (c)[20]

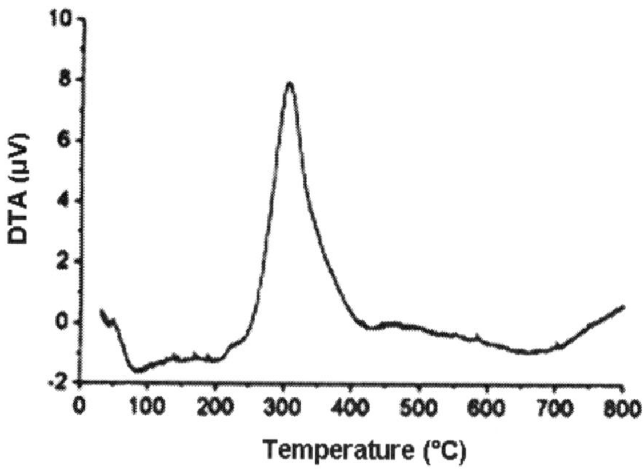

Figure 6 *Differential thermal analysis of an organo-modified LDH sample*[21]

promote the formation of a charred layer that enhances the thermal stability of the composite at high temperature.[21]

Thermal degradation of epoxy resin is characterized by a double stage process: in the first stage, at about 370°C, the main loss in weight occurs (about 75%), while in the second one, at 550°C, almost all the exothermic effect is developed, although the weight loss is only 25%.[22] Such a trend is also shown in the presence of ATH and organo-modified montmorillonite. When an organo-modifier is added, the thermal degradation is highlighted by the splitting of the main exothermic peak and by the consequent delay of more than 50°C of

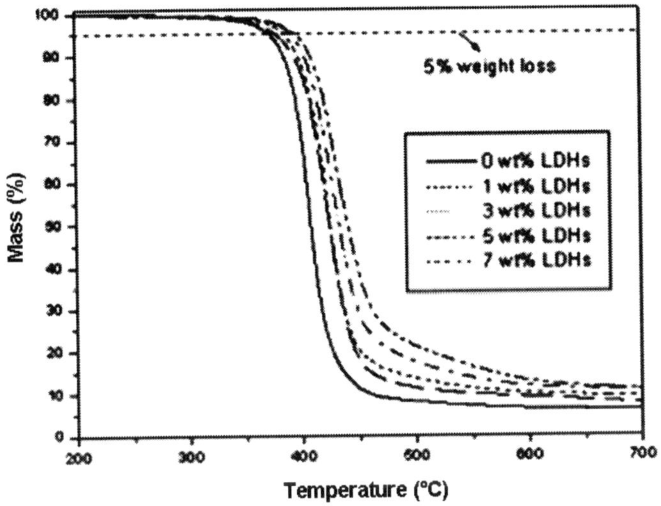

Figure 7 *Thermogravimetric analysis (under nitrogen at 20 °C min*$^{-1}$*) of a LDH/epoxy nanocomposites with various contents of LDH*[15]

the complete development of combustion heat and by the decrease in the rate of weight loss (Figure 7).[23,24]

Li *et al.*[20] have synthesized nanocomposites by exfoliation of ZnAl layered double hydroxides in nonpolar LLDPE solution. The nanomaterials obtained exhibit improved thermal stability compared with pure LLDPE (Figure 8).

The degradation rate for pure LLDPE becomes very quick after 350°C, while all the LLDPE/ZnAl LDH exfoliated nanocomposites show much slower

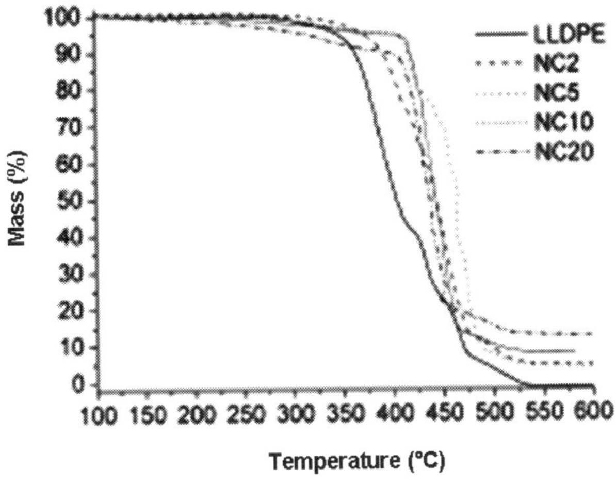

Figure 8 *Thermogravimetric curves under air of LLDPE and LLDPE/ZnAl LDH exfoliated nanocomposites containing increasing amount of LDH*[25]

degradation rates before 400°C. There is around 10% weight loss with each LLDPE/ZnAl LDH exfoliated nanocomposite in the range 200–400°C due to dehydration of ZnAl hydroxide sheets, thermal degradation of alkyl chains, and volatilization of thermo-oxidative product of LLDPE.

3.6 Flame Resistance of LDHs/Polymer Composites

LDHs inorganic fillers are described in the literature as agents that reduce the flammability of polymeric matrices. They are supposed to act by producing a refractory oxide on the surface of the material and by releasing aqueous vapour and carbon dioxide during its decomposition. This endothermic effect and the release of dilution gases increase the time to ignition and reduce the heat released during combustion.[1,23]

Zammarano *et al.* have studied the fire retardant properties of LDHs/epoxy nanocomposites and compared them with classical formulations containing aluminum trihydroxide, montmorillonite and ammonium polyphosphate.[24] From the UL94 horizontal burning test, organo-modified LDHs based nanocomposites show higher fireproof properties than conventional microcomposite formulations containing montmorillonite, ATH and neat LDH. This phenomenon may be explained by the formation of an intercalated nanostructure of metal oxides and char (LDHs maintain their layered structure after thermal decomposition[26]). Cone calorimeter testing of organo-modified LDH containing epoxy shows a reduction of the peak of heat release of about 51% compared with the virgin epoxy resin, while this reduction is about 27% for organo-modified clays and 62% for APP containing resins. In terms of weight loss, organo-modified clays and organo-modified LDHs have the same efficiency (9% residual weight) while APP leads to a final weight of about 90%. However, notably, like clays, LDHs do not modify the time to ignition of the material, which is 30 s higher than in the case of formulation containing APP.[24]

The smoke suppressing action of LDHs in polymeric matrices is the main topic discussed in the literature. Flame retardancy effects of LDH nanocomposites on epoxy resin (EP), poly(vinyl chloride) (PVC) and polyethylene (PE) and polyamide-6 (PA6)/polypropylene (PP) blends were studied. This has been highlighted using classical fire and smoke emission tests.

The oxygen index and the smoke density of LDHs/EP nanocomposite materials in conditions of no flaming combustion have been investigated.[27] The oxygen index of EP increased slightly, but the smoke suppression effect was remarkable when LDH was added in 20–60 parts per 100 parts of resin. The porous thermal decomposition products of LDH with large specific surface area gives smoke suppression effects by adsorbing the smoke and gases produced during combustion. Influences of LDHs on inflaming and smoking behavior of PVC have been studied by a NBS smoke density chamber.[28,29] Effects of loading of LDHs and its dispersion in PVC on smoke suppression were also investigated. The experimental results show that the smoke suppressing efficiency of PVC induced by LDHs powder doesn't increase as the contents of LDHs increases. However, better dispersed LDHs have a better smoke suppressing effect. LDHs have obviously different effects on smoke suppressing of PVC by different inflaming test means.[28] The results show that only 3–5% of LDHs can make the smoke-suppressing efficienly for the PVC/LDHs system reach up to 50%.[29] In addition, the smoke generation rate and maximum density of smoke of LDHs/PVC nanocomposite decrease by 40% with about 20–40 phr of LDH in comparison with pure PVC.[30] The combustion and thermal degradation behavior of polyamide 6 (PA-6)/polypropylene (PP) blends containing nano-layered double hydroxides (LDH) and ammonium polyphosphate (APP) were studied.[31] The flame-retardancy was improved by the synergistic LDH/APP addition. LDHs promote crosslinking and char formation during thermal degradation of the blends. The flame-retarding mechanism is related to the chemical and physical effects of the reaction products of LDHs and APP.

These results tend to show that LDH materials have a strong ability to modify the processes of degradation and combustion of polymers, leading to both condensed phase and gas phase mechanisms.

3.7 Conclusions

Polymer-layered silicate nanocomposites have been widely studied and present improved mechanical and fire properties. Layered double hydroxydes present, like clays, a layered structure and show the benefits of containing hydroxyls groups and water molecules. The synthesis of LDHs and LDHs/polymer nanocomposites have been investigated but few studies describe the contribution of LDHs in term of fire behavior and mechanical properties.

The possibility of synthesizing organo-modified LDHs allows improved compatibility between the mineral load and the polymer. This is due to improved mechanical properties of the composite and, in particular, to an increase in tensile strength and Young's modulus. However, as with clays, LDHs must be used with relatively weak contents of about 5 wt%.

Because of its layered structure, its hydroxyls groups and its water molecules, LDHs are supposed to be very efficient flame retardant. Moreover, it satisfies absolutely the global desire to develop environmentally friendly, cheap flame retardant additives. The few published studies show that nanocomposites systems LDHs/polymer have a reduced flammability compared with virgin polymer. This improvement of fire behavior results mainly in a significant decrease in the release of smoke during combustion. Testing using a cone calorimeter also reveals that LDH/polymer nanocomposites exhibit a lower peak of heat released than the virgin polymer. Possible synergistic effects, in particular with ammonium polyphosphate, have been shown. This association enables much improved fire properties.

LDHs/polymer nanocomposites can offer useful new properties compared to conventional materials and can lead to the emergence of a new class of multifunctional hybrid materials that may be suitable for various applications as diverse as protonic conductors, photochromic materials or fire retardant materials.

3.8 References

1. G. Camino, A. Maffezzoli, M. Braglia, M. De Lazzaro and M. Zammarano, *Polym. Degrad. Stab.*, 2001, **74**, 457–464.
2. L. Delfosse, C. Baillet, A. Brault and D. Brault, *Polym. Degrad. Stab.*, 1989, **23**, 337–347.
3. J. Yeh, M. Yang and S. Hsieh, *Polym. Degrad. Stab.*, 1998, **61**, 465–472.
4. S. Sinha Ray and M. Okamoto, *Prog. Polym. Sci.*, 2003, **28**, 1539–1641.
5. F. Dabrowski, M. Le Bras, S. Bourbigot, J.W. Gilman and T. Kashiwagi, in *Proceedings of Euro-fillers'99*, Lyon-Villerbanne, France, 6–9 September 1999.
6. S. Bourbigot, M. Le Bras, F. Dabrowski, J.W. Gilman and T. Kashiwagi, *Fire Mater.*, 2000, **24**, 201–208.
7. J.W. Gilman, T. Kashiwagi, E.P. Giannelis, E. Manias, S. Lomakin, J.D. Lichtenhan and P. Jones, 'Flammability properties of polymer-layered silicate nanocomposites' in *Chemistry and Technology of Polymer Additives*, S. Al-Malaika, A. Golovoy and C.A. Wilkie (eds.), Blackwell Science Pub., Oxford, UK, 1999, Chapter 14.
8. A. Vaccari, *Appl. Clay Sci.*, 1999, **14(4)**, 161–198.
9. B.R. Shaw and K.E. Creasy, *J. Electroanal. Chem.*, 1988, **243(1)**, 209–217.
10. J. Qiu and G. Villemure, *J. Electroanal. Chem.*, 1988, **428(1–2)**, 165–172.
11. F. Leroux and J.P. Besse, *Chem. Mater.*, 2001, **13**, 3507–3515.
12. F. Malherbe, and J.P. Besse, *J. Sol. State Chem.*, 2000, **155(2)**, 332–341.
13. M. Badreddine, A. Legroui, A. Barroug, A. De Roy and J.P. Besse, *Mat. Lett.*, 1999, **38(6)**, 391–395.
14. H.B. Hsueh and C.Y. Chen, *Polymer*, 2003, **44**, 1151–1161.
15. H.B. Hsueh and C.Y Chen, *Polymer*, 2003, **44**, 5275–5283.
16. M. Lakraimi, A. Legroui, A. Barroug, A. De Roy and J.P. Besse, *J. Mater. Chem.*, 2000, **10(4)**, 1007–1011.
17. S. Miyata, *Clays Clay Miner.*, 1975, **23(5)**, 369–375.
18. F. Kooli, I.C. Chisem, M. Vucelic and W.Jones, *Chem. Mater.*, 1996, **8(8)**, 1969–1977.
19. F. Leroux, M. Adachi-Pagano, M. Intissar, S. Chauviere, C. Forano and J.P. Besse, *J. Mater. Chem.*, 2001, **11(1)**, 105–112.
20. B. Li, Y. Hu, R. Zhang, Z. Chen and W. Fan, *Mater. Res. Bull.*, 2003, **38**, 1567–1572.
21. W. Chen and B. Qu, *Chem. Mater.*, 2003, **15**, 3208–3213.
22. R. Kotsilkova, V. Petkova and Y. Pelovski, *J. Therm. An. Cal.*, 2001, **64**, 591–598.
23. M. Zammarano, *Modifica delle Proprieta di ritardo alla fiamma del poli-(etilene-co-acetato di vinile) con cariche inorganiche.* Degree Thesis, University of Lecce, Italy, July 2000.
24. M. Zammarano, M. Franceschi, F. Mantovani, A. Minigher, M. Celotto and S. Meriani, "Flame resistance properties of layered-double-hydroxides/epoxy nanocomposites." in *Proceed. 9th european meeting on Fire*

Retardancy and Protection of Materials, M. Le Bras *et al.* (eds.), USTL Pub., Villeneuve d'Ascq, France, 17–19 September 2003.
25. W. Chen, L. Feng and B. Qu, *Chem. Mater.*, 2004, **16(3)**, 368–370.
26. W.T. Reichle, S.Y. Kang and D.S. Everhardt, *J. Catal.*, 1986, **101(2)**, 352–359.
27. Y. Zhao, F. Li, D.G. Evans, X. Duan and J.W. Hao, *Yingyong Huaxue*, 2002, **19(10)**, 954–957.
28. X. Zheng, D. Wu, Y. Liu and F. Zhu, *Zhong. Suliao*, 2003, **17(8)**, 36–38.
29. X. Zheng, Y. Liu, D. Wu, O. Meng, X. Duan and H. Wang, *Suliao Gongye*, 2002, **30(5)**, 40–41.
30. B.S. Huang, F. Li, H. Zhang, O.Z. Jiao, X. Duan and J.W. Hao, *Yingyong Huaxue*, 2002, **19(1)**, 71–75.
31. J. Xu, J. Hao, Y. Zhao and X. Duan, *Xiandai Huagong*, 2002, **22(3)**, 34–37.

CHAPTER 4

Effect of a Small Amount of Flame Retardant on the Combustion of PC, PBT and PET

TOMOHIRO OHKAWA,[1] TOMOYUKI ISHIKAWA[2] AND KUNIHIKO TAKEDA[2]

[1]Department of Materials, Physics and Energy Engineering, Graduate School of Engineering, Nagoya University, Furo-cho, Chikusa-ku, Nagoya 464-8603 Japan
[2]Institute for Advanced Research, Nagoya University, Furo-cho, Chikusa-ku, Nagoya 464-8603 Japan tak@leaf.ocn.ne.jp

4.1 Introduction

It has long been known that a small amount of red phosphorous is effective in flame retardancy.[1] Gilman and his co-workers discovered some nanocomposites that can be used as flame retardants when blended by 2–3% to polymers such as polyamide and polystyrene.[2] These compounds are noteworthy because the necessary amount of additives is small, whereas almost all other flame retardants need to be added to plastics by 10% or more to clear the V-0 level of the UL-94 test.[3,4]

Generally, it is very desirable for industry to decrease the amount of flame retardants added to plastics. This has especially been expected in the fields of electronic equipments and other sophisticated industries. When a flame retardant is freely soluble in polymer, the thermal durability, such as the heat distortion temperature, drops. Conversely, if it is insoluble, it reduces the mechanical strength, in particular, the impact strength. Researchers at General Electric (G.E.) and we have observed that 0.1 wt% or less of several compounds such as fluoroalkane sulfonic acid salts and polyphenols retard the combustion of plastics.

In particular, this study focused on the flame retardancy of polycarbonate (PC) because PC polymerized bisphenol-A is one of the most useful engineering plastics due to its toughness, transparency, easy processability and flame retardancy.[5–7] Thermal degradation, light stability and other characteristics of PC have been well studied because of its high practical use.[8–10] Many researchers have also studied the flame retardancy of PC and have proposed the mechanism and the method to enhance the flame retardancy.[11–13]

We have studyed the thermal degradation of engineering plastics and the causal relationship between thermal degradation routes and flame retardancy.[14–17] The thermal degradation routes and the computer simulation by which the routes can be quantitatively analyzed were partially successful, but the relation between the above two phenomena were not always recognized clearly. In particular, the flame retardancy cannot always be improved even when the polymer is successfully stabilized.[18,19] In this chapter, we analyze the degradation route and the kinetics problems of the blended-PC with the strong acid salt compare with that of red phosphorous.[20–22]

4.2 Experimental

PC, polybutylene terephthalate (PBT) and polyethylene terephthalate (PET) used in this study were manufactured by GE Co., Japan (Lexan-121), by Teijin Co., Ltd. C7000 and Kurare (KS-750RC). Perfluorobutane sulfonic acid potassium salts (PPFBS), trifluoromethane sulfonic acid potassium salts (PTFMS), polyhydroxyphenol (PPh), red phosphorous (Red-P) and metal oxides such as ZnO, La_2O_3, Cr_2O_3 were used as flame retardants.

These metal oxides were adsorbed on very fine porous silica particles as supporting materials that had been prepared as follows. An SiO_2/inorganic salt mixture solution was dried at 80°C and, subsequently, the dried lumps were calcined at 600°C. After cooling, the samples were dipped into hot water to leach the inorganic salt. The metal nitrates were dissolved in water and the porous silica was dipped in the solution. Subsequently, the slurry was dried by a rotary evaporator. The thin layer of the metal oxides on the silica particles was formed after calcination at 600°C. The metal oxides were analyzed and observed by SEM-EDX. (Scanning electron microscope/energy dispersive X-ray spectroscopy).

Figure 1 shows the X-ray diffraction diagram of Zn/silica powder. The signals of the specimens coincided with that of the authentic sample. The broad peak at 21° was assigned to the cristobalite-type crystal of silica. Very fine silica particles were observed by SEM.

The chemicals and the silica particle were processed with PC and PBT by a twin rotor manufactured by Custom Scientific Instruments (CSI) Co., Max Mixing Extruder CS-194A. The blended-PC with each flame retardant (FR) is written here as FR/PC. For example, Red-P/PC means the blended-PC with red phosphorous.

Py-GC/MS (manufactured by Shimadzu, Py: PYR-4A, and GC/MS: QP-5000) and TGA (Shimadzu: TGA-50) were used in analysing pyrolysis

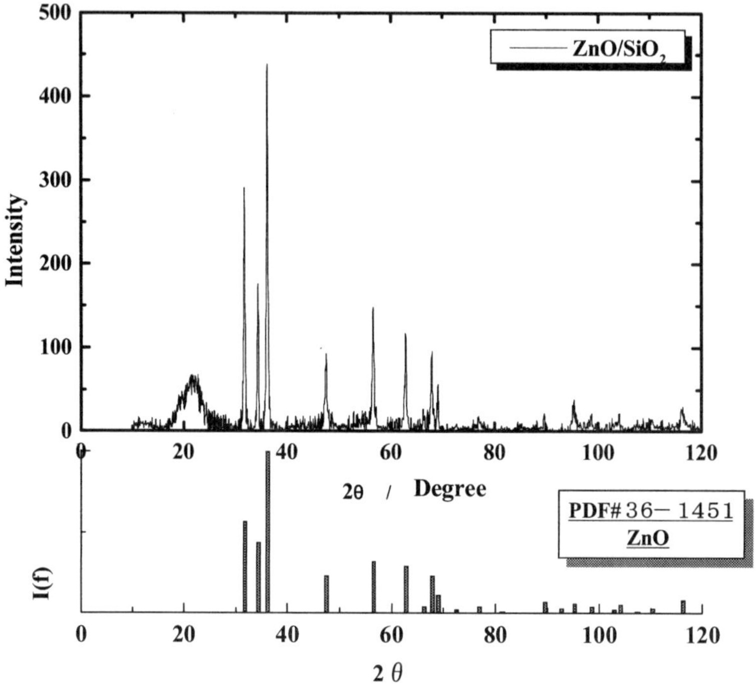

Figure 1 *X-Ray diffraction diagram of Zn/silica*

products and thermogravimetry, respectively. UL-test and a cone calorimeter (CC: manufactured by Toyo Seiki Seisaku-sho: C3 type) were used to measure flame retardancy.[23–25] Ten specimens per kind of samples were prepared in the UL-test, five of which were for measuring the flaming combustion time of the UL (Underwriter laboratory)-94 method, which is well-known as a burning test for plastics. The other five were used for "ignition time", which is defined as the time at which the observer judged the specimen began to fire by the same style of UL-test. It was difficult to decide the ignition time because two kinds of fires (a burner and a flaming combustion fire of specimen) could not be exactly recognized. Thus, the average of three specimens was calculated after the longest and the shortest times were eliminated. The Molic mouse method, developed by us, was applied for the computer simulation.[26,27]

4.3 Results

4.3.1 Combustion Data of Blends with PPFBS, PTFMS and PPh

The flaming combustion times of the UL-94V test as a function of the amount of additives (PPFBS in PC and PPh in PET) are plotted in Figure 2. Both examples restrained remarkably the flammability at a very low concentration level.

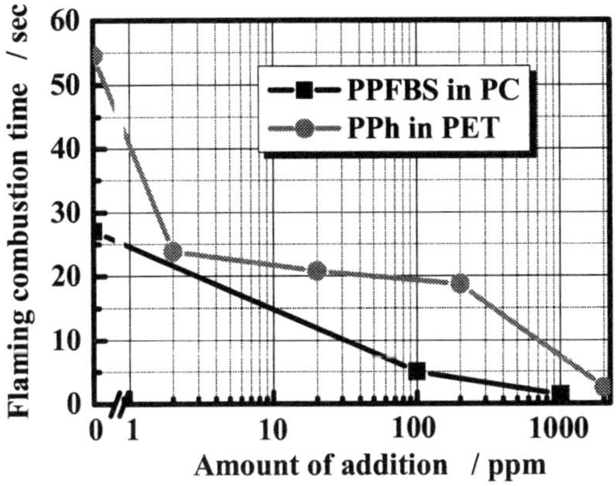

Figure 2 *Flaming combustion time as a function of the amount of additives*

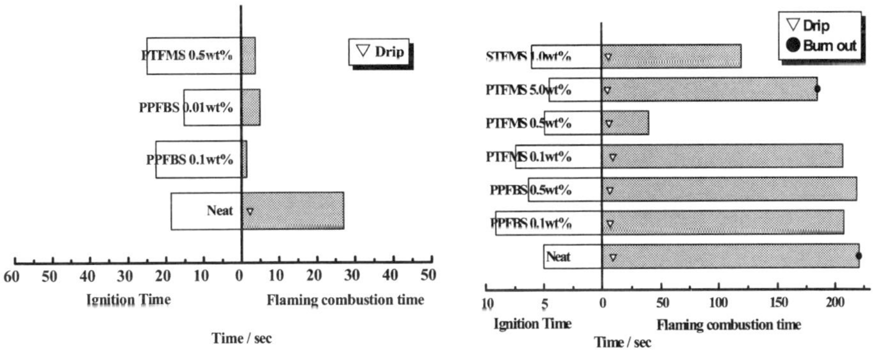

Figure 3 *Ignition and flaming combustion times of PPFBS/PC and PTFMS/PC (left) and FR/PBT (right)*

The flaming combustion times of PPFBS/PC and PTFMS/PC are compared in Figure 3 where the ignition and flaming combustion times can be simultaneously understood. A shift of the bar to the left means that the ignition time is longer and the flaming combustion time is shorter. A small amount of PTFMS as well as PPFBS to PC was effective on the flaming combustion time whereas the ignition times did not change clearly [Figure 3 (left)]. The results with PBT blends were more complex than those of PC. Only 0.5% of PTFMS was effective in shortening the flaming combustion time.

The tendency of the PET and PBT blends, both of which are classified as polyester group, was different (Figure 4). The flaming combustion time of the PPh/PET decreased with increasing PPh but that of PPh/PBT did not show such a clear tendency. The ignition times of various blends of these two polymers did not depend on the amount of the additives (Figures 3 and 4).

The heat fluxes of the specimens were observed by a cone calorimeter and the results are plotted in Figure 5 (left). Neat-PC ignited after 100 s and the heat

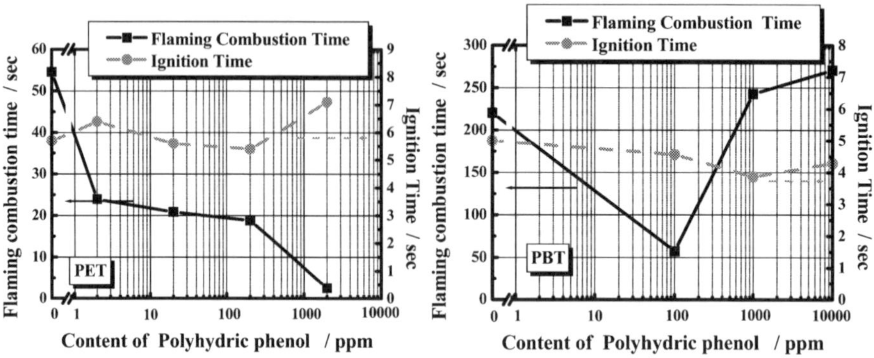

Figure 4 *Flaming combustion times of blends with PPh of PET (left) and PBT (right)*

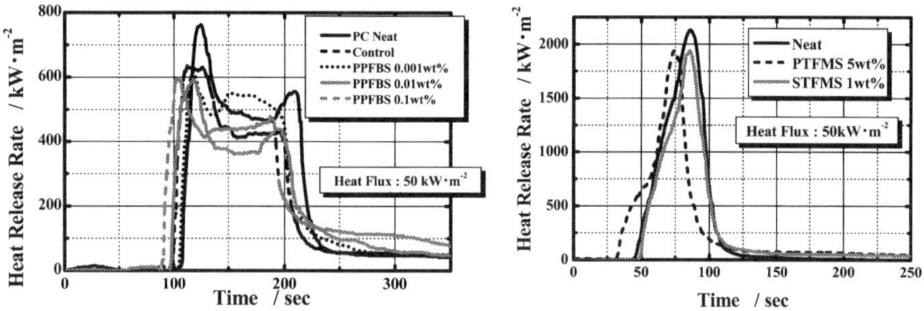

Figure 5 *Heat release rate of the blends of PC (left) and PBT (right) with fluoroalkane sulfonic acid salts*

release rate curve rised rapidly to 780 kW m^{-2}. A carbon layer formed on the surface and the heat release rate dropped after 130 s and was restrained to about 500 kW m^{-2}. Carbon dioxide and other scission products broke the surface layer at 200 s. The blended-PC had a faster ignition time than neat-PC. The additives retarded the flammability but the effects were not remarkable compared with the results of the UL-test.

The profile of the blend with fluoroalkane sulfonic acid salts to PBT [Figure 5 (right)] was much simpler than that of PC. The ignition time of 5% of PTFMS was 30 s, which was shorter than that of neat-PBT. After ignition, a vigorous fire was observed and the heat release rate indicated 2000 kW m^{-2} or more.

4.3.2. Combustion of Blends with Metal Oxides, Red Phosphorous

To compare the effects of other compounds in the region of small additive amount, the effects of metal oxides on porous silica and red phosphorous were studied. The ignition times of the blends were shorter than that of neat-PC [Figure 6 (left)] and the flaming combustion times were also clearly shorter. Generally, a shorter ignition time means that the specimen tends to burn easily.

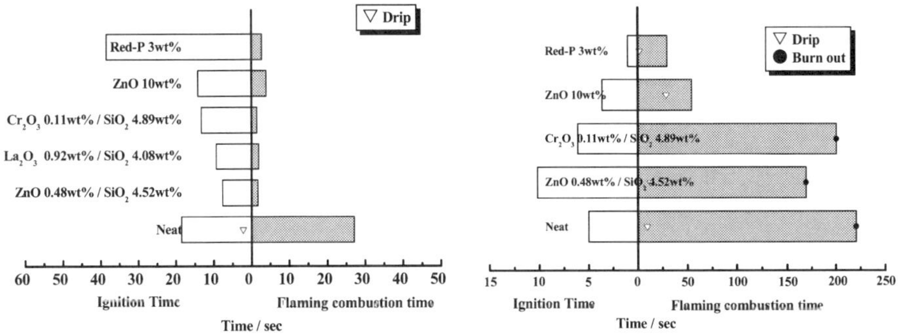

Figure 6 *Effect on UL-test of metal oxides on porous silica (PC: left, PBT: right)*

As the results in Figure 6 (left) differ from the general tendency it of interest to study the mechanism of the flame retardancy.

Although, many metal oxides were tried to improve the flame retardancy of the blended PBT, only ZnO was effective, when added by 10% [Figure 6 (right)]. As ZnO is amphoteric and is expected to coordinate with O-containing compounds, it is seemed that it interacts with the silica surface or polycarbonate.

4.3.3 TGA and Elemental Analysis of PC

Thermal degradation is important to consider in flame retardancy. Both ignition and flaming combustion times are expected to depend on degradation because the combustion heat is decided by the amount or the species of scission products. Each degradation profile of the blends of PC with fluoroalkane sulfonic acid salts (Figure 7 left) and of PBT did not shift from those of neat plastics. TGA curves of the blends with oxides differed [Figure 7 (right)]. The 50% degradation temperature of Red-P/PC shifted to 620°C and the residue increased from 26% to 42% at 800°C. The results supported the idea that the

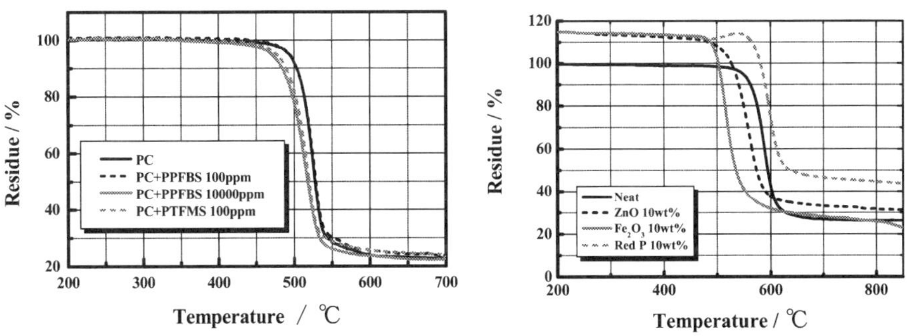

Figure 7 *TGA profiles of neat- and blended-PC (left: blended with fluoroalkane sulfonic acid salts, right: metal oxide)*

Table 1 *Results of elemental analysis of PC and its blends*

Additive	*Measured temperature*	*C/H*	*O/H*	*Nearest "n"*
Neat	As processing	1.12	0.23	1
	At 500 °C	1.78	0.15	2
	800 °C	5.51	0.07	5
PPFBS	As processing	1.13	0.22	1
	500 °C	1.15	0.21	1
	800 °C	5.99	0.11	6
Red-P	As processing	1.06	0.18	1
	500 °C	1.07	0.19	1
	800 °C	15.4	0.21	15

effect on flame retardancy of Red-P depends on the stabilization and the char forming. Conversely, ZnO/PC decomposed at lower temperature.

The results of elemental analysis (Table 1) are very interesting when considering flame retardancy. For example, the C/H ratio of neat-PC at 800°C is 5.51, which is much higher than that of neat-PC before heating, showing dehydrogenation by dehydration or by simple elimination of hydrogen. The dehydrogenation is accelerated by blending red-P and the C/H ratio is 15.4. Conversely, the C/H ratio of PPFBS/PC is 5.99. The C/H ratio and char formation will be discussed in the next section.

If a typical structure of the aromatic compounds can be depicted and "n" is defined as in Figure 8, Equation (1) shows the relation of the C/H ratio and "n".

$$\frac{C_n}{H_n} = \frac{4n^2 - 1}{4n - 1} \tag{1}$$

4.4 Discussion

Numerous flame retardants have been developed since the 1950s. The blocking effect of the radical chain reaction by a halogenated compound in a gaseous phase is the most powerful method. Char and a porous layer on a polymer

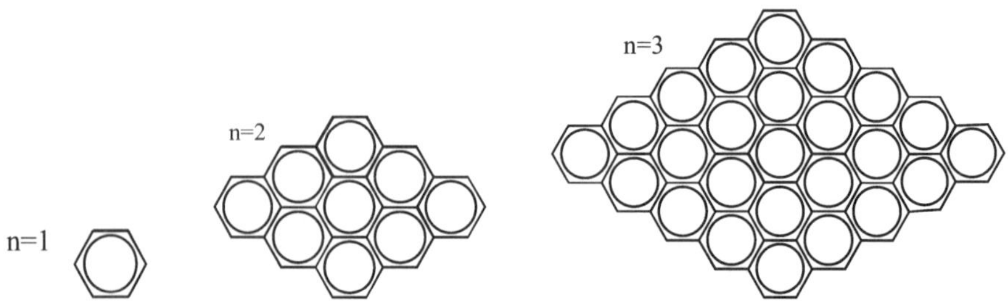

Figure 8 *Model of aromatic compounds after heating (left; n = 1, center; n = 2, right; n = 3)*

surface, which can be increased by phosphorous compounds and the blending of inorganic compounds, are other known methods. A satisfactory level can be obtained when those flame retardants are added by 10 wt% or more.

Conversely, the surface temperature of the polymer is approximately equal to the decomposition temperature of polymer in the initial stage of the combustion because the entire heat generation by oxidation reactions in a gaseous phase is not sufficient to elevate the temperature to 800°C or more. So, the temperature of the polymer is the same as the degradation temperature. Degradation paths of condensed-type polymers such as PC and PBT in the region can be divided into three categories. The first is the inverse reactions of the polymerization or similar degradations and the second is the more complicated paths through radical transfer of direct scission. The third path is the subsequent degradation of the scission products. The following discussion on the flame retardancy and the mechanism of neat- and blended-PCs is based on the above recognition on flame retardancy and thermal degradation.

4.4.1 Degradation at Different Temperatures

The amounts of the scission products of neat-PC and PPFBS/PC are plotted in Figure 9. Abbreviated symbols of the scission products are listed in Table 2.

The thermal degradation started at 500°C and the rate reached the maximum at 600°C at which point the scission products belonging to the A4 group was major. The ratio of A4 of neat-PC in the entire scission products was about 50% and that of B2 was the second largest. However, the distribution of the blend was more complicated. A4 and B3 were major and the ratios were about 30%. B2, whose ratio was 30% of neat-PC, dropped to 12%. In particular, the ratio of B3, most of which was bisphenol-A, changed drastically as a result of adding PPFBS to PC.

4.4.2 Degradation Paths of Neat-PC and Blends

Many studies of PC thermal degradation have been conducted and some of the results have been reported by Tsuge,[28] Montaudo *et al.*,[29] and McNeil *et al.*,[8] in

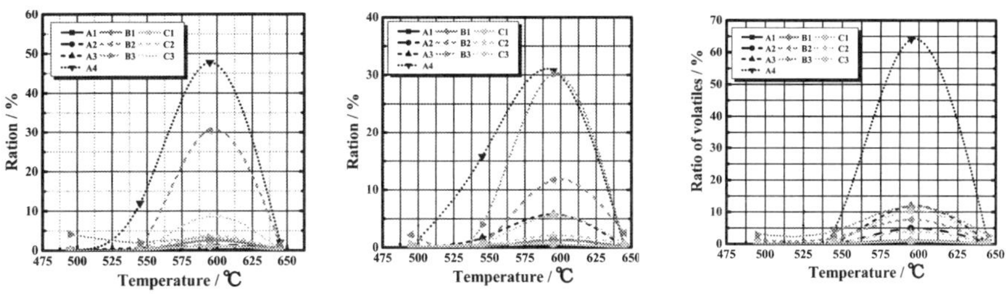

Figure 9 *Scission product distribution of neat-PC (left), PPFBS/PC (middle) and red-P/PC (right) as a function of temperature*

Table 2 *Abbreviated symbols of PC scission products*

	A	*B*	*C*
1	A-1 	B-1 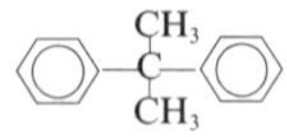	C-1 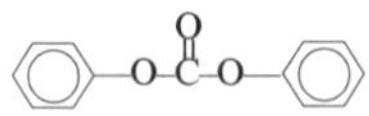
2	A-2 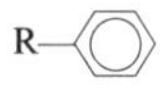	B-2 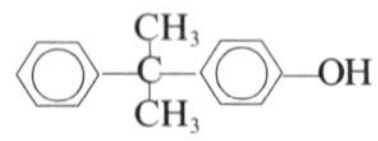	C-2 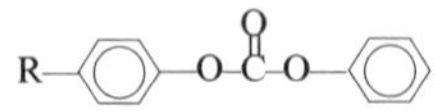
3	A-3 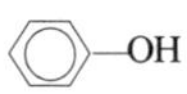	B-3 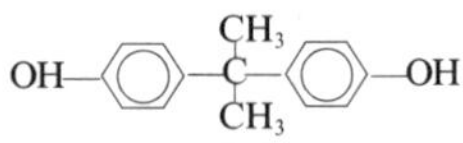	C-3 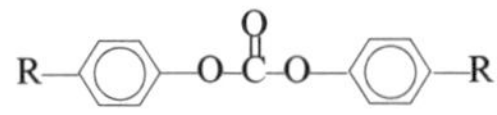
4	A-4 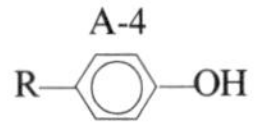	B-4 —	C-4 —

which the scission products were assigned. De-carbonation, dehydrogenation, and Fries rearrangement are typical reactions in the initial stage. Successively, cross-linking, cyclization and many other side reactions occur. Although there are many routes and the reactions are very complicated, the final products are limited. Some major products are bisphenol-A, alkyl phenols, and a cyclic compound. Thus, only the main routes were selected among these known mechanisms in this study (Figure 10).

Two major paths can be considered. One is decarbonation (route-C) and the other is the scission between two benzene rings prior to dehydrogenation from the isopropylidene groups (route-P). Successively, four routes, the C1-, C2-, P1- and P2-routes, might be considered to produce more small scission products as shown in Figure 10. After the two-step degradation, the probability is divided into 16 kinds. For example, the probability of route-C of neat-PC was 94.5% and that of route-P was only 5.5% for neat-PC at 495°C measured by TGA-PyGC-MS.

The scission probabilities of neat- and blended-PC at various temperatures have been calculated (Table 3).

4.4.3 Estimated Char Structures

The Fries rearrangement reaction has been considered as a typical reaction at the initial stage of the chemical structure of the PC main chain. Subsequently, a dehyroxy-reaction occurs and several complicated paths can be reasoned by comparing with other polymer degradations. The C/H ratio of neat-PC is 1.17 and those of the two intermediates in Figure 11 are 1.17 and 1.33, which are much lower than the experimental data at 800°C. The final structure of Figure 11 is presumed to be the char structure of neat-PC after degradation. That of the blend with red phosphorous was much larger because the C/H ratio and the "*n*" was about 15.

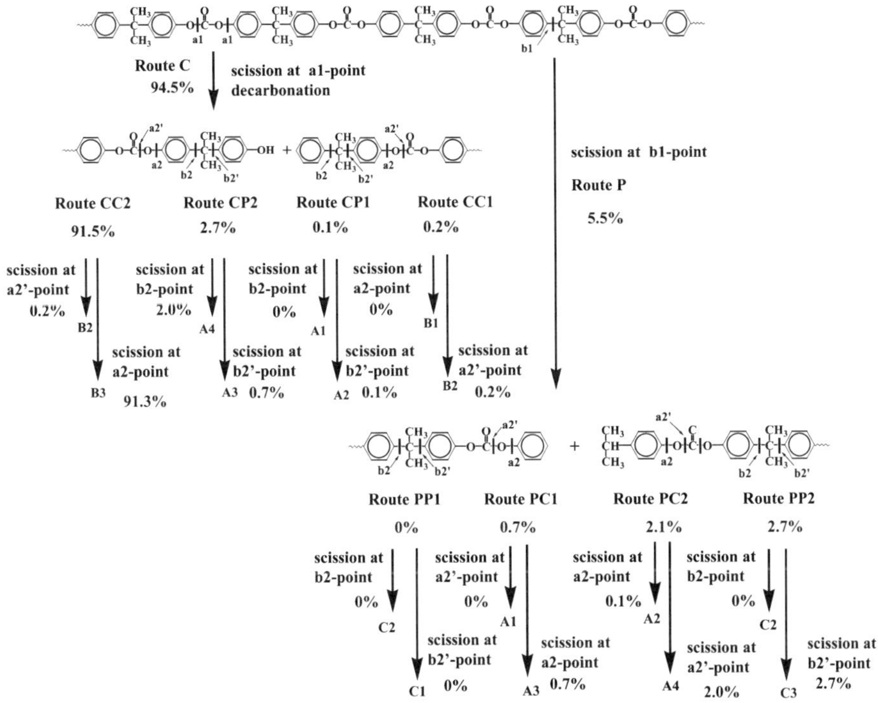

Figure 10 *Simplified degradation paths of PC and the degradation ratio of each path of neat-PC at 495 °C*

4.4.4 Degradation Routes and Flame Retardancy

The amount of the strong acid potassium salt in this study is about 100 times smaller than that of red phosphorous and nano-composites. The result of this study should be discussed from the viewpoint of the degradation paths in PC and the mobility of the additives in the polymer network.

The degradation of PC at just below the degradation temperature is important for the control of combustion. The degradation paths of PC in the region can be divided into two categories. One is relatively simple and can be grouped into two reactions types. The de-carbonation due to the scission of the carbonate bond made phenol type and phenyl type ends [Figure 12 (left)]. There was little possibility for these to recombine with other fragments. So the products seemed to be simple if the degradation is dominated by decarbonation.

On the contrary, dehydrogenation from the isopropylidene group tended to lead to subsequent reactions such as cross-linking, transfer of the radicals, and the generation of a cyclic compound. The volatile products could not be expected because the scission product had a high boiling point.

Carbon-rich materials were generated at higher temperature and Red-P accelerated it. Conversely, the C/H ratio of PPFBS/PC was about the same as that of neat-PC. Adding flame retardants is considered to restrain the flammability

Table 3 *Scission probability table*

Degradation routes	*Probability*											
	Neat				*PPFBS*				*Red P*			
	495°C	545°C	595°C	645°C	495°C	545°C	595°C	645°C	495°C	545°C	595°C	645°C
C	94.5	58.1	61.3	66.0	76.8	57	69.5	70.9	77.5	70.2	52.5	71.3
CC2	91.5	15.5	18.4	20.3	41.8	18	40.2	36.8	55.9	41.2	13.4	53.6
CP2	2.7	40.2	24.4	25.6	3.1	36.8	20.4	3.2	8.6	19.1	32.4	11
CP1	0.1	0.1	0.8	6.8	0.0	0.1	0.9	0.8	2.2	0.9	2.4	3.9
CC1	0.2	2.2	17.7	13.3	31.9	2.1	8	30	10.8	9.3	4.3	2.8
P	5.5	51.9	38.7	34.0	23.2	43	30.5	29.1	22.5	29.8	47.5	28.7
PP1	0.0	0.4	2.9	0.3	2.6	2.6	1.8	7.7	1.3	1.7	2.3	2.8
PC1	0.7	2.0	0.6	9.0	0.3	3.9	3.4	1.2	3.2	2.3	5.4	4.4
PC2	2.1	38.3	24.6	23.4	2.8	33	17.9	2.8	7.6	18.4	29.4	10.5
PP2	2.7	0.1	10.6	1.3	17.5	3.5	7.4	17.5	10.4	7.4	10.4	11

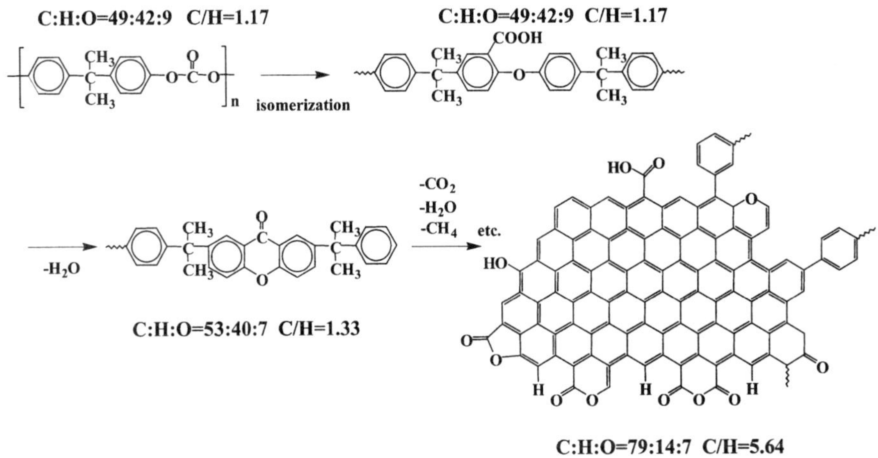

Figure 11 *Estimated char formation process at the initial stage of degradation*

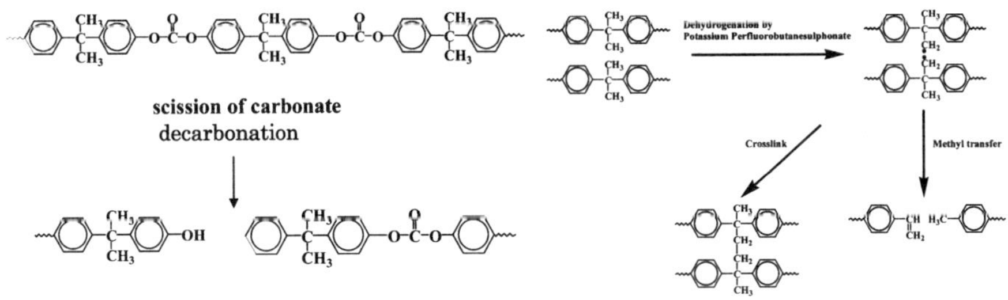

Figure 12 *Simple de-carbonation (left) and cross-linking reaction (right)of PC main chain*

of PC by accelerating the formation of char-like materials on the surface. The result of PPFBS/PC was quite different, whereas it showed excellent flame retardancy. This is very interesting because the flame retardancy of PPFBS/PC and other blends in this study depend on a different mechanism. The same tendency was observed with ZnO.

Another interesting problem is how to react with the degradation site with such a small amount of additives. The reaction time of combustion is very short and the flames of blended-PC are unstable and not uniform. The reaction includes several processes such as (1) degradation of the polymer chain, (2) diffusion of the decomposed product into melted polymer, (3) diffusion in gaseous phase, (4) reaction with oxygen, (5) heat conduction in the polymer. Thus, the contact time of the flame retardants with the degradation site of polymer is very short.

As the strong acid potassium salts changes the degradation path, it reacts with carbonate or the isopropylidene groups on the chain of PC in the short interval. When the concentration is 100 ppm to PC, the molar ratio of the acid to the groups on the PC chain are 7300 in PPFMS/PC and about 13300 in that of PPFBS/PC.

It is very difficult to find a reasonable mechanism to explain why the small amount of acid can react with the enormous groups on the PC chain in such a short time. PC is a rigid plastic and the diffusion constant is about 10–13 $m^2 s^{-1}$. The radius of the sphere in which about 10000 groups are included is calculated to be about 10 nm. The half-time calculated by equation (2) in which the strong acid diffuses in the sphere is 10 ms or less,

$$\tau_{50} = 0.0075\frac{d}{D} \tag{2}$$

where τ_{50} is the half-time of the completed reaction, d is the diameter, and D is diffusion constant.[30]

Consequently, these contradictory phenomena could be explained by considering that the interaction of the PC chain is too weak to construct solid materials in such a small space whose dimension is about 10 nm or less. The molecular force between the two chains of PC is supposed to be very small, like other polymers. PC is in a solid state is not because of the mutual interaction between the adjacent chains but because of the entanglement between several chains. So, the mobility of the strong acid salt is high even in the solid state.

The carbon content of the residue is smaller than those of neat-PC and red-P/PC. Many researchers studying flame retardancy recognize that char forming on the surface of PC can restrain combustion. So, the additives that decrease the char should not be effective. The strong acid salt creates a different kind of circumstance in a very small space of PC polymers and accelerates the degradation of all PC chains in it, as observed in the cone calorimeter test. The decomposed PC changes into a different structure as compared to the original. The cross-link or other reactions are supposed to occur mainly by dehydrogenation instead of dehydration and the structure after degradation in the early stage changes without generating few volatile products.

Simply speaking, four major effects have been recognized to reduce the flammability of polymers. They are the radical trap in a gaseous phase, char formation on the surface of polymer, the blending of inorganic compounds, and the so-called "intumescent" method. The mechanism described in this chapter differs from these four types.

4.5 Acknowledgement

The authors sincerely appreciate the assistance given by Dr S. Nikkeshi who kindly offered the data on polyphenols, which was one of his excellent studies when he worked at the Tohoku Munekata Co. Ltd., IM Dept.

4.6 References

1. M.T. Huggard, *Recent Adv. Flame Retardancy Polym. Mater.*, 1992, **3**, 192–207.

2. J.W. Gilman and T. Kashiwagi, *SAMPE J.*, 1997, **44(3)**, 40–46.
3. J.W. Lyons, *J. Fire & Flamm.*, 1970, **1(9)**, 302–311.
4. S. Bourbigot, M. Le Bras, R. Leeuwendal, K. Shen and D. Schubert, *Polym. Degrad. Stab.*, 1999, **64**, 419–425.
5. W.H. Carothers and F.J. van Natta, *J. Am. Chem. Soc.*, 1930, **53**, 314.
6. P.V. Kozlov, *Vysokomol. Soedin.*, 1961, **3**, 462.
7. R.A. Ekvall and J.R. Low, Jr., *J. Appl. Polymer Sci.*, 1964, **8**, 1677.
8. I.C. McNeil and A. Rincon, *Polym. Degrad. Stab.*, 1993, **39**, 13–19.
9. A. Factor, *Polym. Durability*, 1996, 59–76.
10. C.A. Pryde, *ACS Symp. Ser.*, 1985, **280**, 329.
11. R.P. Kambour, *J. Appl. Polym. Sci.*, 1981, **26**, 861–877.
12. M.R. Mclaury and A.L. Schroll, *J. Appl. Polym. Sci.*, 1985, **30**, 461–472.
13. C.P. Fenimore and F.J. Martin, *Combustion Flame*, 1965, **10**, 135–139.
14. S. Takayama and K. Takeda, *Polym. Degrad. Stab.*, 1995, **50(4)**, 277–284.
15. M. Kinoshita and K. Takeda, *Polym. Degrad. Stab.*, 1995, **68**, 437–443.
16. S. Takayama and K. Takeda, *J. Appl. Polym. Sci.*, 1999, **70**, 2521–2526.
17. K. Takeda and T. Nemoto, in *Proceedings of the 10th Annual Symposium of BBC*, Business Communications Company, INC., Norwalk, 1999, 1–9.
18. K. Takeda, *Polym. Degrad. Stab.*, 2000, **69**, 191–196.
19. K. Takeda, K. Minamitani and T. Nemoto, in *Proceedings of the 11th Annual Symposium of BBC*, Business Communications Company, INC., Norwalk, 2000, 11–18.
20. K. Takeda, F. Amemiya and A. Nanasawa, *J. Appl. Polym. Sci.*, 1997, **64**, 1175–1183.
21. T. Ohkawa and K. Takeda, *Soc. Polym. Sci.*, 2001, **10**, 163–164.
22. Y. Ikeda and K. Takeda, *Polym. Preprints*, 2001, **50(3)**, 579.
23. J. Rycchly, *Polym. Degrad. Stab.*, 1996, **54**, 249–254.
24. M. Foley, *Fire Mater.*, 1994, **18**, 385–387.
25. H.W. Emmons, *Fire Mater.*, 1990, 67–83.
26. T. Koshiduka, T. Ohkawa and K. Takeda, *Polym. Degrad. Stab.*, 2003, **79**, 1–12.
27. T. Ishikawa, T. Ohkawa, M. Suzuki, T. Tsuchiya and K. Takeda, *J. Appl. Polym. Sci.*, 2003, **88**, 1465–1472.
28. S. Tsuge, T. Okumoto, Y. Sugimura and T. Takeuchi, *J. Chromatogr. Sci.*, 1969, **7** (April) 253–256.
29. A. Ballistreri, G. Montaudo and E. Scamporrino, *J. Polym. Sci. Part A*, 1988, **26**, 2115–2121.
30. K. Takeda and K. Morita, *Separat. Sci. Technol.*, 1996, **31(19)**, 2665–2670.

CHAPTER 5

Intumescent Silicates: Synthesis, Characterization and Fire Protective Effect

CHRISTINE PÉLÉGRIS, MURIELLE RIVENET AND MICHEL TRAISNEL

Laboratoire PERF, ENSCL, USTL, BP 108, F-59652 Villeneuve d'Ascq Cedex, France (michel.traisnel@ensc-lille.fr)

5.1 Introduction

Soluble alkali silicates are well-known as fire-proofing materials in building construction.[1] They have long been used to impregnate wood to increase its resistance against fire.[2] More recently, they have been introduced in paintings, enamels, polymers[3,4] or used as coating for fire resistant glasses.[5] The extensive use of alkali silicates is due to their property of swelling when heated or exposed to flame, forming a solid, rigid insulating foam. Moreover, no toxic fumes are released when they are dried, they are environmentally compatible and are in abundant supply. However, their poor water resistance and hygroscopic nature limit their application as coatings.

The mechanisms by which silicates act as fire retardant have been identified as

- Diluting effect due to water vaporisation.
- Cooling effect due to endothermic phenomena related to water vaporisation.
- Thermal and diffusion barrier effects owing to the formation of a mineral foam when heated.[4,6]

The literature also evidenced that, if the water contained in silicates, ionic water has a drastic effect on volume expansion.[4,6,7] Intumescence is allowed by the formation of a viscous phase that forms at the same temperature as the vaporisation temperature of ionic water molecules.[8] Owing to the influence of ionic water, the SiO_2/M_2O (M = alkali) molar ratio and the nature of M have a significant effect on intumescent properties.[7]

The present work aims to compare different silicate precursors on silicate properties and investigate the blending of silicate powder in a polymer matrix to enhance its fire resistance. The influence of silicate synthesis and composition on water loss, solubility and intumescent properties were studied on the raw silicate materials. The fire protective effect was measured using ethylene vinyl acetate (EVA)/silicate materials (90:10).

5.2 Silicate Solution Chemistry

Liquid alkali silicates are commonly produced by dissolving glasses that are made by fusing, or heating under pressure, varying proportions of sand (SiO_2) and soda or limestone. A third method can consist of direct silica attack with soda or potash concentrated solutions.

Soluble silicates (or water glass) are differentiated by their ratio SiO_2/M_2O, defined as the molar proportion of silica to alkali. This ratio determines the physical and chemical properties of silicates.

Silicate solutions are alkaline and, in general, show a pH of 10 to 14. These solutions have a variable viscosity according to the amount of dissolved solids, the molar ratio, and the temperature. They can range in viscosity from very fluid, slightly sticky consistencies to thick substances.

Aqueous solutions are complex and composed of a mix of anions and polysilicate anions.[9] The fundamental building block is the tetrahedral monomer $(SiO_4)^{4-}$, which can be linked through a shared oxygen in two- or three-dimensional structures where the negative charges are balanced by the cations Na^+ or K^+. Figure 1 shows examples of the types of anions that can be present in aqueous silicate solution.

Two main factors influence the distribution of anions: the ratio and the concentration of solids (SiO_2 and M_2O). Hydration and dehydration of silicates depend greatly on this ratio. The more water a solid product contains, the faster it will dissolve. Also, the higher the ratio, the faster the drying rate because alkalis tend to hold moisture readily.

In this study, the soluble silicates are potassium and/or sodium silicates with a high SiO_2/M_2O (molar ratio equivalent to 3.9). They were made from

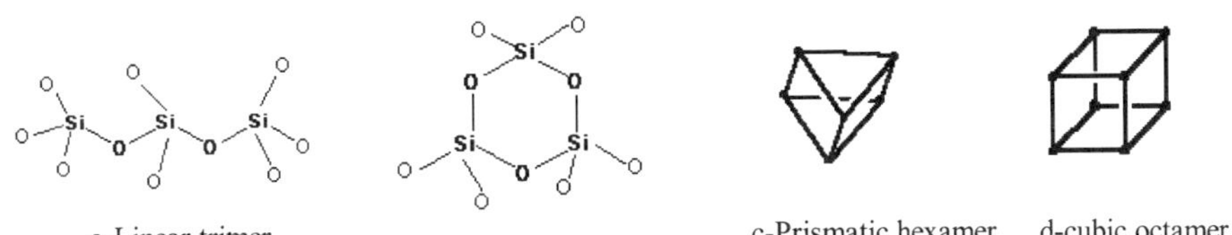

a-Linear trimer b-Cyclic trimer c-Prismatic hexamer d-cubic octamer

Figure 1 *Examples of silicate anion structures: linear, planar cyclic and three-dimensional structures. For* c *and* d *structures, dots represent the silicon atoms; the oxygen linking the silicon atoms are not shown*

commercial solutions of either potassium or sodium silicate with fumed silica, or by depolymerization of silica in concentrated alkaline solutions.

5.3 Experimental

5.3.1 Sample Preparation

5.3.1.1 Aqueous Silicates

Silicate solutions were prepared with a molar ratio equivalent to 3.9. Their reference and composition are reported in Table 1.

K-100 and Na-100 were prepared from two commercial solutions, respectively, Kasil®#6 from PQ Corporation and a sodium trisilicate solution from Riedel-de-Haën; their characteristics are summarized in Table 2. Fumed silica (380 m^2 g^{-1} – Aldrich) was incorporated in the two solutions to increase the molar ratio until 3.9. A mixture of potassium and sodium silicate was also

Table 1 *Silicate materials synthesis*

Silicate materials references	*Raw materials*	*Amount for synthesis* (g)
K-100	K_2O,SiO_2 (Kasil®#6) PQ Corporation	100
	Fumed silica (380 m^2 g^{-1}) Aldrich	5
Na-100	NaOH,SiO_2 Riedel-de-Haën	100
	Fumed silica (380 m^2 g^{-1}) Aldrich	2
K-Na 50/50	K_2O,SiO_2 (Kasil®#6) PQ Corporation	50
	NaOH,SiO_2 Riedel-de-Haën	53
	Fumed silica (380 m^2 g^{-1}) Aldrich	4.5
PK	Fumed silica (380 m^2 g^{-1}) Aldrich	40
	KOH solution, 37.7%	50
PNa	Fumed silica (380 m^2 g^{-1}) Aldrich	40
	NaOH solution, 37.7%	68

Table 2 *Chemical and physical properties of commercial silicate solutions*

	Kasil®#6	*Sodium trisilicate solution*
%K_2O	12.7	
%NaOH		10
%SiO2	26.5	27
Molar ratio	3.29	
Weight ratio	2.1	
pH	12.7	
Viscosity	1050 ctp	
Specific gravity	1.39 g cm^{-3} (20 °C)	1.39 g cm^{-3} (20 °C)
M		242.23 g mol^{-1}

studied through the K-Na 50/50 sample. In each case, samples were synthesized at ambient temperature in Teflon beakers, and were stirred manually. To prepare PK and PNa samples, fumed silica was incorporated, respectively, in a cold solution of KOH (37.7%) and NaOH (20%), and stirred in beakers placed in ice. Depending on the cation, samples appear as a viscous liquid in the case of sodium silicate or a paste with the potassium precursors.

5.3.1.2 Dried Silicates

Silicates were dried at 100°C before characterization and incorporation in the polymer matrix. For intumescence tests, samples were deposited on aluminium boards and dried overnight at 100°C. For TGA, infrared spectroscopy, lixiviation tests and measure of fire protective effect, samples were dried at 100°C for 24 hours on Teflon supports, then powdered.

5.3.2 Blending of Dried Silicates Powders and Ethyl Vinyl Acetate (EVA-19%) Polymer

A 20 g charge of silicate and polymer in the respective proportions 2:18 g was fed into the preheated (160°C) mixer. All blends were mixed using the Brabender Laboratory mixer measuring head (type 350/EH, roller blades, mixing conditions were checked using the data processing torque rheometer system Brabender Plasticorder PL2000, constant shear rate: 50 rpm). Then, each system was compression moulded into plates of $100 \times 100 \times 3$ mm^2 under 40 kN at 160°C.

5.3.3 Characterisation

5.3.3.1 Intumescence Test

Each dried sample was submitted to a thermal shock for 1 hour. The thickness of each sample was measured before (e_i) and after (e_f) the thermal treatment to determine the degree of intumescence, which was defined by Equation (1)

$$\mathrm{I} = \frac{e_f - e_i}{e_i} \tag{1}$$

Tests were performed at 200, 400 and 600 °C.

5.3.3.2 TGA Studies

Dried samples were powdered in a mortar and then thermogravimetrically analysed using a Setarum TGA 92 analyser. TGA tests were conducted in the 20–600°C range at a heating rate of 10°C min^{-1}.

5.3.3.3 *Lixiviation Test*

After drying for 24 h at 100°C, each sample was powdered in a mortar and immersed in a tray of 500 mL distilled water at 22°C for 1 h. For each sample, about 6 g was used. The water was stirred manually every 5 mins. Then the sample was removed, filtered and allowed to dry overnight before weighing to determine the mass loss.

5.3.3.4 *Infrared Spectroscopy*

Infrared spectroscopy was used to gain information on the structure of the silicate at the end of the intumescence test. The intumesced samples were powdered and diluted with KBr (10% of silicate), then pressed at 5 bar. The analysis was performed with an FTIR Nicolet Impact 400D apparatus over the 4000–600 cm^{-1} region.

5.3.3.5 *Fire Protective Effect*

The fire protective effect of silicates was evaluated on the silicate-EVA systems. Each plate was placed under a radiant heat source at an appropriate distance to raise and keep the temperature at the surface of the plate to 350°C. A thermocouple placed underneath the plate followed the temperature on the other side of the sample *versus* exposure time. The heat source was an epiradiator made of viteous silica, with a radiant heat surface of 100 mm diameter and a nominal power of 500 W.

5.4 Results and Discussion

Intumescence tests are reported in Figure 2. Whatever the temperature, the sodium-based silicates exhibit the highest swelling properties. The degree of

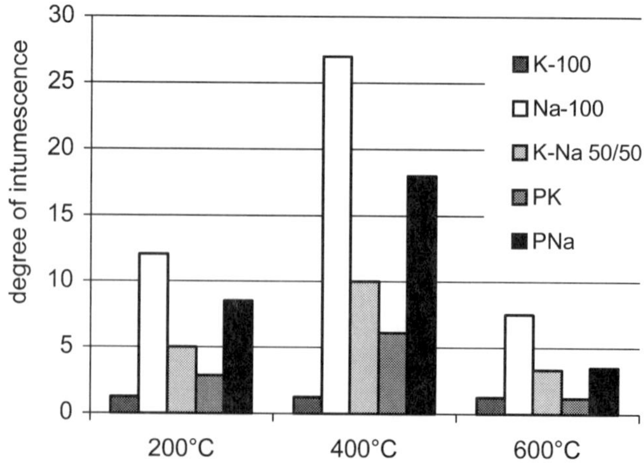

Figure 2 *Influence of silicates composition and temperature on the degree of intumescence*

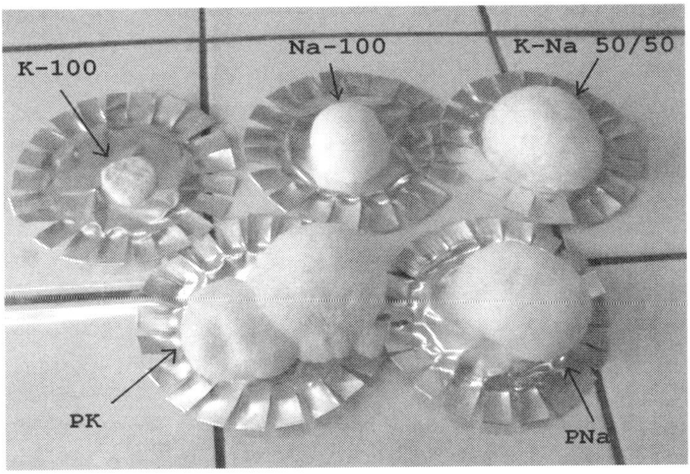

Figure 3 *Aspect of silicates at the end of intumescence test at 400 °C*

intumescence follows the pattern Na-100 > PNa > K-Na50/50, with the highest values measured for the 400°C temperature test. Potassium-based silicates were less intumescent, in particular K-100, as seen in the photograph taken at the end of the 400°C thermal shock (Figure 3).

The total mass loss after drying (100°C) and intumescent tests (400 and 500°C) are reported in Figure 4 for K-100 and Na-100. The mass loss is a direct result of the evolution of water since no combustibles are present in the samples. Both in the sodium and the potassium-based silicate, about 30% of water are eliminated at 100°C. At 400°C, the intumescence of Na-100 involves less than 10% of water, whereas the mass loss is only 3% with K-100. These results confirm the direct link between the release of water and intumescence. However, the increase of mass loss observed for K-100 at higher temperature (500°C) suggests that differences in terms of intumescence between Na-100 and K-100 are not only due to the amount of water but also to its nature.

Thermal degradation of silicates was investigated by TGA (Figure 5). The TGA curves evidence the relationship between intumescence (Figure 2) and

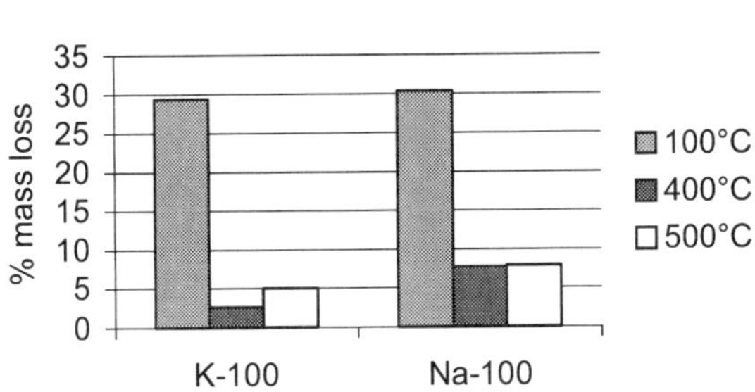

Figure 4 *Total mass loss (%) of K-100 and Na-100 after drying (100 °C) and the intumescence tests at 400 and 500 °C*

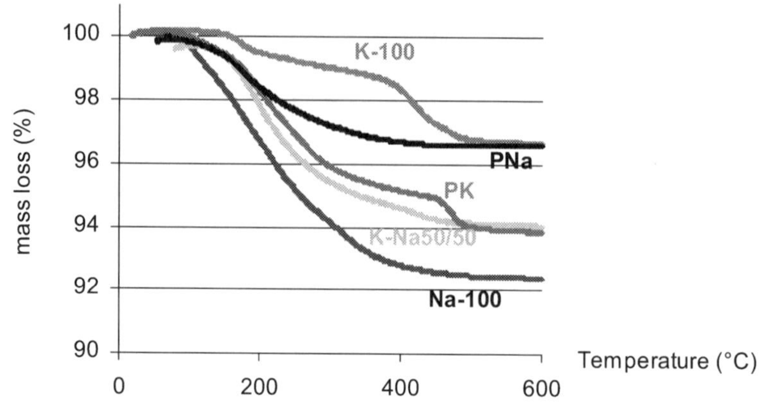

Figure 5 *TGA traces of the degradation of the silicates under air flow*

release of water when we compare Na-100 and K-100. Moreover, they indicate that the water evolution is function of the nature of the cation. Indeed, the evolution of water takes place in one step with the sodium based silicates (Na-100, PNa and K-Na50/50), in the wide range from 100 to 500°C, whereas for the potassium silicates K-100 and PK it involves two steps: one in the interval 180–400°C, the other in the narrow range 400–500°C. According to the literature, the water contained in silicates is described as free water and bound water.[7] Free water corresponds to the water physically adsorbed by hydrogen-bonding to silanol groups. As shown in Figure 5, free water has been totally eliminated by drying at 100°C. The mass loss recorded from 100°C by TGA corresponds to bound water, which is identified as ionic water in the temperature range 100 to 200°C,[7] and is due to the various hydrated species of the silicate anions. Clearly, from Figure 5, ionic water is present in Na-100 whereas for K-100 the release recorded from 180°C corresponds to more strongly bonded water.

The rapid evolution of the water vapour in this region reportedly resulted in the initial intumescence of the alkali silicates.[7] In this study, maximum of intumescence is observed at higher temperature and so cannot be only related to ionic water but to water resulting from the condensation of silanol groups by the following reaction, with the formation of siloxane bonds.

$$\equiv Si—OH + HO—Si\equiv \Rightarrow \equiv Si—O—Si\equiv + H_2O$$

Evolution of water in the sodium-based silicates occurs gradually over a wide range of temperature (Figure 5). This phenomenon is assigned to the dehydration of silanol having an irregular arrangement, associated to the irregular polysilicate ions.[7] In contrast, for K-100 and PK, the mass loss observed in the narrow temperature range (400–500°C) involves the loss of more strongly bonded water. This water has been identified in potassium-based silicates as structural water, and could be the result of the reaction of dehydration of $KHSi_2O_5$:

$$2KHSi_2O_5 \Rightarrow K_2Si_4O_9 + H_2O$$

evidenced by X-Ray diffraction as a significant component in the silicate matrix. The structural water would result from the reaction of silanol groups having an orderly arrangement (*i.e.* belonging to $KHSi_2O_5$). Infrared analysis was performed on K-100 after the intumescence tests at 200, 400 and 600°C. The corresponding spectra (4000–600 cm^{-1}) are reported in Figure 6.

Although the spectra show great similarity in the 600–1100 cm^{-1} region, the trend of the band at 1000 cm^{-1}, assigned to the Si–O–Si stretching frequency, is different when K-100 is heated at 600°C. The broadness of this band observed in the corresponding spectrum is indicative of an irregular structure.[7] In contrast, the narrow band observed at 200 and 400°C suggests an orderly structure. Therefore, if we combine the IR results with the TGA data, the K-100 silicate precursor would lead to the formation of an organized phase in the temperature range of interest for intumescence (200–400°C), whereas for Na-100 an amorphous phase would be obtained.

K-100 was also submitted to irradiance at 350°C (*cf.* the test for fire protective effect), then analysed by IR spectroscopy. The corresponding spectrum is

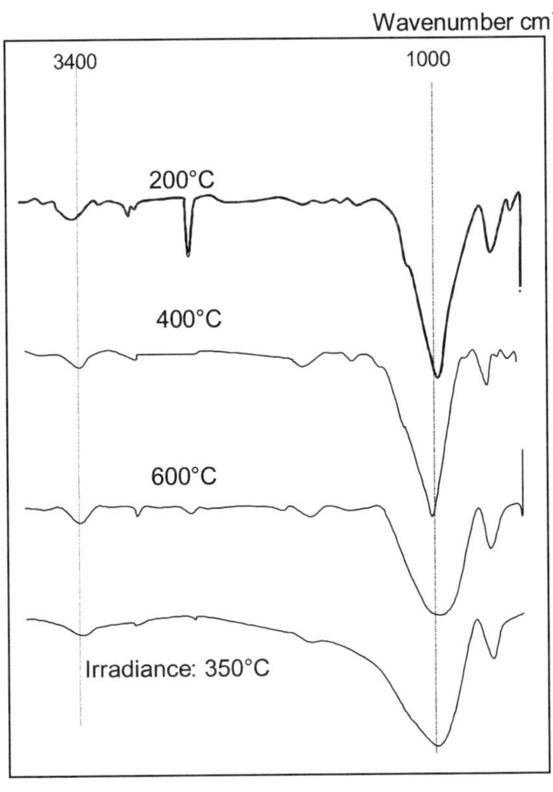

Figure 6 *Infrared spectra of K-100 at the end of intumescent tests at 200, 400 and 500 °C, and after exposure to irradiance at 350 °C (cf. measure of fire protective effect)*

reported in Figure 6 and compared with spectra resulted from the intumesced silicates. A significant broadness of the band at 1000 cm^{-1} is observed, as for the intumesced silicate at 600°C. This shows that the heating conditions are important and influence the structure of the potassium silicate.

Thus, water content and its release are the driving force of the intumescence of the alkali silicates. So it is important to consider, at the same time, the phenomena of dehydration and hydration of these materials. Hydration has been studied through the solubility test in water. The water resistance of dried silicates was evaluated by measuring the mass loss of sample after immersion in water at ambient temperature. The results of the lixiviation tests are reported in Figure 7. As for the intumescent test, differences are observed as a function of the nature of the cation.

Whatever the mode of synthesis, potassium silicates exhibit the highest water resistance whereas the sodium silicates dissolve faster. In contrast, the combination K-Na gives a poor resistance. The water mass loss of each silicate after drying at 100°C is reported in Figure 8. The less hydrated silicate (*i.e.* the

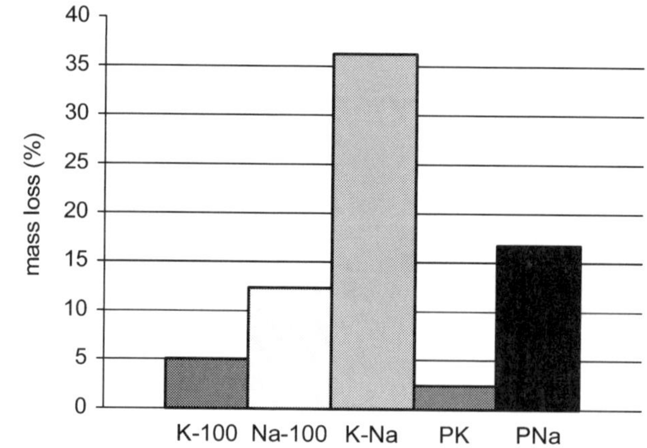

Figure 7 *Results of lixiviation tests in water (T = 22 °C)*

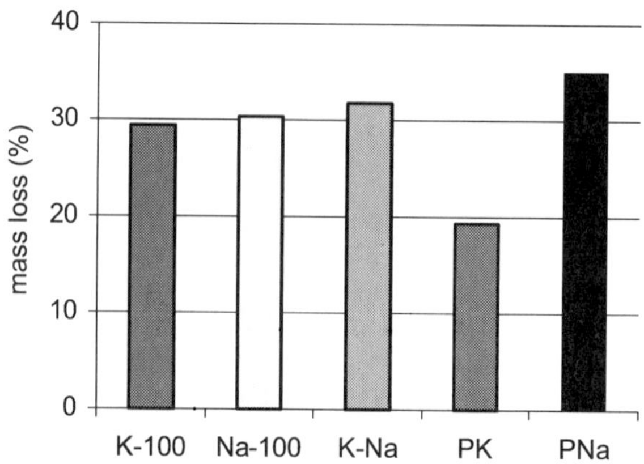

Figure 8 *Mass loss after drying at 100 °C*

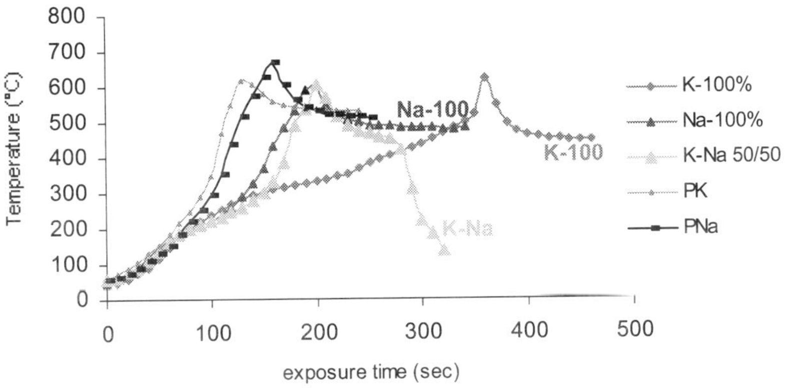

Figure 9 *Measurement of the silicates fire protective effect in an EVA matrix*

potassium silicate PK), appears as the more water resistant. The intumescence related to the amount of water could also be associated to the solubility in water.

After the study of silicates properties, we tested each of them as fire retardant additive in a polymer matrix. The influence of the various silicate precursors on thermal barrier was investigated with blending of silicate powder in ethyl vinyl acetate polymer at concentration of 10%. Compression moulded plates were submitted to an irradiance regulated at 350°C. The temperature of the unexposed side sample was followed during exposure and is reported in Figure 9. The samples swelled immediately after the exposure. As can be seen in Figure 9, the mode of synthesis has a significant influence on fire retardant performance. The mixes PK and PNa give poor fire protective effects whereas a good thermal shield is obtained using K-100, the mix of Kasil®#6 and fumed silica.

Comparison of Figures 2 and 9 evidences that the protective effect of K-100 can not be related directly to the intumescence properties, but involves either chemical reactions between the silicate and EVA matrix or the physical and structural properties of the silicate foam. Thus further analyses are needed to understand the thermal behaviour of the two antagonist samples K-100 and Na-100.

5.5 Conclusion

Our goal was to study the influence of various alkali silicates as potential fire retardant additives in polymer matrix. First, we studied the influence of synthesis and composition (nature of the alkali) on intumescence, and the solubility in water of the silicates materials with formulations corresponding to a molar ratio SiO_2/M_2O equal to 3.9. Two modes of synthesis were compared: silicates prepared from commercial alkali silicates in which fumed silica was added, and silicates obtained by depolymerization of silica in concentrated alkaline solution.

Whatever the mode of synthesis, the sodium-based silicates exhibit the highest swelling properties. Intumescence occurred in the temperature range 200 to

400°C, with a maximum expansion observed at the end of the test at 400°C for all silicates. The best results are obtained with Na-100: the mix of the commercial sodium trisilicate solution and fumed silica. In contrast, the equivalent potassium silicate K-100, obtained using the mix of Kasil®#6 and silica, shows the lowest intumescent behaviour whatever the temperature test.

Intumescence of the sodium-based silicates is related to the amount of water, and is associated with the release of bound water resulting from the condensation of silanol groups contained in an amorphous phase.

For the potassium-based silicates, however, in particular K-100, the release of more strongly bounded water above 400°C would be associated with the condensation of silanol units contained in a crystalline phase. Also, the heating conditions influence the structural arrangement in this silicate.

The sodium-based silicates dissolve faster than the potassium-based in water. Thus, intumescence could be related to the water content and a high solubility in water.

Fire behaviour was evaluated with blending of silicate powder and the EVA polymer matrix. A good thermal shield was obtained using K-100. The protective effect of K-100 can not be explained on the basis of intumescent properties but involves either chemical reactions between the potassium silicate and EVA or either the physical properties of the mineral foam.

5.6 References

1. W. Becker, *Fire Mater.*, 1991, **15**, 169–173.
2. Slimack, *US Patent, 6,303,234*, 2001.
3. E. Metcalfe, Z. Feng and D. Kendrick, *Recent Adv. Flame Retard. Polym. Mater.*, 1997, **8**, 129–135.
4. T.M. Liu, W.E. Baker, K.B. Langille, D.T. Nguyen and J.O. Bernt, *J. Vinyl Addit. Technol.*, 1998, **4**, 246–258.
5. H. Horacek, and S. Pieh, *Polym. Int.*, 2000, **49**, 1106–1114.
6. E.M. Bulewicz, A. Pelc, R. Kozlowski and A. Miciukiewicz, *Fire Mater.*, 1985, **9**, 171–175.
7. K.B. Langille, D.T. Nguyen, J.O. Bernt, D.E. Veinot and M.K. Murthy, *J. Mater. Sci.*, 1991, **26**, 695–710.
8. G. Yakovlev and V. Kodolov, *Int. J. Polym. Mater.*, 2000, **47**, 107–115.
9. J.L. Bass, G.L. Turner and M.D. Morris, *Macromol. Symp.*, 1999, **140**, 263–270.

Use of Nanocomposite Materials

CHAPTER 6

Flammability of Nanocomposites: Effects of the Shape of Nanoparticles

TAKASHI KASHIWAGI

Fire Research Division, National Institute of Standards and Technology, Gaithersburg, MD 20899-8665, USA (takashi.kashiwagi@nist.gov)

6.1 Introduction

There is a high level of interest in using nanoscale reinforcing fillers for making polymeric nanocomposite materials with exceptional properties.[1–3] Nanocomposites are particle-filled polymers where at least one dimension of the dispersed particle is on the nanometer scale. When all three dimensions are of the order of nanometers, we are dealing with true nanoparticles, such as spherical silica, having an aspect ratio of 1. Another type of nanocomposite is characterized by particles having only one dimension on the nanometer scale. In this case, the filler is present as sheets/layers, such as layered silicate or graphite, which are one to a few nanometers thick and hundreds to thousands of nanometers in the other two dimensions. At present, the most common approach to improving flammability is the use of layered silicates having large aspect ratios. When two dimensions are on the nanometer scale and the third is larger, forming an elongated structure, we speak of nanotubes, whiskers, or rods with a high aspect ratio.

Flammability properties of polymers have been improved with nanoscale additives and these filled systems provide an alternative to conventional flame retardants. It is important to explore how the asymmetry (aspect ratio) and other geometrical effects of nanoparticle additives influence the flammability properties of polymer nanocomposites. This chapter describes flammability properties of nanocomposites based on the three different shapes of nanoscale additives, such as nanosilica, clay, and carbon nanotube. Dispersion of the nanoscale particles, sample behavior during gasification, and the shape and the structure of the sample residues after a gasification test are described, as are the flammability characteristics of each nanocomposite based on the different shapes of nanoscale particles.

6.2 Flammability Measurement

Evaluation of flammability properties was achieved using the Cone calorimeter, which was designed and built at NIST (ASTM E 1354-92). Aluminum foil was wrapped around the sample, except on the irradiated surface, to form a sample container instead of using the standard, heavy metal container. Tests were performed either at an incident radiant flux of 40 or at 50 kW m^{-2} in air. Heat release rate and mass loss rate are reproducible to within ±10%. Another device, a radiative gasification instrument similar to the Cone calorimeter, was used to observe the gasification behavior and to measure mass loss rate of the sample in a nitrogen atmosphere (no burning) at 40 or at 50 kW m^{-2}. A more detailed discussion of the device is given in our previous study.[4] The unique advantages of this device are twofold: first, the results obtained are based only on the condensed phase processes due to the absence of any gas phase oxidation reactions; second, it enables visual observation of gasification phenomena under a heat flux similar to that of a fire without interference from a flame.

6.3 Polymer-Nanosilica Nanocomposites

Detailed preparation of the poly(methyl methacrylate) (PMMA)/nanosilica nanocomposites is described in reference 5. The average diameter of the nanosilica particles used in this study was *ca.* 12 nm. The sample was made by *in situ* polymerization of methyl methacrylate in the presence of nanosilica particles. The disk shaped sample were *ca.* 8 cm diameter and 0.6 cm thick. The number averaged molecular weight of this PMMA, measured by size-exclusion chromatography, was 147000 ± 1000[a] and that of the PMMA in the nanosilica composite was 183000 ± 8000. The polydispersities of both samples were 1.9 ± 0.1. The actual content of silica particles in the PMMA/nanosilica nanocomposite was determined by pyrolysing the sample in air at 900°C in a muffle furnace. By weighing the white powdery residue, a value of 13 ± 1% by mass was found rather than the originally intended value of 10%.

The two polymerized samples, the PMMA/nanosilica nanocomposite and the pristine PMMA (polymerized by the same procedure as for the nanocomposite), were transparent (Figure 1). Although the transparency of the samples with nanosilica particles suggests reasonably good dispersion of the particles in the PMMA, TEM and AFM images were taken to examine dispersion at the silica scale. TEM analysis of the PMMA/nanosilica nanocomposite at low magnification shows well dispersed areas and also areas of greater silica particle

[a]According to ISO 31-8, the term "molecular weight" has been replaced with "relative molecular mass," symbol Mr. The conventional notation, rather than the ISO notation, has been employed here.

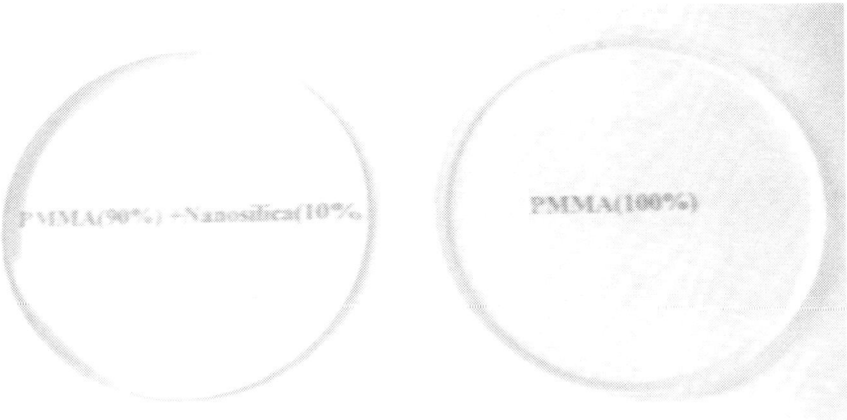

Figure 1 *Pictures of the PMMA/nanosilica nanocomposite sample (left) and the PMMA sample (right)*

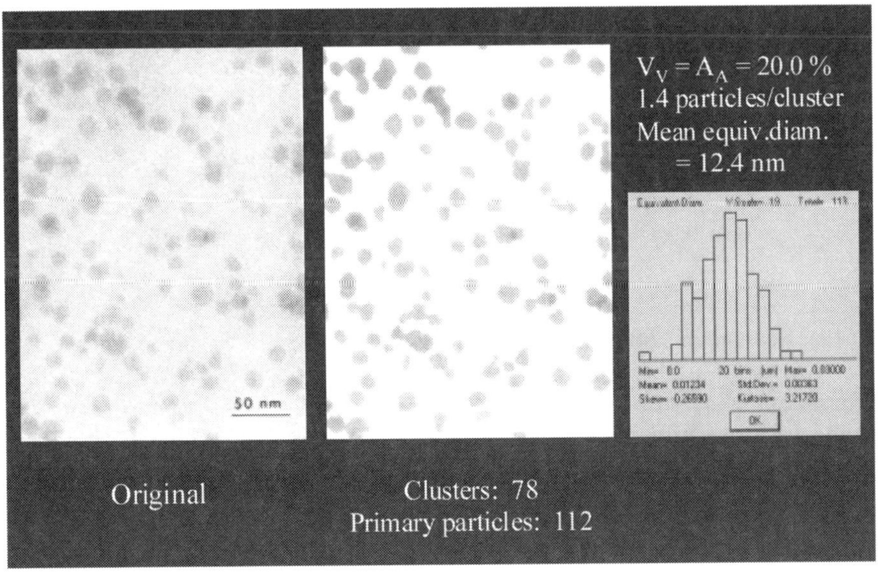

Figure 2 *TEM image of the PMMA/nanosilica nanocomposite (left), analyzed image (middle) and a histogram distribution of diameter (right)*

concentration without clustering. The high magnification TEM image in Figure 2 shows well-dispersed silica spheres. The image analysis of the figure shows a histogram of the particle diameter distribution from 10 to 30 nm and an average diameter of 12.4 nm.

Observation of the sequence of events in the gasification of the pristine PMMA sample in a nitrogen atmosphere at a radiant flux of 40 kW m^{-2} (no burning) first revealed the appearance of small bubbles bursting at the sample

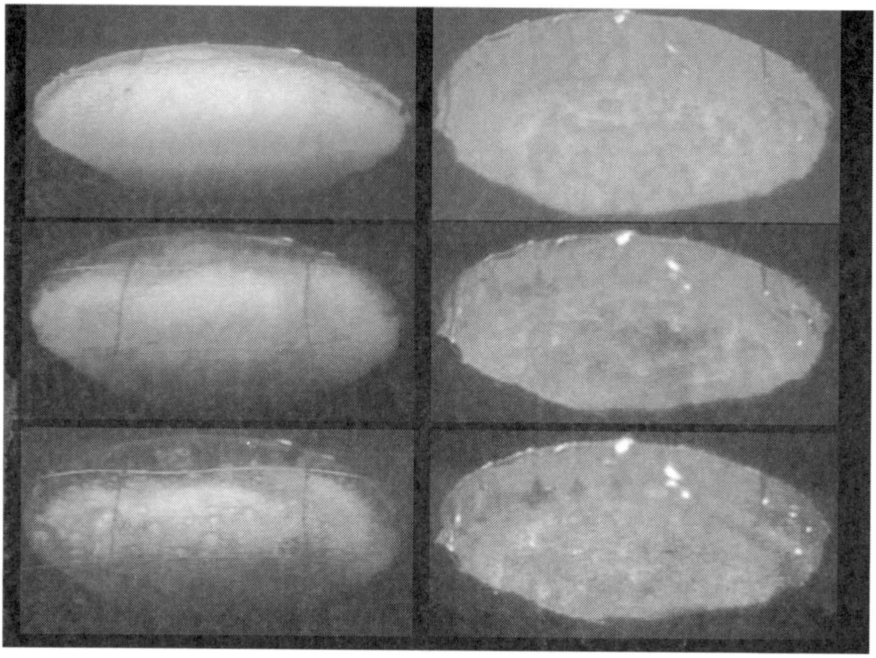

Figure 3 *Selected sequence of video images of gasification phenomena of PMMA and PMMA/nanosilica samples in N_2 at 40 kW m^{-2} (Left column: PMMA, and right column: PMMA/nanosilica, time at 120 s, 240 s, and 300 s from top. The container of the PMMA sample was held by four small wires to avoid its movement)*

surface around 15 s after the start of irradiation, followed by a rapid increase in the number of bubbles bursting so that they covered the entire sample surface after about 30 s. Around 120 s, the sample surface acquired the appearance of a fluid with larger bursting bubbles and with slight swelling, as shown in Figure 3. More vigorous bubbling appeared for times greater than 120 s and the sample became less viscous (more fluid in appearance). After 240 s, the surface was covered by large bursting bubbles and vigorous bubbling with a very fluid sample continued. At the end of the test, no significant amount of residue was left except a thin, black coating on the container surface.

The gasification behavior of the PMMA/nanosilica nanocomposite was quite different from that of the above PMMA sample. Many small, bubbles were observed initially, but at about 60 s many white islands appeared on the sample surface with vigorous bursting of small bubbles around the islands. Around 120 s, the islands became darker and irregular (Figure 3, top right). The islands appeared to be made of coarse, granular particle clumps. The fractional coverage of the sample surface by the islands continued to increase and a random motion of the granular particle clumps on the sample surface was often observed. At about 300 s, the sample surface was completely covered by coarse, granular particle clumps (Figure 3, bottom right). Also, the sample surface was

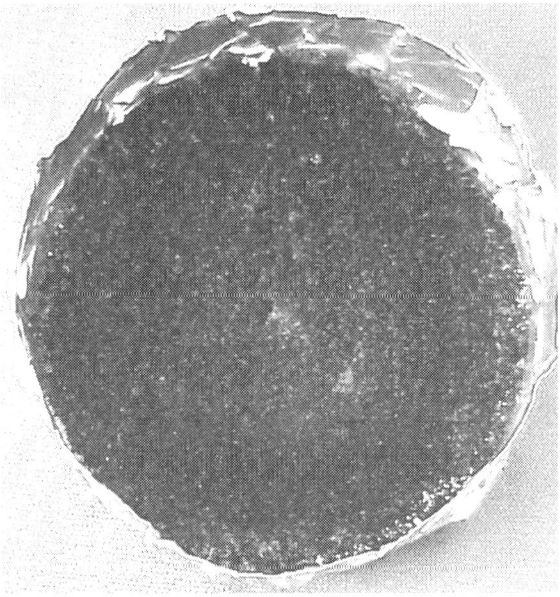

Figure 4 *Picture of the residue of the PMMA/nanosilica nanocomposite consisting of granular, coarse particles after the gasification test in nitrogen at 40 kW m^{-2}*

slowly receding. At the end of the test, a dark, coarse powdery layer was left in the sample container (Figure 4). The highly fluid behavior observed for the PMMA sample was not seen for this sample. The mass of the residue was almost the same as that of the initial weight of nanosilica and the thickness of the residual layer at the end of the test was roughly half of the initial sample thickness.

The calculated mass loss rates from the measured sample masses of pristine PMMA and PMMA/nanosilica samples are plotted in Figure 5. The peak mass loss rate of the PMMA/nanosilica nanocomposite is roughly 40% less than that of pristine PMMA. However, the mass loss rate up to 50 s and the total sample mass loss (integrated values of the mass loss curve) are about the same for both samples. These trends are very similar to those of a low molecular weight PMMA/silica gel sample.[6]

The heat release rates of pristine PMMA and the PMMA/nanosilica nanocomposite are shown in Figure 6. The addition of the nanosilica reduced the peak heat release rate of the PMMA sample to roughly 50% of the pristine PMMA value, but the ignition delay time and the total heat release (integrated values of the heat release rate curve) are about the same for both samples. The trends of the measured mass loss rate (burning rate) curves (not shown) are very close to those of the heat release rate curves and thus the calculated specific heat of combustion (measured heat release rate divided by measured mass loss rate) is 24 ± 2 MJ kg^{-1} for both types of sample. Furthermore, the trends of the heat release rate curves are very similar to those of the mass loss rate in nitrogen, as

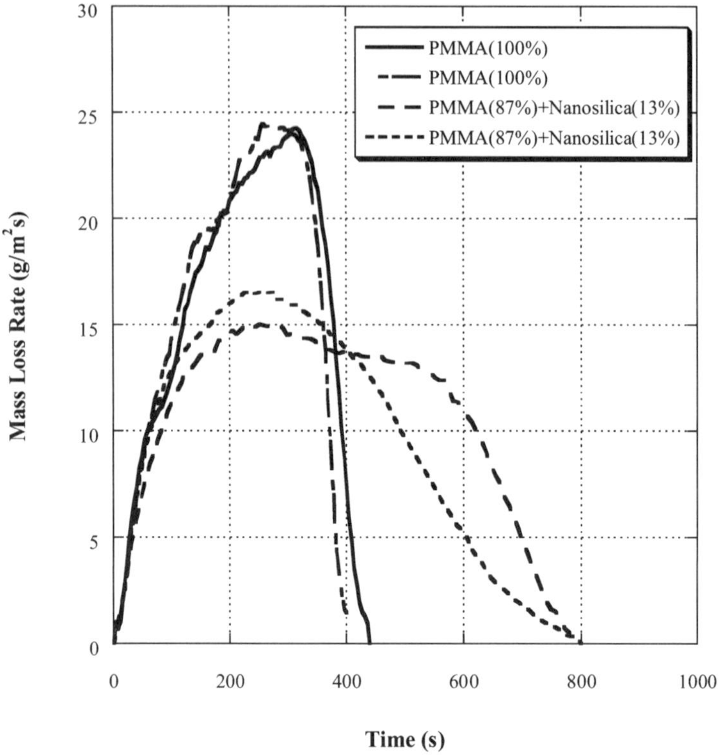

Figure 5 *Effects of nanosilica addition on mass loss rate of PMMA at 40 kW m^{-2} in nitrogen*

shown in Figure 5. The PMMA/nanosilica nanocomposite residue after the cone calorimeter tests was a gray layer consisting of coarse, granular powder accumulated at the bottom of the sample container, which is very similar to that shown in Figure 4.

6.4 Polymer-Clay Nanocomposites

Polyamide 6, PA6, was selected as a resin for this study and commercially available PA6/clay samples were used. They were PA6 homopolymer (molecular mass (M_w) of about 15000 g mol^{-1}, UBE 1015B[b]), PA6 ($M_w \approx$ 15,000) with montmorillonite (MMT) of 2% by mass (UBE 1015C2), and PA6 ($M_w \approx$ 18,000) with MMT of 5% by mass (UBE 1018C5). They were selected due to their exfoliated clay dispersion in PA6. Sample disks (75 mm diameter and 8 mm thick) were injection molded. TEM images of the original sample show that the clay

[b]Certain commercial equipment, instruments, materials, services or companies are identified in this article to specify adequately the experimental procedure. This in no way implies endorsement or recommendation by the National Institutes of Standards and Technology (NIST).

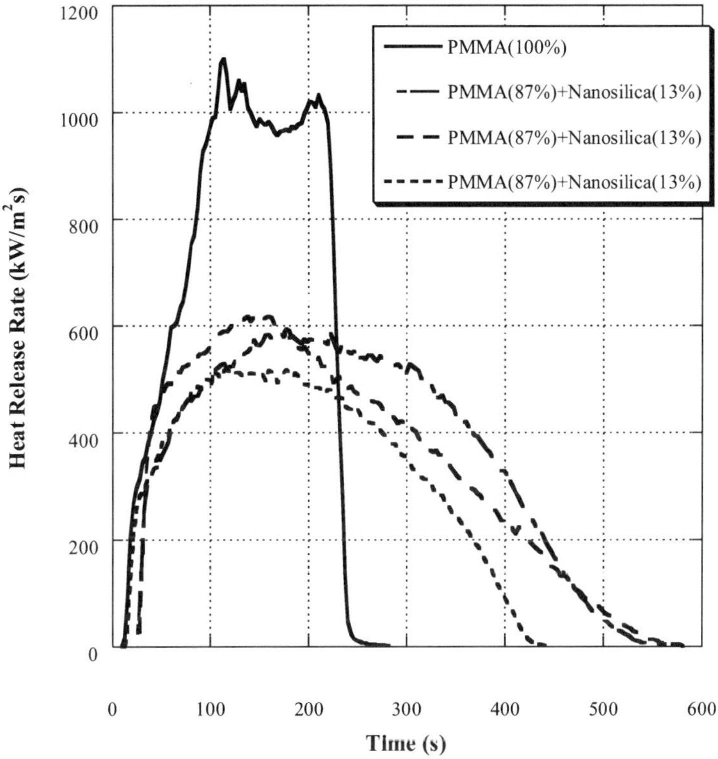

Figure 6 *Effects of nanosilica addition on heat release rate of PMMA at 40 kW m^{-2} (Dashed lines were the results of three replica of nanocomposites made at three different times)*

platelets were fully exfoliated (Figure 7). Note at the position labeled "A" in the figure, two platelets spaced approximately 1.5 nm apart can be seen clearly. However, such close platelets were rare. Wide-angle XRD measurements were conducted to obtain the clay particle structure in the sample. A comparison of the XRD data between clay and PA6/clay(5%) sample is shown in Figure 8. XRD data of the original Na-clay shows many peaks with a *d*-spacing of about 1.19 nm (at 2θ of about 7.44°). However, the PA6/clay(5%) sample shows a sharp peak at 2θ of about 21.4°, corresponding to the γ crystalline phase of PA6, without any peak corresponding to the clay. The results indicate that clay platelets were fully exfoliated, in agreement with the TEM images. More detailed discussion of the dispersion of clay platelates in PA6 and further XRD data of collected residues are given in our previous publication.[7]

Observation of the sequence of events in the non-flaming gasification of the PA6 sample without clay first revealed small bubbles of evolved degradation products at the sample surface, followed by the appearance of many large bubbles. About 60 s after the start of irradiation, the bubbles became gradually smaller and around 120 s many small bubbles, with few larger bubbles, appeared

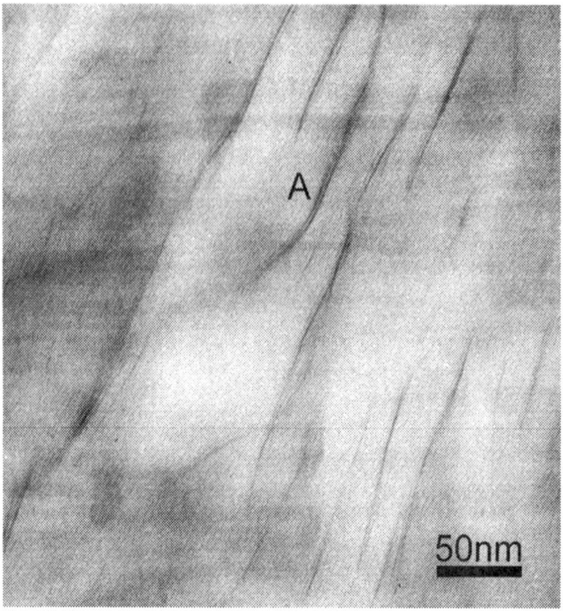

Figure 7 *TEM image of the PA6/clay(5%) sample*

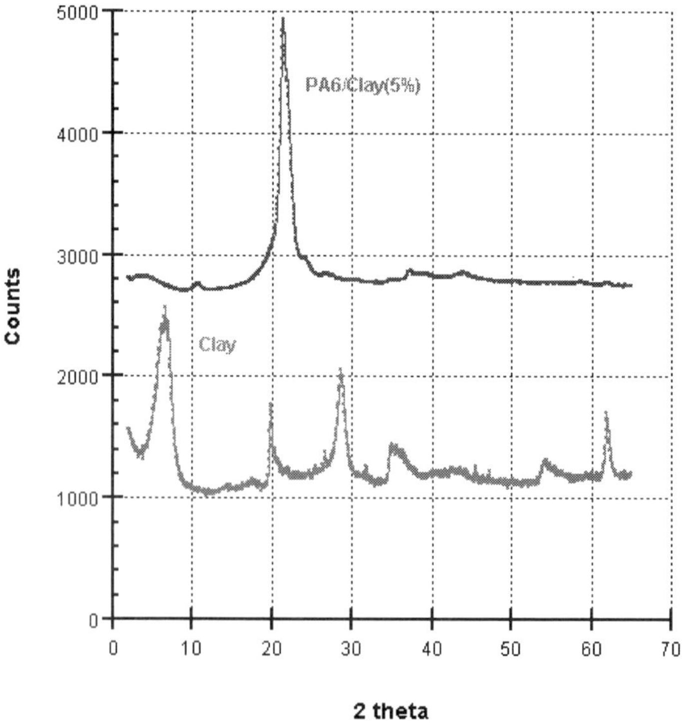

Figure 8 *XRD of Na$^+$ clay and PA6/clay(5%)*

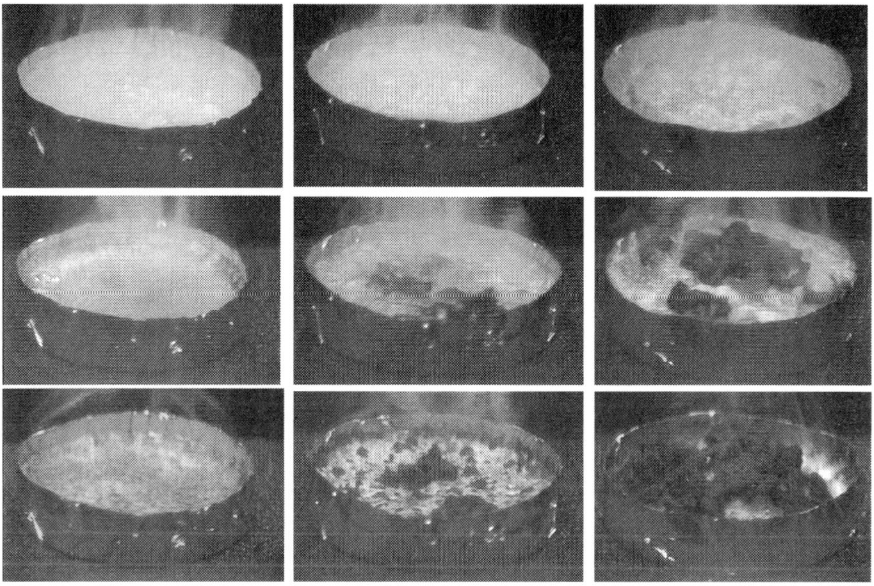

PA6 PA6/Clay(2%) PA6/Clay(5%)

Figure 9 *Selected video images of the three samples at 100, 200, and 400 s in nitrogen at 50 kW m^{-2}*

(Figure 9). The sample appeared less viscous (more fluid-like) with numerous small bubbles. Shortly after 200 s, some swelling of the sample was observed, giving it the appearance of a highly viscous mound. Vigorous bubbling in the very fluid-like upper layer of the sample continued, and the sample surface gradually darkened after 400 s. A very thin, black coating over the bottom of the container was left at the end of the test (Figure 10). The amount of the residue at the end of the test was less than 1% of the initial sample mass.

PA6 PA6/Clay(2%) PA6/Clay(5%)

Figure 10 *Residue pictures at the end of the gasification tests in N_2 at 50 kW m^{-2}*

The gasification behavior of the PA6/clay(2%) nanocomposite was initially similar to that of the PA6 sample, except that it appeared to be more viscous; it still had the appearance of a viscous fluid. Around 150 s, several small, dark floccules appeared on the surface and these grew with time, as shown in Figure 9. However, they never covered the entire sample surface. Numerous small dark floccules were formed, together with a few large floccules. The dark crust-like floccules were left at the bottom of the container at the end of the test. The mass of the residue was about 2% of the initial sample mass. The PA6/clay(5%) nanocomposite appeared to be much more viscous than the PA6 sample during the gasification test but it still formed numerous larger bubbles. Around 100 s after the start of irradiation, a thin, black ring (not continuously connected) appeared at the perimeter of the sample and this ring moved toward the center of the sample then collapsed to form a large black clump around 150 s. More black floccules appeared near the perimeter of the sample and moved gradually toward the center and formed larger rough-surface floccules. This can be seen in the images at 200 s in Figure 9.

Vigorous bubbling of evolved degradation products was observed over that portion of the sample surface not covered by the black floccules. The floccules gradually grew and were left at the bottom of the container at the end of the test. The mass of the residue was about 5% of the initial sample mass. A picture of the residue collected after the test is shown for each sample in Figure 10. These residues look like a carbonaceous char and are brittle and fragile. The PA6/clay(5%) nanocomposite generated more residue of the black floccules than the PA6/clay(2%) nanocomposite. Similar black floccules were also observed in the residues of the burned samples tested in the cone calorimeter. These pictures indicate that the formation of the protective, black floccules and their coverage over the sample surface are desirable as a means of reducing the exposure of the molten polymer to external radiant flux or to heat feedback from am: flame. For most effective FR performance, they need to cover the entire sample surface to fully shield/protect the polymer melt.

The ideal structure of a protective surface layer (consisting of clay particles and some char) is net-like and has sufficient physical strength not to be broken or disturbed by bubbling. The protective layer should remain intact over the entire burning period. Although the PA6/clay nanocomposites studied here formed such a protective layer covering a part of the sample surface, it was reported that the polystyrene (PS)/clay nanocomposite sample formed such a protective layer, covering the entire sample surface (Figure 11).[8] This could be due to enhanced formation of char from PS by the addition of clay. However, several large cracks were observed in the residues of this particular PS/clay nanocomposite sample.

Heat release rate curves of the three PA6-based samples are shown in Figure 12. The results show that the nanocomposite samples have a slightly increased ignition delay time and a significantly peak heat release rate in comparison to pristine PA6. The greater the clay content the lower the heat release rate corresponding to the surface coverage by the floccule layer. There is no significant reduction in total heat release due to the nanocomposites for the

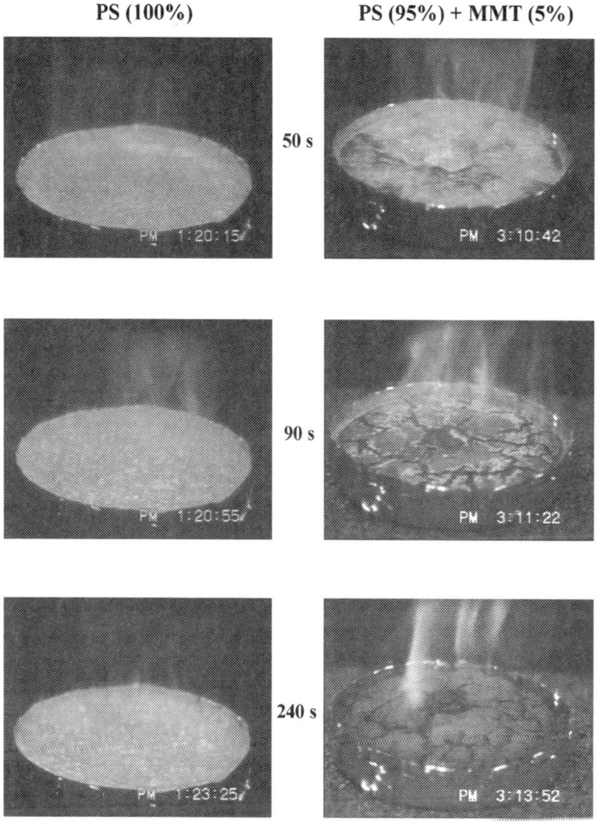

Figure 11 *Selected video images of gasification behavior of PS and PS/MMT(5%) nanocomposite in nitrogen at a flux of 50 kW m^{-2}. Extensive carbonaceous char formation can clearly be observed in the nanocomposite (right) samples*

levels of clay content used in this study. This indicates that the nanocomposites burn more slowly, but they burn nearly completely.

The mass loss rate curve of each sample (not shown) is proportional to the heat release rate curve. Thus, the specific heat of combustion obtained from the heat release rate divided by mass loss rate is 30 ± 2 kJ g^{-1} for all three samples. This unchanged specific heat of combustion implies that the observed reduction in heat release rate (and mass burning rate) tends to be due to chemical and physical processes mainly in the condensed phase rather than in the gas phase.

6.5 Polymer–Carbon Nanotube Nanocomposites

Reference 9 gives a detailed description of the preparation of poly(propylene) (PP)/multi-walled carbon nanotube (MWNT) nanocomposites. The nanocomposites were prepared by melt blending without any compatibilizer or organic treatment. All samples were compression molded to make 75 mm diameter by

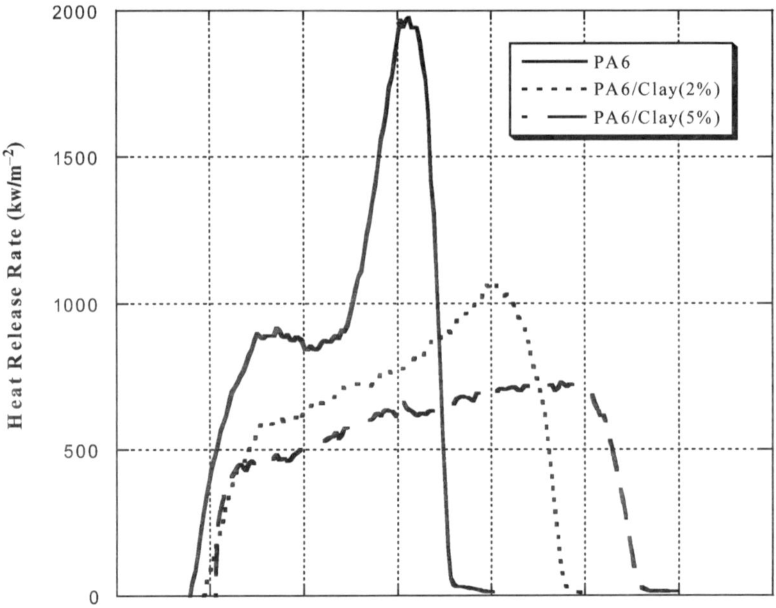

Figure 12 *Effects of clay content on heat release rate of PA-6 (8mm thick at 50 kW m^{-2})*

Figure 13 *SEM picture of MWNT dispersion in the PP/MWNT(1%) nanocomposite after solvent removal of PP*

8 mm thick disks. The distribution of the nanotubes in the sample was examined by two different methods and magnifications. One used scanning electron microscopy (SEM). A SEM picture of the recovered MWNTs after solvent removal of PP from a PP/MWNT (1%) nanocomposite is shown in Figure 13. Although it shows well dispersed MWNTs, *implying* good dispersion in the PP/MWNT nanocomposite, more direct observation and a larger observation area of the dispersion of MWNTs in the PP/MWNT samples are preferred. The second method used optical microscopy. The image of PP/MWNT(1%) in Figure 14 shows globally well-dispersed nanotubes in PP, along with a wide range of diameters and lengths of nanotubes (Figure 13).

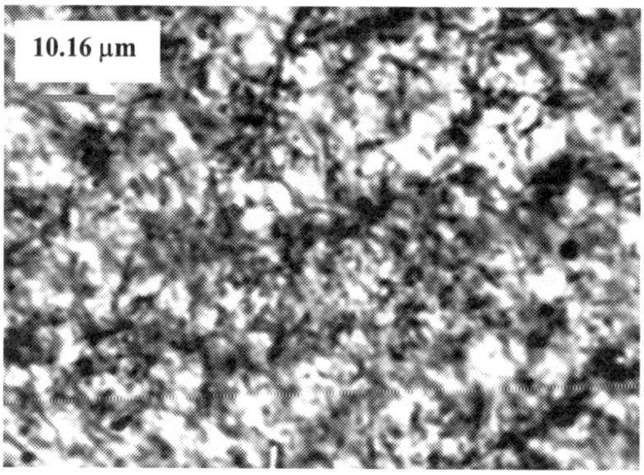

Figure 14 *Optical microscopy image of MWNT/PP(1%) nanocomposite in the melt*

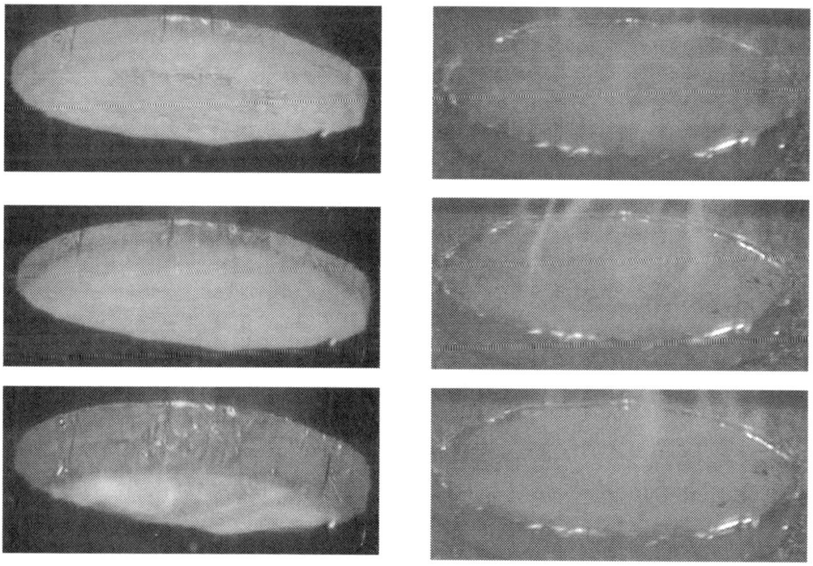

(a) (b)

Figure 15 *Sample behavior in the gasification test of* (a) *PP, and* (b) *PP/MWNT(1%) at 50 kW m*$^{-2}$ *in nitrogen*

The physical behavior of the PP/MWNT nanocomposites was significantly different from that of pristine PP during the gasification test. As shown in Figure 15(a), the pristine PP sample behaved like a liquid with a fine froth layer generated by the bursting of numerous small bubbles at the sample surface. No char was left at the end of the test. However, all PP/MWNT samples tested in this study behaved like a solid without any visible melting, except at the very beginning of the test, and the shape of the sample did not significantly change

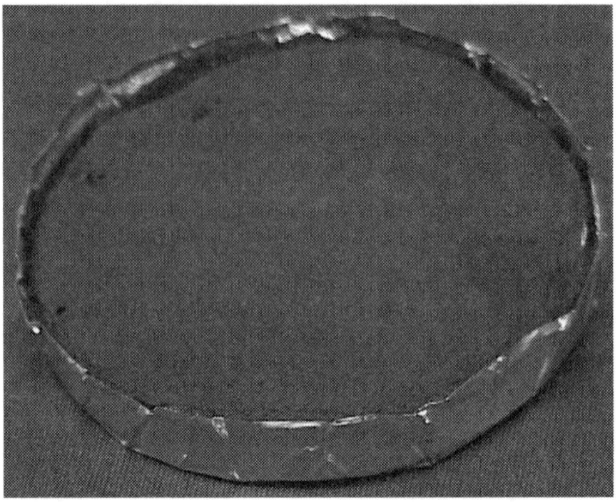

Figure 16 *Residue of PP/MWNT(1%) after the gasification test in nitrogen at 50 kW m*$^{-2}$

during the test. A picture of the residue of the PP/MWNT(1%) nanocomposite is shown in Figure 16. The shape of the residue was nearly the same as the original sample except for slight shrinkage. No cracks were observed in any residue of the PP/MWNT nanocomposites studied here. The networked floccule residues of the PP/MWNT samples covered the entire sample surface and extended to the bottom of the residue (Figure 17). The floccule residue was porous, but had physical integrity and did not break when lightly picked at by one's fingers. The mass of the floccule residue was very close to the initial mass of carbon nanotubes in the original nanocomposite. This indicates that the networked floccule did not enhance char formation from PP.

Figure 18 compares the heat release rate curves among the three samples. The results show that the heat release rates of the PP/MWNT nanocomposites

Figure 17 *Cross section of the residue of the PP/MWNT (1%) nanocomposite shown in Figure 15(b)*

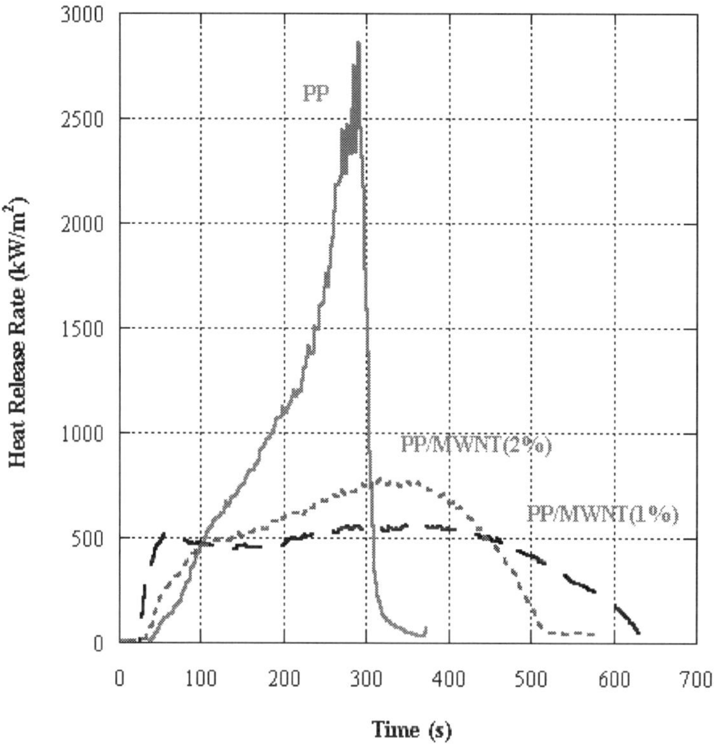

Figure 18 *Comparison of heat release rate curves among PP and PP/MWNT nanocomposites at 50 kW m*$^{-2}$

are much lower than that of pristine PP even though the amount of MWNT in PP is quite small. The total heat release, the integral of the heat release rate curve over the duration of the experiment, is about the same for the three samples. The curves of the mass loss rate per unit surface area for the three samples are very similar to those of the heat release rate. Since the specific heat of combustion value is calculated by dividing measured heat release rate with measured mass loss rate, this indicates that the specific heat of combustion is about the same for the three samples. The calculated specific heat of combustion of each sample is 43 ± 1 MJ kg^{-1}. The above results indicate that the PP/MWNT nanocomposites burn much slower than PP, but they all burn nearly completely.

These observations are similar to those made with clay-nanocomposites and with composites made by the addition of nanoscale silica in PMMA, as described above. This indicates that the observed FR performance of the PP/MWNT nanocomposite is mainly due to chemical and/or physical processes in the condensed phase instead of in the gas phase.

6.6 Discussion

Heat release rate curves of the three different types of nanocomposites show that a large reduction in heat release rate is achieved in the following decreasing order: polymer–MWNT nanocomposites, polymer–clay nanocomposite, and polymer–nanosilica nanocomposites. The residues of these nanocomposites show (1) a low density network-like protective layer covering the entire sample surface without any cracks for the PP/MWNT nanocomposite, (2) floccular islands consisting of clay and char for the PA6/clay nanocomposites, and (3) the formation of coarse, granular particulate clumps for the PMMA/nanosilica nanocomposites. This suggests that the formation of a continuous network-like protective layer that covers the entire sample surface is critical to significantly reduce the heat release rate with only a small mass of these nanoscale particles. Some polymer–clay nanocomposites can form such a layer with the PS/MMT nanocomposites[8] and the PP/polypropylene-grafted maleic anhydrate (PP-g-MA)/MMT nanocomposites.[10] However, their residues were slightly brittle and were cracked. Therefore, it appears that a higher clay content (5–10%) in polymers is needed to obtain a similar amount of reduction in heat release rate as that of the PP/MWNT (1%) nanocomposite.

The formation of a continuous, low-density network-structured protective layer is easiest with high aspect ratio nanoscale particles. The aspect ratio of the nanosilica in PMMA is about 1, and coarse, granular particulate clumps, probably coagulated silica particles, were formed instead of a network structure. The effects of aspect ratio for plate-like clay particles on the heat release rate of polymer/clay nanocomposites were determined using three different clays; synthetic mica (aspect ratio of roughly one thousand), MMT (about 100), and synthetic hectolite (few hundreds). The PP/PP-g-MA/clay nanocomposites (all had 7.7% clay by mass) were prepared by melt mixing and the mass loss rate of each sample was measured in nitrogen in the gasification test at 50 kW m^{-2}. The sample was a $10 \times 10 \times 0.3$ cm thick plate. The results are shown in Figure 19.

The lowest mass loss rate was observed for the sample with synthetic mica, which generated a significantly lower mass loss rate than those with MMT and synthetic hectolite. There was little difference in mass loss rate between the sample with MMT and with synthetic hectolite. Figure 20 shows pictures of the residues of the four samples after the tests. The surface of the residue of the sample with synthetic mica is relatively smooth without any large cracks. However, the two samples with the other two clays show large cracks (in particular, with synthetic hectolite). Melting and vigorous bubbling were observed in the large cracks during the test. Similar behavior was observed at the exposed sample surface of PA6/clay (Figure 9). The different types of clay in the PP/PP-g-MA/clay nanocomposites formed a slightly different structure of the protective layer during the test. This appears to be caused by the difference in aspect ratio among the three clays. Nanoscale particles with higher aspect ratio tend to form a network-like structure of a protective layer, covering the entire surface without any significant cracks.

Dispersion of the clay particles in a polymer/clay nanocomposite has significant effects on the reduction in heat release rate. If the clay particles are not well dispersed in the nanocomposite, no significant reduction in heat release rate has been reported for the PS/MMT sample.[8] A similar trend was recently

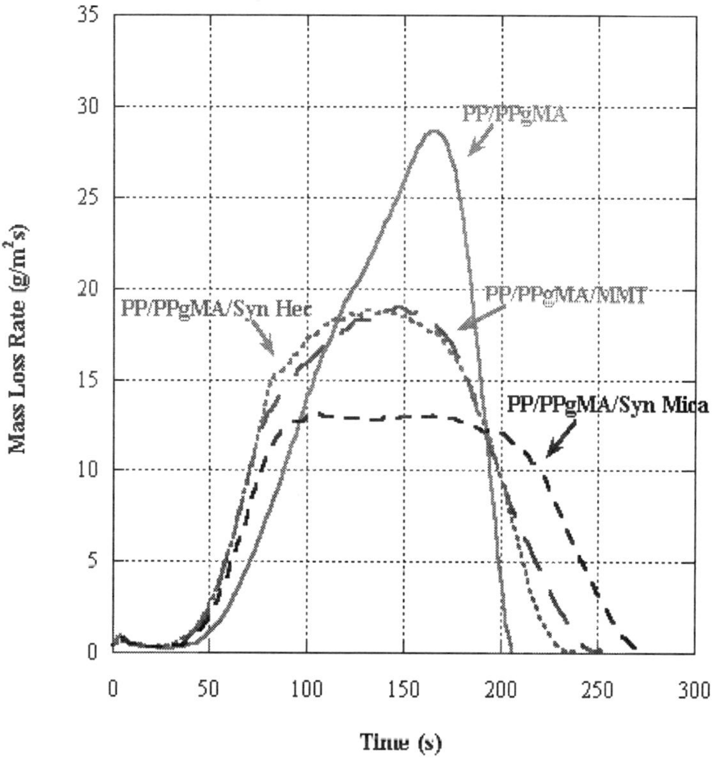

Figure 19 *Effects of clay type on mass loss rate of PP(92.3%)/PPgMA(7.7%) and PP(84.6%)/PPgMA(7.7%)/clay(7.7%) samples in nitrogen at 50 kW m^{-2}*

observed with polymer/MWNT nanocomposites. Therefore, nanoscale particles in a polymer nanocomposite should be well dispersed to obtain the maximum performance in reducing heat release rate.

Nanocomposites generally reduce heat release rate, in particular the peak heat release rate. However, they tend to have slightly shorter ignition delay times (or no significant increase in ignition delay) and about the same total heat release as those of pristine polymers. As described here, nanocomposites tend to burn slowly and nearly completely. Although the use of a small quantity of nanoscale particles in polymers to form nanocomposites could be one of the alternatives to conventional flame retardants, nanocomposites need further improvements to increase ignition delay time and reduce total heat release.

6.7 Conclusion

Nanocomposites based on three different shapes of nanoscale particles, sphere (silica), plate (clay), and tube (carbon nanotube) were prepared and the dispersion of the particles in the nanocomposites was confirmed by various techniques (TEM, SEM, optical microscopy, and XRD). Their flammability properties

Figure 20 *Effects of clay type on the surface pattern of the residues of PP(92.7%)/PPgMA(7.7%) and PP(84.7%)/PPgMA(7.7%)/clay(7.7%) samples (after gasification test in nitrogen at 50 kW m^{-2}).*

were measured by using a cone calorimeter and a radiative gasification apparatus in nitrogen. Residues of these nanocomposites after the gasification test were collected and their shape and structure examined. The results show that the reduction in heat release rate is achieved in the order: carbon nanotubes, clay platelets, and silica spheres, providing these particles are well dispersed in the sample. It appears that the particles having higher aspect ratio tend to form an effective protective layer consisting of network-structured floccule that covers the entire sample surface without forming any cracks during burning. The formation of such a layer is critical to obtain low heat release rate from nanocomposites.

6.8 Acknowledgement

This chapter is based on results obtained from many different projects, as described in references 6–8, with the collaboration of many people. The author thanks Richard Harris, John Shields, Alexander Morgan (currently at Dow

Chemical Co.), Kathryn Butler, and Jeffrey Gilman at the Fire Research Division for their collaboration, Xin Zhang and Robert Briber at the Department of Materials Science and Engineering of University of Maryland for collaboration with the PA6/clay work, Jenny Hilding and Eric Grulke at the Department of Chemical and Materials Engineering of University of Kentucky for their collaboration with the PP/MWNT work, Joseph Antonucci, Semen Kharchenko and Jack Douglas of the Polymers Division of NIST for the preparation of PMMA/nanosilica nanocomposites and valuable discussions about the properties of polymer nanocomposites, and, finally, Koichi Iwasa at Sekisui Chemical Co. for collaboration with PP/PP-g-MA/clay work.

6.9 References

1. Y. Kojima, A. Usuki, M. Kawasumu, A. Okada, Y. Fukushima, T. Kurauchi and O. Kamigaito, *J. Mater. Res.*, 1993, **8**, 1185–1189.
2. E.P. Giannelis, *Adv. Mater.*, 1996, **8**, 29–35.
3. Z. Wang and T.J. Pinnavaia, *Chem. Mater.*, 1998, **10**, 1820–1826.
4. P.J. Austin, R.R. Buch and T. Kashiwagi, *Fire Mater.*, 1998, **22**, 221–237.
5. T. Kashiwagi, A.B. Morgan, J.A. Antonucci, .M.R. VanLandingham, R.H. Jr Harris, W.H. Awad and J.R. Shields, *J. Appl. Polym. Sci.*, 2003, **89**, 2072–2078.
6. T. Kashiwagi, J.R. Shields, R.H. Jr Harris and R.D. Davis, *J. Appl. Polym. Sci.*, 2003, **87**, 1541–1553.
7. T. Kashiwagi, R.H. Jr Harris, X. Zhang, R.M. Briber, B.H. Cipriano, S.R. Raghaven, W.H. Awad and J.R. Shields, *Polymer*, 2004, **45**, 881–891.
8. A.B. Morgan, R.H. Jr Harris, T. Kashiwagi, L.J. Chyall and J.W. Gilman, *Fire Mater.*, 2002, **26**, 247–253.
9. T. Kashiwagi, E. Grulke, J. Hilding, R.H. Jr Harris, W.H. Awad and J. Douglas, *Macromol. Rapid Commun.*, 2002, **23**, 761–765.
10. A.B. Morgan, T. Kashiwagi, R.H. Harris, J.R. Campbell, K. Shibayama, K. Iwasa and J.W. Gilman, in *Fire and Polymers*, G.L. Nelson and C.A. Wilkie (eds.), ACS Sym. Series, 2001, Volume 797, pp. 9–23.

CHAPTER 7

Thermal Degradation and Combustibility of Polypropylene Filled with Magnesium Hydroxide Micro-Filler and Polypropylene Nano-Filled Aluminosilicate Composite

SERGEI M. LOMAKIN,[1] GENNADY E. ZAIKOV[1] AND ELENA V. KOVERZANOVA[2]

[1]Institute of Biochemical Physics of Russian Academy of Sciences, 4 Kosygin Street, Moscow, 119991, Russia (chembio@sky.chph.ras.ru)
[2]Institute of Chemical Physics of Russian Academy of Sciences, 4 Kosygin Street, Moscow, 119991, Russia

7.1 Introduction

Mineral fillers, metals, and fibers have been added to thermoplastics and thermosets for decades to form composites. Compared to the neat resins, these composites show improved properties including tensile strength, heat distortion temperature, and modulus. Thus, for structural applications, composites have become very popular and are sold in billion pound quantities. These filled thermoplastics are sold in even larger volumes than neat thermoplastics. Furthermore, the volume of fillers sold is roughly equal to that of thermoplastic resin sold. Clearly, the idea of adding fillers to thermoplastics and thermosets to improve properties, and in some cases decrease costs, has been very successful for many years. More recently, advances in synthetic techniques and the ability to readily characterize materials on an atomic scale have lead to interest in nanometer-size materials. Since nanometer-size grains, fibers and plates have dramatically increased surface area compared to their conventional micrometer-size materials, the chemistry of these nanosized materials is altered compared to conventional materials.[1] Emerging nanotechnologies offer the potential for revolutionary new polymer materials with enhanced physical features: reduced flammability, thermal expansion coefficients, increased stiffness and strength, barrier properties, and heat resistance, without loss of impact strength.

Nanocomposites, which contain nanometer-scale particles that are homogeneously dispersed throughout traditional polymers, can provide a stiffness and strength approaching that of metals, but with significant reductions in weight. Reinforcing polymers at the molecular level with inorganic fillers can bring about property improvements in polymeric materials.

The commercial importance of polypropylene (PP) has driven the investigation of PP composites reinforced by particulates, fibers, and layered inorganic fillers. Specifically, with respect to layered inorganic fillers, the aluminosilicate minerals talc and mica have been of greatest interest. However, recent advances in polymer/clay and polymer/silicate nanocomposite materials have motivated efforts to disperse fillers in PP based on montmorillonite, a naturally occurring mineral in the 2 : 1 aluminosilicate family. Because of the PP nonpolar (aliphatic) nature, it has proved challenging to develop a clay-based filler that is directly miscible with neat (*i.e.* nonfunctionalized) PP.

General interest in the flammability reduction and thermal degradation of PP arose in the second half of the 20th century. Halogen-based flame retardants are the most commonly used due to their high efficiency, but this use is now reversing due to the presumed high toxicity and corrosiveness of its breakdown products. Nowadays, the main interest in flame retardancy research is focused in halogen-free flame retardants. Both magnesium hydroxide and nanocomposite compositions with PP present a flame retardancy effect in addition to total harmlessness. However, for PP composition with $Mg(OH)_2$ the utilization of this kind of filler involves important changes in the mechanical properties of the polymer due to the high filling level required to obtain a good flame retardancy (weight concentrations up to 60% may be required). In general, dispersion in PP of fillers such as $Al(OH)_3$ and $Mg(OH)_2$ provokes a decrease in tensile yield strength and fracture toughness measured at high rate on the homopolymer based compounds or measured at low strain rate on the block copolymer PP compounds.

The ability of nanoclay incorporation to reduce the flammability of polymeric materials has been a major theme of publications.[2,3] Gilman *et al.* demonstrated the extent to which flammability behaviour could be restricted in polymers such as PP with only 2% nanoclay loading. In particular, heat release rates were found to decrease substantially with nanoclay incorporation.[4] Although conventional microparticle filler incorporation, together with the use of flame retardant and intumescent agents, would also minimise flammability behavior, this is usually accompanied by reductions in various other important properties. With the nanoclay approach, this is usually achieved whilst maintaining or enhancing other properties and characteristics.

In this work, different compositions of thermal degradation of polypropylene filled with magnesium hydroxide and a hybrid nanocomposite with polypropylene (maleic anhydride-modified PP matrix reinforced with 10 wt% of organically modified montmorillonite (Cloisite 15A)) are studied, focusing on the thermal stability and flammability characteristics.

7.2 Experimental

7.2.1 Materials

Polypropylene (PP; BO677MO, Bolearis) and polypropylene-*graft*-maleic anhydride (PPgMA, Aldrich, 0.4% by mass fraction MA) were dried for 2 h at 70°C in an air-flow oven and then stored over silica gel before use.

PP composition with magnesium hydroxide (Aldrich) was blended using a using a Brabender mixing chamber at 210°C. Composites with mineral levels in the range 1–60 wt% were prepared.

Organically treated layered alkylammonium montmorillonite, Cloisite 15A, was supplied by Southern Clay Products (San Antonio, TX) Hybrid nanocomposite with polypropylene (maleic anhydride-modified PP) matrix reinforced with 10 wt% fraction of organically modified montmorillonite (Cloisite 15A) was prepared *via* melt intercalation in a Brabender chamber at 210°C for 10 min.

Pyrolysis of polypropylene compositions was performed in a pyrolytic cell at 300, 500 and 700°C in air (flow rate of 30 ml min^{-1}). The products of pyrolysis were dissolved in hexane at 0°C. The oven temperature was monitored with a thermocouple and a stability of ±5°C was verified.

7.2.2 Thermal Analysis

Vertical TG balance Derivatograph 950Q was used for TGA of samples pyrolyzed at different heating rates under 100 ml min^{-1} nitrogen or air flow. The average weight of samples was 5 mg.

7.2.3 Gas Chromatography/Mass Spectrometry Analysis (GC-MS)

Degradation products have been analyzed by gas chromatography using a "Zvet 500M" with an electron capture detector. A glass column (3 mm × 4 m) filled with OV-17 (phenylmethyl silicon) was used at 230°C.

GC/MS analysis of samples was performed using a "Varian 3300" gas chromatograph connected to a mass spectrometer detector (ion trap), "Finnigan MAT ITD 800". A DB-5 fused capillary column (0.32 mm × 30 m) temperature programmed from 50 to 270°C at 10°C min^{-1} was used in GS/MS analysis. Mass spectra detection (from 40 to 650 Da) were obtained in electron impact mode (energy of 70 eV). All mass spectra were assigned using the Wiley275 mass spectral Library.

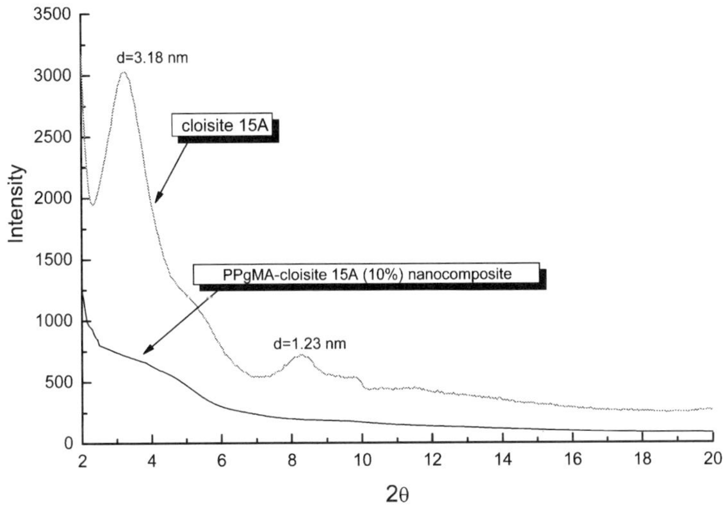

Figure 1 *XRD of Cloisite 15A and PPgMA-nanocomposite*

7.2.4 Clay and Composite Characterization

Microstructures of Cloisite 15A and PPgMA-Cloisite 15A nanocomposite were characterized using XRD on a Philips diffractometer using Cu K_α radiation ($\lambda = 0.1540562$ nm). XRD patterns of Cloisite 15A and PPgMA-Cloisite 15A nanocomposite, which reveal the intercalated structure of Cloisite 15A and exfoliated structure of PP nanocomposite, are shown in Figure 1. The interlayer spacing (d_{001}) for the Cloisite 15A is 3.18 nm.

Generally, XRD data gives an initial picture of the clay distribution. In intercalated nanocomposites the dimension and distribution of tactoids could be very variable and a complete characterization of the nanocomposite morphology could be achieved using transmission electron microscopy. TEM images of PP nanocomposite samples (cooled at −70°C and then microtomed with a diamond knife at *ca.* −50°C) are obtained at 80 kV, with a ZEISS EM 900. An image of a PP nanocomposite (10 wt%) is presented (Figure 2). The micrograph shows the layers of the clay (dark lines in the micrograph represent an aluminosilicate layer): a part of them being well dispersed with the presence of tactoïds. This proves a partial exfoliation of the clay throughout the PP matrix.

7.3 Results and Discussion

As mentioned in previous studies, the thermal degradation behavior of polymer compositions directly influences their combustibility. The first routine procedure to evaluate this behavior is a formal kinetic approach based on TGA data. Improved thermal stability of PP compositions of PPgMA-Cloisite 15A nanocomposite and PP-$Mg(OH_2)$ (60%) was demonstrated by thermogravimetric analysis.

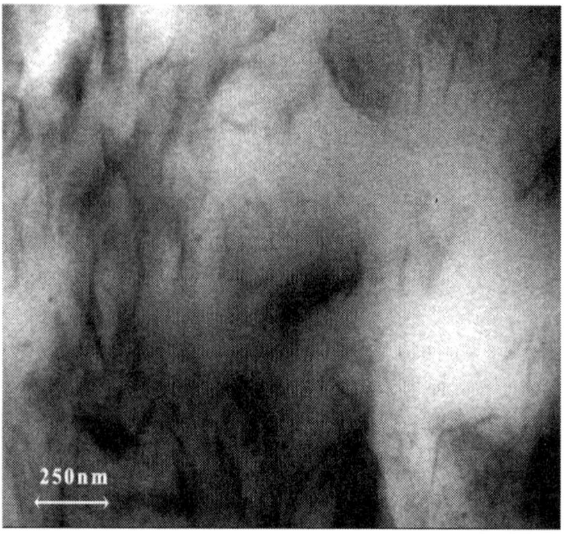

Figure 2 *TEM image of PPgMA-nanocomposite (10 wt% of Cloisite 15A)*

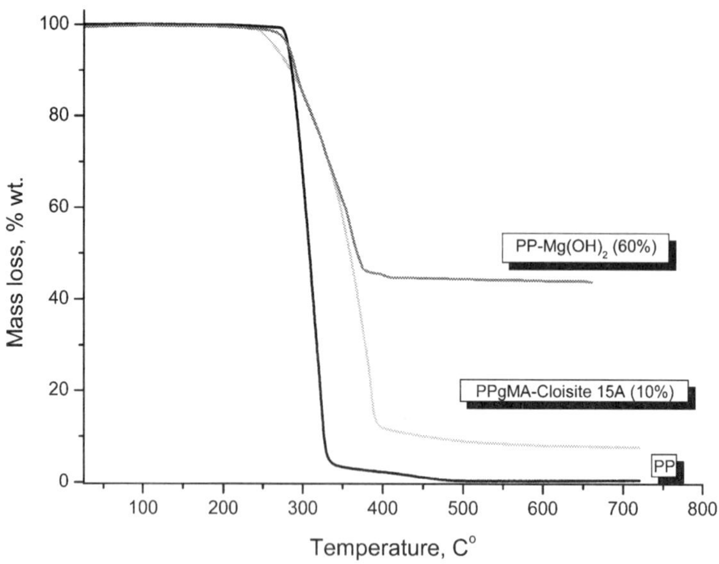

Figure 3 *TGA of poly(propylene) formulations (10 °C min^{-1}, in air)*

The thermal stability at 50% weight loss was increased by 60°C (Figure 3). However, the extent of thermal stability is related to the chemical nature and the degradation mechanism of the pristine polymer. TG traces of a PP conventionally filled with 60% of $Mg(OH)_2$ and an exfoliated nanocomposite of PPgMA-Cloisite 15A (10 wt% clay fraction) in air and under nitrogen are given in Figure 3.

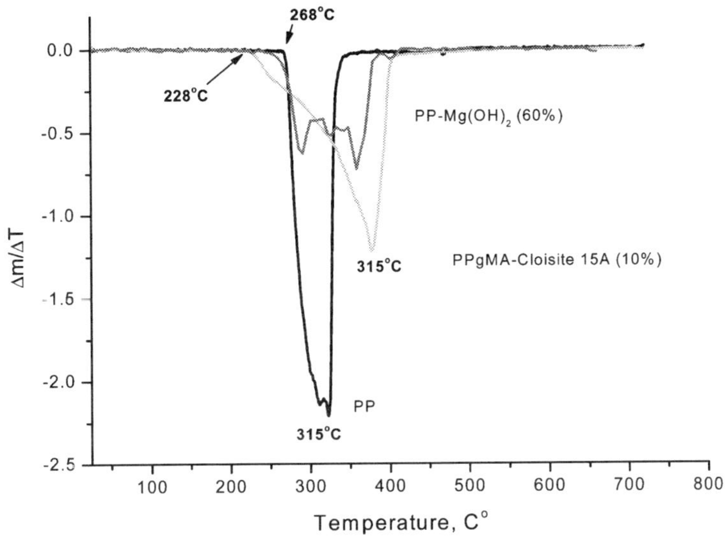

Figure 4 *DTG of polypropylene compositions (10 °C min^{-1}, in air)*

At the initial stage of thermal degradation in air, the nanocomposite showed poorer performance than the conventional PP and $Mg(OH)_2$-filled PP, which may be explained by the low thermal stability of the compatibilising agent (octadecylammonium ions). Therefore, the choice of the compatibilising agent is obviously important. Generally, an increase in the maximum of the mass loss rate of PPgMA-Cloisite 15A in comparison with PP can be described by a diffusion process that limits the evolution of the gaseous products.

Whereas thermal degradation process of PP and PPgMA-Cloisite 15A (10%), PP + 50% $Mg(OH)_2$ proceeds in the stage the degradation of PP filled with 60% of $Mg(OH)_2$ is a multistage process (Figure 4).

As mentioned above, an excess of organo-modifier (alkylammonium ions) destabilizes PP at the initial stage of thermal degradation. The initial temperature of thermal degradation of PPgMA-Cloisite 15A is 228°C, whereas the pure PP starts to decompose at 268°C (Figures 3 and 4).

The complexity of TGA data does not allow easy reliable mechanistic conclusions. Thus, GC-MS analysis became a very important procedure to evaluate the thermal degradation mechanism of the studied formulations. Such analysis has shown that at 300 and 500°C the basic products of the thermal degradation are aliphatic saturated and unsaturated hydrocarbons, ketones and alcohols with siding methyl side groups.

Identification of these products is quite complicated because of aliphatic hydrocarbons and alcohols $C_2H_5^+$, the ions with odd and even mass number of electrons are formed under ionization: $C_2H_5^+$, $C_3H_7^+$, $C_4H_9^+$, etc. Intensities of peaks of ions are the greatest in the field of mass numbers 43, 57, 71. Intrinsic mass spectra of methyl ketones have ions with mass 43, 57, 71 and 85, which

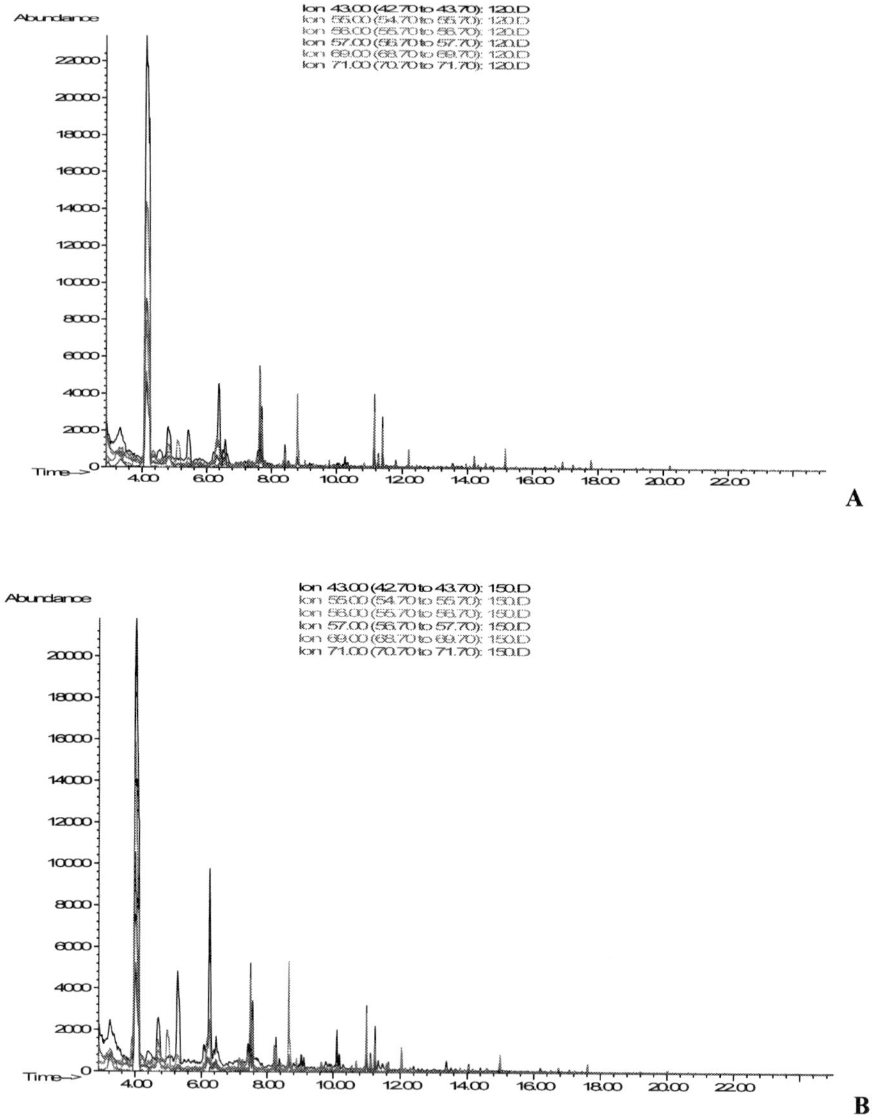

Figure 5 *GC-MS chromatograms of typical pyrolysis products of PP* (A) *and PP-Mg(OH)*$_2$ *(60%)* (B) *[alcohols (at 500 °C) with m/z = 43, 55, 56, 57, 69 and 71]*

are formed by the breaking of the bond in α to the carbonyl group. GC-MS chromatograms of alcohol products of pyrolysis with *m/z* 45, 59, 73 and 87 and ketones with *m/z* 58, 72 and 86, respectively, are presented in the Figures 5 and 6.

Retention times and concentrations of the pyrolysis products for various PP samples are given in Table 1. The most abundant product (retention time 4 min)

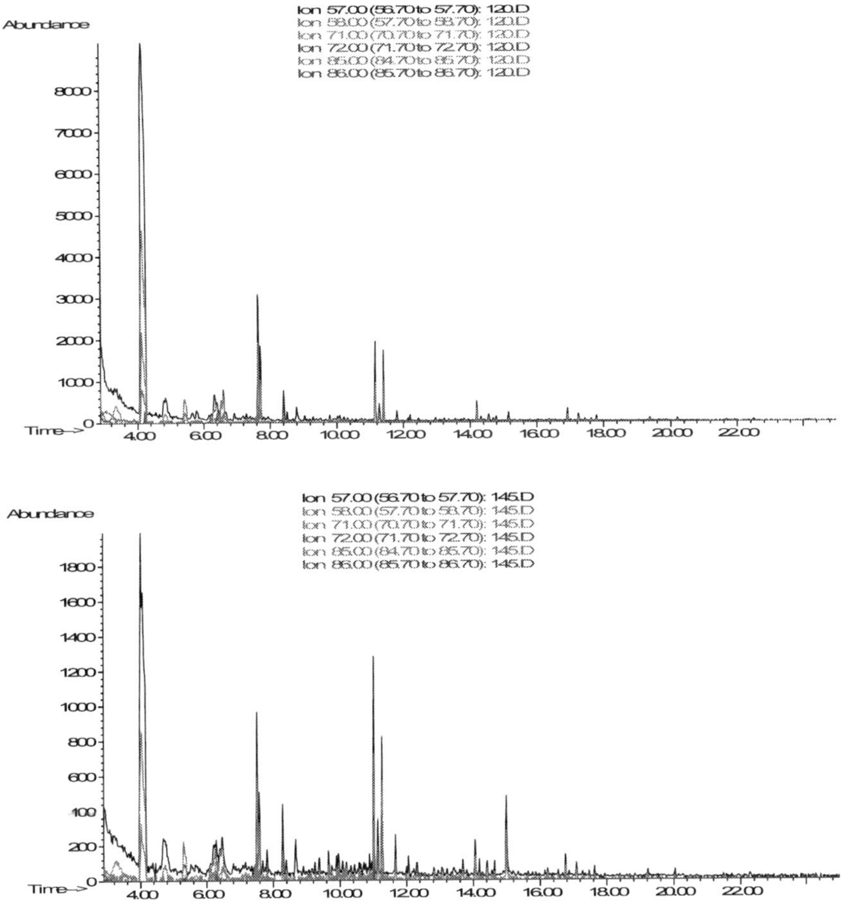

A

B

Figure 6 *GC-MS chromatograms of typical pyrolysis products of PP* (A) *and PP-Mg(OH)$_2$ (60%)* (B) *[ketones (at 500 °C) with m/z = 58, 71, 72, 85, 86]*

may be identified as 2,4-dimethyl-1-heptene. At 700°C some fused aromatic compounds were identified.

The products of PP and PP-$Mg(OH)_2$ (60%) are practically identical at 300°C. However, at higher temperatures (700°C) some differences in product distributions and concentrations have been found, which are explained by the presence of isomers of methylnaphthalene, azulenes and biphenyls.

This phenomenon was explained by means of solid-phase catalysis, which leads to the formation of condensed aromatics and, ultimately char. TGA indicated about 5% char yield under conditions of thermal-oxidative degradation of PP with 60% of $Mg(OH)_2$ (Figure 3). A mechanism of char formation was proposed for the PP composition that consists of a high amount of $Mg(OH)_2$ additive. It starts to decompose above 300°C according to:

Table 1 *Thermal degradation products of polypropylene formulations vs. temperature. Glossary: tr: retention time, PP/Mg : PP/Mg(OH)$_2$(60%)*

		Concentrations (wt%)					
		300		*500*		*700*	
Pyrolysis T (°C) *Products*	t_r (min)	*PP*	*PP/Mg*	*PP*	*PP/Mg*	*PP*	*PP/Mg*
Methylbenzene	3:16	–	–	–	–	16.2	19.3
	3:31	–	1.8	1.8	–	–	–
2,4-Dimethyl-1or3-heptene	4:01	7.0	4.6	33.2	18.2	7.9	–
Ethylbenzene	4:23	–	–	–	–	6.9	9.9
	4:26	–	1.0	0.6	1.4	–	–
2,4-Dimethylnonane	4:41	1.6	2.2	1.7	2.3	–	–
Dimethylbenzene	4:42	–	–	–	–	5.3	7.6
	4:59	1.1	1.4	1.6	1.3	–	–
4-Methyl-2-heptanone	5:17	7.5	6.8	1.2	0.9	–	–
Isomers of methyethyllbenzene	5:42	–	–	–	–	1.4	2.0
	6:01	–	–	–	–	1.2	1.3
2,4,6-Trimethyl-2-nonane	6:05	5.6	3.8	–	–	–	–
	6:12	–	–	2.8	1.5	3.8	–
2,6-Dimethyl-4-heptanone	6:15	10.2	8.5	4.4	3.6	–	–
	6:26	3.2	1.4	1.4	0.9	0.6	0.4
	6:42	–	–	–	–	0.6	–
	7:10	0.8	–	0.7	0.9	0.5	4.7
	7:16	0.8	0.5	0.5	0.6	0.4	–
Isomers of dimethyloctanol	7:26	4.7	2.0	–	0.3	1.0	–
	7:30	5.1	3.5	9.8	7.4	2.8	–
	7:34	4.5	3.4	5.9	3.9	1.5	–
	7:40	–	–	0.5	0.8	1.0	0.8
	7:52	0.6	0.8	–	0.3	–	0.7
	8:13	2.2	–	1.9	–	1.4	–
3-Methy-3,5-hexadiene	8:16	1.1	2.0	0.4	1.0	0.2	–
	8:22	0.2	1.2	–	2.5	0.7	–
2,5,5-Trimethyl-1,6-heptadiene	8:39	3.8	3.8	5.7	0.5	2.1	–
4,8-Dimethyl-1,7-nonadiene	8:41	1.6	–	1.3	6.0	–	–
	8:53	0.4	–	0.5	–	0.2	2.5
Azulene	8:58	–	–	–	–	–	1.1
	9:01	1.8	1.8	0.2	–	0.4	–
	9:06	1.7	1.2	0.4	0.4	0.5	–
Naphthalene	9:17	–	–	–	–	6.8	14.6
	9:38	–	–	0.7	1.1	1.4	–
	9:46	1.1	1.1	–	0.5	–	0.4
4,6-Dimethyl-5-hepten-2-one	10:06	6.0	5.1	0.4	0.9	0.9	0.4
	10:11	2.9	2.8	0.2	0.3	0.4	–
	10:21	–	–	0.1	0.6	–	–
	10:22	–	–	–	–	–	–
	10:31	–	–	–	0.8	–	0.8
	10:42	0.6	–	0.5	1.3	1.7	–

Table 1 *Continue*

		Concentrations (wt%)					
		300		*500*		*700*	
Pyrolysis T (°C) *Products*	t_r (min)	*PP*	*PP/Mg*	*PP*	*PP/Mg*	*PP*	*PP/Mg*
Methylnaphthalene	10:59	–	–	–	–	–	5.5
2,4,6,8-Tetramethyl-1-undecene	11:00	5.1	4.3	6.7	11.5	18.0	–
	11:07	2.5	2.2	1.4	2.4	2.4	–
	11:15	3.5	3.5	4.9	7.8	10.2	–
Dimethylnaphthalene	11:15	–	–	–	–	–	3.3
	11:20	1.9	0.5	0.1	0.2	–	–
	11:28	1.1	0.3	0.1	0.1	–	–
	11:39	0.9	0.7	0.7	1.6	1.6	0.1
	12:02	1.7	2.3	1.6	3.5	3.6	0.4
1,1-Biphenyl	12:12	–	–	–	–	–	3.2
	13:22	1.9	2.1	0.3	0.6	–	–
	13:32	1.0	0.7	0.1	0.2	–	–
	13:40	1.1	–	0.1	0.4	–	2.1
	13:47	0.1	–	0.5	0.6	0.5	0.2
	14:03	0.3	0.8	1.3	1.5	1.0	0.9
1-(2-Phenylethylnyl)-benzene	14:11	0.4	1.1	0.3	1.7	–	0.8
	14:36	–	–	0.3	0.9	0.4	0.3
	14:59	0.7	11.4	1.9	2.5	0.8	2.9
1-(3-Phenylpropyl)-benzene	15:51		6.2		0.4	–	–
	16:12	0.8	1.0	0.1	0.3	–	–
	16:32	–	–	0.3	0.4	–	–
	16:49	0.3	1.0	0.8	1.0	0.7	0.2
	17:04	0.2	0.4	0.5	0.6	0.2	–
	17:37	0.3	0.6	1.0	0.9	0.6	–
	19:12	–	–	0.2	0.3	0.4	–
	20:01	0.2	–	0.4	0.4	0.3	–
Total Σ (prod.)		100.0	100.0	100.0	100.0	100.0	100.0

$$[Mg(OH)_2 = MgO + H_2O]$$

Magnesium oxide is a well-known catalyst for the dehydrogenetion of aliphatic and aromatic substitutes due to its low-basic properties and high surface area.[5,6] During high temperature pyrolysis or under combustion conditions, PP-$Mg(OH)_2$ undergoes a solid-phase catalytical condensation, with polycyclization followed by char formation on the activated surface of MgO (Scheme 1). The concentrations of methylnaphthalenes, azulenes and biphenyls, as well as methylbenzene, ethylbenzene, naphthalene and substituted biphenyls tend to grow significantly in comparison with conventional PP (Table 1). From the above results and discussion, the incorporation of $Mg(OH)_2$ clearly leads to flame retardancy of polypropylene, initiated in the gaseous phase, followed by a solid-phase catalytic char formation.

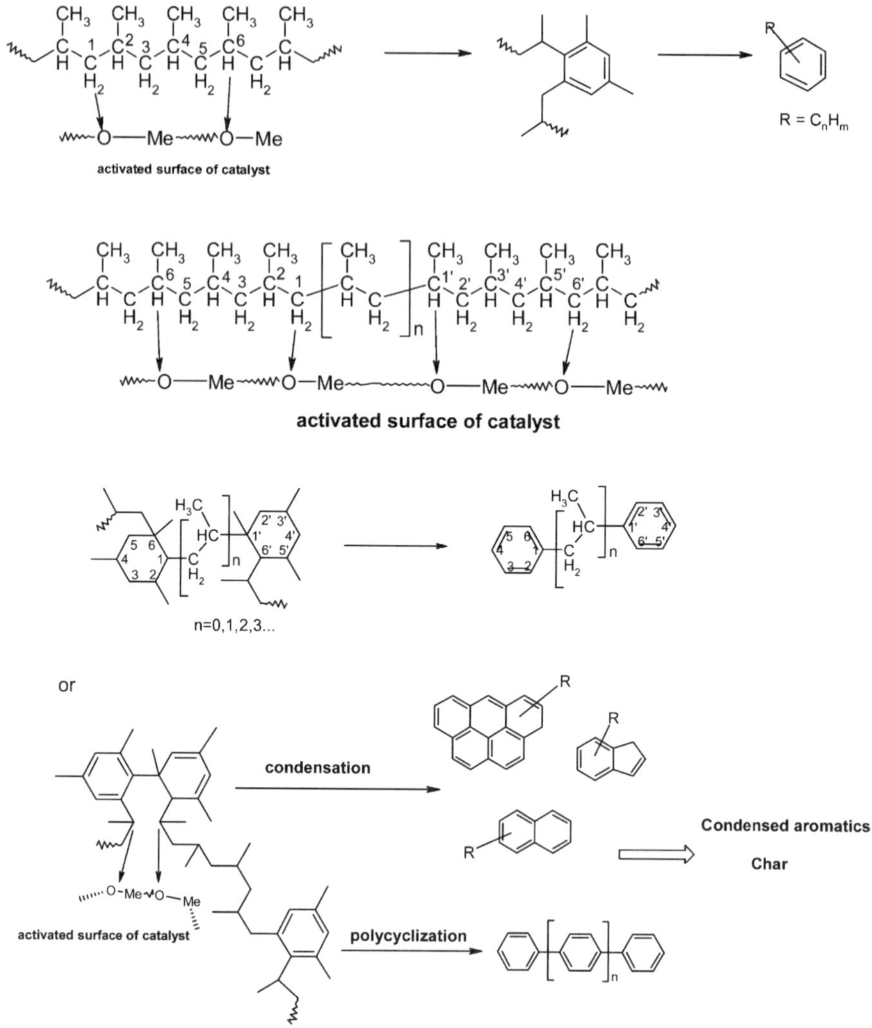

Scheme 1 *Solid-phase catalytic char formation of PP-60% $Mg(OH)_2$*

The combustibility of PP compositions was estimated by cone calorimetry (under an external heat flux 35 kW m^{-2}). Figures 7 to 10 show the heat release rate, heat of combustion, CO evolution and total hydrocarbons unburnt during combustion, respectively.

For a PPgMA nanocomposite with 10 wt% Cloisite 15A, there is a 40% relative reduction in flammability compared to the unfilled polymer (Figure 7, Table 2). The flame retardant performance arises from the formation of a carbonaceous-char layer, which develops on the outer surface during combustion. This surface char has a high concentration of aluminosilicate layers and becomes an excellent insulator and a mass transport barrier (slowing the oxygen

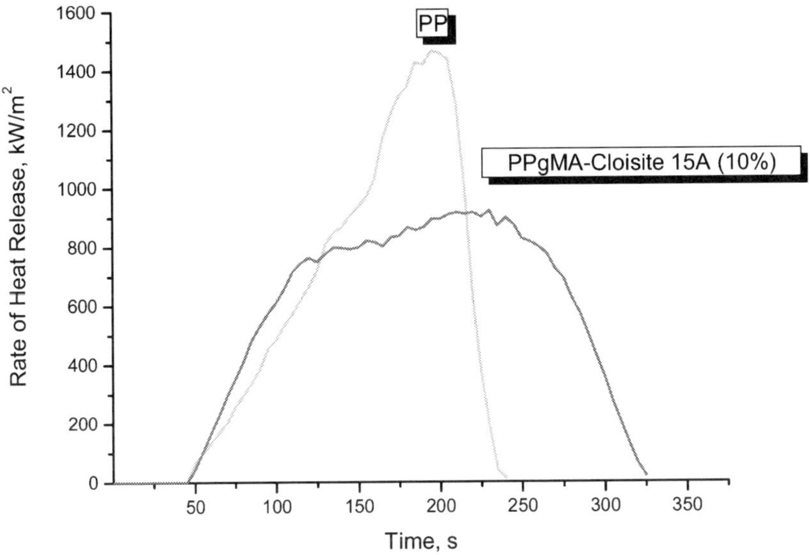

Figure 7 *Maximum heat of release for PP and PPgMA-Cloisite 15A (cone calorimeter; external heat flux: 35 kW m^{-2})*

Table 2 *RHR values from cone calorimetry of PP formulation*

Samples RHR (kW m^{-2})	*PP*	*PP-60%Mg(OH)$_2$*	*PP-g-Ma / OMC 15A-10%*
PkRHR	1480	593	907
Average RHR	577	401	472

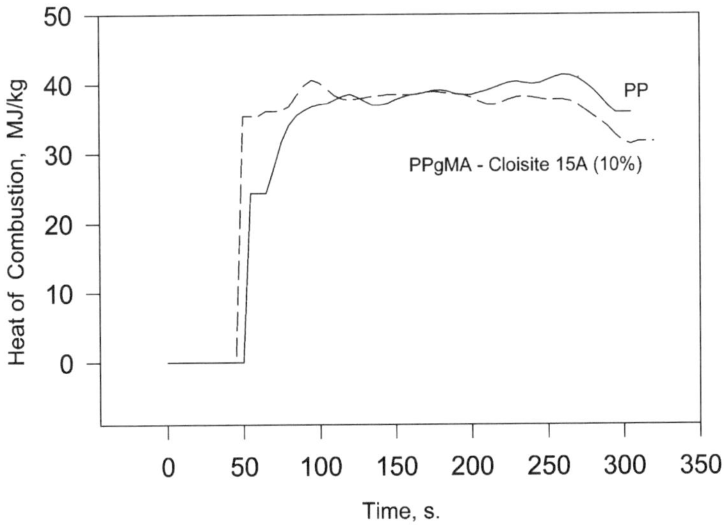

Figure 8 *Heat of combustion of PP and PPgMA-Cloisite 15A vs. time (external heat flux: 35 kW m^{-2})*

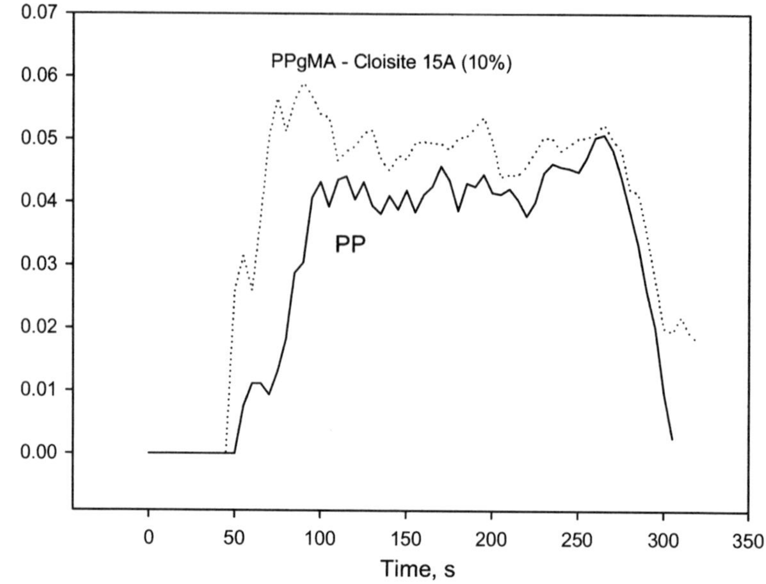

Figure 9 *Carbon monoxide yield for PP and PPgMA-Cloisite 15A vs. time (external heat flux: 35 kW m^{-2})*

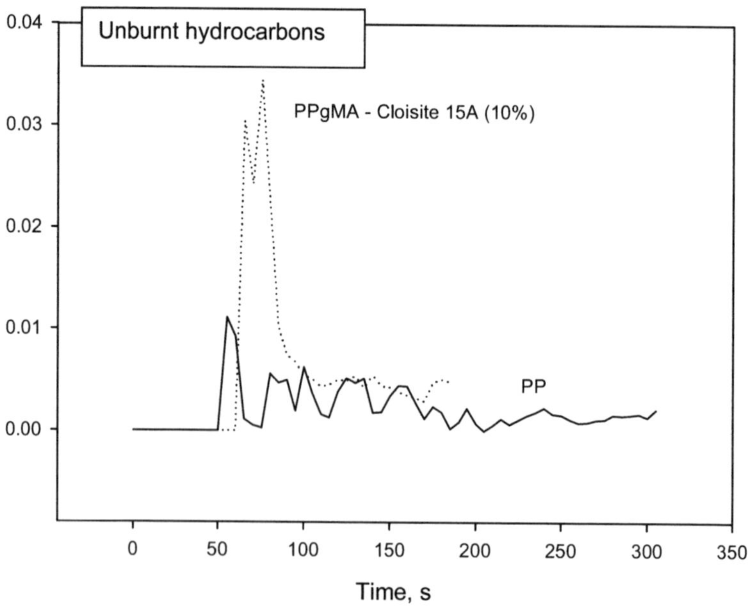

Figure 10 *Unburnt hydrocarbons for PP and PPgMA-Cloisite 15A vs. time (external heat flux: 35 kW m^{-2})*

supply as well as the escape of the combustion products generated during decomposition). Conversely, the reduction in flammability for PP-Mg(OH_2) (60%) over the PP is equal to 60%, which is more significant than for the nanocomposite (Table 2).

However, from a practical viewpoint, because only a few percent of inorganic fillers are needed in the PP-layered silicates nanocomposites, the resulting hybrids are lightweight and tend to preserve the mechanical properties of polymeric materials.

7.4 References

1. A.B. Morgan and J.D. Harris, *Polymer*, 2003, **44**, 2313–2320.
2. E.P. Giannelis, *Adv. Mater.*, 1996, **8(1)**, 29.
3. J.W. Gilman, T. Kashiwagi, M. Nyden, E.T.J. Brown, C.L. Jackson, S.M. Lomakin, E. P. Giannelis and E. Manias, "Flammability studies of polymer layered silicate nanocomposites: polyolefin, epoxy, and vinyl ester resins", in *Chemistry and Technology of Polymer Additives*, S. Al-Malaika, A. Golovoy and C.A. Wilkie (eds.), Blackwell Pub., Oxford, UK, 1999, pp. 249–265.
4. J.W. Gilman, C.L. Jackson, A.B. Morgan, R. Harris, E. Manias, E.P. Giannelis, M. Wuthenow, D. Hilton and S.H. Phillips, *Chem. Mater.*, 2000, **12**, 1866–1873.
5. A.Tkáč, I. Špilda, *J. Polym. Sci. Polym. Chem. Ed.*, 1981, **19**, 1495–1508.
6. N.K. Jha, A.C. Misra and P. Bajaj, *J. Macromol. Sci.-Rev., Macromol. Chem. Phys.*, 1984, **C24.1**, 69–116.

CHAPTER 8

Effect of the Processing Conditions on the Fire Retardant and Thermomechanical Properties of PP–Clay Nanocomposites

ABDEL BENDAOUDI,[1] SOPHIE DUQUESNE,[1] CHARAFEDDINE JAMA,[1] MICHEL LE BRAS,[1] RENÉ DELOBEL,[1] PHILIPPE RECOURT,[2] JEAN-MICHEL GLOAGUEN,[3] JEAN-MARC LEFEBVRE[3] AND AHMED ADDAD[3]

[1]Laboratoire des Procédés d'Elaboration de Revêtements Fonctionnels, Ecole Nationale Supérieure de Chimie de Lille, BP 108, F-59652 Villeneuve d'Ascq, France (sophie.duquesne@ensc-lille.fr)
[2]Processus et Bilans des Domaines Sédimentaires, Université des Sciences et Technologies de Lille, F-59655 Villeneuve d'Ascq Cedex
[3]Laboratoire de Structure et Propriétés de l'Etat Solide, UMR 8008, Université des Sciences et Technologies de Lille, F-59655 Villeneuve d'Ascq Cedex

8.1 Introduction

Nanocomposite polymers, and in particular polymer/clay systems, have been of interest for around 15 years but were first referenced as early as 1950.[1] This interest is linked with the research carried out by Toyota,[2] which demonstrated the superior performance that can be achieved using a nano-dispersion of the filler (nanocomposite) when compared to a micro-dispersion (conventional composites). The barrier properties to gas diffusion (in particular to oxygen and carbon dioxide) of nanocomposite led to the development of some applications in the food packaging industries. The improvement in mechanical properties for low loading systems may find advantages in the automotive industry. Another property that can be improved using these materials is their flame retardant properties.[3–5] These enhanced properties are closely linked to the degree of dispersion of the nano-filler.

Several methods exist to obtain a good dispersion of the clay layers in the polymer matrix.[6] *In -situ* polymerization leads generally to a well exfoliated structure. However, this method is difficult to adapt in an industrial scale to all the families of polymers. It is also possible to process nanocomposite using a solution method. This method consists first in a exfoliation of the clay in a solvent and then the polymer is solubilised. After extraction of the solvent, a nanocomposite is obtained. This method presents the inconvenient to use solvent, which is usually an environmental and safety problem. Finally, a melt processing method can also be employed to prepare nanocomposite. The dispersion of the clay is obtained in the melted polymer using the shear developed in the process (such as, for example, mixing or extrusion). The direct melt blending process is most attractive because of its low cost, high productivity and compatibility with current polymer processing techniques. In this method, the processing parameters (temperature, rotor or screw speed (rotor speed), mixing duration, presence of oxidative atmosphere *etc.*) as well as the chemical nature of the clay and the intercalating agent are key parameters.[7]

The effect of the type of the clay,[8–10] of the clay loading[11,12] and of the intercalating agent[13–17] (also called surfactant, or compatibilizer) of the clay have been widely investigated. It is generally accepted that modification of the clay by organophilic compounds is needed to provide a good dispersion or an exfoliation of the clay platelets into the polymer matrix. In polypropylene (PP) matrix, exfoliation is more difficult due to the non-polar chemical structure of the polymer. In general, PP/clay nanocomposites form an exfoliated structure only when a compatibilizer such as maleic anhydride functionalised polypropylene (PP-g-MA) is added. Recently, a novel approach to make exfoliated PP/clay nanocomposites without adding PP-g-MA has been developed, applying a large electric field to the exfoliated structures.[18–20]

The aim of this study is to investigate the effect of processing parameters on the fire retardant performance and on the thermal and thermomechanical properties of PP/clay systems. To optimize those properties, different materials have been prepared using different processing conditions defined by an experimental design. The nanocomposites were characterized using X-ray diffraction analyses and transmission electronic microscopy (TEM). The properties of the materials have been finally correlated with the processing conditions.

8.2 Experimental

8.2.1 Materials

Raw materials were PP [polypropylene supplied by Atofina – PPH7060 MFI = 12 g/10 min] PP-g-MA (maleic anhydride grafted polypropylene supplied by Crompton – Polybond 320–2% MA, MFI = 110 g/10 min) and

Table 1 *Processing conditions of experimental design n°1 (mixing duration = 15 min)*

Experiment code	*Temperature* (°C)	*Occupied volume* (vol%)	*Rotor speed*(rpm)
ED1-1	230	90	80
ED1-2	190	90	80
ED1-3	230	70	20
ED1-4	190	70	20

Table 2 *Processing conditions of experimental design n°2 (occupied volume in the mixing chamber = 90%, rotor speed = 80 rpm)*

Experiment code	*Temperature* (°C)	*Mixing duration* (min)
ED2-1	170	30
ED2-2	190	15
ED2-3	170	15
ED2-4	190	30

organically modified montmorillonite (Cloisite 20A, Southern Clay Product, organic modifier = dimethyl dihydrogenatedtallow quaternary ammonium salt). The study was carried out using the ratio PP/PP-g-MA/20A = 90:5:5 (wt/wt). Mixtures were prepared using a Brabender mixer measuring head (type 350/EH, roller blades), monitoring the mixing conditions using a data processing torque rheometer system Brabender Plasticorder PL2000. Processing conditions (temperature, rotor speed, mixing duration, and percentage of occupied volume in the mixing chamber) are presented in Tables 1 and 2.

Sheets ($100 \times 100 \times 3$ mm^3) or bars ($40 \times 4 \times 1$ mm^3) were then obtained using a Darragon press at $T = 190$°C and a pressure of 3 MPa.

8.2.2 Cone Calorimetry

The Stanton Redcroft Cone Calorimeter was used to carry out measurements on samples following the procedure defined in ASTM 1354-90. The method is based on oxygen consumption calorimetry.[21] The standard procedure used involves exposing specimens measuring $100 \times 100 \times 3$ mm^3 in horizontal orientation. An external heat flux of 50 kW m^{-2} was used for running the experiments. This flux was chosen because it is a common heat flux in mild fire scenarios.[22] When measured, HRR (heat release rate) values are reproducible to within ±10%. The cone data reported here are the average of three replicated experiments.

8.2.3 Thermogravimetry

TG analyses were performed using a Setaram MTB 10–8 thermobalance at 10°C min^{-1} from 20 to 800°C under air flow (Air Liquide grade, 5×10^{-7} m^3 s^{-1}

measured in standard conditions). Samples (powder, about 10 mg) were placed in vitreous silica pans. The precision of temperature measurements was ±1.5°C.

8.2.4 Dynamic Mechanical Analysis

Dynamic mechanical properties were studied on compression-molded bars ($40 \times 4 \times 1$ mm^3) with a Metravib dynamic mechanical analyzer in the tensile mode. Samples were tested from –40 to 140°C at a heating rate of 3°C min^{-1} and a frequency of 1 Hz.

8.2.5 Characterization of Nanocomposites

The degree of the nano-dispersion, intercalation and/or exfoliation, of the blended PP/PP-g-MA/20A was investigated by X-ray diffraction measurements and TEM.

X-Ray scattering measurements were performed with a Philipps PW 1729 diffractometer with CuKα radiation ($\lambda = 1.5418$ Å), a step size of 0.02° 2θ and count time of 1 s. The *d*-spacing experimental standard uncertainty was ±0.2 nm. Cloisite 20A was analyzed as received. The polymer nanocomposites were melt-pressed before analysis into 1.6 mm sheets.

Transmission electron microscopy (TEM) samples of PP-clay nanocomposites were prepared with cryoultramicrotome (LEICA UltracutE FC4) at –100°C and cut into 50 nm thick sections with a 35° diamont knife. The sections were transferred onto Cu grids of 200 mesh. TEM images were obtained at 200 kV with a Jeol 200 CX electron microscope at magnifications of 20,000 and 150,000.

8.2.6 Experimental Design

Two experimental designs based on a Hadamard matrix (two or three factors, at two levels) were employed to obtain the correlation between the processing parameters and the fire retardant performance, using the Nemrodw software.

8.3 Results and Discussion

8.3.1 Fire Retardant Performance of PP Nanocomposites

The fire retardant performance of materials set up according to experimental design n°1 are reported in Figure 1. The cone calorimeter, simulating the conditions of a fire, is the most adapted tool to evaluate the fire retardant properties of nanocomposite. Indeed, other tests such as LOI or UL94 are less sensitive to the improvement of the fire retardancy of polymer when clay is added. For example, LOI values of PP nanocomposites have been reported between 18.6 and 19.2 vol% in comparison with 17.8 vol% for virgin PP.[23] Considering this low variation, it is very difficult to obtain a good discrimination between the materials.

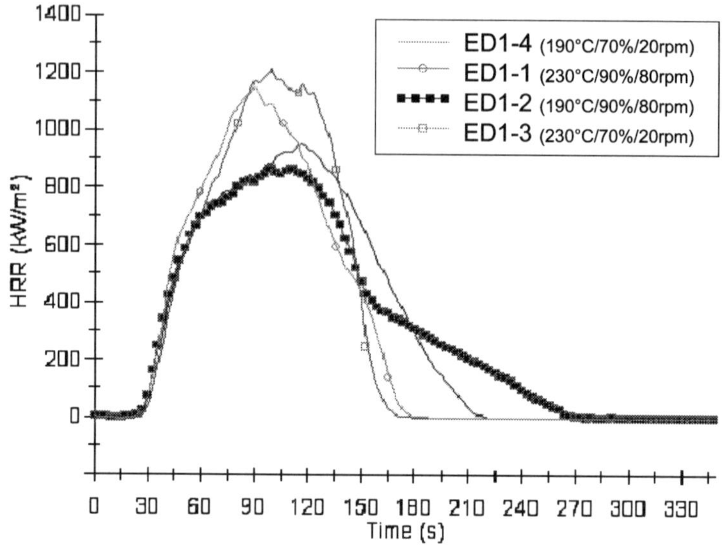

Figure 1 *HRR curves of PP/PP-g-MA/20A – Experimental design n°1*

It is generally known that the degree of delamination and dispersion of layered silicate by melt compounding using an extruder is affected by the type of extruder and its screw design and the processing conditions. A mechanism of delamination and dispersion has been proposed in the literature: shear intensity is required to start the dispersion process, by shearing particles apart into tactoids. Then, a long residence time is required to allow polymer to enter into the clay galleries and peel the platelets apart.[24]

Figure 2 shows the processing parameters obtained from the experimental design n°1. In our study, the shear has been modified varying the temperature, the occupied volume in the mixing chamber and the rotor speed. Clearly, the most important parameter is the temperature of the process. The lower temperature (190°C) leads to higher performance, *i.e.* the peak of heat release rate (PHRR) reaches its minimum around 870 kW/m^2. Even if compounding in the conditions of internal mixer and extruder are not comparable, the results are in good agreement with the proposed mechanism of delamination since a decrease

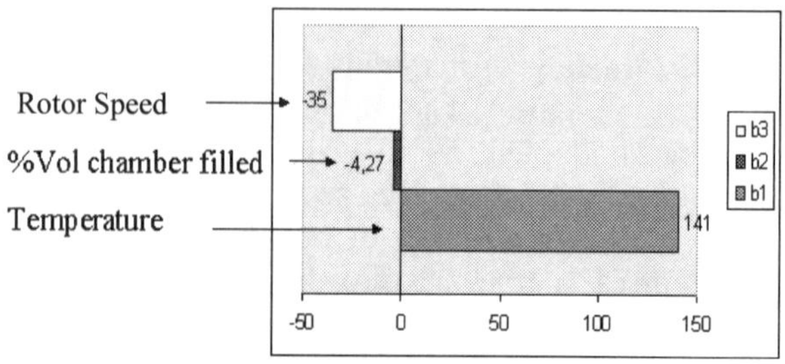

Figure 2 *Processing parameters (experimental design n°1)*

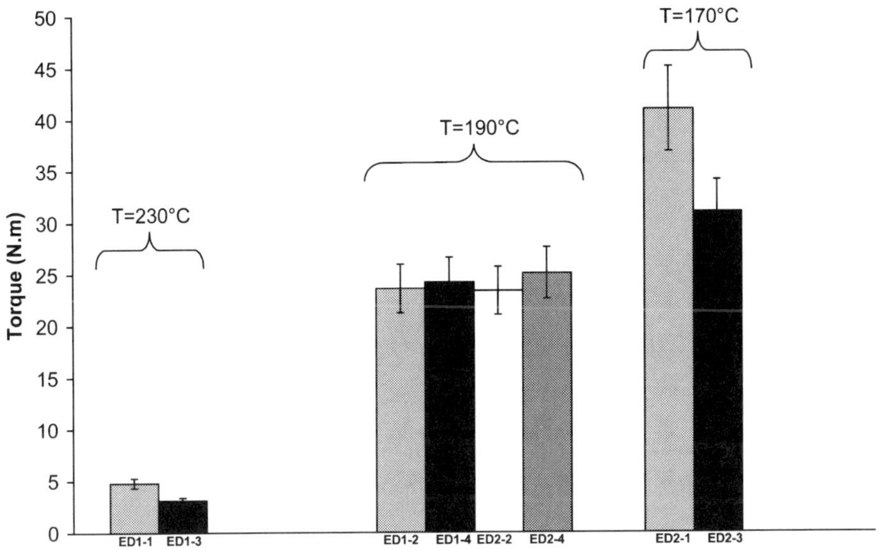

Figure 3 *Variation of the torque versus time during processing of ED 1 and -2*

in the processing temperature leads to an increase in the viscosity of PP and thus to an increase in the shear intensity of the polymer, as demonstrated by the variation of the Brabender torque during the experiment (Figure 3).

The influence of rotor speed is also in agreement with the results reported in the literature. The higher the rotor speed, the lower the PHRR. Finally, the percentage of the occupied volume in the mixing chamber has no influence on the fire retardant properties of PP nanocomposites.

To optimize the processing conditions, a second experimental design has been set. The rotor speed and the occupied volume in the mixing chamber were fixed according to experimental design n°1, respectively to 80 rpm and 90%. The studied parameters are the temperature and residence time. Increasing the mean residence time in the extruder generally improves the delamination and dispersion, and thus leads to an improvement in performance of the materials.[24] However, this subject is controversial.[25]

Figures 4 and 5 report, respectively, the fire retardant performance and the effect of the processing conditions on those performances. They confirm that the lower the temperature of the process, the better the fire performance. The residence time has little influence on the fire retardant properties of PP nanocomposites. This may reasonably explained by the fact that the residence time in mixer is very long in comparison with that of an extruder.

8.3.2 Thermal Stability of PP/PP-g-MA/20A Nanocomposites

Figure 6 presents the thermogravimetric curves of PP nanocomposites prepared according to experimental design n°2. TGA data corresponding to the temperatures at which 10% and 50% mass loss occur are reported in Table 3. Whatever the materials, an important increase in thermal stability of the

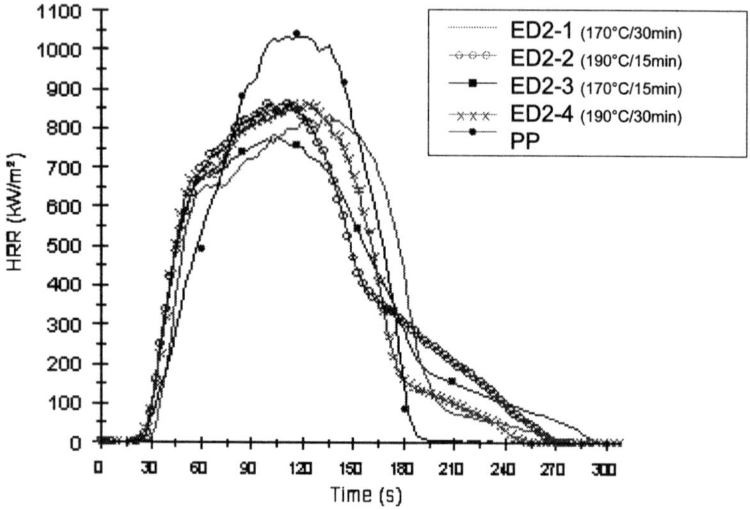

Figure 4 *HRR curves of PP/PP-g-MA/20A – Experimental design n°2*

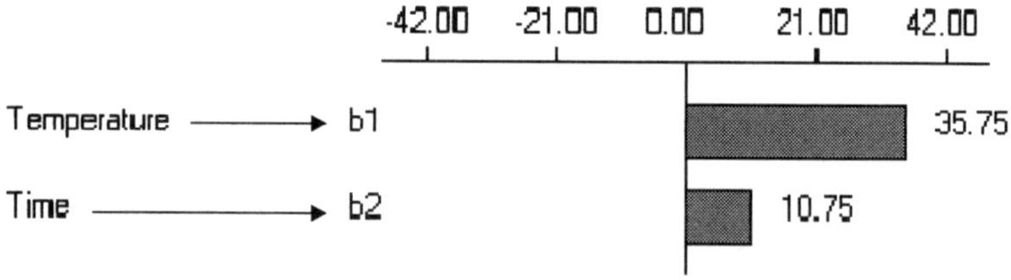

Figure 5 *Processing parameters (experimental design n°2)*

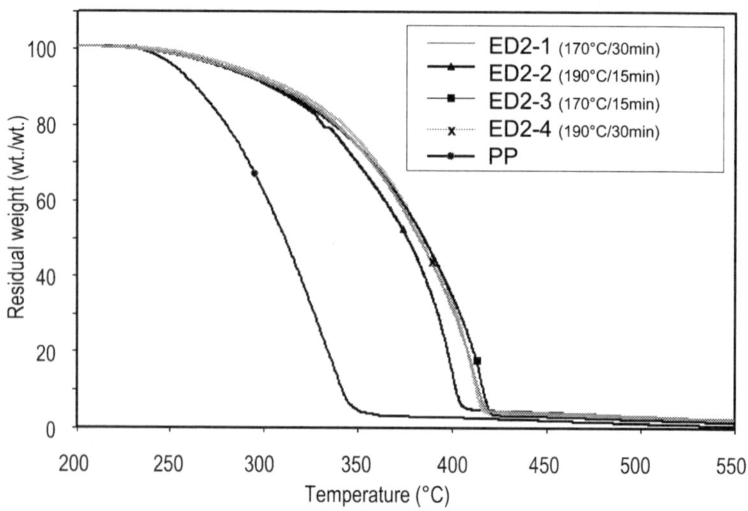

Figure 6 *TG curves of PP/PP-g-MA/20A – experimental design n°2*

Table 3 *TGA data under air of PP/PP-g-MA/20A nanocomposites*

Experiment code	$T_{10\%}$ (°C)	$T_{50\%}$ (°C)
PP	263	310
ED2-1	307	383
ED2-2	303	376
ED2-3	311	385
ED2-4	304	385

polymer is observed when clay is added. An increase of about 40°C and of about 70°C is, respectively, observed for the temperature at which 10% and 50% mass loss occurs.

PP degrades in one step. Thermal decomposition of PP in the presence of oxygen occurs *via* initiation, propagation and termination sequences.[26] For PP-clay nanocomposites, degradation also occurs in a single step, and a slight decrease in degradation rate is observed at the early stages.

In the literature, an improvement in the thermal stability of PP upon adding clay has been reported, and attributed to a difference in chain structure that restricted thermal motion or to the formation of a diffusion barrier that delays the decomposition process.[13,27] This increase in thermal stability of PP in the presence of clay may indicate that a good dispersion is obtained since no significant change in degradation temperature is generally observed for immiscible microcomposites.

8.3.3 Dynamic Thermo-Mechanical Properties of PP Nanocomposites

Dynamic mechanical analysis results are shown in Figure 7 and Table 4. At –40°C the storage modulus, which correlates directly with the stiffness or flexural modulus, increases from 3.3 GPa for PP to 4.2±0.1 GPa for PP/PP-g-MA/20A nanocomposites. Moreover, as the temperature increases up to 20°C, the clay has a low influence in preserving the stiffness of PP. The processing parameters appears to a have little effect on the response to DMA of the nanocomposites, indicating that the increased mixing time, in the range of the tested temperatures, has no influence on the dispersion. The increase in storage modulus is higher below the glass–rubber relaxation of the amorphous portion of PP (T_g) in the PP-clay nanocomposite. It may be assumed that the links between the clay and the polymer are relatively poor since strong links (such as, for example, covalent chemicals bonding) should lead to a sharp increase in the storage modulus above T_g.

8.3.4 Characterization of PP Nanocomposites

WAXS patterns of PP/PP-g-MA/20A nanocomposites in the range of $2\theta = 3 - 25°$ for the two experimental designs are shown in Figures 8 and 9. The mean interlayer spacing of the (001) plane d_{001} for the Cloisite 20A obtained by WAXS

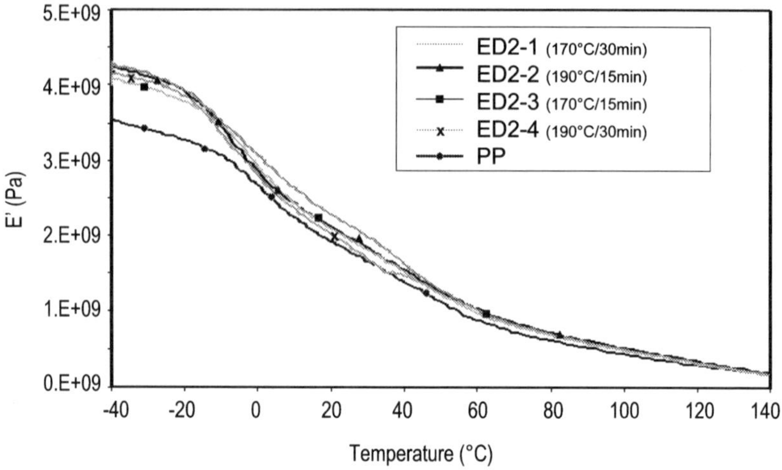

Figure 7 *Thermomechanical properties of PP/PP-g-MA/20A – experimental design n°2*

Table 4 *Thermomechanical data under air of PP/PP-g-MA/20A nanocomposites*

Experiment code	$E_{-40°C}$ (GPa)	$E_{20°C}$ (GPa)	$E_{140°C}$ (GPa)
PP	3.3	1.9	0.18
ED2-1	4.3	2.3	0.18
ED2-2	4.2	2.1	0.19
ED2-3	4.1	2.1	0.17
ED2-4	4.2	2.0	0.18

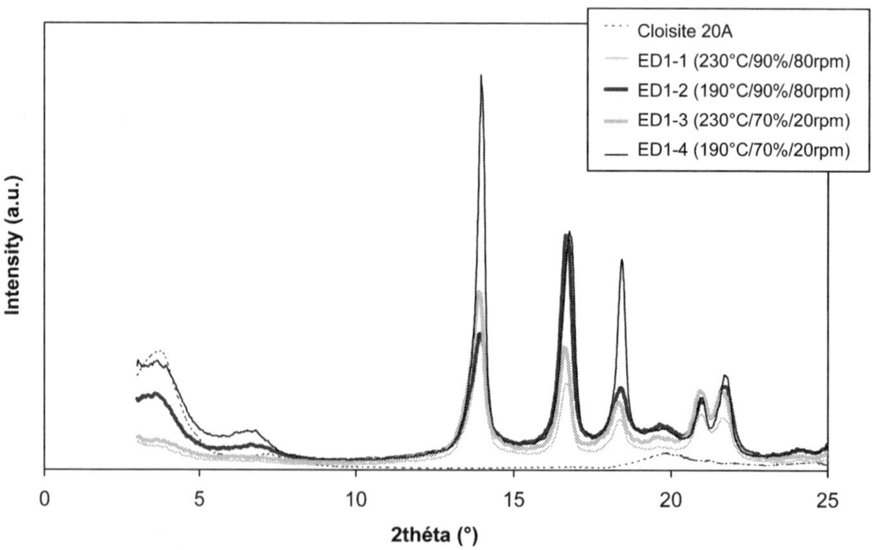

Figure 8 *WAXS patterns of PP/PP-g-MA/20A nanocomposite (experimental design n°1)*

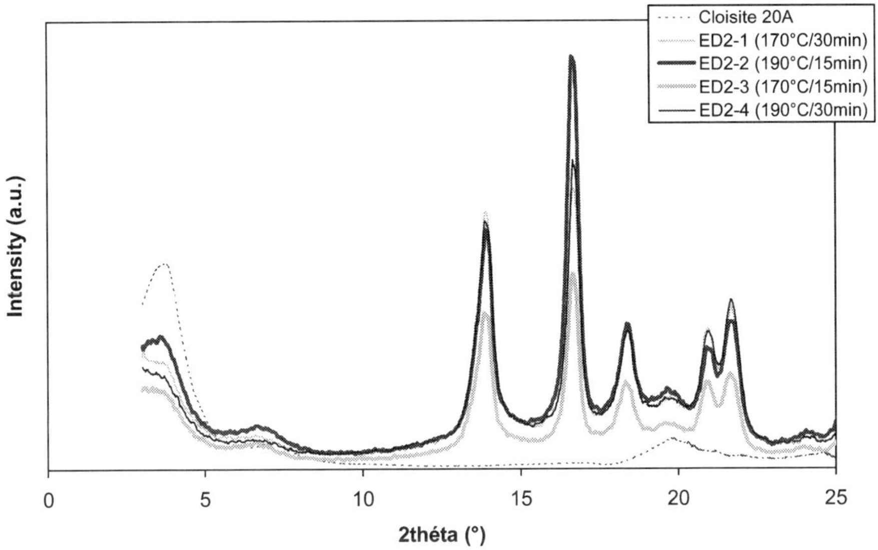

Figure 9 *WAXS patterns of PP/PP-g-MA/20A nanocomposite (experimental design n°2)*

measurements is 2.4 nm ($2\theta = 3.65°$). A small remnant shoulder is observed for ED 1–1 and ED 1–3 nanocomposites samples, and a smaller peak, in comparison to Cloisite 20A, is still detected for ED 1–2, ED 1–4 and all the samples from the second experimental design. These observations correspond to the (001) plane d_{001} of silicate layers due to the intercalation of polymer chains in the silicates galleries. Clearly, with increasing processing temperature the d_{001} peak is smaller, suggesting that the extent of intercalation increases at higher temperature.

To check the dispersion of the clay particles in the PP/PP-g-MA/20A nanocomposites, TEM has been carried out. A uniform dispersion of the clay particles is seen [Figure 10(a)] at a meso-structural scale. At higher magnification, an intercalated structure is obvious. In particular, Figure 10(b) demonstrates that the distance between the platelets increases and an ordered structure is maintained.

8.4 Conclusion

This study has investigated the effect of the processing parameters (temperature, rotor speed, residence time and occupied volume of the mixing chamber) on the fire retardant, thermal and thermomechanical stability of PP-clay nanocomposites. Whatever, the processing conditions, an intercalated nanocomposite is obtained.

The temperature is a key factor of the process. A decrease in the processing temperature allows improvement of the fire properties of the PP-clay materials whereas the thermal and thermomechanical properties are slightly affected.

a)

b)

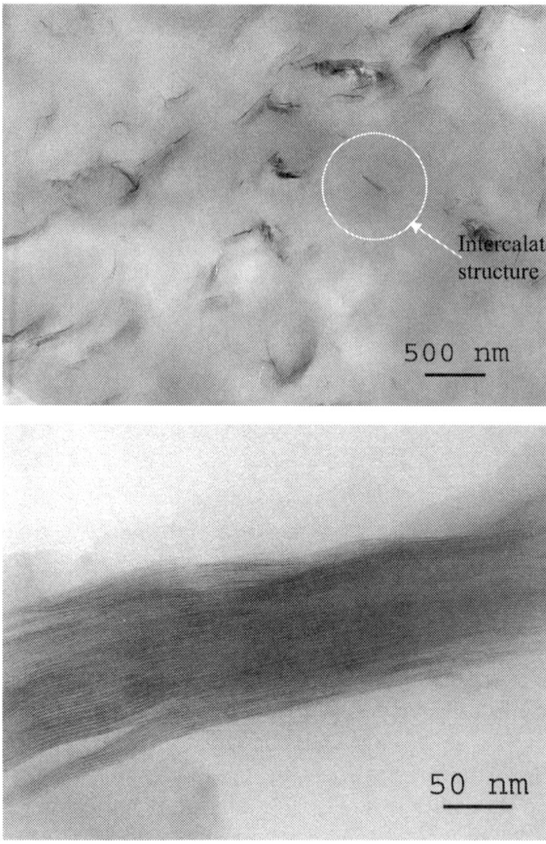

Figure 10 *TEM image of PP/PP-g-MA/20A nanocomposite*

8.5 References

1. L.W. Carter, J.G. Hendricks and D.S. Bolley, *US Patent 2531396*, 1950.
2. Y. Fukushima, A. Okada, M. Kawasumi, T. Kurauchi and O. Kamigaito, *Clay Mineral*, 1988, **23**, 27–34.
3. J.W. Gilman, A. Morgan, E.P. Giannelis, M. Wuthenov and E. Manias, *11th Conference on Recent Advances in Flame Retardancy of Polymeric Materials*, Stamford, USA, 2000.
4. E.P. Giannelis, *Adv. Mater.*, 1996, **8(1)**, 29–35.
5. Gy. Marosi, P. Anna, A. Marton, Sz. Matko, A. Szep, S. Keszei, B. Csontos and B. Marosfoi, *12th International Conference on Additives*, San Francisco, April 2003.
6. M. Alexandre and P. Dubois, *Mater. Sci. Eng.*, 2000, **R28(1–2)**, 1–63.
7. H.R. Dennis, D.L. Hunter, D. Chang, S. Kim, J.L. White, J.W. Cho and D.R. Paul, *ANTEC 2000*, Orlando, Florida, 8–9 May 2000.
8. C.A. Wilkie, in *Proceed. 9th European Meeting on Fire Retardancy and Protection of Materials*, M. Le Bras *et al.*, (ed.) USTL Pub., 17–19th September 2003, Lille, France, p. 49.

9. Y. Tang, Y. Hu, L. Song, R. Zong, Z. Gui, Z. Chen and W. Fan, *Polym. Degrad. Stab.*, 2003, **82(1)**, 127–131.
10. D. Chaiko, *PCT Int. Appl.*, 2002, 24 pp *WO 2002044101 A2 WO 2001-US51210 20011113*. Priority: *US 2000-717590 20001121*.
11. A. Pozsgay, L. Papp, T. Frater and B. Pukanszky, *Progr. in Colloid Polym. Sci.*, 2001, **117**, 120–125.
12. H. Wang, M. Elkovitch, L.J. Lee and K.W. Koelling, *Annu. Tech. Conf. – Soc. Plastics Eng.*, 2000, **58th(Vol. 2)**, 2402–2406.
13. M. Zanetti, G. Camino, P. Reichert and R. Mulhaupt, *Macromol. Rapid Commun.*, 2001, **22(3)**, 176–180.
14. J.W. Lee, Y.T. Lim and O.O. Park, *Polym. Bull.*, 2000, **45(2)**, 191–198.
15. D. Merinska, Z. Malac, J. Hrncirik, J. Simonik, J. Trlica, M. Pospisil, P. Capkova, Z. Weiss, *Annu. Tech. Conf. – Soc. Plastics Eng.* 2001, **59th(Vol. 2)**, 2166–2170.
16. L. Wu and Y. Hua, *Abstracts of Papers, 226th ACS National Meeting*, New York, September 7–11, 2003 (2003), PMSE-366, Conference; Meeting Abstract, Pub. American Chemical Society, Washington, D.C.
17. G.D. Barber, C.M. Carter and R.B. Moore, *Annu. Tech. Conf. – Soc. Plastics Eng.*, 2000, **58th(Vol. 3)**, 3763–3767.
18. D.H. Kim, J.U. Park, K.H. Ahn and S.J. Lee, *Annu. Tech. Conf. – Soc. Plastics Eng.*, 2003, **61st(Vol. 2)**, 2215–2218.
19. K.H. Ahn, D.H. Kim, J.U. Park, J. Hong and S.J. Lee, *Annu. Tech. Conf. – Soc. Plastics Eng.*, 2002, **60th(Vol. 2)**, 1457–1460.
20. K.H. Ahn and S.J. Lee, *PCT Int. Appl.*, 2003, *W 2003016208 A1 20030227 WO 2002-KR1511 20020808*.
21. C. Huggett, *Fire Mater.*, 1980, **4(2)**, 61–65.
22. V. Babrauskas, *Fire Mater.*, 1984, **8(2)**, 81–95.
23. U. Wagenknecht, B. Kretzschmar and G. Reinhardt, *Macromol. Symp.*, 2003, **194**, 207–212.
24. H.R. Dennis, D.L. Hunter, D. Chang, J.L. Kim, J.W. Cho and D.R. Paul, *Polymer,* 2002, **42**, 9513–9522.
25. C.H. Davis, L.J. Mathias, J.W. Gilman, D.A. Schiraldi, J.R. Shields, P. Trulove, T.E. Sutto and H.C. Delong, *J.. Polym. Sci.: Part B: Polym. Phys.*, 2002, **40**, 2661–2666.
26. T.J. Henman, *Develop. Polym. Stabil.* 1979;**1**, 39–99.
27. A. Tidjani, O. Wald, M.M. Pohl, M.P. Hentschel and B. Schartel, *Polym. Degrad. Stab.*, 2003, **82**, 133–140.

CHAPTER 9

Fire Retardancy of Polystyrene –Hectorite Nanocomposites

DONGYAN WANG, BOK NAM JANG, SHENGPEI SU, JINGUO ZHANG, XIAOXIA ZHENG, GRACE CHIGWADA, DAVID D. JIANG AND CHARLES A. WILKIE

Department of Chemistry, Marquette University, PO Box 1881, Milwaukee, WI 53201, U.S.A. (charles.wilkie@marquette.edu)

9.1 Introduction

Polymer nanocomposites have become an area of extensive research in recent years. The properties of polymer nanocomposite are expected to be improved significantly in the presence of layered silicate materials.[1–6] Amongst these layered silicate materials, montmorillonite is the most popular one studied, but some attention has also been paid to magadiite,[7–12] bentonite[13–16] and hectorite.[17–22]

While montmorillonite is an aluminosilicate, magadiite and hectorite contain only silicates. The chemical formula for hectorite is $Na_{0.3}(Mg,Li)_3Si_4O_{10}(OH)_2$; a specimen of hectorite, fresh from the mine, has a soft, greasy texture; it is one of the more expensive clays, due to its unique thixotropic properties. The major uses of hectorite are in cosmetics and in chemical and industrial material production.

Polyolefin microcomposites and layered silicate nanocomposites have been prepared by Dubois *et al.*[17] *via in situ* polymerization, using both montmorillonite and hectorite that had been treated with trimethylaluminum-depleted methylaluminoxane. The encapsulated filler particle within the (co)polyolefinic matrix formed polymers ranging from thermoplastics to elastomers. The obtained "homogeneous" (nano)composites exhibit improved mechanical properties, as compared to more conventional melt blends for the same filler content. Sandi[18] has synthesized a series of polymer–clay nanocomposites based on synthetic lithium hectorite and different mass ratios of poly(ethylene oxide) and tested these as candidates for polymeric electrolytes in lithium ion cells.

Transparent films with excellent mechanical strength were obtained with a conductivity that is comparable to more traditional polymer electrolytes made with added lithium salts.

Coumarin dye molecules were first intercalated into the gallery of hectorite; extensive shaking and sonication of this water suspension leads to exfoliation, which is confirmed by both atomic force microscopy (AFM) and transmission electron microscopy (TEM).[19] The resulting nanocomposite films were transparent and displayed fluorescence centered at around 470 nm.

Polystyrene–clay and poly(methyl methacrylate)–clay nanocomposites[20] have been prepared using cetyltrimethylammonium-modified hectorite by solution blending in toluene. Wide-angle X-ray diffraction (WAXD) as well as 2D ^{1}H–^{29}Si and ^{1}H–^{1}H correlated solid-state NMR confirmed the dispersion of the intercalated clay stacks in the polymer matrix.

Multinuclear solid-state NMR (two-dimensional ^{1}H–^{29}Si heteronuclear correlation (HETCOR) NMR with spin diffusion and refocused ^{29}Si detection for enhanced sensitivity) revealed that during the intercalation of poly(styrene-ethylene oxide) block copolymers (PS-b-PEO) into hectorite, the PS block is not intercalated but the PEO segment is intercalated.[21] In PS-rich samples, a small amount of PEO is intercalated and a significant fraction of PEO is not intercalated. Intercalated PEO exhibits reduced mobility, most prominently for the PEO nearest to the silicate surface. *In situ* small-angle X-ray scattering studies were conducted to monitor the structural changes of polymer nanocomposites upon heating.[22]

These silicates usually have excess negative charge, which is balanced by the exchangeable cations in the gallery space. Like montmorillonite clay, the cation exchangeability offers the possibility for the modification of pristine hectorite by organic cations, which can increase the organophilic character of the gallery space so that it is compatible with an organic polymer. Because of the outstanding performance of montmorillonite clay in the enhancement of barrier properties and in fire retardancy, there is interest in examining hectorite, and other clays, to determine how different clays behave with respect to nanocomposite formation and in fire performance.

In this chapter, pristine hectorite was modified with two different quaternary ammonium salts, one of which is known to give intercalated and the other to give exfoliated nanocomposites with montmorillonite, and polystyrene nanocomposites were prepared by bulk polymerization.

9.2 Experimental

9.2.1 Materials

Dimethylhexadecylamine (≥98%) was acquired from Fluka. Most of the other chemicals used in this study, including vinylbenzyl chloride (97%), monomeric styrene, benzoyl peroxide (BPO) 97% and tetrahydrofuran (THF) (99+%), were purchased from the Aldrich Chemical Company. The polymerization inhibitor was removed from the monomer by passing it through an inhibitor-remover column, also acquired from Aldrich. The quaternary ammonium salt known as 10A was kindly provided by Akzo-Nobel, while the salt known as VB was synthesized in this laboratory following a previously published procedure.[23] Distilled water was used throughout as needed. Hectorite slurries were kindly provided by Elementis Specialties, Inc.; the iron contents of the clays and the lot numbers by which they are reported herein are: 66A, 0.053%; 66B, 0.53%; and 66G, 2.57%.

9.2.2 Organic Modification of Hectorite

The method used for the organic-modification of hectorite was quite similar to that used to modify montmorillonite clay, as reported previously.[24] The cationic exchange reaction occurs between pristine hectorite and a quaternary ammonium salt, in this case styryldimethylhexadecylammonium chloride (VB16, denoted VB) and dimethylbenzylhydrogenated tallow chloride (10A) were utilized. Hydrogenated tallow contains ~65% C18, ~30% C16 and ~5% C14. A 10% mole excess of the quaternary ammonium salt (based on the CEC of the hectorite) was added to the hectorite slurry for the cationic exchange reaction. After overnight stirring, the reaction was stopped, then the organically-modified hectorite was dried in a vacuum oven at room temperature.

9.2.3 Preparation of Nanocomposites

A bulk polymerization technique was utilized in the preparation of the polystyrene (PS) hectorite nanocomposite. This procedure, which has been used for montmorillonite, has been previously described.[24]

9.2.4 Instrumentation

X-Ray diffraction (XRD) patterns were obtained using a Rigaku Geiger Flex, 2-circle powder diffractometer equipped with Cu Kα generator ($\lambda = 1.5404$ Å). Generator tension was 50 kV and generator current was 20 mA.

Thermogravimetric analysis (TGA) was performed on a TA Instruments, model SDT 2960 Simultaneous DTA-TGA unit under a 40 mL min^{-1} flowing nitrogen atmosphere at a scan rate of 10°C min^{-1} from room temperature to 700°C; temperatures are reproducible to ±3°C, and the fraction of nonvolatile materials is reproducible to ±3%.

Cone calorimetry was performed on an Atlas CONE2 according to ASTM E 1354-92 at an incident flux of 35 kW m^{-2} using a cone shaped heater. Exhaust flow was set at 24 L s^{-1} and the spark was continuous until the sample ignited. Cone samples were prepared by compression molding the sample (about 30 g) into square plaques. Typical results from Cone calorimetry are reproducible to within about ±10%. These uncertainties are based on many runs in which thousands of samples have been combusted.[25]

9.3 Results and Discussions

9.3.1 X-ray Diffraction

XRD can provide information about the *d*-spacing of hectorite according to the Bragg equation. The *d*-spacing of pristine hectorite is 1.1 nm ($2\theta = 7.9°$); after organic-modification with the 10A salt, the *d*-spacing increased to 2.0 nm ($2\theta = 4.4°$), indicating that ion-exchange occurred. After bulk polymerization of styrene with the clay, the XRD trace shows a sharp, strong peak at about 3.5 nm ($2\theta = 2.5°$), clearly indicating the formation of an intercalated nanocomposite; XRD traces for the PS-hectorite 10A nanocomposites are shown in Figure 1.

For the VB system, the *d*-spacing after ion exchange is also at 2.0 nm ($2\theta = 4.4°$). Peaks in the XRD traces (Figure 2) are much weaker for the VB system than for the 10A system. This may be attributable to either a greater extent of exfoliation or disorder of the clay and an immiscible system. This last possibility is rejected because VB invariably gives better exfoliation than does a non-functionalized organic-modification such as 10A.[23]

Identical results were obtained for all of the hectorites examined in this study. There is little doubt, based on the XRD results, that these are intercalated and exfoliated nanocomposites.

9.3.2 Transmission Electron Microscopy

The XRD results very strongly suggest that good nanodispersion has been achieved for all of these nanocomposites, but the only proof of this assertion lies

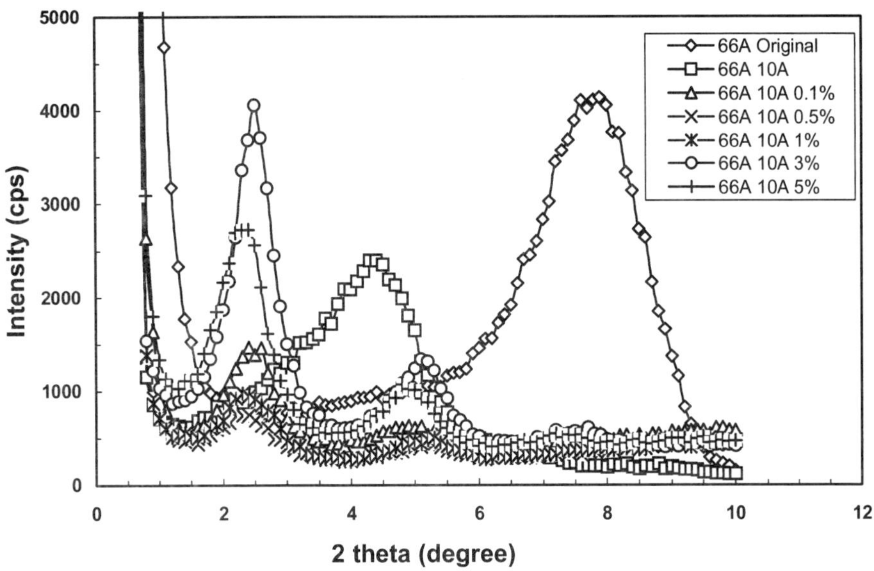

Figure 1 *XRD traces for PS-hectorite 10A systems*

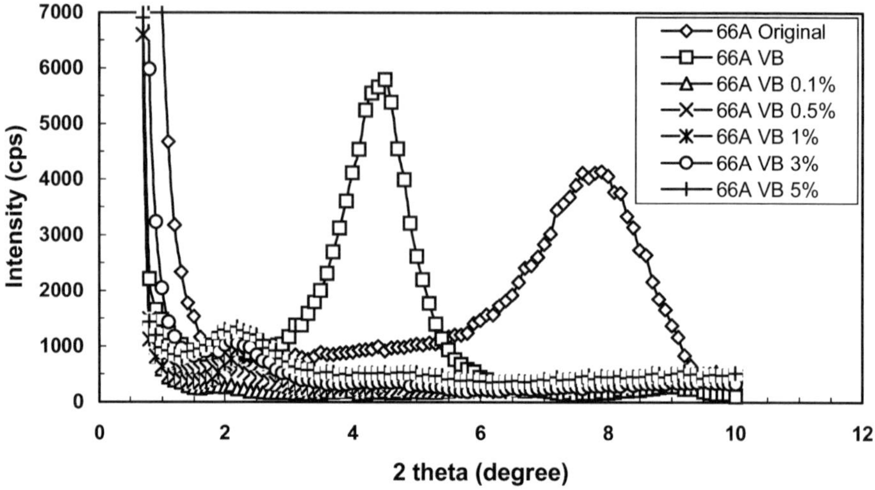

Figure 2 *XRD traces for PS-hectorite VB systems*

in TEM data. TEM images at both low and high magnification for the VB-PS nanocomposites are shown in Figure 3 while those for the 10A-PS system are shown in Figure 4. For both systems one can see that good nanodispersion has been obtained and the high magnification images enable one to see the individual clay layers and to specify that the VB system is more exfoliated than is the 10A system.

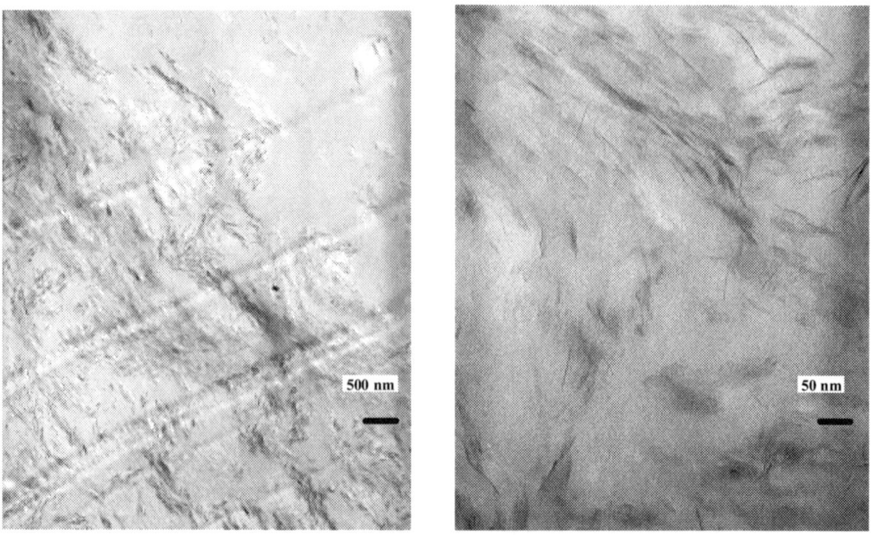

Figure 3 *TEM images at low magnification (left) and high magnification (right) for the VB-PS nanocomposite*

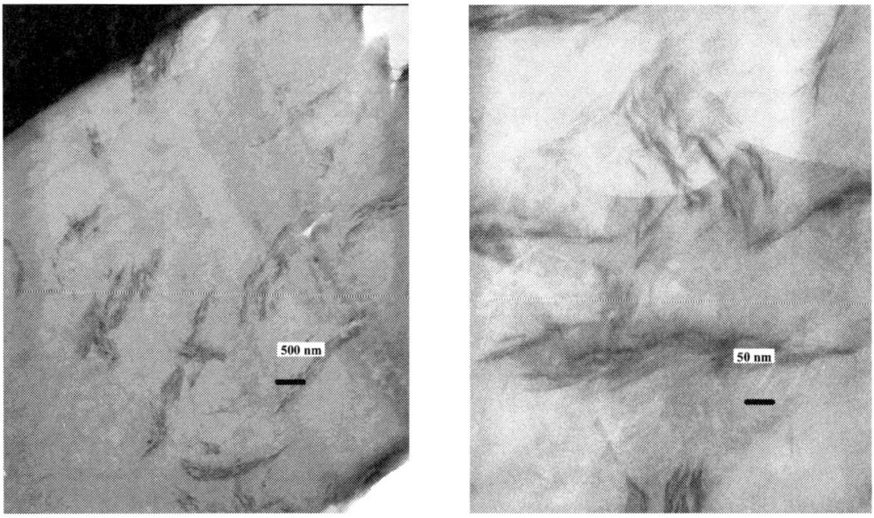

Figure 4 *TEM images at low (left) and high magnification (right) for the 10A-PS nanocomposite*

9.3.3 Thermogravimetric Analysis

Parameters extracted from the TGA include the temperature at which 10% of the mass has been lost, $T_{0.1}$, a measure of the onset of degradation, the temperature at which 50% of the mass is lost, $T_{0.5}$, the mid-point of the degradation, and the fraction of material that is not volatile at 600°C (denoted as char). With montmorillonite clays we found previously that the onset temperature, as well as the mid-point temperature, of iron-containing clay increases by about 50°C compared to virgin polymer.[26] Results for the hectorite clay systems (Table 1) show comparable results for both temperatures; a representative set of TGA curves for one of these systems is shown in Figure 5.

There appears to be some difference between lower amounts of clay and the results at 3 or 5%, especially for the mid-point of the degradation, with larger increases in temperature at these clay levels. With the montmorillonite, if one compared iron-containing with iron-free clays, there was a significant temperature difference between the two clays at low amounts of clay, but this difference became smaller as the amount of clay increased and became negligible at 3 or 5% clay. With hectorite, there is no difference that can be attributed to the presence or absence of iron. The tentative conclusion is that hectorite and montmorillonite exhibit similar effects according to TGA analysis, but the amount of iron is important for montmorillonite but not for hectorite.

9.3.4 Cone Calorimetry

Cone calorimetry enables the evaluation of the fire parameters for a system; the data that may be obtained includes the heat release rate curve, the total heat

Table 1 *TGA parameters for hectorite-PS nanocomposites*

Sample	$T_{0.1}$ (°C)	$T_{0.5}$ (°C)	*Char* (wt%)
PS	351	404	0
66A-10A-PS, 0.1%	382	425	0
66A-10A-PS, 0.5%	399	436	0
66A-10A-PS, 1%	392	435	0
66A-10A-PS, 3%	390	444	2
66A-10A-PS, 5%	389	446	3
66A-VB-PS, 0.1%	387	428	0
66A-VB-PS, 0.5%	391	427	0
66A-VB-PS, 1%	399	436	0
66A-VB-PS, 3%	389	428	1
66A-VB-PS, 5%	414	457	5
66B-10A-PS, 0.1%	365	417	0
66B-10A-PS, 0.5%	380	426	0
66B-10A-PS, 1%	404	437	0
66B-10A-PS, 3%	389	443	3
66B-10A-PS, 5%	399	448	3
66B-VB-PS, 0.1%	337	402	0
66B-VB-PS, 0.5%	356	410	0
66B-VB-PS, 1%	364	422	2
66B-VB-PS, 3%	404	445	2
66B-VB-PS, 5%	408	452	5
66G-10A-PS, 0.1%	382	424	0
66G-10A-PS, 0.5%	403	436	0
66G-10A-PS, 1%	400	440	1
66G-10A-PS, 3%	400	444	3
66G-VB-PS, 0.1%	383	421	0
66G-VB-PS, 0.5%	394	421	0
66G-VB-PS, 1%	393	425	0
66G-VB-PS, 3%	400	445	2

released, mass loss rate, time to ignition and smoke evolution, known as the specific extinction area. For montmorillonite-polystyrene nanocomposites, the time to ignition is decreased, the total heat released is unchanged but the peak heat release rate, PHRR, is significantly decreased, typically by 50–60%, the mass loss rate is also reduced and there is little change in smoke evolution. In a study of iron-containing versus iron-free polystyrene-montmorillonite nanocomposites, we found a significant difference in the PHRR for iron-containing clays at low amounts of clay, but this difference vanishes as the amount of clay increases.[26] The commonly accepted mechanism by which nanocomposite formation reduces the PHRR is through barrier formation, which can act both as an insulator and a barrier to mass transport.[27] Based upon these observations on the effect of iron, it was proposed that some radical trapping may occur and that this is an effective mechanism at low clay content but, at high clay content, the barrier effect becomes dominant.

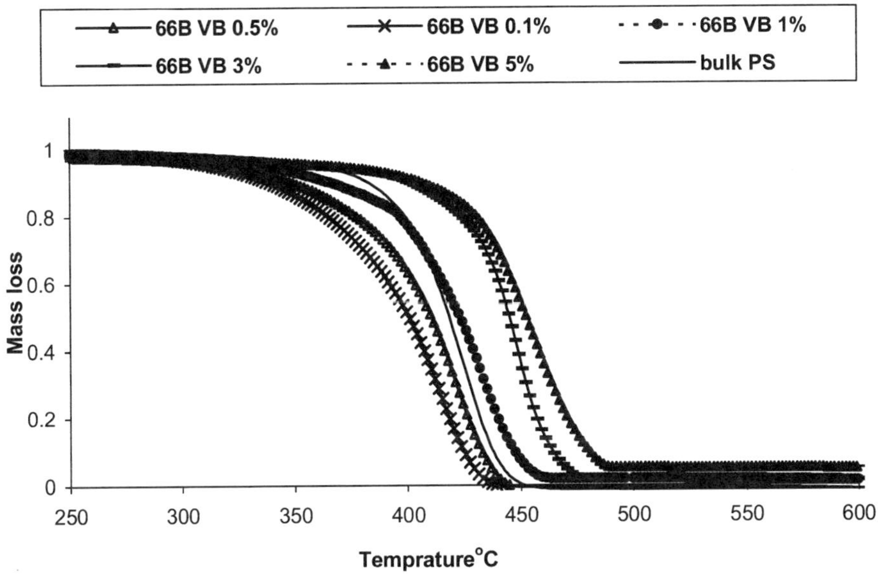

Figure 5 *TGA curves for 66B-VB-PS nanocomposites*

For hectorite-polystyrene nanocomposites the results are quite different; the cone calorimetric results are shown in Table 2, while Figure 6 shows a representative plot of the heat release rate for one of the polystyrene-hectorite nanocomposites. There is essentially no reduction in PHRR when the clay content is 1% or less and in some cases there is no effect at 3% clay, an amount at which montmorillonite is very effective. In most cases, at 5% clay one sees a reasonable reduction in PHRR. The other parameters recorded in Table 2 are typical values and confirm the PHRR observation. For instance, when there is no reduction in PHRR, there is also no change in mass loss rate.

One can examine the data as a function of morphology, the 10A series *versus* the VB series, or as a function of iron content. From the XRD traces, those nanocomposites made with the 10A salt show a peak and are presumed to be intercalated while those made with VB16 show no peak and thus are assumed to be exfoliated; the TEM data confirms these suggestions. For the very low iron content clay, 66A, the one discontinuity appears at 3% clay where the intercalated material, 10A, gives a reduction while the exfoliated VB system gives no reduction.

For the intermediate iron content clay, 66B, a similar trend is seen in which the intercalated system gives a slight reduction at low amounts of clay and there is a major difference at 3% clay. This trend is not continued at higher amounts of iron, 66G; here there is no reduction for the intercalated system but a better reduction for the exfoliated system. These discrepancies cannot be attributed to changes in the dispersion of the clay within the polymer matrix and must be due to something else.

Table 2 *Cone calorimetric data for the hectorite-polystyrene nanocomposites*

Sample	T_i, (s)	*PHRR* (kW m^{-2})	*THR* (mJ m^{-2})	*WLR*, (g s^{-1} m^{-2})	*SEA$_{av}$*, (g s^{-1} m^{-2})
Polystyrene	59	1200	89	31.7	952
66A-10A-PS, 0.1%	58	1149(4)	94	27.8	1138
66A-10A-PS, 0.5%	60	1223(0)	98	29.8	1117
66A-10A-PS, 1%	62	1440(0)	120	29.8	1342
66A-10A-PS, 3%	51	634 (31)	91	20.8	1285
66A-10A-PS, 5%	59	771 (36)	94	20.1	1322
66A-VB-PS, 0.1%	54	1505 (0)	112	29.2	1325
66A-VB-PS, 0.5%	56	1350 (0)	107	27.9	1321
66A-VB-PS, 1%	59	1240 (0)	104	28.0	1226
66A-VB-PS, 3%	54	1147 (4)	97	25.9	1340
66A-VB-PS, 5%	53	888 (26)	100	21.1	1394
66B-10A-PS, 0.1%	56	1027 (14)	66	31.9	776
66B-10A-PS, 0.5%	50	1093 (9)	76	32.8	950
66B-10A-PS, 1%	40	972 (19)	75	30.3	936
66B-10A-PS, 3%	32	679 (43)	64	22.9	1025
66B-10A-PS, 5%	44	598 (50)	62	22.2	1104
66B-VB-PS, 0.1%	58	1361 (0)	92	31.0	1092
66B-VB-PS, 0.5%	50	1329 (0)	85	31.0	1151
66B-VB-PS, 1%	56	1337 (0)	99	34.0	1134
66B-VB-PS, 3%	47	1237 (0)	94	29.0	1214
66B-VB-PS, 5%	41	547 (54)	59	21.8	1157
66G-10A-PS, 0.1%	58	1587(0)	126	29.7	1319
66G-10A-PS, 0.5%	55	1481 (0)	132	26.7	1419
66G-10A-PS, 1%	48	1338 (0)	123	24.0	1496
66G-10A-PS, 3%	51	1242 (0)	120	24.1	1458
66G-VB-PS, 0.1%	54	1015 (15)	57	32.9	776
66G-VB-PS, 0.5%	43	926 (23)	62	31.2	825
66G-VB-PS, 1%	50	947 (21)	65	31.9	802
66G-VB-PS, 3%	41	894 (26)	66	30.7	849

Glossary: t_i: time to ignition; THR: total heat released; WLR: mass loss rate; SEA$_{av}$: Average specific extinction area.

When the data are examined from the point of view of the iron content, it is possible to suggest that, as the iron content increases, the PHRR also increases. This is not in accord with work with montmorillonite, in which there is an iron effect at low amounts of clay but when the amount of clay is 3% or larger, the PHRR is unaffected by the iron content.

One can summarize the cone calorimetry results for various clay-polystyrene nanocomposites as follows: montmorillonite gives a 50–60% reduction in PHRR at 3% clay;[23] while magadiite[12] and fluorohectorite[27] give no reduction and hectorite gives a reduction of up to 50%, but only at 5% clay; the reduction with hectorite at 3% clay is lower than is seen for montmorillonite at this clay level. These variations require an explanation. TEM information is available for all systems and there is excellent dispersion for montmorillonite,

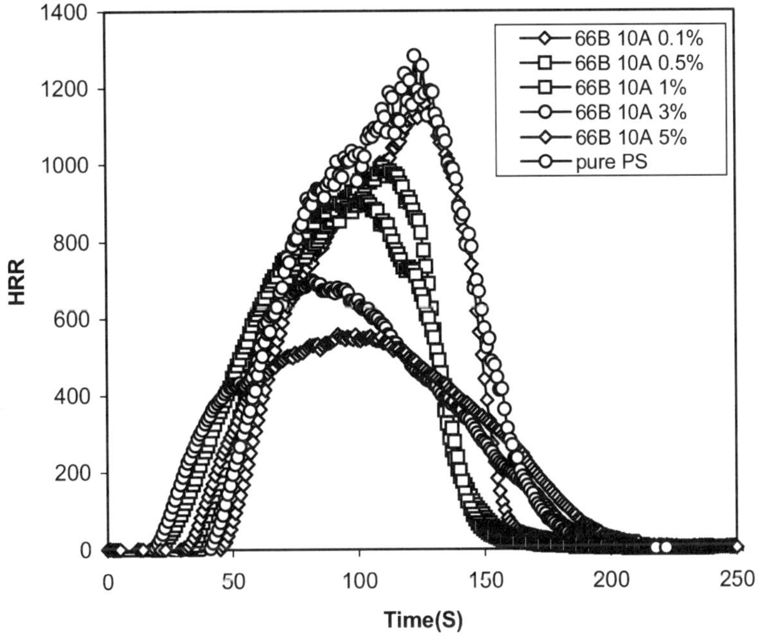

Figure 6 *Heat release rates for the 66B-10A-hectorite-polystyrene nanocomposites*

fluorohectorite and hectorite, but there is some question on the dispersion for magadiite. From previous work in this and other laboratories, there is a correlation between nano-dispersion and reduction in PHRR; good nanodispersion leads to significant reduction in PHRR while no reduction is seen if the clay is not well dispersed.[25,28] For magadiite, the dispersion is not as good as one might like but the enhanced mechanical properties are suggestive of good dispersion. There is a potential sampling problem with TEM in that the amount of material examined is quite small and may not be representative of the whole sample.

We hope to use this information to begin to identify what is important in a clay for fire retardancy. To that end, we must begin by identifying the differences between these various clays; differences that are under consideration include: composition, morphology, charge location, and size. Montmorillonite is an aluminosilicate material while fluorohectorite, hectorite and magadiite are all-silicate materials; since the all silicates give different results, one cannot attribute the differences in PHRR changes to composition.

Previous work showed that there is no difference in PHRR of styrene-montmorillonite nanocomposites for intercalated and exfoliated systems; thus we tentatively decide that changes in morphology, as long as there is good nano-dispersion, do not influence the reduction in PHRR.

Clays consist of octahedral and tetrahedral layers and the substitution of one ion for another may occur in either layer. Differences in charge location might be important, but this information for the clays that have been used is not available and thus cannot be evaluated.

The last difference that has been considered is size. Hectorite is lath-like, while fluorohectorite is much more floppy and tends to fold onto itself to reduce the aspect ratio, and magadiite is very monolithic. The plate diameter and aspect ratios of the clays under consideration are: magadiite, plate diameter ~40 μm, (this is an average reported value obtained from scanning electron microscopy);[30] fluorohectorite, plate diameter, ~4–5 μm,[27] 5 μm,[30] aspect ratio, 500:1 to 4000:1;[27] montmorillonite, plate diameter, ~0.1–1 μm,[27] 0.3–0.6 μm,[30] 0.25 μm,[31] aspect ratio, 100:1 to 1000:1;[27] hectorite, 0.05 μm,[31] ~0.02–0.03 μm.[32] Clearly, there is a great variation in the sizes of the clay particles and it is possible that the differences may be attributed to changes in size. Figure 7 shows plot of the size parameter, peak heat release rate and mass loss rate.

The most accepted process by which the heat release rate is affected by nanocomposite formation is barrier formation.[27] This may occur by loss of the polymer due to thermal degradation so that the clay platelets fall over and come into contact. If the platelet is too large, it may not fall flat but may stick up, leaving a gap in the barrier. However, if the particles are too small, it may require more material to form this impermanent barrier, so the poorer barrier will lead to a smaller reduction in PHRR; this suggests that magadiite and fluorohectorite are too large and do not form a suitable barrier while hectorite, the smallest material, requires additional material to permit the complete reduction in PHRR. One may ask if the larger clays would form a good barrier at higher amounts, which could lead to a substantial reduction in PHRR; this is under investigation.

9.4 Conclusions

Hectorite has been organically-modified with two different ammonium salts and these salts show the same behavior seen with montmorillonite; with an

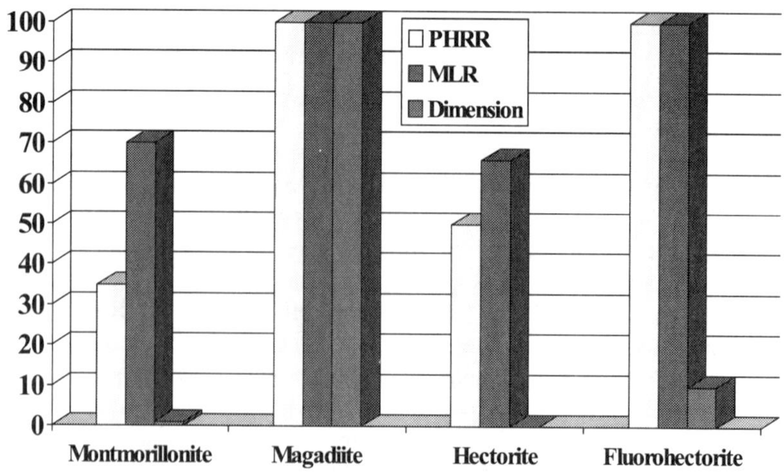

Figure 7 *Comparison of peak heat release rate (PHRR), mass loss rate (MLR), and the dimension of the clay for four different polystyrene-clay nanocomposites*

ammonium salt that contains a styryl unit, exfoliation is observed. When this polymerizable unit is absent, the result is intercalation. Hectorite appears to offer similar thermal properties to montmorillonite but at higher amounts of clay. The significant differences between various clays have been tentatively attributed to differences in size. This is only a tentative conclusion and further work is underway to further elucidate the reason for the differences. Based on this and other work from this laboratory and others, it seems that nanocomposite formation alone will never lead to fire retardancy. Instead, it is felt that nanocomposite formation may be one part of a combination of materials that can be used to achieve fire retardation. The role of the clay is likely to change the heat release rate curve but, more importantly, it will help achieve excellent mechanical properties that may be compromised by the addition of other components of the fire retardancy package. The choice of the clay will probably be made on the basis of enhanced mechanical properties rather than because of some inherent fire retardant properties. Further work is underway to evaluate additional clays and combinations of clays with conventional fire retardants.

9.5 Acknowledgement

This work was performed under the sponsorship of the US Department of Commence, National Institute of Standards and Technology, Grant Number 70NANB6D0119.

9.6 References

1. M. Alexandre and P. Dubois, *Mater. Sci. Eng.*, 2000, **R28**, 1.
2. E.P. Giannelis, R. Krishnamoorti and E. Manias, *Adv. Polym. Sci.*, 1999, **138**, 107.
3. E.P. Giannelis, *Adv. Mater.*, 1996, **8**, 29.
4. R.A. Vaia, K.D. Jandt, E.J. Kramer and E.P. Giannelis, *Chem. Mater.*, 1996, **8**, 2628.
5. D.A. Brune and J. Bicerano, *Polymer*, 2002, **42**, 369.
6. R.K. Bharadwaj, *Macromolecules*, 2001, **34**, 9189.
7. G. Lagaly, K. Beneke and A.Weiss, *Am. Mineral.*, 1975, **60**, 642.
8. K. Beneke and G. Lagaly, *Am. Mineral.*, 1977, **62**, 763.
9. K. Beneke and G. Lagaly, *Am. Mineral.*, 1983, **68**, 818.
10. Y. Sugahara, K. Sugimoto, T. Yanagisawa, Y. Nomizu, K. Kuroda and D. Kato, *Yogyo Kyokai Shi*, 1987, **95**, 117.
11. Z. Wang, T. Lan and T.J. Pinnavaia, *Chem. Mater.*, 1996, **8**, 2200.
12. D. Wang, D.D. Jiang, J. Pabst, Z. Han, J. Wang, and C.A. Wilkie, *Polym. Eng. Sci.*, 2004, **44**, 1122.
13. D. Garcia-Lopez, O. Picazo, J.C. Merino and J.M. Pastor, *Eur. Polym. J.*, 2003, **39**, 945.
14. C. Decker, K. Zahouily, L. Keller, S. Benfarhi, T. Bendaikha and J. Baron, *J. Mater. Sci.*, 2002, **37**, 4831.

15. X. Tong, H. Zhao, T. Tang, Z. Feng and B. Huang, *J Polym. Sci.: Part A: Polym. Chem.*, 2002, **40**, 1706.
16. Z. Shen, G.P. Simon and Y-B. Cheng, *Polymer*, 2002, **43**, 4251.
17. P. Dubois, M. Alexandre, and R. Jerome, *Macromolecular Symposia* (Eurofillers'01 Conference, 2001), 2003, **194**, 13.
18. G. Sandi, K.A. Carrado, H. Joachin, W. Lu and J. Prakash, *J. Power Sources*, 2003, **119–121**, 492.
19. D.W. Kim, A. Blumstein, J. Kumar and S.K. Tripathy, *Polym. Mater. Sci. Eng.*, 2001, **84**,182.
20. S-S. Hou and K.Schmidt-Rohr, *Chem. Mater.*, 2003, **15**, 1938.
21. S-S. Hou, T.J. Bonagamba, F.L. Beyer, P.H. Madison and K. Schmidt-Rohr, *Macromolecules*, 2003, **36**, 2769.
22. G. Sandi, H. Joachin, R. Kizilel, S. Seifert and K.A. Carrado, *Chem. Mater.*, 2003, **15**, 838.
23. J. Zhu, A.B. Morgan, F.J. Lamelas and C.A. Wilkie, *Chem. Mater.*, 2001, **13**, 3774.
24. D. Wang, J. Zhu, Q. Yao and C.A. Wilkie, *Chem. Mater.*, 2002, **14**, 3837.
25. J.W. Gilman, T. Kashiwagi, M. Nyden, J.E.T. Brown, C.L. Jackson and S. Lomakin, in *Chemistry and Technology of Polymer Additives*, S. Al-Maliaka, A. Golovoy and C.A. Wilkie (eds.), Blackwell Scientific, London, 1998, pp. 249–65.
26. J. Zhu, F.M. Uhl, A.B. Morgan and C.A. Wilkie, *Chem. Mater.*, 2001, **13**, 4649.
27. J.W. Gilman, C.L. Jackson, A.B. Morgan, R. Harris, Jr., E. Manias, E.P. Giannelis, M. Wuthenow, D. Hilton and S.H. Phillips, *Chem. Mater.*, 2000, **12**, 1866.
28. D. Wang, D.D. Jiang, J. Pabst and C.A. Wilkie, 2004, **10**, 44.
29. S. Su, D.D. Jiang and C.A. Wilkie, *J. Vinyl Add. Technol.*, in press.
30. J.S. Dailey and T.J. Pinnavaia, *Chem. Mater.*, 1992, **4**, 855.
    H.O. Pastore, M. Munsignatti and A.J.S. Mascarenhas, *Clay Clay Miner.*, 2000, **48**, 224. K. Isoda, K. Kuroda and M. Ogawa, *Chem. Mater.*, 2000, **12**, 1702. K. Kikuta, K. Ohta and K. Takagi, *Chem. Mater.*, 2002, **14**, 3123.
31. J. Ren, B.F. Casanueva, C.A. Mitchell and R. Krishnamoorti, *Macromolecules*, 2003, **36**, 4188.
32. S.-S. Hou and K. Schmidt-Rohr, *Chem. Mater.*, 2003, **15**, 1938.
33. T. Kasawa, T. Murakami, N. Kohyama and T. Watanabe, *Am. Mineral.*, 2001, **86**, 105.

CHAPTER 10

Pyrolysis and Flammability of Polyurethane–Organophilic Clay Nanocomposite

GENNADY E. ZAIKOV,[1] SERGEI M. LOMAKIN[1] AND ROMAN A. SHEPTALIN[2]

[1]Institute of Biochemical Physics of Russian Academy of Sciences, 4 Kosygin Street, Moscow, 119991, Russia (chembio@sky.chph.ras.ru)
[2]D.I. Mendeleyev Russian Chemical-Technological University, Moscow, Russia

10.1 Introduction

Polymer layered silicate (clay) nanocomposites are materials with unique properties when compared with conventional filled polymers. Polymer nanocomposites, especially polymer-layered silicates, represent a new alternative to conventionally filled polymers.

New, more effective, and environmentally friendly flame resistance polymers are needed. Recent data on the combustion of polymer nanocomposites indicate that they could be employed for this purpose.[1]

There are several proposed mechanisms as to how the layered silicate affects the flame retardant properties of polymers.[1] The first is an increased char layer that forms when nanocomposites are exposed to flame. This layer is thought to inhibit oxygen transport to the flame front, as well as gaseous-fuel transport from the polymer and, therefore, reduces the heat release rate of the burning surface. This may interrupt the burning cycle as radical species are needed to break polymer chains into fuel-fragments. The disordered nanocomposites also inhibit oxygen and combustible "fuel" species transfer by increasing the path length of these species to the flame front. There is also a high possibility of alumina-silicate solid-phase catalysis of polymer decomposition, which can dramatically change the overall scheme of thermal degradation process kinetics.

Polyurethane foam (PU) is a unique and most useful commercial polymeric material. In the present study, a polyurethane–organophilic montmorillonite nanocomposite (PU-OM) was synthesized by a known two-stage procedure.[2] Simulated heat release analysis was applied to provide information on the flame-resistance properties of PU-OM.

10.2 Experimental

10.2.1 Materials

The original sodium montmorillonite fraction having a cation-exchange capacity of 80 mEq $(100\ g)^{-1}$ was obtained from natural bentonite clay (Ural Region, Russia).

10.2.2 Preparation of Organophilic Montmorillonite (OM)

Montmorillonite was gradually added to a solution of alkyltrimethylammonium chloride (C16–C18) – Quartamin 60L, Kao Corporation S.A. and resultant suspension was stirred vigorously for 3 h at 50°C. After filtration the product was placed in a vacuum oven at 100°C for 24 hours.

10.2.3 Synthesis of Propylene Oxide-OM (PO-OM)

OM and propylene oxide glycol (Laprol 5003-2B-10), provided by Nizjnemkamskneftkhim co., whose average molecular weight is about 5000 (Scheme 1) were mixed together, (1 : 1) with rapid stirring at 40°C for 1 hour to give the colloidal PO-OM intercalated hybrid.

10.2.4 Synthesis of Polyurethane–Organophilic Montmorillonite Nanocomposite (PU-OM)

PU-OM was synthesized by polycondensation *in situ* of component A: hybrid polyol (PO-OM) and plain polyol (1 : 10), catalyst (triethylenediamine) – Tegostab 100, by Goldschmidt, water and stabilizer and component B: toluene diisocyanate – Voranate T-80, by Dow Chemical co. and polyisocyanate, PITZ-B by NPO Korund co. (1 : 2). Components A and B were mixed with stirring. This mixture was then poured out in the special form followed by its solidification and foaming during 15 minutes.

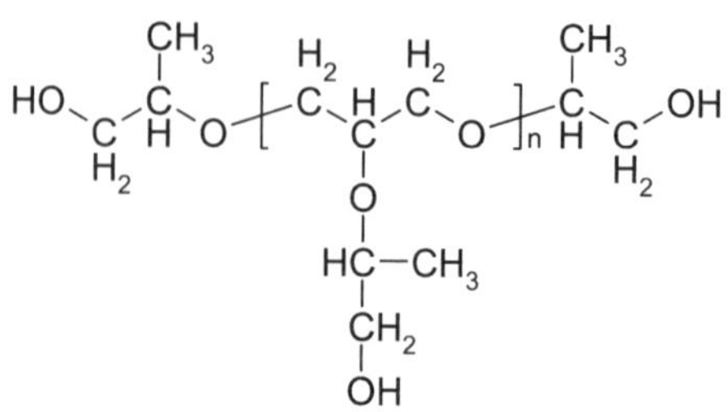

Scheme 1 *Propylene oxide glycol (Laprol 5003-2B-10)*

10.2.5 XRD Characterization

X-Ray diffraction (XRD) analysis of OM, PO-OM and PU-OM was performed on a Philips diffractometer using CuK_α radiation, ($\lambda = 0.1540562$ nm). The microstructure of PU-OM nanocomposite was characterized using XRD. The XRD patterns of OM and PU-OM, which reveal the intercalated structure of OM and delaminated structure of PU-OM nanocomposite, are shown in Figure 1. The interlayer (basal) spacing (001) for the OM is 3.5 nm.

10.2.6 Pyrolysis

Extensive pyrolysis of PU and PU-OM/nanocomposite samples over a one minute period was carried out in a laboratory model pyrolizer in an air atmosphere at 250 and 500°C. The oven temperature was monitored with a thermocouple and a stability of ±5°C was attainable. Pyrolysis products were dissolved in hexane at 0°C.

10.2.7 Gas Chromatography/Mass Spectrometry (GC-MS) Analysis

GC-MS analysis of samples was performed using a "Varian 3300" gas chromatograph connected to a mass spectrometer detector (ion trap), "Finnigan MAT ITD 800". A DB-5 fused capillary column (0.32 mm × 30 m) temperature programmed from 50 to 270°C at 10°C min^{-1} was used. Mass spectra detection

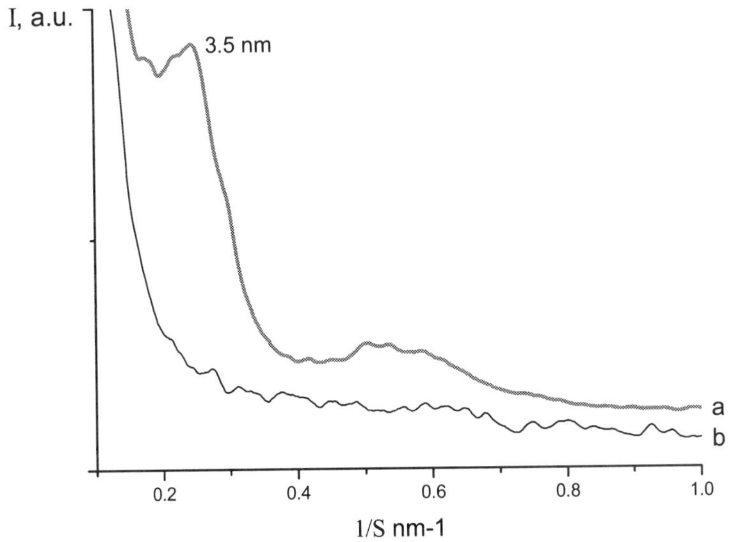

Figure 1 *XRD patterns of OM and PU-OM delaminated nanocomposite:* a – *OM,* b – *PU-OM*

were obtained in the electron impact mode scanned from 40 to 650 Da with a energy of 70 eV. All mass spectra were assigned using the Wiley275 mass spectral library.

10.2.8 Combustion Tests

Simulated heat release analyses were performed on spherical samples of foamed PU and PU-OM with OM clay fraction of 10 wt% using a laboratory thermal analysis balance (Mettler T300). Samples were ignited in air at ambient conditions. The external heat flux was assumed to be zero.

The heat of combustion of polymers was calculated using the principle of molar additivity of the heats of formation of the combustion products and reactants [PhysProps 1.6 by G & P Engineering Software Co.]. The concept derives from the fact that the enthalpy (*H*) is a state function and, therefore, its change in any process is independent of the path from reactants to products. Thus, the overall enthalpy of a reaction is simply the sum of the enthalpies of the component reactions. In practice, the heat of combustion of the reaction can be calculated by subtracting the heat of formation of the products from the heat of formation of the reactants [Equation (1)].

$$\Delta h_c = \sum_i n_p \Delta h^o_{f,p} \pm \sum_j n_r \Delta h^o_{f,r} \tag{1}$$

For polymeric reactants the molar heat of formation can be estimated from the tabulated molar contributions of the chemical groups that constitute the monomer or repeat unit.

10.3 Results and Discussion

Thermochemical calculations of the gross heat of combustion from molar group additivity of the heats of formation of products and reactants achieves better accuracy than calculations based on oxygen consumption for the polymers examined in this study. This is not surprising since the group contributions to the heats of formation used in this study were originally determined from the gross heats of combustion of materials with known composition. The net heats of combustion of PU polymers of known chemical composition were measured and calculated.

The thermochemical calculation is based on complete combustion of PU:

$$C_{50}H_{56}O_{11}N_7 + 71O_2 \longrightarrow 50CO_2 + 28H_2O + 7NO_2$$

The heat of combustion of PU, as well as this of PU-OM, was calculated as 5.36 MJ kg^{-1} using the PhysProps 1.6 [Equation (2)].

$$\Delta H_c = \Sigma \Delta H_{producls} - \Sigma \Delta H_{reactants} \tag{2}$$

Bench-scale fire calorimeters have since been developed that use the oxygen consumption principle to determine the chemical heat release rate of burning materials. This principle is based on the observation that combustion of a wide range of organic compounds and common polymers produces 13.1 ± 0.7 kJ of heat per gram of diatomic oxygen consumed, independent of the chemical composition of the organic material. The gases evolved during polymer decomposition are usually unknown and do not burn to completion in real fires. Oxygen consumption is a means of measuring heat release without detailed knowledge of the fuel species. Oxygen consumption calorimetry measures the heat released by the burning of volatile polymer decomposition products, the net heat of complete combustion of which can be written as in Equation (3),

$$\Delta h^o_{c,v} = \frac{\Delta h^o_{c,p} \pm \mu \Delta h^o_{c,\mu}}{1 \pm \mu} \tag{3}$$

where $\Delta h^{\circ}_{c,v}$, $\Delta h^{\circ}_{c,p}$, $\Delta h^{\circ}_{c,\mu}$ are the heats of complete combustion for the volatiles, polymer and char, respectively, and μ is the char fraction. The heat of combustion of the volatile fraction can differ significantly from that of the polymer and the char, so polymer heats of combustion should not be used to calculate flaming combustion efficiency of materials.

The experimentally observed minimum irradiance level, below which no sustained burning occurred, was approximately 20 kW m^{-2}.[3] The critical irradiance level derived from the experimental correlation was 12.6 kW m^{-2}. The difference between the minimum and critical irradiance for this type of material has been explained by Janssens.[4] At low irradiance levels, fuel volatiles are exhausted before the lower flammability limit is reached in the gas phase. At higher irradiance levels, the minimum mass flux of volatiles to create flammable mixture is generated before fuel exhaustion. The ignition times for the polyurethane foam were short (2–6 s). In Cone calorimeter tests the heat of combustion can be obtained as the ratio of heat release rate and mass loss rate, both measured in the Cone calorimeter. Our approach was based on simulated heat release analysis, which operates with experimental mass loss data and the calculated heat of combustion.

An empirical correlation is needed to relate bench-scale data to full-scale results. It is normally dealt with by normalizing the results by the exposed surface area of the specimen. We have chosen the normalizing surface area as 1 cm^3 [Equation (4)],

$$\dot{q}_{full\text{-}scale}(t) = \dot{q}''_{bench\text{-}scale} \times S(t) \tag{4}$$

where $\dot{q}_{full\text{-}scale}$ is the total rate of heat release at any time t; $\dot{q}''_{bench\text{-}scale}$ is the bench-scale heat release rate; $S(t)$ is the area of the full-scale specimen that is at any time ignited, covered with flames, but not yet burned out.

Figures 2 and 3 compare the results obtained for PU bench-scale combustion tests. The spherical samples of PU were ignited and mass loss data were measured during flaming out.

Similar results were obtained for PU-OM nanocomposite. The maximum mass loss rate at 1 cm^2 surface area of sample was obtained *via* approximating the maximum loss rate (PU and PU-OM) *vs.* time to an exponential equation (Figures 4 and 5).

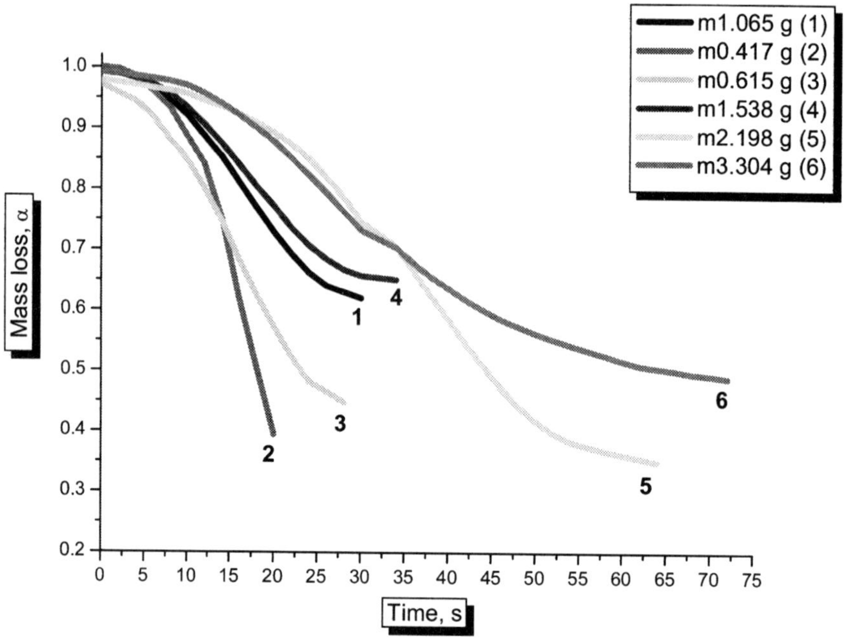

Figure 2 *Bench-scale data of PU spheres combustion tests: mass loss (conversion degree) vs. time*

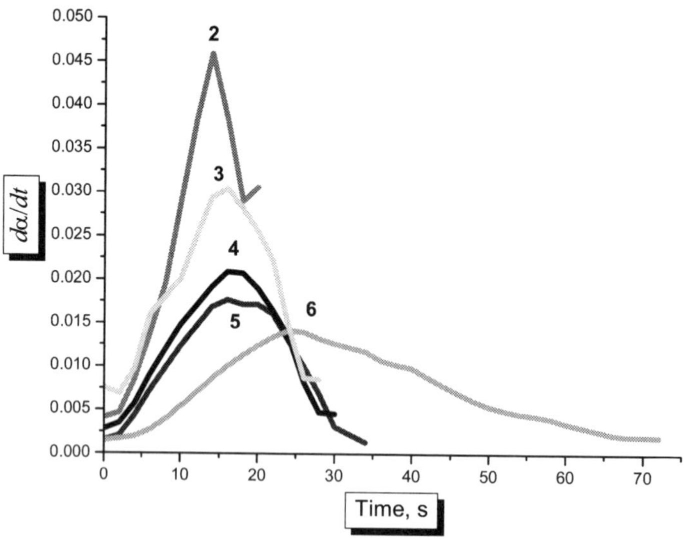

Figure 3 *Bench-scale data of PU spheres combustion tests: rate of mass loss rate (conversion degree) vs. time*

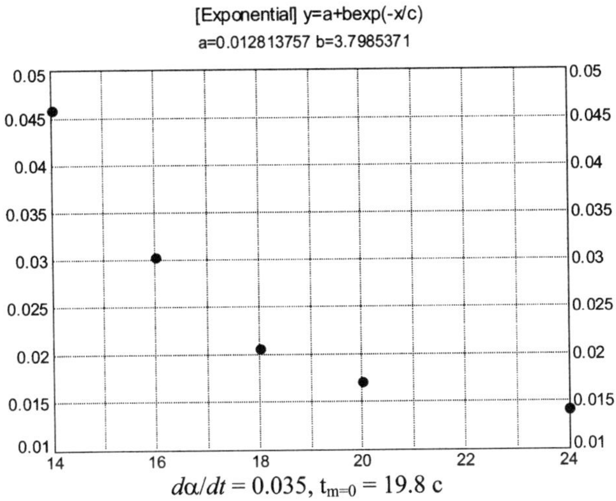

Figure 4 *Approximation of maximum loss rate (PU) vs. time to an exponential equation* $y = a + b\exp(-x/c)$

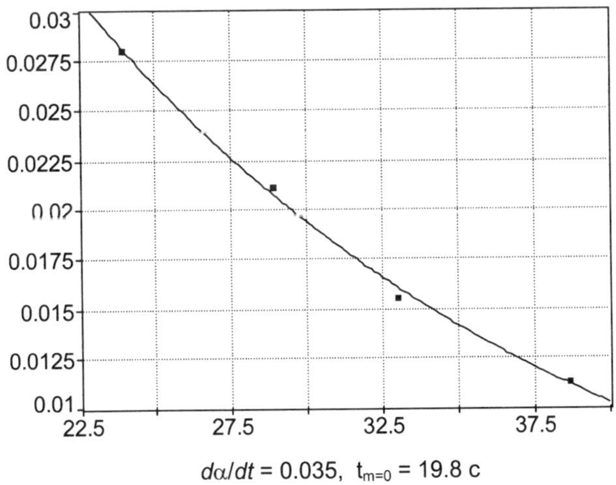

Figure 5 *Approximation of maximum loss rate (PU – OM nanocomposite, 10%) vs. time to an exponential equation* $y = a + b\exp(-x/c)$

The maximum heat release rates were estimated using Equation (5),

$$q''_{bench\text{-}scale} = \dot{m}''_{bench\text{-}scale} \times \Delta H_{combustion} \tag{5}$$

where $\dot{m}''_{bench\text{-}scale}$ is the bench-scale mass loss rate and $\Delta H_{combustion}$ is the effective heat of combustion (kJ kg^{-1}). Effective heats of combustion for PU and PU-OM nanocomposite were calculated using the PhysProps 1.6 software. Figure 6 presents the maximum heat of release of PU and PU-OM nanocomposite. These

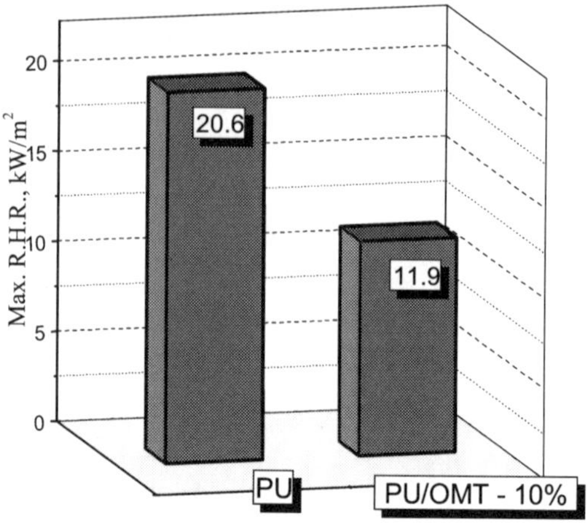

Figure 6 *Maximum heat of release of PU and PU-OM nanocomposite*

data indicate about 40% relative improvement of PU-OM (10%) (nanocomposite) over the native PU maximum heat of release rate.

During the combustion test of the nanocomposite specimen, a carbon layer formed on its surface from the start, grew over time and resisted the heat. Formation of a carbonized layer on the surface of the polymer is a feature of all nanocomposites studied so far: the pattern illustrated decomposition, which can dramatically change the overall scheme of thermal degradation process kinetics by way of solid-phase catalysis or by shifting the reaction mechanism from radical random-scission degradation to aromatization followed by carbonization.

10.4 References

1. J.W. Gilman, C.L. Jackson, A.B. Morgan and R. Harris Jr., *Chem. Mater.*, 2000, **12**, 1866–1873.
2. Y. Hu, L Song, J. Xu, L. Yang, Z.Chen and W. Fan, *Colloid Polym. Sci.*, 2001, **279**, 819.
3. O. Grexa, M. Janssens, "Wood & fire safety", in *Proceed. of 3rd International Scientific Conference*, Ed. By A. Osvald, CSc., Nikara Publ., Krupina, Slovak Republic, 1996, pp. 139.
4. M. Janssen, "Improved method of analysis for the LIFT apparatus – Part 1: ignition", in *Proceed. 2nd Fire & Materials Conference*, Arlington, VA, September 23–24, 1993, Interscience Communication, London, pp. 37.

CHAPTER 11

Thermal Degradation Behaviour of Flame-Retardant Unsaturated Polyester Resins Incorporating Functionalised Nanoclays

BALJINDER K. KANDOLA, SHONALI NAZARE AND A. RICHARD HORROCKS

Centre for Materials Research and Innovation, Bolton Institute, Deane Road, Bolton, BL3 5AB, UK (bk1@bolton.ac.uk)

11.1 Introduction

Recent interest in the reported char-promoting behaviour of functionalised dispersed nanoclays at levels of 2–5%, to yield nanocomposite structures having improved fire properties, has prompted investigation of their potential as fire retardants.[1–4] The nanocomposite flame retardant mechanism is believed to be a consequence of high-performance carbonaceous-silicate char build-up on the surface during burning.[1] This insulates the underlying material and slows the mass loss rate of decomposition products. Unfortunately, however, these nanocomposites on their own are not sufficient to reduce flammability of low char-forming polymers like polyesters to a significant and specified level. However, when used with conventional flame retardants, their action may be synergistic in a way that less flame retardant is required as compared to the situation where nanoclays are not used. The other advantage is that nanoclays maintain and sometimes increase matrix mechanical properties while it is well known that any conventional additive in a polymer often reduces its mechanical performance.[1–3,5]

We have explored the effect of incorporating different types of organically modified clays in polyester resin with and without flame retardants (FR), such as ammonium polyphosphate, melamine phosphate with and without dipentaerythritol and alumina trihydrate. For polyester resins significant flame retardancy is observed only at FR concentrations greater than 20% (w/w).[6,7] In the present work, to observe the significance of nanocomposite structures, the clays have been introduced at a typical 5% loading level and the FR level has been kept 20%. X-ray diffraction studies have shown that the clays are well exfoliated. Thermal degradation behaviour of these samples has been studied by simultaneous DTA-TGA and results have been analysed to assess possible effects of nanoclays on resin thermal stability with and without flame retardants.

11.2 Experimental

11.2.1 Materials

The resin is a polyester resin (orthophthalic, Crystic 471 PALV, supplied by Scott Bader). Clays are Cloisite Na^+, 10A, 15A, 25A and 30B (supplied by Southern clay Products, USA). Properties of these clays are given in Table 1.

The following commercially available flame retardants (FR) were used without further purification:

(i) APP–Ammonoium polyphosphate (Antiblaze MCM, Rhodia Specialities)
(ii) NH–Melamine phosphate (Antiblaze NH, Rhodia Specialities)
(iii) NW–Dipentaerythritol/melamine phosphate (Antiblaze NW, Rhodia Specialities)
(iv) ATH–Alumina trihydrate (Martinal, Martinswerk, GmbH).

Table 1 *Treatment/ properties of organically modified clays*

Clay	*Commercial name*	*Organic modifier*[a]	*Modifier conc.* (meq/100 g clay)	*d spacing* (Å)
Inorganic	Cloisite Na^+	–	93	11.7
Cl 1	Cloisite 10A	$CH_3-N^+(CH_3)(HT)-CH_2-C_6H_5$	125	19.2
Cl 2	Cloisite 15A	$CH_3-N^+(CH_3)(HT)-HT$	125	31.5
Cl 3	Cloisite 25A	$CH_3-N^+(CH_3)(HT)-CH_2CH(CH_2CH_3)CH_2CH_2CH_2CH_3$	95	18.6
Cl 4	Cloisite 30B	$CH_3-N^+(CH_2CH_2OH)(CH_2CH_2OH)-T$	90	18.5

[a]T is tallow and HT is hydrogenated tallow (~65%C1 8; ~30%C1 6; ~5%C1 4); Anion : Chloride in Cloisite 10A, 15A and 30A; sulphate in 25A.

Table 2 *Weight percentages of various components in the formulations*

Sample		*Resin*	*FR*	*Clay*
Res	– Resin	100	–	–
Res/Cl	– Resin + Clay	95	–	5
Res/FR	– Resin + FR	83	17	–
Res/Cl/FR	– Resin + Clay + FR	79	17	4

11.2.2 Preparation of Polyester–Clay Nanocomposites

Polyester–clay nanocomposites incorporating flame retardants have been prepared by *in situ* intercalative polymerisation; 5% (w/w) clay was gradually added to the resin polyester resin, while stirring with a mechanical mixer under high shear (900 rpm). The mixing was carried out for 60 min at room temperature. For samples incorporating flame retardants, 20% (with respect to resin-clay mixture) of the respective flame retardant was added to the mixture of resin and clay after 20 min of mixing. The percentages of various components in the formulations are given in Table 2. Small amounts of samples were taken from the mixture for simultaneous DTA-TGA analysis before the catalyst was added and laminates were cast and cured at room temperature for further flammability testing (to be discussed in further publications).

The nanocomposite structures were characterised by X-ray diffraction, XRD, in the laboratories of the National Institute of Standards and Testing (NIST), USA.

11.2.3 Equipment

Simultaneous DTA-TGA analysis was performed using SDT 2960 TA instruments under flowing air (100 ml min^{-1}) and at a heating rate of 10 K min^{-1} on 25 mg sample masses.

11.3 Results and Discussion

The nanocomposite formation has been studied by X-ray diffraction measurements. In the diffraction curves for pure clays, a prominent peak in each curve corresponding to basal spacing of respective clays occur at *d*-spacings as shown in Table 1. This reflection is missing in the scattering curves of all the polyester-clay nanocomposites, irrespective of the presence of flame retardant, confirming the presence of nanocomposite structures. While detailed XRD analysis will be presented in a separate publication, here representative diffraction curves for clays Cloisite Na$^+$ and 25A, polyester resin with APP and resin-Cloisite 25A with and without APP are given in Figure 1. Cloisite 25A shows a peak at 2θ of 4.8° [Figure 1(a)], representing a *d*-spacing of 18.6 Å (Table 1), which is missing in Res/Cl 3 and Res/Cl 3/APP formulations in Figure 1(b).

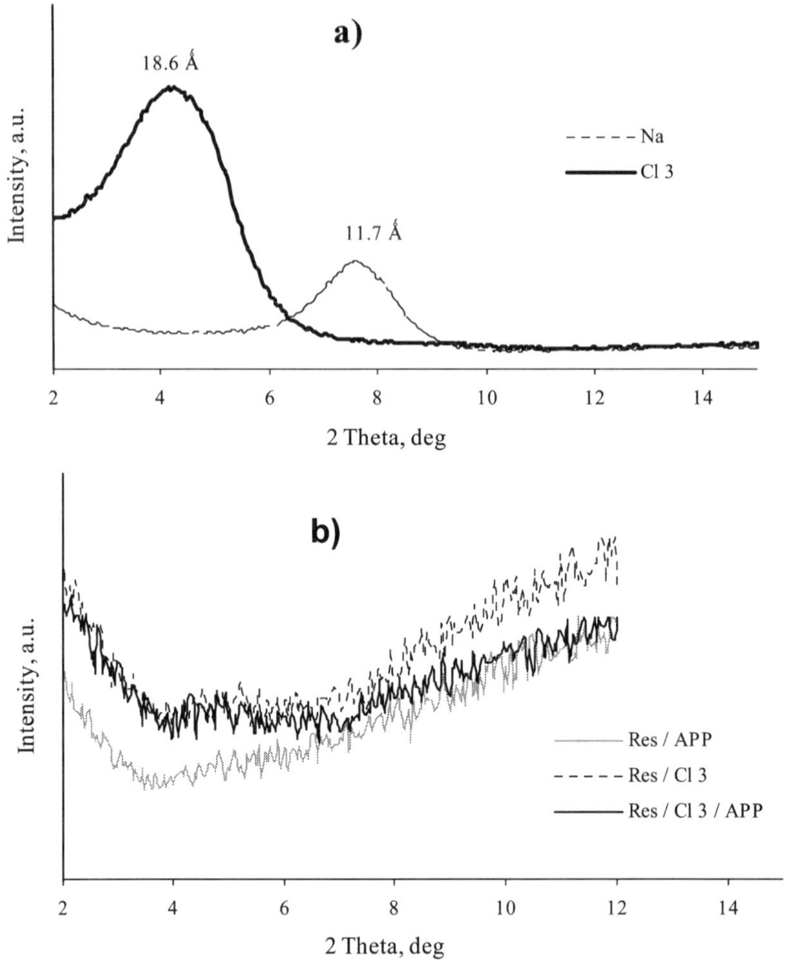

Figure 1 *XRD data for* (a) *Cloisite Na⁺, 25A (Cl 3) clays and* (b) *polyester resin with APP, resin-Cl 3 nanocomposite with and without APP*

11.3.1 Thermal Degradation of Clays

DTA and DTG peak maxima for all organically modified clays used in this work are given in Figure 2 and Table 3. Although Na-montmorillonite was not dispersed in polyester resin, its thermal analytical behaviour is discussed here for comparison with other organically modified clays. Na-montmorillonite shows very little weight loss [Figure 2(a)] and high residues at 800°C. In the main the DTA response is featureless [Figure 2(b)] showing inertness of inorganic clay, however there are two very broad and small endotherms with maxima at 78 and 663°C. All organically modified clays, however, show two stages of weight loss, the first represented by double peaked (in the temperature

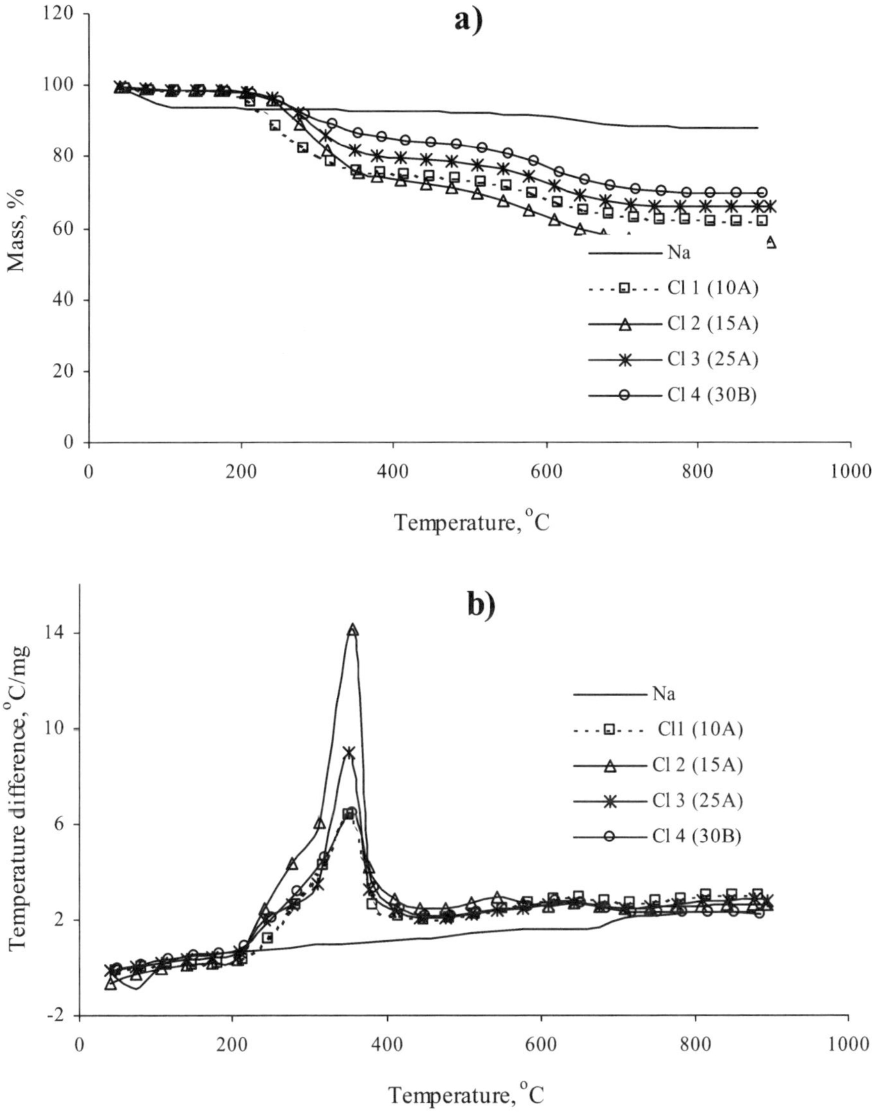

Figure 2 *TGA* (a) *and DTA* (b) *responses for all clays in air*

range of 235–293 and 307–348°C) and the second by single peaked DTG maxima (575–605°C) shown in Table 3. The first stages are most probably due to decomposition of the respective organic components of the clays and the second single one due to dehydroxylation of the clay layers.[8] Residues at 800°C in Table 3 represent residual silica contents. These two stages of weight loss are also supported by DTA curves, where all organically modified clays show exotherms. The first exotherm is double peaked with the first maxima in the

Table 3 *Analysis of DTA responses (peak temp °C), DTG maxima (°C) and % mass residue of unsaturated polyester resin formulations under flowing air*

Samples	*DTA peaks(°C)*	*DTG peaks(°C)*	*% Mass residue* 600°C	800°C
Clays				
Na^+	78 En(b); 663 En(b)	76, 665	91	88
Cl 1	347 Ex; 633 Ex (s,b)	235, 348; 597	68	62
Cl 2	264 (s), 354 Ex(d); 545 Ex (s,b)	260, 307; 575	63	56
Cl 3	258 (s), 351 Ex (d); 626 Ex (s,b)	293, 341; 603	73	66
Cl 4	270 (s), 345 Ex (d); 630 Ex (s,b)	263, 335; 605	77	67
Resin	131 Ex(s,b); 313 (s), 365 Ex (d); 552 Ex	201, 360, 554	1.1	1
Resin /clays				
Res/Cl 1	163 En (s); 307 Ex; 338 En; 532 Ex	161, 338, 567	4	4
Res/Cl 2	158 En (s), 310 Ex; 338 En; 524 Ex	156, 337, 533	4	4
Res/Cl 3	159 En (s); 302 Ex; 340 En; 519 Ex	157, 338, 531	6	5
Res/Cl 4	154 En; 292 Ex; 333 En; 528 Ex	147, 330, 533	6	5
Flame retardants				
APP	96 En (s); 173 En (s); 330 En; 669 Ex	331, 655	62	8
NH	267 En; 302 En; 386 En; Exo baseline deviation – 529, 756 (b)	267, 305, 383, 472, 573, 740	41	18
NW	118 En (s); 221 En; 241 En: 328 Ex; 393 En; 492 Ex; 665 Ex	120, 238, 341, 389, 653	32	1
ATH	93 En (s); 242 (s), 308 En (d); 520 En (s,b)	237, 302, 520	66	65
Resin/FRs and/or clays				
Res/APP	320 En, 346 Ex (s), 382 Ex; 696Ex	317, 372, 699	30	3
Res/Cl 3/APP	146 En (s); 290 Ex; 313 En; 351, 455, 530 Ex(t); 712 Ex(b)	155, 312, 745	23	15
Res/NH	266 En; 297 Ex; 313 En; 323 Ex; 528 Ex; 795 Ex; 846 Ex	69, 306, 800	16	7
Res/Cl 3/NH	174 En; 257 En; 312 En, 323 Ex; 588 Ex	169, 310, 610	16	9
Res/NW	204 En; 250 En; 312 En; 331 Ex; 530 Ex; 815 Ex	135, 311, 537, 801	16	7
Res Cl 3/NW	155 En; 251 En; 306 En; 329 Ex; 458, 588 Ex (d)	151, 310, 613	15	7
Res/ATH	164 En; 243 En; 330 En; 382 Ex (s); 518 Ex	66, 160, 326, 348, 518	12	12
Res/Cl 3/ATH	150 En (s); 247 En; 288 Ex; 326 En; 377 (s); 389(s); 490 Ex	146, 150, 324, 350, 374, 504	15	15

En = endotherm; Ex = exotherm; b = broad; d = double peaked; s = small.

range 258–270°C, and the second, more prominent one around 345–355°C. The second small, broad exotherm appears in the temperature range 545–633°C.

11.3.2 Thermal Degradation of Resin

The TGA curve of polyester resin shows three stages of weight loss [Figure 3(a)], the first occurring up to about 250°C, the second over the range 250–400°C and the third smaller weight loss from 400 to 600°C. DTA [Figure 3(b)] and DTG peaks at 365 and 360°C, respectively, for polyester resin in Table 3 probably represent release of styrene and other volatile products. Resin starts to decompose above 200°C, whereas the main step of weight loss occurs between 300 and 400°C,[9] as shown by the DTG peak at 360°C in Table 3. Above 400°C, solid phase oxidation reactions predominate.[10] The detailed mechanism of these reactions is discussed elsewhere.[11]

11.3.3 Effect of Different Clays on Thermal Degradation of Resin

One of the most important property enhancements expected from formation of a polymer nanocomposite is that of thermal stability either of the initial stages or final carbonaceous residues. However, for polyester nanocomposites, Figures 3(a) and 3(b) show that thermal stability of the resin is reduced below 400°C. Furthermore, the main DTA decomposition peak of the resin at 365°C is replaced by endotherms in Res/Cl nanocomposite samples. The initial endothermic DTA peaks (Table 3) in the 154–163°C range are most probably due to decomposition of the organic component of the clay. The onset of degradation of resin temperature is lowered on addition of nanoclays, as reflected in shifts in DTA peaks from 313°C to as low as 292°C for the Res/Cl 4 combination [Table 3 and Figure 3(a)]. Similar effects of nanoclays in cross-linked polyester resin thermal analysis responses were also seen by Bharadwaj *et al.*,[12] who introduced clay at 1, 2.5, 5 and 10% (wt/wt). Above 600°C, thermal stability is increased slightly [Figure 3(a)].

Figure 3(c) presents mass difference versus temperature plots, which show the difference between TGA experimental residual masses and calculated (from weighted average component responses) masses at each temperature.[13] Below 400°C, char formation in all nanocomposites is much less and hence volatilizations greater than expected, which may be because clays are interfering with the cross-linking of the resin. At 400–600°C, char formations are again less than expected. Above 600°C, char formations are similar to that expected from respective calculated values. However, the type of clay has no effect on residual char formation at high temperatures. This clearly indicates that nanoclays on their own are not effective in increasing char formation and, hence, in reducing flammability of unsaturated polyester resins. However, Gilman *et al.* have also observed this behaviour with vinyl ester and epoxy nanocomposites, where there is little improvement in carbonaceous char yield once the presence of silica is accounted for, but these samples showed a reduction in peak heat release when tested with cone calorimetry.[14]

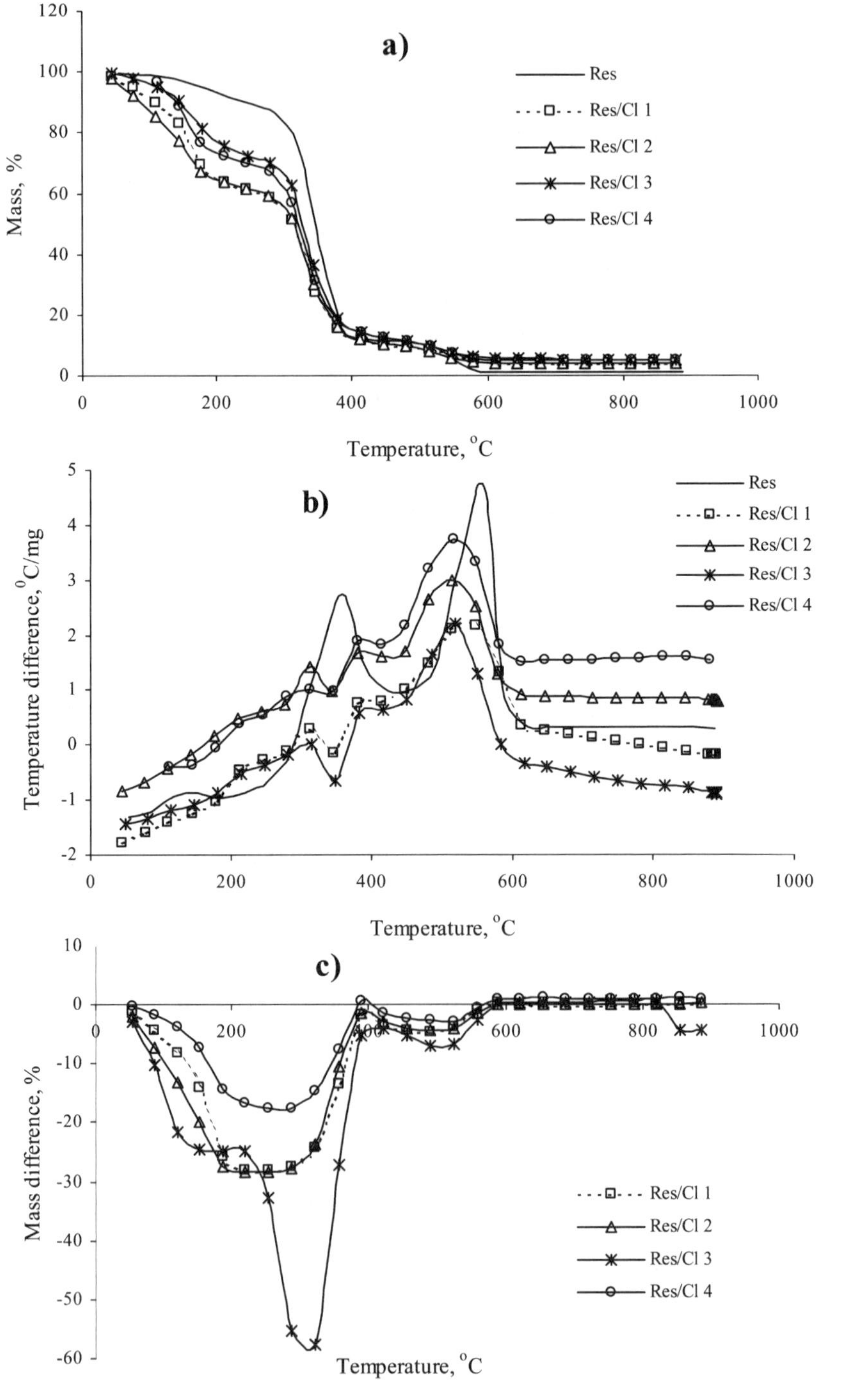

Figure 3 *TGA* (a) *and DTA* (b) *responses in air; percentage residual mass differences (actual – calculated) as function of temperature* (c) *for different resin – clay nanocomposites*

11.3.4 Effect of Flame Retardants on Thermal Degradation of Polyester Resin

TGA curves of resin with all flame retardants are given in Figure 4(a); DTA and DTG peaks are given in Table 3. All flame retardants affect the thermal degradation mechanism of the resin. Thermal degradation mechanisms of

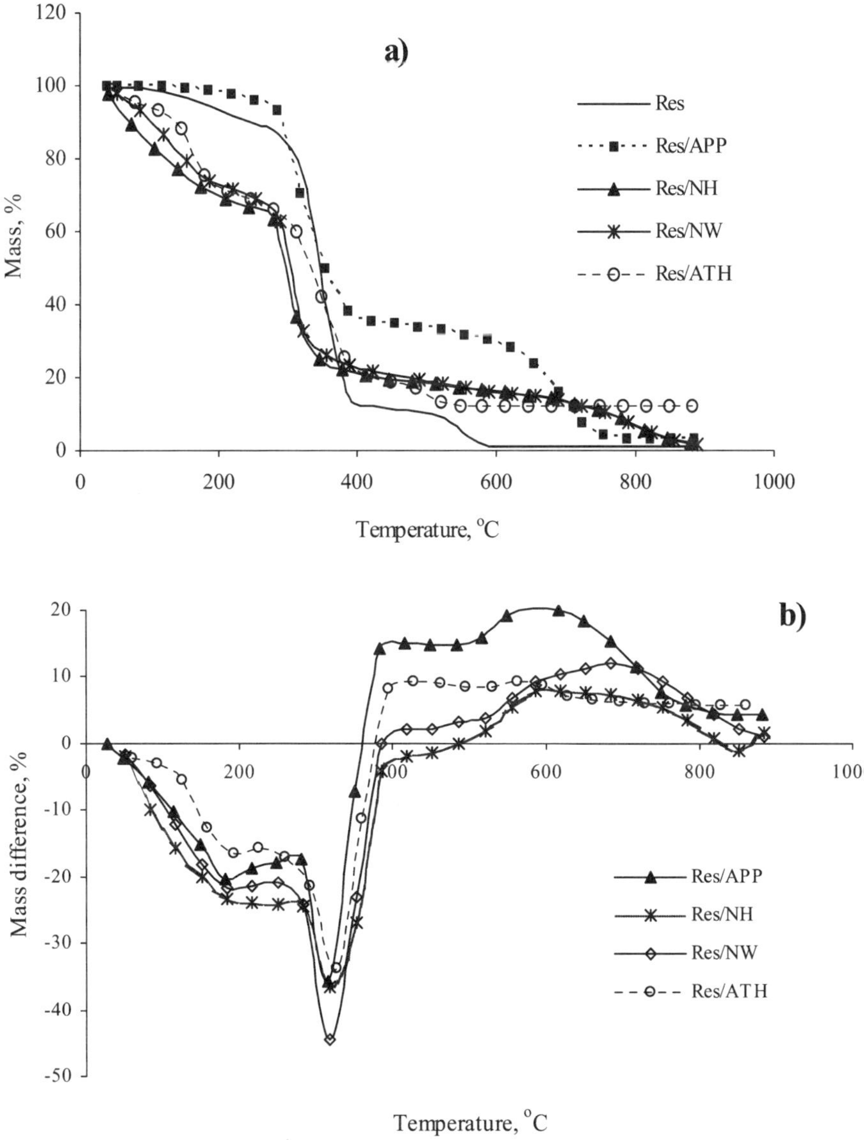

Figure 4 *TGA responses in air* (a) *and percentage residual mass differences (actual–calculated) as a function of temperature* (b) *for resin with different flame retardants*

ammonium polyphosphate, melamine phosphate with and without dipentaerythritol have been discussed in previous communications,[15] where all of them show endothermic decomposition peaks. APP starts decomposing just after melting at 210°C, releasing ammonia and phosphoric acid, and then polymerises to polyphosphoric acid, which at higher temperatures decomposes to P_2O_5. These reactions are represented by an endothermic peak at 330°C and exothermic peak at 669°C, respectively (Table 3). Melamine phosphate decomposes over the temperature range 250–380°C, forming melamine pyrophosphate and polyphosphate at about 280 and 310°C,[15] decomposing further in the temperature range 330–410°C, releasing melamine, ammonia and water,[16] as shown by endothermic peaks at 267, 302 and 386°C. In Antiblaze NW, dipentaeythritol melts at about 125°C and reacts with melamine phosphate, forming polyol phosphate. Antiblaze NW also shows a series of endothermic peaks at 118, 221, 241, 393°C, followed by exothermic peaks at 492 and 665°C. ATH decomposes at 180–340°C with a series of endothermic peaks at 93°C, double peaked at 242 (small) and 308°C (main peak), followed by small, broad peak at 520°C (Table 3), due to release of water and subsequent decomposition.

All of these flame retardants are, therefore, decomposing in the temperature range 200–300°C, and so offer the chance of interacting with the cross-linking polyester resin. This is seen by changes in weight loss in this temperature range by TGA curves in Figure 4(a) and DTA peaks in Table 3. Ammonium polyphosphate enhances the thermal stability of resin, whereas, melamine phosphate with and without dipentaerythritol and alumina trihydrate, decreases its thermal stability, showing more weight loss in this temperature range. However, above 400°C, all flame retardants enhance residual levels and hence thermal stability. APP is seen to be the most effective char promoter up to 700°C, after which the char oxidises. Alumina trihydrate shows superior behaviour above this latter temperature and even at 800°C 12% char is left behind (Table 3), which is expected to be residual alumina. When the char difference between actual and calculated values are plotted in Figure 4(b), all the observations discussed above are clearly seen, except that APP does not produce more than expected char formation below 400°C, as previously observed in Figure 4(a). ATH also produces more than expected char above 400°C, showing that this is not acting just like a filler, but as a reactive flame retardant.

11.3.5 Effect of Clays on Thermal Degradation of Flame Retarded Resin

Although all the resin clay nanocomposites were studied with different flame retardants present, only the results of Cloisite 25A (Cl 3) are presented here. Figure 5(a) shows that clay is effective with the resin-APP mixture in changing thermal degradation, making it less stable than the Res/Cl mixture at lower temperatures and producing more char above 500°C, as seen from Table 3 and in the absence of nanoclay in Figure 4(a). Clay addition has less effect on resin containing melamine phosphate with and without dipentaerythtritol, as can be seen by comparing Figures 4(a) and 5(a), whereas, with alumina trihydrate,

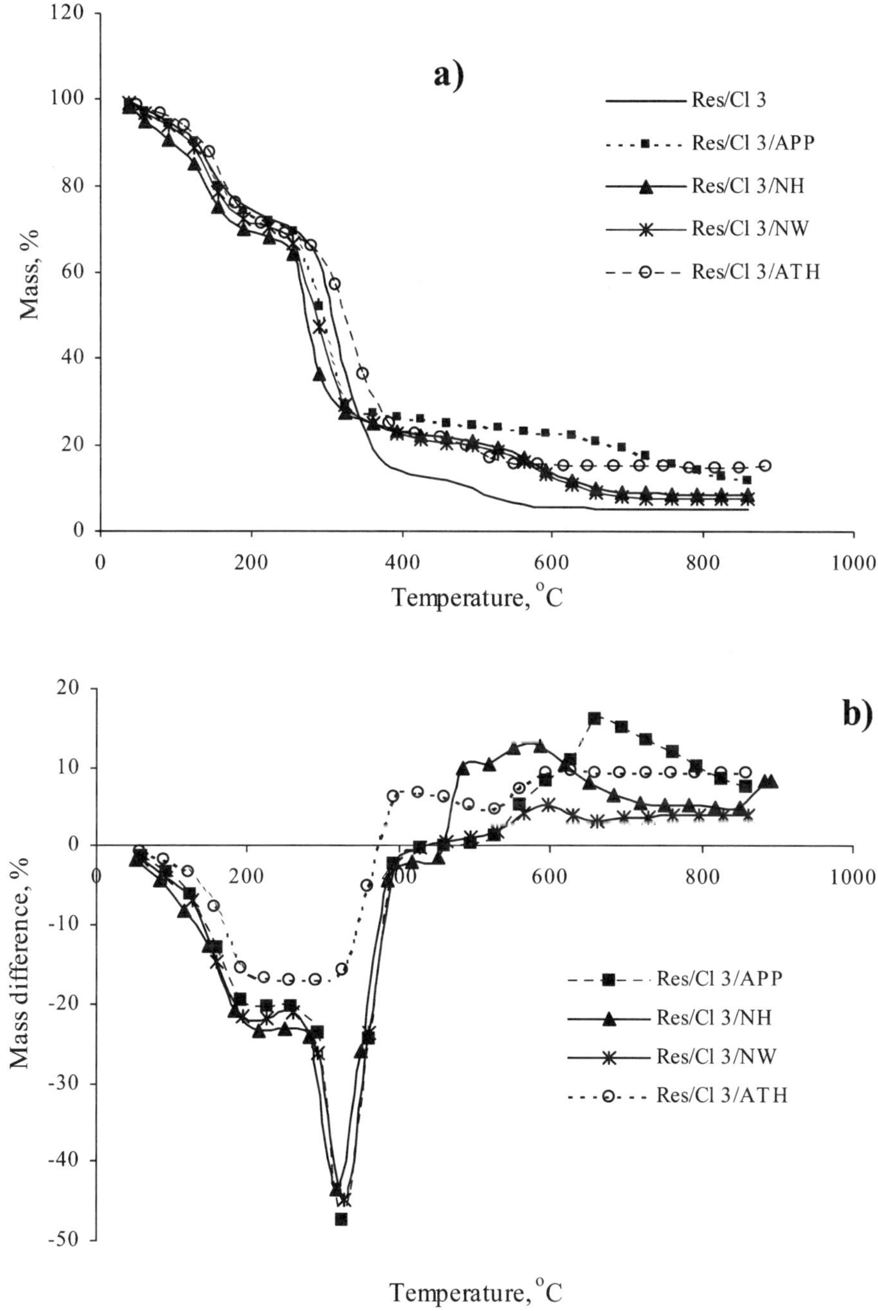

Figure 5 *TGA responses in air* (a) *and percentage residual mass differences (actual–calculated) as a function of temperature* (b) *for resin nanocomposite with different flame retardants*

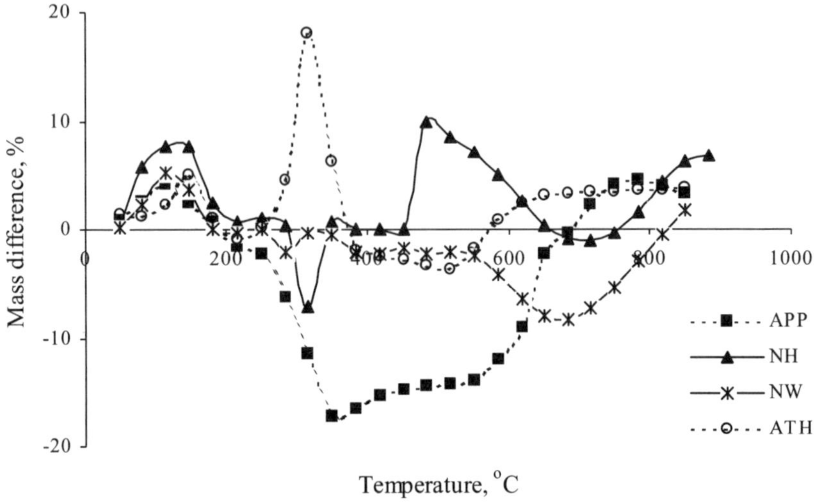

Figure 6 *Effect of Cloisite 25 A clay: percentage residual mass differences between Res/Cl 3/FR [Figure5(b)] and Res/FR samples [Figure 4(b)]*

char formation is increased above 600°C. This effect can also be seen in Figure 5(b), where differential curves are very similar in the range 100–400°C. Clearly, clay with ammonium polyphosphate is the most effective of the samples studied to enhance char formation, yielding above 10% at 700°C. However, this enhanced char is less than that seen for APP alone in Figure 4(b). This is in contrast to what is expected from such structures for other polymer – nanocomposites.[2]

Figure 6 shows the effect of clay on the Res/FR systems by subtracting Res/FR char difference values [Figure 4(b)] from Res/Cl 3/FR [Figure 5(b)]. It can be seen that clay reduces char formation tendency of Res/APP system up to 700°C and then it is increased up to 5%. In the presence of melamine phosphate clay further increases char at 450–700°C. However, when both melamine phosphate and dipentaerythritol are present, char formation tendency is reduced. Clay addition enhances the charring tendency of resin in the presence of ATH in the temperature range 200–400°C and above 600°C.

Most work in the area of polymer nanocomposite flame retardancy has been in thermoplastics, where it is proposed that a reduction in heat release (flammability) and, hence, an increase in thermal stability is due to formation of a protective surface barrier/insulation layer consisting of accumulated silica platelets with a small amount of carbonaceous char.[17,18] The accumulation or precipitation of silicate layers on the surface is due to gradual degradation and gasification of the polymer. However, according to Lewin[19] the layered silicates are dispersed and not dissolved in the polymer and should not precipitate as a consequence of the progressive gasification of the polymer. Lewin has postulated that this migration is driven by the lower surface free energy of the montmorillonite and by convection forces, arising from the temperature gradients, perhaps aided by movement of gas bubbles present during melting of the thermoplastic polymers. Alternative or even additional mechanisms include their presence as dispersed barriers to diffusion of molten polymer and decomposing products including release of combustible gases as well as diffusion of air and heat by convection. Each of these permeabilities contributes to the FR effectivity of the charred surface.[19] Clearly, such complex mechanisms could interact, positively or negatively, with flame retardants also present and the evidence above suggests that both possibilities exist with the polyester resin used.

11.4 Conclusions

In unsaturated polyester resin nanocomposites, nanoclays reduce thermal stability and char formation tendency of the resin up to 600°C and after that there is no observed change. Different condensed phase active flame retardants increase char formation of the resin above 400°C. When nanoclays are added, char formation is not greatly affected and in fact for APP-containing resins it is reduced. Clearly, for unsaturated polyester resin some char consolidating agent/group is required and work is ongoing in this area in our laboratories.

However, thermal analysis techniques are not representative of the real fire situation. Consequently, the effect of these nanoclays on resin fire performance with and without flame retardants in cone calorimetry is presented in a separate publication,[20] where their mechanical properties are also evaluated.

11.5 Acknowledgements

The authors acknowledge the financial support from the Engineering and Physical Science Research Council during this work. They also thank the National Institute of Standards and Technology (NIST), USA and in particular Dr Jeffrey W. Gilman for technical and financial support.

11.6 References

1. J.W. Gilman, and T. Kashiwagi, in *Polymer–clay Nanocomposites*, T.J. Pinnavaia and G.W. Beall (eds.), John Wiley & Sons Ltd, New York, 2000, Chapter 10.
2. B.K. Kandola, in *Fire Retardant Materials*, A.R. Horrocks and D. Price (eds.), Woodhead Publishing, Cambridge, U.K., 2001, Chapter 6.
3. J. Lee, T. Takekoshi and E.P. Giannelis, *Mater. Res., Soc., Symp. Proc.*, 1997, **457**, 513–518.
4. E.P. Giannelis, *Adv. Mater.*, 1996, **8(1)**, 29.
5. Y. Kojima, A. Usuki, M. Kawasumi, A. Okada, Y. Fakushima, T. Kurauchi and O. Kamigaito, *J. Mater. Res.*, 1993, **8**, 1185–1189.
6. F. Le Lay and J. Gutierrez, *Polym. Degrad. Stab.*, 1999, **54**, 397–401.
7. A. Hernangil, M. Rodriguez, L.M. Leon, J. Ballestero and J.R. Alonso, *J. Fire Sci.*, 1999, **17**, 281–306.

8. K.P. Pramoda, T. Liu, Z. Liu, C. He and H.-J. Sue, *Polym. Degrad. Stab.*, 2003, **81**, 47–56.
9. S.V. Levchik, in *Fundamentals, International Plastics Flammability Handbook*, 3rd Edn, M. Le Bras, S. Bourbigot and J. Troitzsch (eds.), Hanser Pub., Munich, 2004, pp. 83–98.
10. G.S. Learmonth and A. Nesbit, *Br. Polym. J.*, 1972, **4**, 317–325.
11. B.K. Kandola, A.R. Horrocks, P. Myler and D. Blair, in *Fire and Polymers*, G.L. Nelson and C.A. Wilkie (eds.), *ACS Symp. Ser.,* 2001, Volume 797, 344–360.
12. R.K. Bharadwaj, A.R. Mehrabi, C. Hamilton, C. Trujillo, M. Murga, R. Fan, A. Chavira and A.K. Thompson, *Polymer*, 2002, **43**, 3699–3705.
13. B.K. Kandola, S. Horrocks and A.R. Horrocks, *Thermochim. Acta*, 1997, **294**, 113–125.
14. J.W. Gilman, T. Kashiwagi, M. Nyden, J.E.T. Brown, C.L. Jackson, S. Lomakin, E.P. Giannelis and E. Manias, in *Chemistry and Technology of Polymer Additives*, S. Al-Malaika, A. Golovoy and C.A. Wilkie (eds.), Blackwell Science, Oxford, UK, 1999, Chapter 14.
15. B.K. Kandola and A.R. Horrocks, *Polym. Degrad. Stab.* 1996, **54**, 289–303.
16. G. Camino, M.P. Luda and L. Costa, in *Recent Advances in Flame Retardancy of Polymeric Materials*, M. Lewin (ed.), (Proceedings of the 1993 Conference), BCC Company, Stamford, CT, 1993, Volume IV. p. 12.
17. J.W. Gilman, *Appl. Clay Sci.*, 1999, **15**, 31–49.
18. A.B. Morgan, R.H. Jr. Harris, T. Kashiwagi, L.R. Chyall and J.W. Gilman, *Fire Mater.,* 2002, **26**, 247–253.
19. M. Lewin, *Fire Mater.*, 2003, **27**, 1–7.
20. B.K. Kandola, S. Nazare and A.R. Horrocks, presented at 228th ACS National Meeting, Philadelphia, PA, USA, August 22–26, 2004.

CHAPTER 12

Comparative Study of Nano-effect on Fire Retardancy of Polymer–Graphite Oxide Nanocomposites

JIANQI WANG AND ZHIDONG HAN

National Laboratory of Flame Retardant Materials, School of Materials Science & Engineering, Beijing Institute of Technology, 100081 Beijing, China (Bitjq@public.bta.net.cn)

12.1 Introduction

Nanocomposite technology originated in the 1980s at Toyota's central R & D Laboratories in Japan.[1] Research focused on nylon nanocomposites and many patents on related products were published. Updated development work suggests that nanocomposites are a unique class of materials having significant improvements in important properties like modulus, flexural strength and heat distortion temperature.

The ability of nanoclay incorporation to reduce the flammability of polymeric materials was first reported by Gilman *et al.*[2] Particularly, heat release rate, obtained from cone calorimetry experiments, was found to be decreased substantially by nanoclay incorporation at loadings as low as 2–5 wt%.

Cone calorimetry has become the most important methodology in flammability tests, and is very often used in polymer-layered silicate (PLS) nanocomposites and recently carbon nanotube composites.[3] However, the limiting oxygen index (LOI), a different fire model, capable of evaluating the fire extinction, is popularly adopted because of its precision and its excellent reproducibility.

Graphite oxide (GO) samples were first prepared by the Hummers method in 1958.[4] GOs have been prepared and characterized GO[5,6] and polymer/graphite oxide intercalated and exfoliated nanocomposite have been obtained[7–11] with the help of the polar nature of GO attached with various functional groups, *e.g.*

hydroxyl, carbonyl, and ether groups.[6] The main objective of this chapter is to give a brief comparative review of the nano- and micro-dispersive effect on the fire retardancy and flammability properties for polymer/GO composites.[5,6] A short comparison on fire retardancy between PLS and PLGO (polymer-layered GO) is also made. In addition, an XPS/TGA study and flame retardant mechanism are reported.

12.2 Experimental

12.2.1 Sample Preparation

Graphite oxide was prepared by the modified Hummers method.[4–6] Polymer/GO composites have been prepared[7–11] by (i) dispersion/absorption method in aqueous solution (*e.g.* PVA, PEO); (ii) *in situ* polymerisation (*e.g.* acrylic acid); and (iii) suspension polymerisation (*e.g.* PS). Method (i) was used in the work. Prior to the preparation of polymer/GO composite the GO should be well ground in a pulverator. For details concerning sample preparation for nano- and micro-polymer/GO composites, please see references 5 and 6.

12.2.2 Characterization Techniques

X-Ray diffraction (XRD) patterns were carried out on a D/max-RB Japan equipped with a Cu-K_α generator ($\lambda = 0.1540$ nm) operated at 100 mA and 40 kV. Transmission electron micrograph (TEM) experiments were conducted on H-800 HITACHI (acceleration voltage of 200 kV). Thermal gravimetric analysis (TGA) results were performed on a Du Pont 2000 at N_2 flow rate of 50 ml min^{-1} and heating rate of 10°C min^{-1}. X-Ray photoelectron spectra (XPS) were recorded on a PHI 5300 ESCA system (Perkin-Elmer) with MgK_α at 250 W (12.5 kV × 20 mA) under a vacuum better than 10^{-6} Pa, calibrated by assuming the binding energy of adventitious carbon to be 285.0 eV. The pseudo-*in situ* technique is utilized, where the sample is heated outside the XPS chamber under argon protection, keeping the sample orientation intact.

Calorimetry by oxygen consumption: parameters, like heat release rate (HRR), mass loss rate (MLR), CO and CO_2 yield *etc.* were recorded according to ASTM 1356-90 on a Cone calorimeter (Stanton Redcroft) at a heat flux of 15 kW m^{-2} to keep the mechanistic study of the nanostructure effect.

LOI measurement was carried out using a FTA II Instrument (Stanton Redcroft, Polymer Laboratory, UK) according to GB2406-80 (ASTM D 2863).

12.3 Results and Discussion

12.3.1 Morphological Structure

Four systems of polymer-*GO* nano-composites were prepared through the incorporation of GO into *polyacrylic ester* (PAE), polyvinyl alcohol (PVA), *polyurethane* (PU), and poly(vinylidene chloride)[12] (PVDC). Morphological structures were evidenced for each of them by XRD/TEM techniques. For brevity, only the XRD/TEM of PAE, nano-PAE/GO and PVDC, nano-PVDC/GO are presented (Figures 1 to 4).

XRD analysis reveals the disappearance of the layered spacing of GOs ($d = 0.81$ nm at $2\theta = 10.9°$) in both PAE/GO(5%) and PVDC/GO(5%) nanocomposites (Figures 1 and 2). TEM images (Figures 3 and 4) confirm the exfoliated structure of the nanocomposites. PAE-10, -15, and -20 nanocomposites were prepared similarly as nano-PAE-5. Distinct peaks show up in XRD spectra, which verify the formation of nanostructures when adding a certain amount of GO.[12]

12.3.2 Fire Retardancy

LOI measurement – To evaluate the self-extinguishing power, the LOI for nano- and micro-PAE/GO composites was measured as a function of GO level

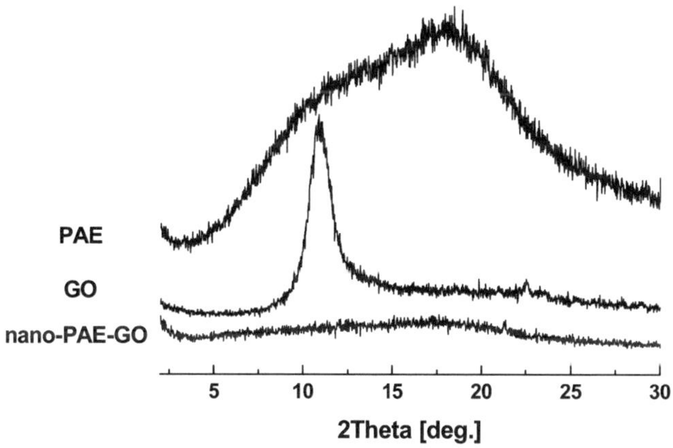

Figure 1 *XRD spectra of PAE, GO and nano-PAE/GO(5%) nanocomposite*

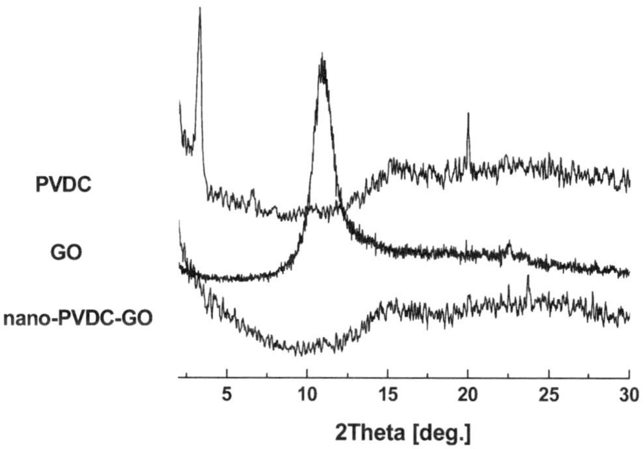

Figure 2 *XRD patterns of PVDC, GO and nano-PVDC/GO(5%) nanocomposite*

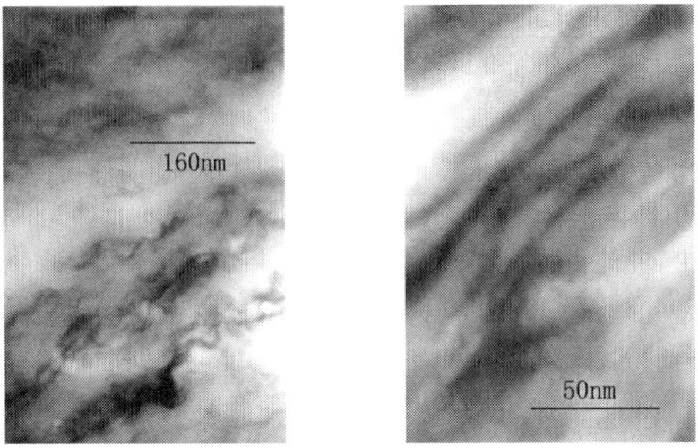

Figure 3 *TEM images of nano-PAE/GO(5%) nanocomposite*

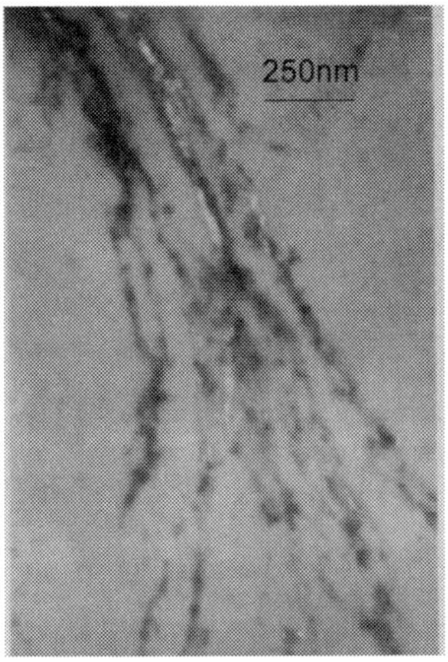

Figure 4 *TEM image of nano-PVDC/GO(5%) nanocomposite*

(Tables 1 and 2). Compared to micro-composite, the LOI of PAE/GO nanocomposites jumps from 18.8 to 24.6% (Tables 1 and 2). We will focus our subsequent discussion on this aspect.

The separation of ~6.0 unit induced here by nano-effect demonstrates a big gap when compared to the polymer–silicate nanocomposites. LOIs of traditional polymer–silicate nanocomposites usually remain unchanged compared to virgin polymers.[12] Our laboratory found that neither polymer layered silicate (*e.g.* MMT) nanocomposites nor nano-particulates (*e.g.* TiO_2)[13] have a really strong dependence of LOI on the level of nano-filler.

The difference in LOI between nano- and micro-composites, *i.e.* $(LOI)_{nano} - (LOI)_{micro}$, (Figure 5) as a function of GO %, can reasonably be attributable to

Table 1 *LOI for nano-PAE/GO composites as function of the GO level*

Sample	*GO* (%)	*LOI* (%)	Δ^a	Δ (GO%)^b
PAE	0	18.8	0	0
Nano-PAE-2	2	22.2	3.4	1.7
Nano-PAE-5	5	23.3	4.5	0.9
Nano-PAE-10	10	24.5	5.7	0.6
Nano-PAE-15	15	24.6	5.8	0.4
Nano-PAE-20	20	24.6	5.8	0.3

[a]LOI increment for nano-PAE/GO relative to PAE; [b]Δ divided by GO (%).

Table 2 *LOIs for micro-PAE/GO composites as function of GO level*

Sample	*GO* (%)	*LOI* (%)	Δ^a	Δ (GO%)^b
PAE	0	18.8	0	0
Micro-PAE-2	2	19.0	0.2	0.10
Micro-PAE-5	5	19.3	0.5	0.10
Micro-PAE-10	10	19.6	0.8	0.08
Micro-PAE-15	15	19.8	1.0	0.07
Micro-PAE-20	20	20.0	1.2	0.06

[a]LOI increment for nano-PAE/GO relative to PAE; [b]Δ divided by GO (%).

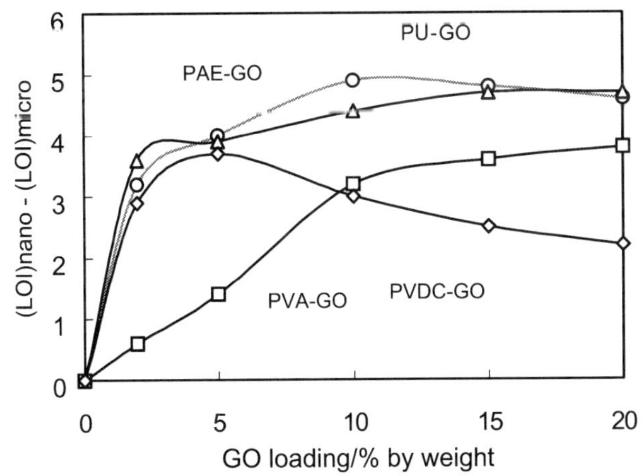

Figure 5 *[$(LOI)_{nano} - (LOI)_{micro}$] vs. GO level for systems of PAE/GO, PU/GO, PVA/GO, PVDC/GO*

the nano-effect. All GO-containing systems exhibit similar behaviour, except for PVDC/GO.

The nano-effect has also been verified by the horizontal burning rate (HBR) test (not shown here, for details see reference 12). It again shows that one can not necessarily predict the coherence among a variety of fire models that impact on the fire retardance of materials. Nevertheless, each fire model gives some information as well as a significance of their own.

Cone calorimetry – Calorimetry data for PAE/GO (5% mass fraction), as an example, are given in Figure 6 and Table 3.

Cone data (p-HRR, TTI, TSR and p-CO yield) shown in Figure 6 and Table 3 reveal that PAE/GO nano-composites have significantly better flammability properties than PAE/GO micro-composites and than the virgin PAE.

Regarding the influence of the nano-effect on the burning behaviour, cone data at higher heat fluxes (25 and 35 kW m^{-2}) were also recoded.[14] The nano-effect of PVDC/GO(5%) based on cone experiments is shown in Table 4.

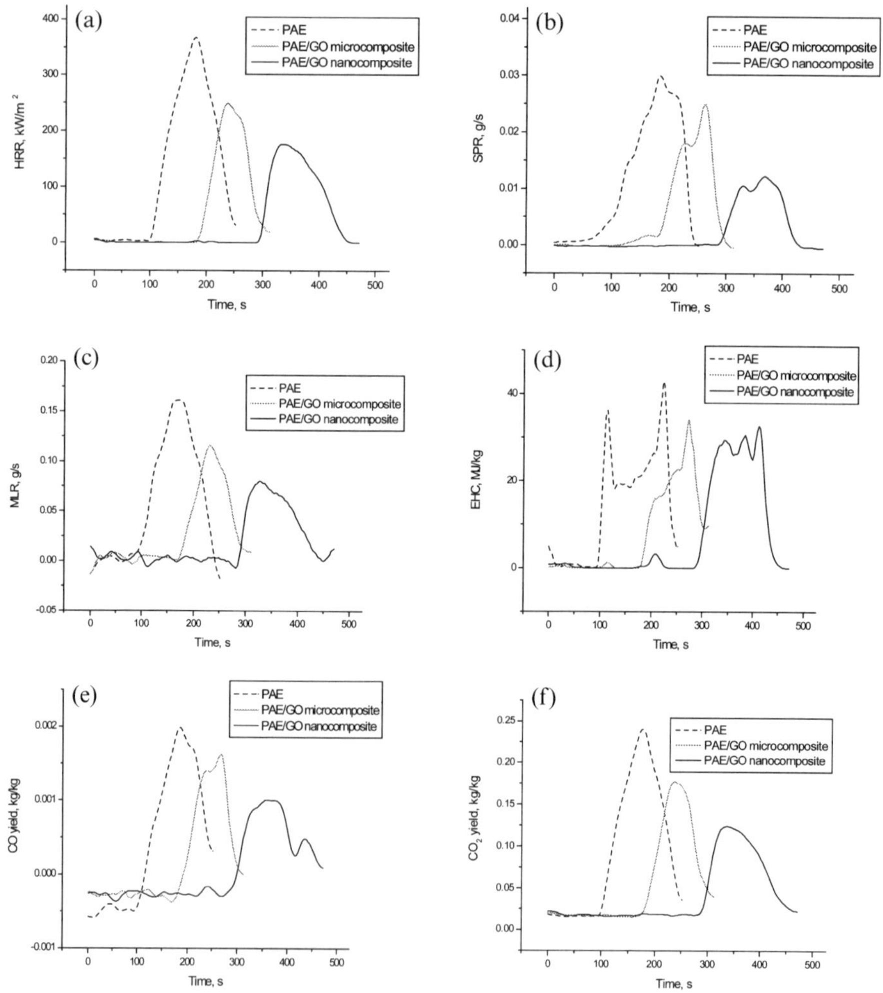

Figure 6 *Cone data of PAE/GO(5%) systems (15 kW m^{-2}):* (a) *Heat release rate;* (b) *smoke production rate;* (c) *mass loss rate;* (d) *efficient heat of combustion;* (e) *CO yield;* (f) *CO_2 yield (external heat flux 15 kW m^{-2})*

Table 3 *Cone data for systems of PAE/GO(5%) (15 kW m^{-2})*

Parameter[a]	*PAE*	*Micro-PAE/GO*	*Nano-PAE/GO*
p-HRR (kW m^{-2})	377	252 (–33.0%)	178 (–52.8%)
TTI (s)	109	181 (+66%)	293 (+168.8%)
TSR (m^2 m^{-2})	323.9	186.8 (–42.3%)	119.5 (–63.0%)
p-CO yield (kg kg^{-1})	0.15	0.12 (–20.0%)	0.05 (–66.7%)

[a]p-HRR: peak-heat release rate; TTI: time to ignition; TSR: total smoke release.

Table 4 *Nano-effect for systems of PVDC/GO (5%): cone data (15kW kW m^{-2})*

Parameter[a]	*Micro-PAE/GO*	*Nano-PAE/GO*
p-HRR (kW m^{-2})	42.6	16.6
TTI (s)	176	235
TSR (m^2 m^{-2})	54.9	16.6
p-CO yield (kg kg^{-1})	0.44	0.0

[a]p-HRR: peak-heat release rate; TTI: time to ignition; TSR: total smoke release.

12.3.3 Mechanistic Study (TGA/XPS)

XPS coupled with TGA has proved to be an instructive combination when studying the fire retardant mechanistic events in the condensed phase.[15] The specimens were prepared as thin films. The 'pseudo-*in situ*' protocol used in this work denotes that only one specimen at a fixed orientation was employed for absolute intensity (cps eV, counts per second eV) measurement from room temperature up to 500°C. Two systems (PAE/GO and PVDC/GO) were used here as models for discussion.

PAE-GO: TGA and XPS data are given in the Figures 7 and 8, and Table 5, respectively. They show that GO decomposes in three steps: the 1st step between

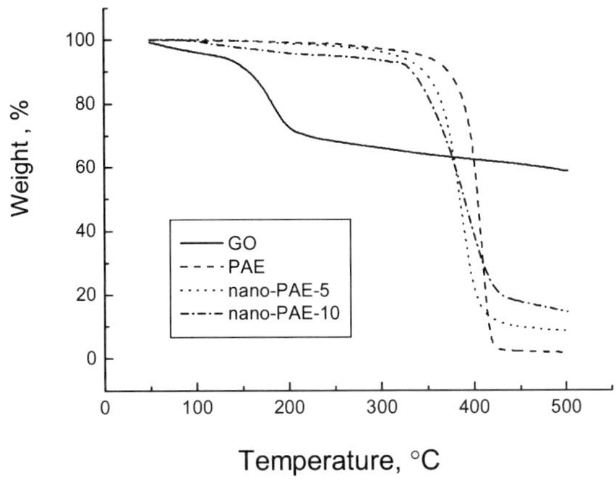

Figure 7 *TG curves of GO, PAE and nano-PAE/GO(5%) and nano-PAE/GO(10%) composites*

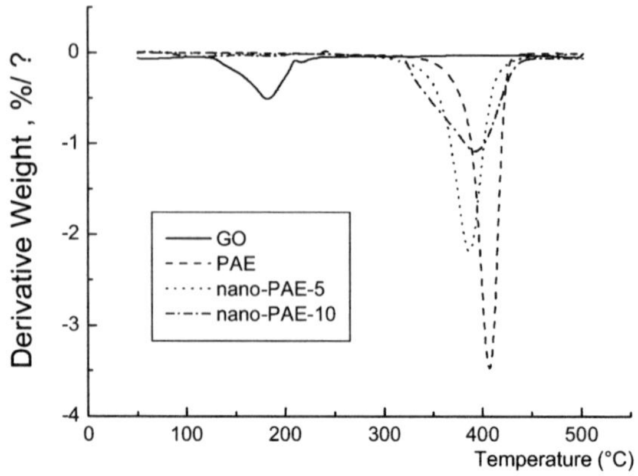

Figure 8 *DTG curves of GO, PAE and nano-PAE/GO(5%) and nano-PAE/GO(10%) composites*

Table 5 *TGA data for nano-PAE/GO composites*

Sample	GO (%)	T_5 (°C)	T_{10} (°C)	*Peak* T_{max} (°C)	*Peak* R_{max} (% °C^{-1})	*Residue at 500 °C* (%)
GO	–	123	155	182	0.51	58.9
PAE	0	341	372	407	3.46	1.8
nano-PAE-5	5	318	347	385	2.18	9.0
nano-PAE-10	10	252	334	392	1.05	15.5

20 and 130°C corresponding to the evolution of adsorbed H_2O, the 2nd step between 130 and 300°C corresponding to the loss of CO_2, CO, H_2O, and the 3rd step between 300 and 500°C corresponding to the loss of CO_2 and small molecules.[16] Consequently, the greater the level of GO incorporated, the earlier degradation of PAE started, resulting in the following points: (i) the onset temperatures (T_5 and T_{10}) decrease by 23 and 25°C for nano-PAE/GO(5%), and 89 and 38°C for nano-PAE/GO(10%) with respect to PAE, respectively, due to the early decomposition of GO; (ii) the degraded rate (R_{max}) for nano-PAE/GO(10%) is about thrice that of PAE; (iii) the char yield at 500°C jumps from 1.8% (PAE) to 15.5% [nano-PAE/GO(10%)].

Figure 9 provides the relative intensity in C1s spectra vs. temperature for both PAE and nano-PAE/GO. Previously,[17–23] we showed that these curves can be rationalized in terms of the relative extent of cross-linking as a function of temperature. The PAE curve indicates that cross-linking commences upon heating at 380°C. This change is reasonably consistent with TGA data. The curve of nano-PAE/GO(5%) composite always lies above the curve of PAE below 400°C, suggesting a higher extent of cross-linking for nano-PAE/GO, which is due to interactions between PAE and GO. Above 400°C the curve of the nanocomposite falls, probably because of the dominant oxygen at the surface (Figure 10).

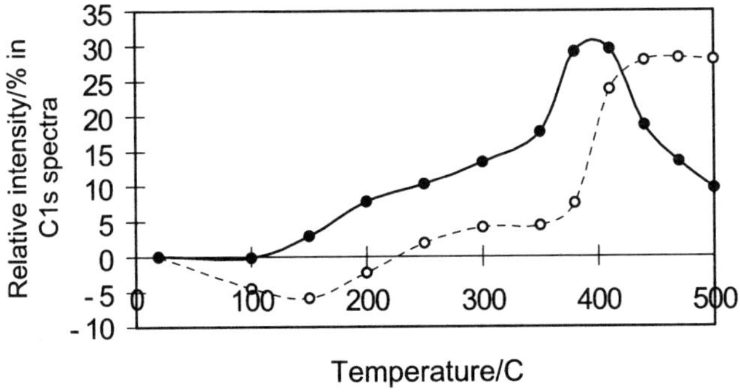

Figure 9 *Relative intensity (%) in C1s spectra for PAE and nano-PAE/GO(5%) vs. temperature (PAE: open circles; nano-PAE/GO: filled circles)*

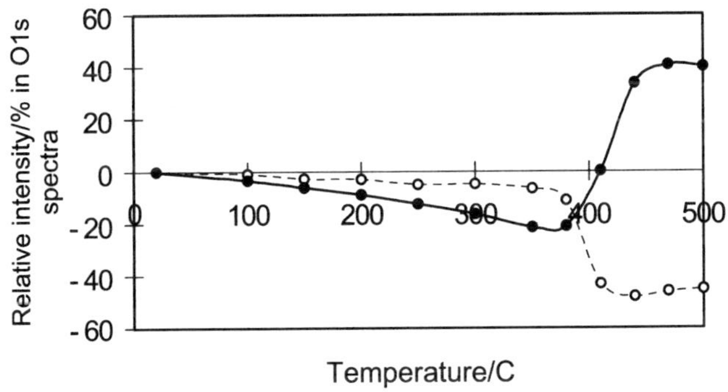

Figure 10 *Relative intensity (%) in O1s spectra for PAE and nano-PAE/GO(5%) vs. temperature (PAE: open circles; nano-PAE/GO: filled circles)*

The two curves in Figure 10 separate at about 380°C. The increase of oxygen content suggests the existence of GO at the surface, probably through two mechanisms: (i) decomposition of the polymer upon heating and/or (ii) migration of GO from bulk towards the surface in the polymer melt.

In other words, the mechanism for nano-PAE/GO in the condensed phase is more like that for nano-PS/MMT,[20] and nano-PMMA/MMT.[21] At the same time, the ratios of C:O are nearly constant before 380°C, and then gradually separates (Figure 11).

PVDC-GO – PVDC and PVDC/GO are halogen-containing systems, another model of interest. PVDC is a well-known char-forming polymer. All TGA data are listed in Figures 12 and 13 and Table 6 for comparison.

Some features can be drawn from Tables 5 (PAE/GO) and 6 (PVDC/GO):

(1) Lesser influence of GO on mass loss rate (R_{max}) and residual yield in nano-PVDC/GO(5%) than in nano-PAE/GO(5%), *i.e.* the nano-effect in PVDC/GO does function but is weaker than that in PAE/GO. The effectiveness may be offset by the voluminous char residue in PVDC/GO.

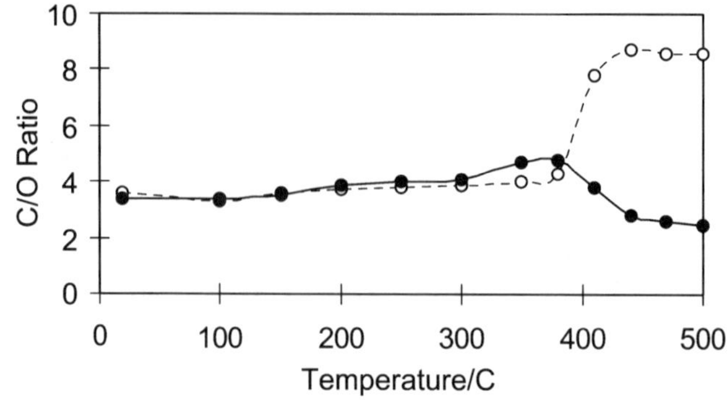

Figure 11 *C:O ratio for PAE and nano-PAE/GO(5%) vs. temperature (PAE: open circles; nano-PAE/GO: filled circles)*

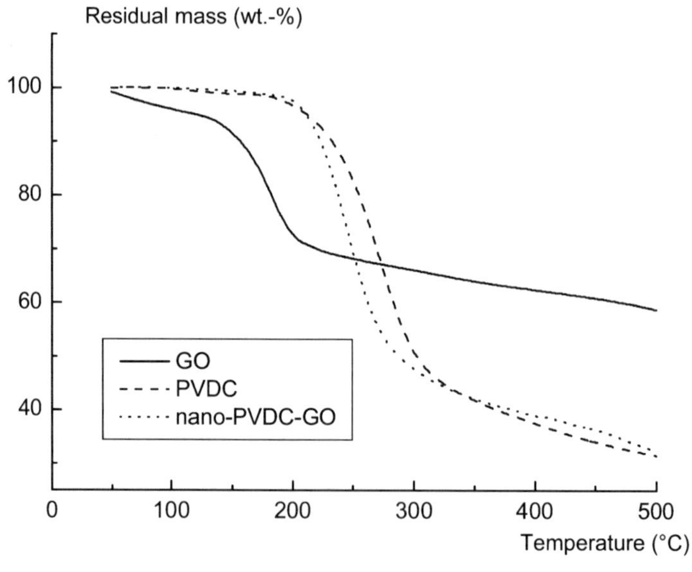

Figure 12 *TG curves of GO, PVDC and PVDC/GO(5%) nanocomposite*

Table 6 *TGA data for nano-PVDC/GO composites*

Sample	T_5	T_{10}	*Peak* T_{max} (°C)	*Peak* R_{max} (% °C^{-1})	*Residue at 500 °C* (%)
GO	123	155	182	0.51	58.9
PVDC	211	232	272	0.75	31.4
Nano-PVDC-GO-5	210	222	244	0.85	32.6

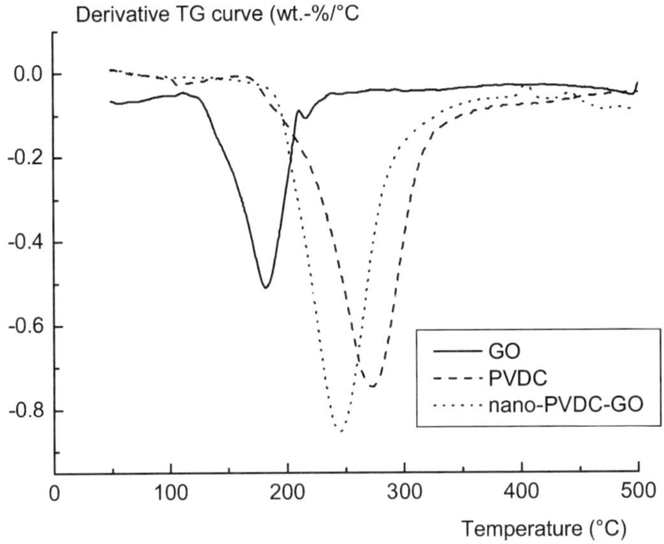

Figure 13 *DTG curves of GO, PVDC and PVDC/GO(5%) nanocomposite*

(2) Both nano-PVDC/GO and nano-PAE/GO composites decompose earlier than the virgin polymer by 28 and 22°C (T_{max}), respectively. This is associated with GO, which decomposes earlier than the virgin polymer.

Figure 14 (Cl2p spectra *vs.* temperature) reveals that the nano-PVDC/GO composite shows evidence of higher stability than pristine PVDC, when taking the chlorine atom as an indicator. This implies an interaction between PVDC and GO.

XPS data for PVDC and nano-PVDC/GO composite (Figures 15 and 16) is very like that for the PVC/MMT nanocomposite,[22] *i.e.* for char-forming polymers the surface on burning is dominated by carbon rather than oxygen, contrary to non-char-forming polymers like PS, PMMA, and PAE.

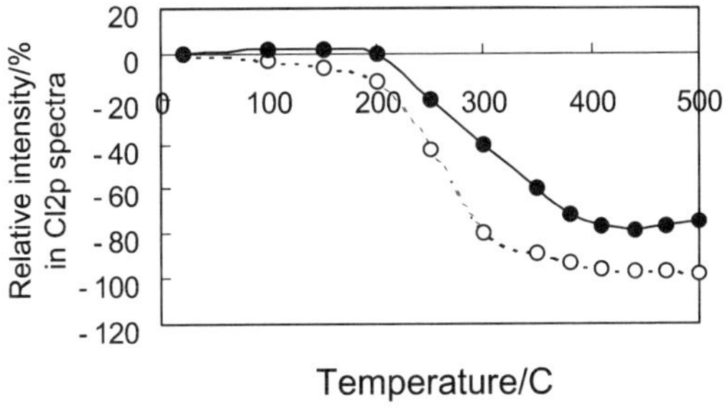

Figure 14 *Relative intensity in C1s spectra for PVDC (○) and nano-PVDC/GO(5%) (●) vs. temperature*

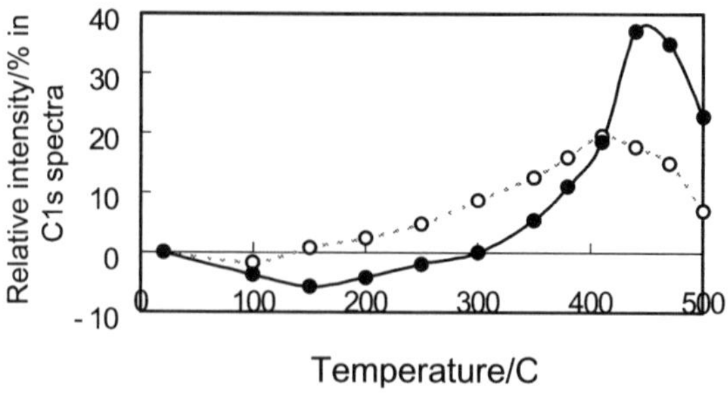

Figure 15 *Relative intensity in O1s spectra for PVDC (○) and nano-PVDC/GO(5%) (●) vs. temperature.*

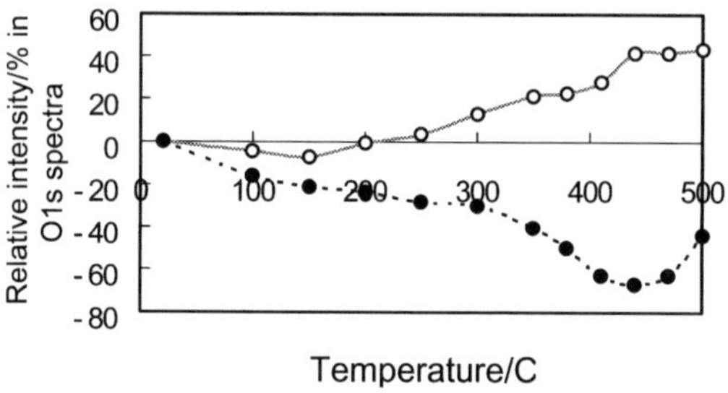

Figure 16 *Relative intensity in Cl2p spectra for PVDC (○) and nano-PVDC/GO(5%) (●) vs. temperature*

As in PVC/MMT, ionic species can be monitored for nano-PVDC/GO on heating (Figure 17). GO indeed accelerates the shift in charring of polymer PVDC to lower binding energies. For example, the emergence of ionic species (say, <199.0 eV) occurs at *ca.* 400°C for nano-state, very much earlier (or lower) than that of PVDC, which can hardly approach the value within this temperature range. The degradation of nano-PVDC/GO develops through two steps, *i.e.* the initial loss of HCl to form allylic species and the degradation of those allylic

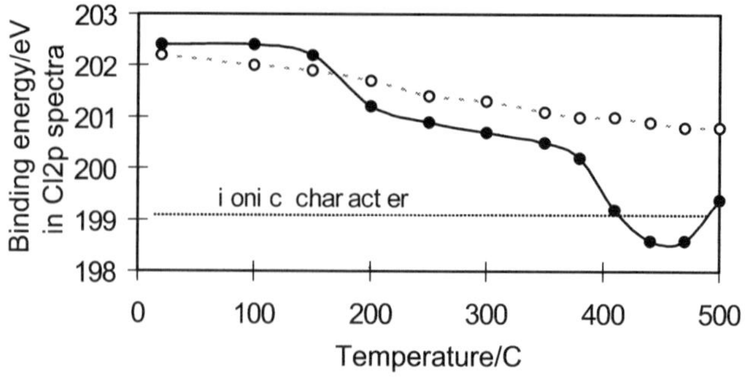

Figure 17 *Binding energy (eV) in Cl2p spectra for systems of PVDC (○) and nano-PVDC/GO(5%) (●) vs temperature (without charging subtraction)*

Table 7 *Comparison for systems of nano- and micro-PAE/GO and -PVDC/GO*

Sample	*Residue %* $TGA_{500°C}$ (CONE)	*O/C* XPS 500°C	*LOI* (%)	*LOI* (Δ)	*p-HRR* CONE (kW m^{-2})	*TTI* CONE (s)
PAE/GO composite						
PAE	1.8(8.9)	0.12	18.8	0.0	377	109
Micro-PAE/GO(5%)	10.8(11.8)	0.43	19.6	0.8	252 (−33.0%)	181 (+66.0%)
Nano-PAE/GO(5%)	15.5(13.5)	0.56	24.5	5.7	178 (−52.8%)	293 (+168.8%)
PVDC/GO composite						
PVDC	31.4(–)[a]	0.48	20.5	0.0	–[a]	–[a]
Micro-PVDC/GO(5%)	32.6(38.4)	–	20.2	−0.3	42.6	176
Nano-PVDC/GO(5%)	32.6(32.6)	0.13	23.9	3.4	16.6	235
Graphite oxide (GO)						
GO	58.9(–)	0.77	–	–	–	–

[a]Not measured.

species. In the first step, only covalent chlorine is lost while in the second both ionic (allylic) and covalent chlorine are lost.[22] Most data on micro- and nano-PAE/GO and -PVDC/GO are summarized in Table 7.

The C1s signal measured at 500°C (Figure 18) seems to offer a likely resemblance between GO and nano-PAE/GO. The nano-PAE/GO looks more like that of GO rather than micro-PAE/GO. In other words, the barrier yielded at the surface at high temperature consists mainly of GO-like species. Figures 19 and

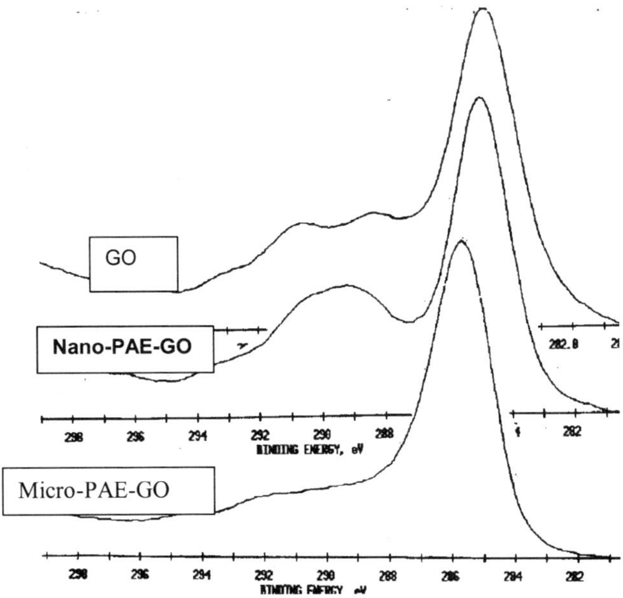

Figure 18 *Comparative C1s spectra of GO, nano-PAE/GO(5%), and micro-PAE/GO(5%) at 500 °C*

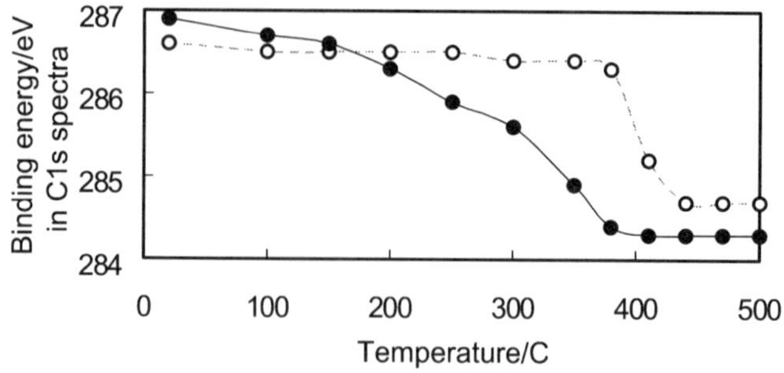

Figure 19 *Binding energy (eV) in C1s spectra for systems of PAE (○) and nano-PAE/GO(5%) (●) vs. temperature*

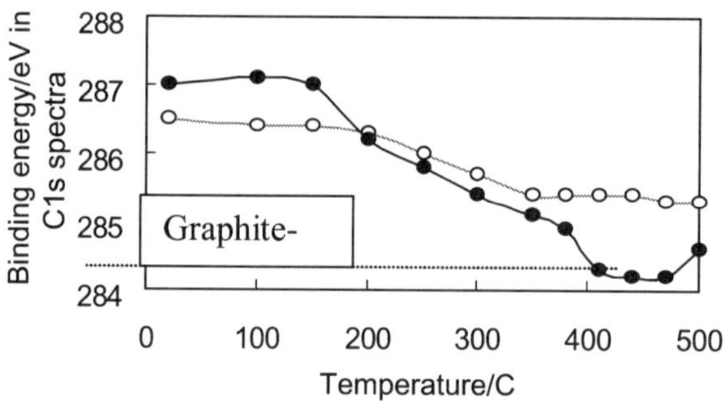

Figure 20 *Binding energy (eV) in C1s spectra for systems of PVDC (○) and nano-PVDC/GO(5%) (●) vs. temperature (without charging substraction)*

20 offer further critical proof that the nano-effect exerts a critical function in forming graphitic structures (~284.3 eV) whatever the PAE or PVDC. Consequently, one may obtain a toughened barrier dominated with a GO-like structure *via* covalent bonding between the polymeric matrix and functional groups attached to GO.

12.4 Conclusions

The use of graphite oxide (GO) as a nanofiller incorporated into polymers (PAE, PVA, PU and PVDC) enables the making of nanocomposites. The finding is that the nano-effect, as revealed by LOI measurement, works pretty well in composites including PAE-, PVA-, PU- and PVDC-/GO. These excellent fire properties are also confirmed by cone calorimetry.

Notably, the systems of nano-polymer/GO composites give promising results based on LOI and cone data in contrast to polymer-layered silicate (*e.g.* MMT) nanocomposites and nano-particulates (*e.g.* TiO_2).

The survival and renascence of graphite-like (sp^2) structure (~284.3 eV in XPS) initiated by GO at 500°C offers a critical proof that a toughened barrier is formed, dominated by a GO-like structure *via* covalent bonding between the polymeric matrix and functional groups attached to the surface of GO.

The nano-effect seems to be partially offset by halogen, which usually plays its role in the gas phase. This may be due to the counter-balanced or compensated influence of the voluminous char residue of PVDC/GO (32.6 wt% at 500°C) *vs.* PAE/GO (9.0 wt% at 500°C).

Our study offers further critical proof, *i.e.* the nano-effect exerts a critical function in forming graphitic structure (~284.3 eV) at high temperature, whatever PAE or PVDC is concerned. A toughened barrier dominated by a GO-like structure *via* covalent bonding between the polymeric matrix and functional groups attached to GO is, presumably, formed on the surface.

12.5 References

1. Y. Kojima, A. Usuki, M. Kawasumi, A. Okada, Y. Fukushima, T. Kurauchi and O. Kamigaito, *J. Mater. Res.*, 1993, **8**, 1185.
2. J.W. Gilman, T. Kashiwagi, E.P. Giannelis, E. Manias, S. Lomakin, J.D. Lichtenhan and P. Jones, in *Fire Retardancy of Polymers: The Use of Intumescence*, M. Le Bras, G. Camino, S. Bourbigot and R. Delobel (eds.), The Royal, Society of Chemistry, Cambridge, 1998; J.W. Gilman, *Appl. Clay Sci.*, 1999, **15**, 31–49.
3. T. Kashiwagi, F. Grulke, J. Hilding, R. Harris, W. Awad and J. Douglas, *Macromol. Rapid Commun.*, 2002, **23**, 761–765.
4. W.S. Hummers, Jr and R.E. Offeman, *J. Am. Chem. Soc.*, 1958, **80**, 1339.
5. Z. Han and J. Wang, *Chin. J. Inorg. Chem.*, 2003, **19(5)**, 459–461.
6. Z. Han and J. Wang, *Chin. J. Inorg. Chem.*, 2003, **19(12)**, 1366–1370.
7. Y. Matsuo, K. Tahara and Y. Sugie, *Carbon*, 1996, **34**, 672.
8. Y. Matsuo, K. Tahara and Y. Sugie, *Carbon*, 1997, **35**, 113.
9. Y. Matsuo, K. Hatase and Y. Sugie, *Chem. Mater.*, 1998, **10**, 2266–2269.
10. K.E. Strawhecker and E. Manias, *Chem. Mater.*, 2000, **12**, 2943.
11. J. Xu, Y. Hu, L. Song, Q. Wang, W. Fan and Z. Chen, *Carbon*, 2002, **40**, 445–467.
12. Z. Han, Doctorial Thesis, *A Comparative Study on Flame Retardant Properties of Polymer/Graphite Oxide Nano-systems*, Beijing Institute of Technology, 2003.
13. X. Mo, Doctorial Thesis, *Study on Nano-effect in Intumescent Flame Retardant Coatings*, Beijing Institute of Technology, 2003.
14. Z. Han and J. Wang, *The 2nd International Symposium on Engineering Plastics (EP'2004)*, Lanzhou, Gansu province, China, August 15–20, 2004.
15. J. Wang, in *Fire Retardancy of Polymers, The Use of Intumescence*, Ed. by M. Le Bras, G. Camino, S. Bourbigot and R. Delobel, The Royal Society of Chemistry, Cambridge, 1998, pp. 154–172.
16. J. Wang and Z. Han, *228th ACS meeting 2004*, Philadelphia, 22–26 August 2004.

17. J. Hao, S. Wu, C.A. Wilkie and J. Wang, *Polym. Degrad. Stab.*, 1999, **66**, 81–86.
18. J. Hao, C.A. Wilkie and J. Wang, *Polym. Degrad. Stab.*, 1999, **71**, 305–315.
19. J. Wang, J. Du, H. Yao and C.A. Wilkie, *Polym. Degrad. Stab.*, 2001, **74**, 321–326.
20. J. Wang, J. Du, J. Zhu and C.A. Wilkie, *Polym. Degrad. Stab.*, 2002, **77**, 249–52.
21. J. Du, J. Zhu, C.A. Wilkie and J. Wang, *Polym. Degrad. Stab.*, 2002, **77**, 377–381.
22. J. Du, D. Wang, C.A. Wilkie and J. Wang, *Polym. Degrad. Stab.*, 2003, **79**, 319–324.
23. J. Du, J. Wang, S. Su and C.A. Wilkie, *Polym. Degrad. Stab.*, 2004, **83**, 29–34.

CHAPTER 13

Styrene–Acrylonitrile Copolymer Montmorillonite Nanocomposite: Processing, Characterization and Flammability

JEFFREY W. GILMAN,[1] SERGE BOURBIGOT,[2] SÉVERINE BELLAYER,[1] HOLLY STRETZ[3] AND DONALD R. PAUL[3]

[1]Fire Science Division, BFRL, National Institute of Standards and Technology, Gaithersburg, MD-20899, U.S.A.
[2]PERF, UPRES EA 1040, École Nationale Supérieure de Chimie de Lille (ENSCL), F-59650 Villeneuve d'Ascq, France (Guest Researcher at Fire Science Division, BFRL, NIST, Gaithersburg, MD-20899, U.S.A.
[3]The University of Texas, Austin, TX-78712, U.S.A.

13.1 Introduction

Interest in polymer clay nanocomposites has increased significantly in recent years. Property improvements include better mechanical properties, better barrier properties, lower water absorption and reduced flammability.[1–6] To achieve these properties, mica-type layered silicates, such as montmorillonite (MMT), are generally dispersed at the nanoscale in the polymer to yield the so-called "nanocomposite".

The pioneering work of Gilman *et al.* demonstrated that the presence of montmorillonite clay produces a substantial improvement in fire performance in polymeric matrices such as polystyrene and polyamides.[5–7] The purpose of this work is to investigate the melt-processing of styrene-acrylonitrile copolymer (SAN) with MMT clays and the influence of the clay on mechanical properties and on fire performance of SAN. Our work is included in a larger project aimed at understanding the flame retardancy of ABS. Computer housings are typically made from acrylonitrile-butadiene-styrene (ABS) polymer and related blends, and these polymers must meet safety standards. The SAN polymer discussed here is a model system for the study of ABS/clay systems, since the multi-phase ABS consists of an SAN matrix embedded with rubber particles. Our concern here is to simplify the system to study the nanodispersion of the clay in SAN. In addition, we will discuss in future publications how the clay dispersion and organoclay compatibilizer affect flammability in SAN and ABS.

The first part of this chapter is devoted to the characterization by X-ray diffraction (XRD), transmission electronic microscopy (TEM) and solid-state nuclear magnetic resonance (NMR) of the nanocomposites. These characterization methods are complementary, but it should be pointed out that solid state NMR characterization has not been previously reported for SAN/MMT, and that the advantage of NMR characterization of nanocomposites is that the sample size probed is much larger (and therefore more representative of a bulk dispersion) than either XRD or TEM. Together, these methods will provide an overview of the dispersion of the clay in SAN. Then, tensile properties and flammability performance evaluated by mass loss calorimetry will be examined and discussed.

13.2 Experimental[a]

13.2.1 Preparation of Nanocomposites

MMT originated from Southern Clay Products, Inc (Gonzales, TX, USA). The starting material, sodium-MMT, was commercially modified using methyl, tallow, bis-2-hydroxyethyl, quaternary ammonium chloride (Cloisite 30B). SAN copolymer, containing 25 mass% acrylonitrile (SAN-25) and formulated for computer housing, was supplied by Dow Chemical.

Polymer melt-direct intercalation is an approach to making polymer layered silicate nanocomposites by using conventional polymer extrusion. The optimized processing temperature for best dispersion was determined to be 220°C.[8] SAN-25 was melt-mixed with the clay using a co-rotating twin-screw extruder (Haake, screw L/D = 10, diameter = 35 mm) at 280 rpm. These conditions on this extruder produce fully exfoliated nylon-6/montmorillonite nanocomposites.[9] The clay loadings (Cloisite 30B) were 2.2, 4.2, 6.2 and 7.9 wt% true silicates. The true silicate concentration is determined by heating the dry, post-extrusion pellets in a muffle furnace to 900°C for 45 min, and correcting for structural rearrangement by dividing the ash percentage by 0.935 (the silicate

[a] This work was carried out by the National Institute of Standards and Technology (NIST), an agency of the U.S. government and by statute is not subject to copyright in the United States. Certain commercial equipment, instruments, materials or companies are identified in this paper in order to adequately specify the experimental procedure. This in no way implies endorsement or recommendation by NIST. The policy of NIST is to use metric units of measurement in all its publications, and to provide statements of uncertainty for all original measurements. In this document, however, data from organizations outside NIST are shown, which may include measurements in non-metric units or measurements without uncertainty statements.

rearrangement results in 6.5% loss of structural water). Organoclay content can then be calculated from the true silicate content as reported here by using:

$$\text{organoclay (\%)} = \frac{0.935 \times \text{MMT}}{\left[1 - \text{LoI}/100\right]}$$

where LoI is the loss on ignition for the dry organoclay (30% for Cloisite 30B), and MMT is the true silicate mass percentage.

13.2.2 NMR Spectroscopy

Measurements were conducted using a Bruker Avance 300 spectrometer (Bruker BioSpin Corp., Billerica, MA) operating at 7.05 T. Proton spectra at 300 MHz were obtained using a 5 mm low proton-background CRAMPS[10] (combined rotation and magic-angle spinning) probe (Doty Scientific of Columbia, SC).

Even though we recognized the strong effect of oxygen (it will be commented in the following),[11] proton longitudinal relaxation times (T_1^H) of oxygen containing samples were obtained using the inversion–recovery sequence with direct proton observation[12] in a ZrO_2 rotor. For purposes of time efficiency, only the delay time, τ_{null}, was determined. The latter is the delay time where, after inversion, magnetization passes through zero on its way back to the Boltzmann equilibrium level. From τ_{null}, a lower limit for T_1^H was calculated *via* the relationship $T_1^H = \tau_{null}$ /ln2. This relationship assumes full initial inversion of the magnetization and single-exponential recovery. The paramagnetic contribution to T_1^H originating from the MMT clay (paramagnetic Fe^{3+} is embedded in the aluminosilicate layers of the MMT) normally produces a slightly accelerated early decay relative to the typically exponential behavior seen at longer times; hence, this relationship systematically yields a lower limit to the T_1^H than would describe this longer time behavior. In any case, all samples had equilibrated with O_2 and had aged for at least one month. Standard uncertainties for τ_{null} measurements are ±2.5% of the given value.

13.2.3 Transmission Electron Microscopy

All samples were ultra-microtomed with a diamond knife on a Leica Ultracut UCT microtome to give sections with a nominal thickness of 70 nm. The sections were transferred from water (room temperature) to Cu grids of 400 mesh. Bright-field TEM images of nanocomposites were obtained at 120 kV under low-dose conditions with a Philips 400T electron microscope, using Kodak SO-163 film. Low-magnification images were taken 22,000×. High-magnification images were taken at 100,000×. The materials were sampled by taking several images of various magnifications over 2–3 sections per grid to ensure that analysis was based on a representative region of the sample.

13.2.4 Tensile Properties

Injection molding of tensile test bars (ASTM D638) was performed on an Arburg Allrounder 305-210-700 injection molding machine. Test specimens were molded at a barrel temperature of 260°C, and a mold temperature of 80°C. Mechanical testing was performed on an Instron model 1137 at a strain rate of 0.51 cm min^{-1}, and moduli data were evaluated using an extensometer with a 2.5 cm gauge length. Standard deviations for 5 or more test bars were under 10% of the mean for moduli. Standard deviations on the tensile strength at break and strain at break increased with clay percentage up to a maximum of 15% of the mean.

13.2.5 Cone Calorimetry by Mass Loss Calorimeter

FTT (Fire Testing Technology) mass loss calorimeter was used to carry out measurements on samples following the procedure defined in ASTM E 906. The equipment is identical to that used in oxygen consumption cone calorimetry (ASTM E-1354-90), except that a thermopile in the chimney is used to obtain heat release rate (HRR) rather than employing the oxygen consumption principle. Mass loss readings are performed simultaneously by ASTM E-1354, and serve as a benchmark of the heat release rate values obtained in this manner. ASTM E 906 is a screening method, and results in an internally representative set of data. For comparison, fully scaleable peak heat release rates from ASTM E-1354 for non-FR ABS grades range from 800[13] to 1000 kW m^{-2} [14] at an external heat flux of 35 kW m^{-2}. The procedure involved exposing specimens measuring 100 × 100 × 3 mm in horizontal orientation. An external heat flux of 35 kW m^{-2} was used for running the experiments. This flux corresponds to common heat flux in mild fire scenario.[15,16] The MLC calorimeter was used to determine heat release rate (HRR). When measured at 35 kW m^{-2}, HRR is reproducible to within ±10%. The cone data reported here the average of three replicated experiments.

13.3 Results and Discussion

13.3.1 Characterization by XRD and TEM

TEM images of SAN-25 MMT nanocomposites containing 2.2, 4.2, 6.2 and 7.9 wt% of true silicate are shown at low magnification (Figures 1a to 4a) and at higher magnification (Figure 1b to 4b).

At low magnification, the images reveal that clay is well and evenly dispersed in all nanocomposites. At higher magnification, all samples exhibit dispersion of multi-layer stacks (tactoids) of individual MMT layers that range from 2 to 10 layers in size. Some individual MMT layers can be also distinguished.

The samples are not all similar; the nanostructure of nanocomposite (size of the tactoids) appears qualitatively to depend on the loading in organo-clay although a quantitative analysis of the distribution has not yet been attempted. We can observe that, as the clay loading is increased, larger tactoids seem to become more prevalent. This trend has been suggested in the literature,[17] *i.e.*

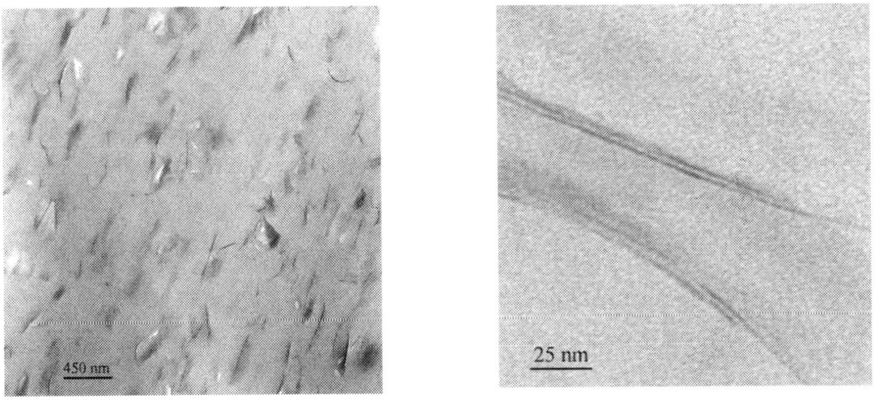

(a) (b)

Figure 1 *TEM images at low* (a) *and high* (b) *magnification of SAN-25 MMT nanocomposites at 2.2 wt% silicate*

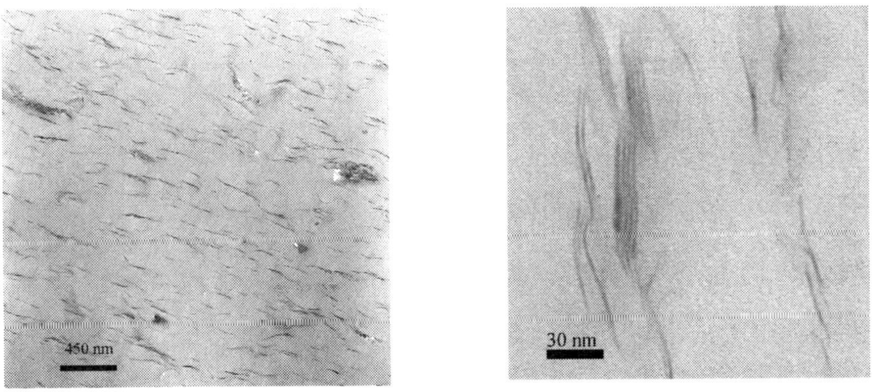

(a) (b)

Figure 2 *TEM images at low* (a) *and high* (b) *magnification of SAN-25 MMT nanocomposites at 4.2 wt% silicate*

relatively high loading in clay favors the agglomeration of MMT platelets. In addition to this, XRD spectra (not shown)[8] reveal diffraction peaks in the low-angle region giving a *d*-spacing falling between 3.1 and 3.25 nm. The *d*-spacing of the pure organo-clay is 1.75 nm and this result suggests that at least a part of the nanocomposite has an intercalated structure. From TEM and XRD, we may conclude that the nanocomposites have a mixed intercalated/exfoliated structure regardless of the loading.

13.3.2 T_1^H of Nanocomposite

An approach has been developed at NIST by VanderHart *et al.* using solid-state NMR to characterize the nanodispersion of clay in polymers.[18–20] The method is

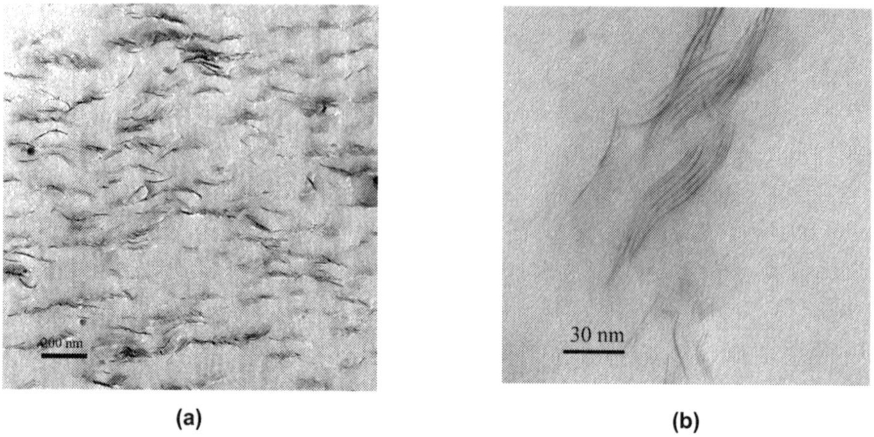

Figure 3 *TEM images at low (a) and high* (b) *magnification of SAN-25 MMT nanocomposites at 6.2 wt% silicate*

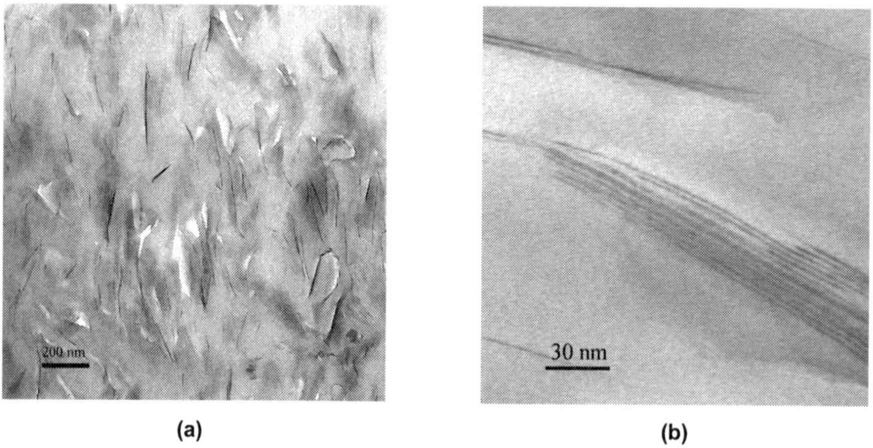

Figure 4 *TEM images at low (a) and high* (b) *magnification of SAN-25 MMT nanocomposites at 7.9 wt% silicate*

based on T_1^H measurement. It utilizes two effects: (1) the paramagnetic character of this MMT, which directly reduces the T_1^H of nearby protons, and (2) spin diffusion, whereby this locally enhanced relaxation propagates to more distant protons. As reported,[11,21,22] the adsorption of paramagnetic oxygen on aromatic polymers causes a major shortening of T_1^H. This was shown previously for polystyrene but we suspect the same effect for SAN polymer because of the presence of phenyl ring in the repeat unit of this polymer. The amount of adsorbed oxygen is a function of the chemical nature of the polymer, its molecular packing, molecular motion and temperature. Thus, in addition to the sensitivity of T_1^H to clay dispersion, any other changes that might affect oxygen solubility or dynamics could also influence T_1^H. The purpose here is to use the shortening of T_1^H measured in SAN-25/MMT nanocomposites to get an estimation of the

Table 1 *NMR, XRD and TEM characterization of oxygenated SAN-25/MMT nanocomposites*

Polymer/MMT nanocomposite	*OrganoMMT concentration* (wt%)	*True silicate concentration* (wt%)	T_1^H ($\tau_{null}/\ln 2$) (s)	*XRD/TEM conclusions*
SAN-25	0	0	1.77	–
SAN-25/3	3.0	2.2	1.14	Well and evenly dispersed, exfoliated/intercalated with small tactoids (2–3 layers)
SAN-25/6	5.6	4.2	0.84	Well and evenly dispersed, exfoliated/intercalated with small tactoids (3–5 layers)
SAN-25/8	8.3	6.2	0.63	Well and evenly dispersed, exfoliated/intercalated with small tactoids (3–8 layers)
SAN-25/11	10.6	7.9	0.54	Well and evenly dispersed, exfoliated/intercalated with small tactoids (3–10 layers)

nanodispersion of the clay in the polymer. In a future paper,[23] we will discuss the effect of oxygen and propose a quantitative approach for determining the nanodispersion using solid-state NMR.

Table 1 shows T_1^H data, obtained from τ_{null}, for our oxygen containing SAN-25/MMT nanocomposites. T_1^Hs of samples are shortened by the paramagnetic character of the clay.

All T_1^H data of SAN-25/MMT nanocomposites fall between 1.14 and 0.54 s (T_1^H of the pure SAN-25 is 1.77 s). T_1^H decreases with increasing MMT concentration for all nanocomposites, and depends on both the MMT concentration and the quality of the dispersion. It decreases noticeably for each sample compared to the pure SAN-25 and, as expected, the higher the MMT concentration, the shorter T_1^H is. By analogy with our previous work with nylon-6/clay nanocomposites[20] and PS/MMT nanocomposites,[11] the T_1^Hs of our samples suggest a modestly good dispersion of the clay. Here, a good correlation is found between T_1^Hs, and XRD and TEM conclusions. However, such a result should be considered only if secondary issues such as the presence of residual monomer do not modify the solubility of oxygen. Indeed, effects like residual monomer, large amounts of organo-modifier, or even small amounts, could potentially lead to an inappropriate conclusion about the dispersion of the clay if one does not properly take account of such phenomena. Nevertheless, we checked the Bloch spectra (not shown), and no degradation of SAN-25 or of the organo-modifier of the clay could be detected, *i.e.* there was no NMR evidence of additional mobile species present in our samples.

13.3.3 Tensile Properties

Figure 5 shows the tensile properties of SAN-25/MMT nanocomposites. Young's modulus clearly increases as a function of the MMT concentration but,

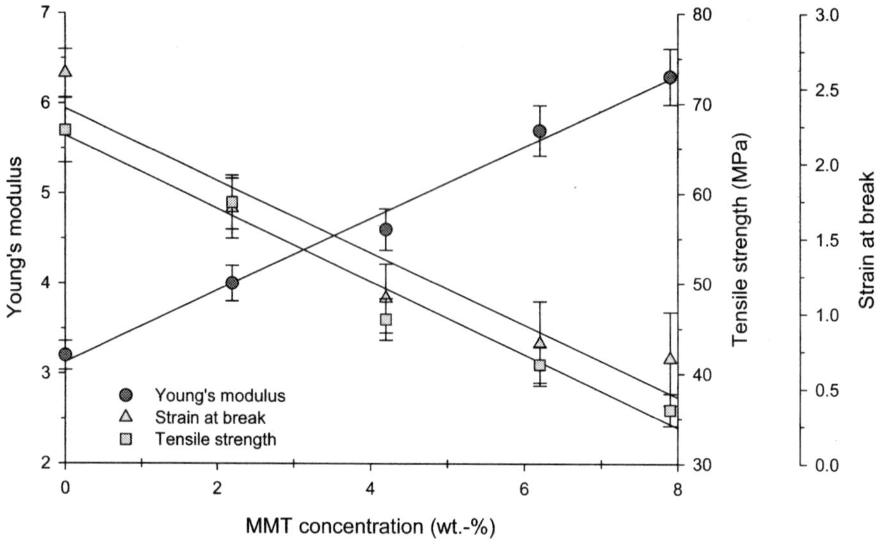

Figure 5 *Young's modulus, tensile strength and strain at break vs. MMT concentration of SAN-25/MMT nanocomposites*

concurrently, tensile strength and strain at break decrease. Young's modulus of SAN-25/MMT nanocomposites is substantially superior to that of neat SAN-25 due to the reinforcement by the clay platelets. The modulus exhibits a linear behavior *versus* the MMT content in the polymer. Surprisingly, tensile strength drops with higher clay loading. Most literature reports show an increase in strength at yield with clay loading,[14] but the SAN/MMT samples broke in a brittle manner before yield, and comparison to a ductile matrix is then not valid in terms of the strain regime. Both tensile strength and strain at break show a linear decrease at low clay loadings, suggesting the presence of small agglomerates in the polymer, as reported elsewhere.[17] This dependence appears to plateau at higher loadings, but the standard deviation also increases with loading, making the trend difficult to evaluate. The nature of the variation of tensile strength with loading for nanocomposites with a brittle matrix (*i.e.* SAN) is not completely understood. All nanocomposites exhibit an intercalated structure with some exfoliation and tactoids of small size. We also observed a linear dependence of the tensile modulus properties *versus* clay content for the range of loading tested. We have, therefore, a "filler effect" increasing the stiffness but decreasing the tensile strength of the nanocomposites.

13.3.4 Flammability Properties

Peak of heat release rate (PHRR) of polystyrene is reduced upon addition of clay;[24,25] however, to our knowledge, no published data are available for melt-processed SAN nanocomposites. Figure 6 shows significant reductions in PHRR in SAN-25 as a function of clay content. This reduction is strongly enhanced at higher clay loading. As an example, PHRR is decreased by 36%

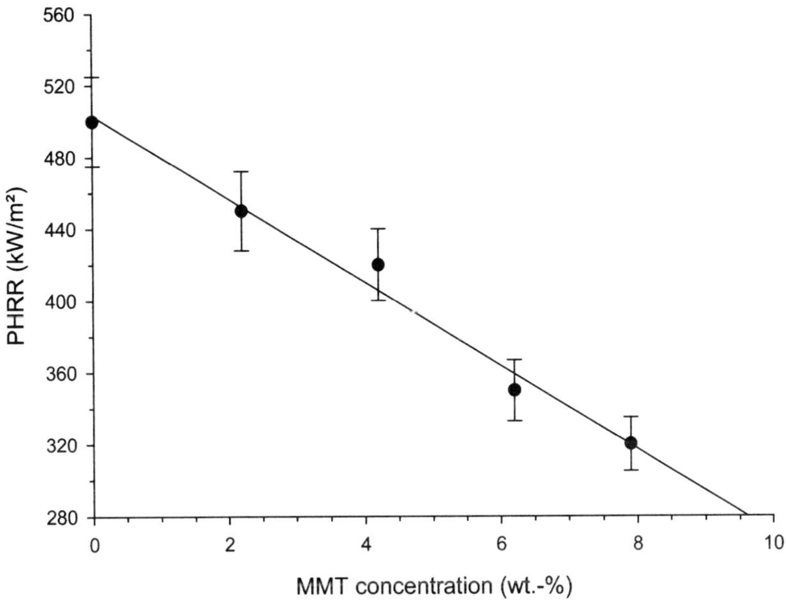

Figure 6 *Peak of heat release rate (PHRR) vs. MMT concentration of SAN-25/MMT nanocomposites*

at 4.6 wt% in MMT concentration. The suggested mechanism by which clay nanocomposites function involves the formation of a surface layer that serves as a potential barrier to both mass and energy transport.[24] The higher the initial concentration of clay, the thicker (and more insulating) is the ceramic-like layer that forms at the decomposing surface of the burning sample, with due consideration for homogeneity and integrity of the forming layer.

13.4 Conclusion

This work has characterized the dispersion of montmorillonite clay in SAN-25 and evaluated the mechanical properties and the reaction to fire of SAN-25/MMT nanocomposites. The nanodispersion has been characterized by XRD, TEM, and solid-state NMR. SAN-25/MMT nanocomposites reveal an intermediate morphology, *i.e.* an intercalated structure with some exfoliation and with the presence of small tactoids whatever the loading in MMT is. The presence of clay in SAN-25 leads to a "filler effect" that increases the stiffness but decreases the tensile strength of the nanocomposites. It also leads to a significant decrease of peak of heat release rate (up to 36%) as measured by mass loss calorimetry.

13.5 References

1. M. Alexandre and P. Dubois, *Mat. Sci. Eng., R.* 2000, **28(1–2)**, 1–63.
2. Y. Kojima, A. Usuki, M. Kawasumi, A. Okada, Y. Fukushima, T. Kurauchi and O. Kamigaito, *J. Mater. Res.*, 1993, **8(5)**, 1185–1189.

3. A. Okada, Y. Fukushima, M. Kawasumi, S. Inagaki, A. Usuki, S. Sugiyama, T. Kurauchi and O. Kamigaito, *US Patent 4,739,007*, 1988.
4. P.B. Messersmith and E.P. Giannelis,. *J. Polym. Sci. A., Polym. Chem.*, 1995, **33(7)**, 1047–1057.
5. J.W. Gilman, T. Kashiwagi and J.D. Lichtenhan, *SAMPE J.*, 1997, **33**, 40–46.
6. J.W. Gilman, *Appl. Clay Sci.*, 1999, **15(1–2)**, 31–49.
7. J.W. Gilman, T. Kashiwagi, E.P. Giannelis, E. Manias, S. Lomakin, J.D. Lichtenhan and P. Jones, in *Fire Retardancy of Polymers – The Use of Intumescence*, M. Le Bras, G. Camino, S. Bourbigot, and R. Delobel (eds.), The Royal Society of Chemistry, Cambridge, 1998, pp. 223–235.
8. H. Stretz, *Preliminary Proposal 2002*, University of Texas, Austin.
9. T.Fornes, P.J. Yoon, H. Keskkula and D.R. Paul, *Polymer*, 2001, **42**, 9929–9940.
10. L.M. Ryan, R.E. Taylor, A.J. Paff and B.C. Gerstein, *J. Chem. Phys.*, 1980, **72**, 508–515.
11. S. Bourbigot, D.L. VanderHart, J.W. Gilman, W.H. Awad, R.D. Davis, A.B. Morgan and C.A. Wilkie, *J. Polym. Sci. B, Polym. Phys.*, in press.
12. T.C. Farrar and E.D. Becker, in *Pulse and Fourier Transform NMR*, Academic Press, New York, 1971, pp. 20f.
13. A. Grand, in *Heat Release Calorimetry Evaluation of Fire Retardant Polymer Systems*, 42nd Intl. SAMPE Proceedings, 1997, Volume 5.
14. M.J. Scudmore, P.J. Briggs and F.H. Prager, *Fire Mater.*, 1991, **15**, 65–84.
15. V. Babrauskas, *Development of Cone Calorimeter – A bench scale rate of heat release based on oxygen consumption*, NBS-IR 82-2611, US Nat. Bur. Stand., Gaithersburg, 1982.
16. V. Babrauskas, *Fire Mater.*, 1984, **8(2)**, 81.
17. J.-H. Chang, Y.U. An and G.S. Sur, *J Polym. Sci, Part B: Polym. Phys.*, 2003, **41**, 94.
18. D.L. VanderHart, A. Asano and J.W. Gilman, *Macromolecules,* 2001, **34**, 3819–3822.
19. D.L. VanderHart, A. Asano and J.W. Gilman, *Chem. Mater.*, 2001, **13**, 3781–3795.
20. D.L. VanderHart, A. Asano and J.W. Gilman, *Chem. Mater.*, 2001, **13**, 3796–3809.
21. D.Capitani, C. de Rosa, A. Ferrando, A. Grassi and A.L. Segre, *Macromolecules*, 1992, **25**, 3874–3880.
22. D. Capitani, A.L. Segre and J.S. Blicharski, *Macromolecules*, 1995, **28**, 1121–1128.
23. S. Bourbigot, D.L. VanderHart, J.W. Gilman, S. Bellayer, H. Stretz and D.R. Paul, manuscript in preparation.
24. J. Zhu, A.B. Morgan, F.J. Lamelas and C.A. Wilkie, *Chem. Mater.*, 2001, **13**, 3774–3780.
25. J. Zhu, F.M. Uhl, A.B. Morgan and C.A., Wilkie, *Chem. Mater.*, 2001, **13**, 4649–4654.

Micro-Sized Fire Retarding Mineral Fillers

CHAPTER 14

Polyhedral Oligomeric Silsesquioxanes: Application to Flame Retardant Textiles

SERGE BOURBIGOT,[1] MICHEL LE BRAS,[1] XAVIER FLAMBARD,[2] MARYLINE ROCHERY,[2] ERIC DEVAUX[2] AND JOSEPH D. LICHTENHAN[3]

[1]Laboratoire des Procédés d'Élaboration des Revêtements Fonctionnels, UPRES EA 1040, École Nationale Supérieure de Chimie de Lille, Université des Sciences et Technologies de Lille, BP 108, F-59652 Villeneuve d'Ascq Cedex, France (serge.bourbigot@ensc-lille.fr)
[2]Laboratoire de Génie et Matériaux Textiles (GEMTEX), UPRES EA2461, École Nationale Supérieure des Arts et Industries Textiles (ENSAIT), F-59056 Roubaix Cedex 01, France
[3]Hybrid Plastics, 18237 Mount Baldy Circle, Fountain Valley, CA - 92708, U.S.A.

14.1 Introduction

Polymeric hybrid materials have attracted great interest recently due to their advantageous performance relative to nonhybrid counterparts.[1] Chemical approaches to nanocomposite plastics have their origins in the mid-1970s with the use of sol-gel technology to form homogeneous dispersions of inorganic domains throughout a polymeric matrix.[2,3] In such systems the inorganic phase may or may not be chemically attached to the organic phase.

This first generation approach has gained some commercial value in coating applications where its complex processing and limited material strength are most amenable. Starting in the early 1980s the field saw the emergence of second generation nanocomposite technology with the resurgence in the use of minerals and clay fillers that have been treated with various surfactants to modify their surface (or interfacial) properties.

The technical goal for second generation nanocomposites continues to be the homogeneous dispersion (*via* exfoliation or intercalation) of these organically modified fillers into common polymer systems. The second generation approach has received considerable interest because the improvements include the unique combination of several properties such as better mechanical properties, better barrier properties, lower water absorption and reduced flammability.[4–7] However, it has suffered from technical drawbacks stemming from limited compatibility between the filler and matrix as well as complex processing requirements. A third generation nanocomposite technology was developed based on low density, nanostructured, silicon-based chemical feedstocks. This technology utilizes nanostructured chemicals as monomers that can be easily incorporated into all known polymeric materials in a rationally controlled manner (*e.g.* atactic, isotactic, syndiotactic) using standard polymerization or processing techniques.

The third generation approach is broadly applicable to all classes of polymers, and these nanocomposites can even be further reinforced with common fillers (*e.g.* organoclays, carbon, glass, fibers). The key technological development behind this nanocomposite technology was the development of nanostructured chemical feedstocks based on POSS.[8–12] The structure of polyhedral oligomeric silsesquioxanes (or POSS) was first reported in 1946,[13] but it is only recently that POSS-based hybrid polymers have received increasing attention because of the unique structure of the POSS macromer, which is a well-defined cluster with an inorganic silica-like core (Si_8O_{12}) surrounded by eight organic corner groups (Figure 1). The nanoscopic size of POSS enables POSS segments to effectively reinforce polymer chains segments and control chain motion at the molecular level through maximizing the surface area and chemical interactions of the nanoreinforcement with the polymer. POSS is thus a candidate to design polymer nanocomposites.[14–16] POSS reinforcement of polymer chains on a molecular level is analogous to the macroscopic reinforcement that fibers provide in composite structures.

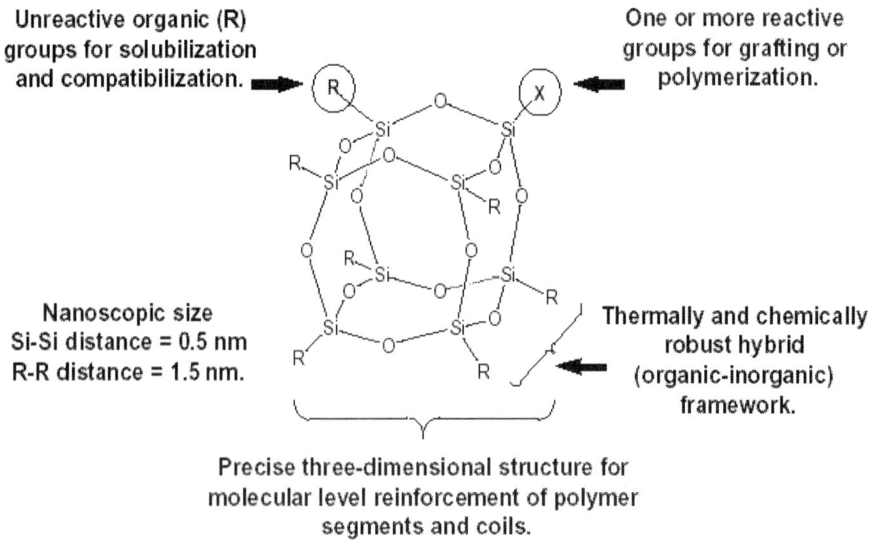

Figure 1 *Anatomy of POSS (from reference 17)*

Increased environmental awareness and fire safety concerns have pushed the plastic industry to look for environmentally friendly alternatives. Replacing traditional halogenated fire retardants with non-halogenated alternatives is a pressing concern and also, important for the image of the company ("eco-label"[18]). One of the approaches currently pursued is the use of inorganic filler (aluminum trihydrate, magnesium hydroxide, phosphate, *etc.*) or the concept of intumescence.

These approaches have led to some success but higher inorganic loadings can deteriorate the mechanical properties, and the increased viscosity makes processing difficult. The use of POSS could be tailored to be compatible with the polymeric matrix, and higher loading could be obtained while maintaining the physical characteristics. Lichtenhan *et al.*[11] have shown the efficiency of using POSS in commodity and engineering polymers. As an example, peak of heat release rate (external heat flux = 35 kW m^{-2}) of polyether block amides polymer (PEBAX) is decreased by 77 when using POSS compared to virgin polymer (Figure 2).[19] The suggested mechanism is char formation at the surface of the material which can act as an insulative barrier.[20] The organic groups on POSS cages undergo homolytic Si–C bond cleavage at ~300–350°C in air. This process is immediately followed by fusion of POSS-cages to form a thermally insulating and oxidatively stable silicon-oxycarbide "blackglass" surface char ("Si-O-C ceramified char").[21]

Nanocomposite textiles have not attracted much attention. Previous work[22] has demonstrated the feasibility of making polyamide-6/clay hybrid yarns, and has shown a significant enhancement of the flammability properties of these yarns as knitted fabrics. In this chapter, our approach is to use POSS as a potential flame retardant for textiles.

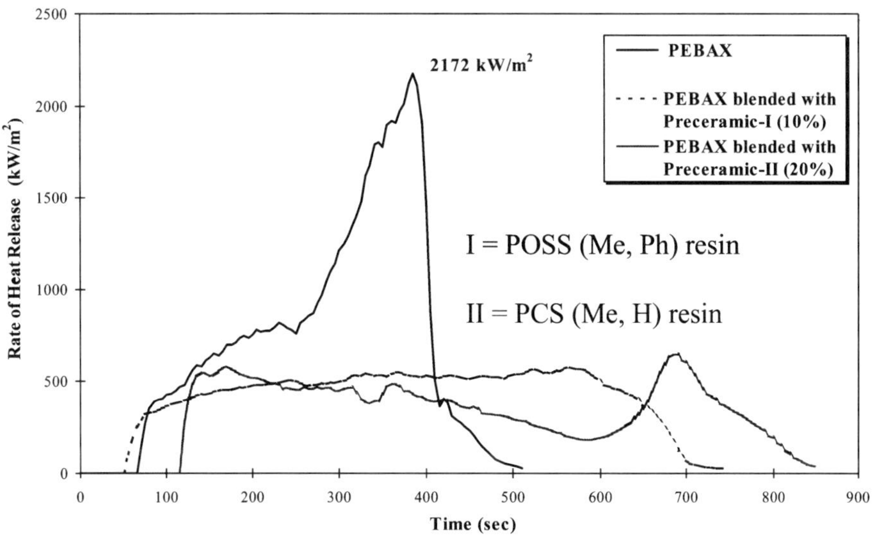

Figure 2 *Rate of heat release (RHR) curves vs. time of PEBAX blended with POSS at 35 kW m^{-2} (from reference 19)*

The feasibility of making polypropylene (PP)-POSS filaments or yarns *via* a melt-spinning process is investigated, and, then, a thermoplastic polyurethane (TPU)–nanocomposite is examined as coating for fabrics. Two kinds of POSS are used. POSS is incorporated in PP *via* melt-processing and PU-POSS nanocomposites are made by organic synthesis. The flammability of the PP/POSS knitted fabric and polyester coated by TPU-POSS is examined and discussed using the cone calorimeter as fire model.

14.2 Experimental

14.2.1 Raw Materials

Raw materials used for the preparation of PP-POSS were PP supplied by AtoFina as pellets [PPH 7060, MFI = 12 g (10 min)$^{-1}$], and dodecaphenyl-POSS (DP-POSS) (Figure 3) and poly(vinylsilsesquioxane) (FQ-POSS) (Figure 4) were supplied by Hybrid Plastics. They were used as received.

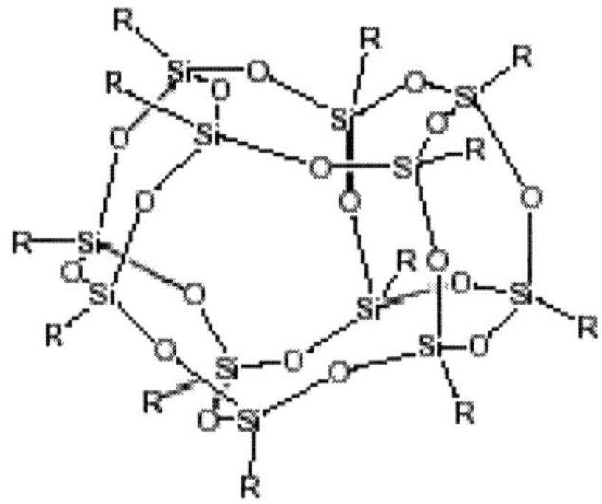

R = phenyl

Figure 3 *Dodecaphenyl-POSS (DP-POSS)*

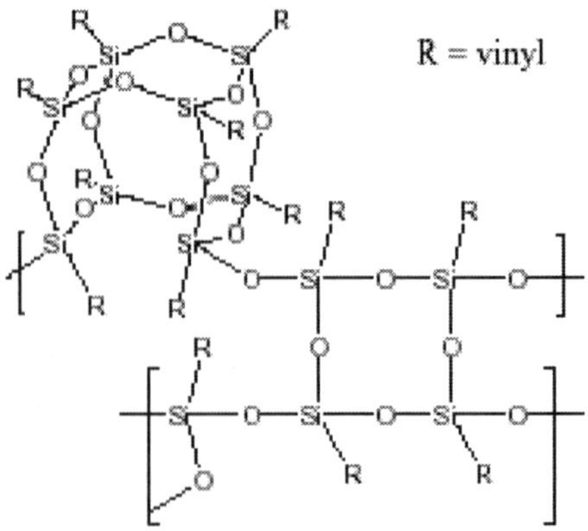

Figure 4 *Poly(vinylsilsesquioxane) (FQ-POSS)*

The ingredients of the synthesis of the polyurethane coating are Poly (oxytetramethylene glycol) (POTM) ($\overline{Mn}$ = 2000 g mol^{-1} = 2000 g mol^{-1}) isophorone diisocyanate (IPDI) and butanediol (BD) supplied by Aldrich.

14.2.2 Processing of Nanocomposite Textiles

PP-POSS. PP was melt-mixed with the clay using a counter-rotating twin-screw extruder (Haake). The rotational speed was 300 rpm in order to have a high shear stress and the temperature was 190°C. The extrudate was then pelletized.

14.2.2.1 PP-POSS Multifilament Yarns

Multifilament yarns were made *via* a melt spinning process using a Busschaert Spinboy 1 melt spinning machine. The spinning apparatus consists of a single-screw extrusion system with five heating zones. The extrusion device is fed with the solid polymeric pellets, and the heating of the screw is regulated at 200°C. The spinning die includes eighty holes with a trilobal shape, and is heated at 210°C. A volumetric pump ensures the injection of molten polymer towards the die with a flow of 100 cm^3 min^{-1}. The molten filaments are air cooled, and then drawn by means of two heated rolls, before being wound. The first roll has a surface speed of 200 rpm and is heated at 80°C, whereas the second turns at 550 rpm and is heated at 100°C. The final winding speed is the same that the second drawing roll. The yarns obtained consist of 80 trilobal continuous filaments.

14.2.2.2 Knitted Fabric of PP-POSS Multifilament Yarns

PP and PP-POSS multifilament yarns have been knitted on an automatic rectilinear machine gauge 7. The texture used is a woven rib. The two samples have a surface weight equaling 1020 g m^{-2} and the thickness is about 2.5 mm.

14.2.2.3 Synthesis of Polyurethane Nanocomposite

The thermoplastic polyurethane (TPU) coating was synthesized in two stages. The first stage was the synthesis of the prepolymer. POTM macrodiol reacts with IPDI for 24 h at 100°C. Using butanediol as chain extender added to the prepolymer, the mixture was then stirred for 2 min at room temperature to give a TPU polymer. The filler was added at 10 wt% during the second step of the synthesis of TPU. The mixture was then stirred at 500 rpm to make nanostructured material.

14.2.2.4 Polyester Fabric Coated with Polyurethane Nanocomposite

Polyester fabric was coated by TPU-nanocomposite using a coating table K Control Coater (Erichsen). The coating paste was spread with a threaded rod.

Coating thickness increases with the threading size of the rod. We used a rod with a medium threading to obtain a coating theoretical thickness of 36 μm. The coated fabric was then placed in a drying oven at 80°C for 4 hours.

14.2.3 Solid state NMR

^{29}Si NMR measurements were performed on a Bruker ASX400 at 79.49 MHz (9.4 T) with magic-angle spinning (MAS) and cross polarization (CP) using a 4 mm probe. A repetition time of 5 s was used for all samples. The reference used was tetramethylsilane (TMS), and the spinning speed was 15000 Hz. Before starting a new experiment, the chemical shift reference was always verified to be within ±0.2 ppm.

14.2.4 Thermogravimetric Analysis

Thermogravimetric analysis (TGA) was carried out using a TA Instruments (TGA 2950) at 10°C min^{-1} from 25 to 1000°C in air flow (60 cm^3 min^{-1}). Samples of 10 mg each were placed in open platinum pans. Typically, three replicates were run for each sample, and the mean was reported. Both the onset (5 mass fraction loss) and peak mass loss rate have an uncertainty of 1.2°C (2σ).

The curves of weight difference, Δ(T), between the experimental and theoretical TG curves are computed as follows:

M_{Poly}(T): TG curve of pure PP or TPU, M_{POSS}(T): TG curve of FQ-POSS or DP-POSS,

M_{exp}(T): TG curve of the blend Polymer/POSS,

M_{sim}(T): simulated TG curve computed by linear combination between the TG curves of polymer and POSS: $M_{sim}(T) = xM_{Poly}(T) + yM_{POSS}(T)$, $x + y = 1$ and x, y are the mass percentage of polymer and POSS in the blend,

Δ(T): curve of weight difference: $\Delta(T) = M_{exp}(T) - M_{sim}(T)$. The Δ(T) curves allow the observation of an eventual increase or decrease in the thermal stability of the polymer.

14.2.5 Cone Calorimetry by Oxygen Consumption

A FFT (Fire Testing Technology) cone calorimeter was employed to carry out measurements on samples. Samples were mounted between two cut steel sheets placed on the usual holder of the cone calorimeter. The surface exposed to the external heat flux was 9 × 9 cm^2. Our method does not correspond to any standard. An external heat flux of 35 kW m^{-2} was used in running the experiments; this flux corresponds to the common heat flux in mild fire scenario.[23,24] The cone calorimeter was used to determine heat release rate (HRR). When measured at 35 kW m^{-2}, HRR is reproducible to within ±10%. The results presented in the following are averages. Cone data reported in this chapter are the average of three replicated experiments.

14.3 Results and Discussion

14.3.1 PP-POSS Multifilament Yarns

Multifilament yarns have been made *via* a melt spinning process after extruding virgin PP and PP/FQ-POSS, and they have knitted to obtain fabrics (Figure 5).

No significant difference is observed between the two fabrics with and without POSS except that the touch of PP/FQ-POSS is softer.

CP-MAS NMR^{29}Si spectra of neat FQ-POSS and of PP/FQ-POSS multifilament yarns are compared on Figure 6. They show only one band at about −80 ppm assigned to R-SiO_3 groups (R is an organic group).[25,26] This is consistent with the chemical structure of FQ-POSS (Figure 1). In addition,

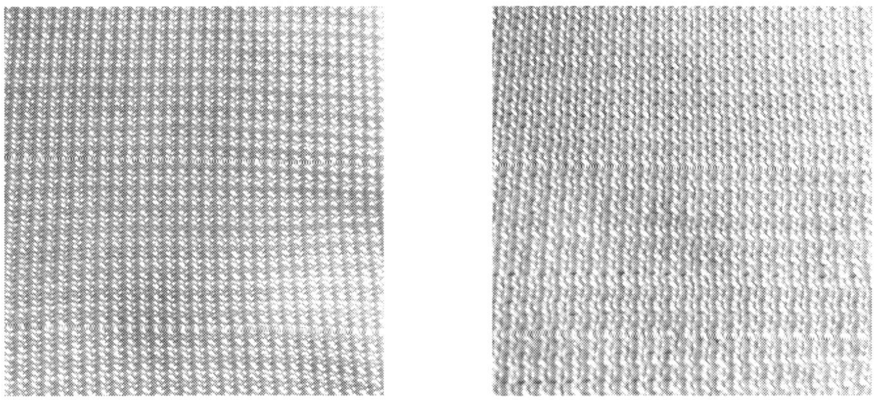

(a) (b)

Figure 5 *PP* (a) *and PP/FQ-POSS* (b) *multifilament yarns knitted as fabric*

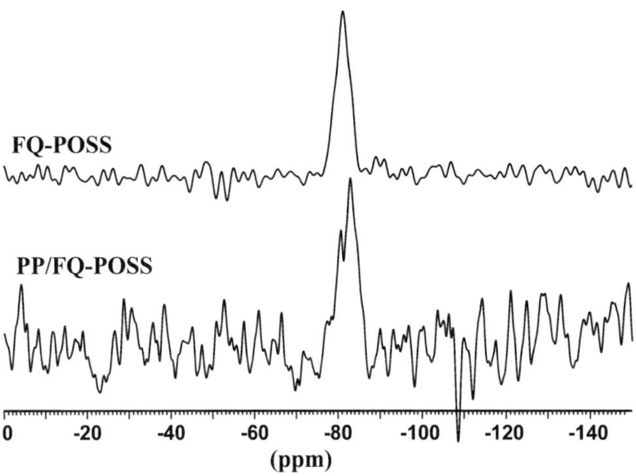

Figure 6 *CP-MAS NMR^{29}Si of FQ-POSS and PP/FQ-POSS multifilament yarns*

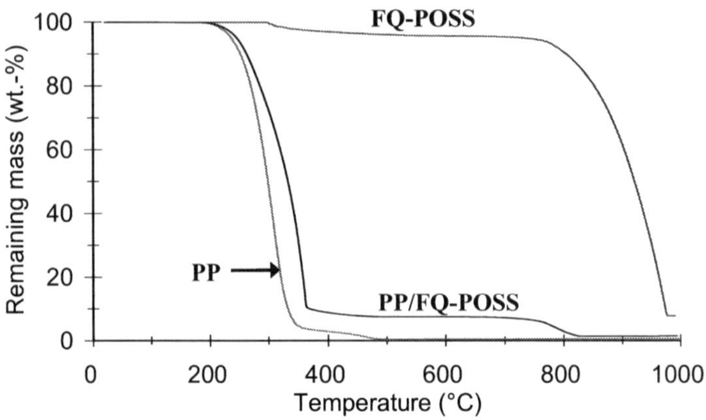

Figure 7 *TG curves of PP and PP/FQ-POSS multifilament yarns (10 °C min^{-1} air flow)*

FQ-POSS and PP/FQ-POSS multifilament yarns exhibit similar spectra, suggesting that no degradation and/or modification of FQ-POSS occurs during extrusion and melt spinning.

The fire behaviour of a material depends on processes occurring in both condensed and gas phase and on the processes of heat and mass transfer. These processes strongly depend on the degradation reactions occurring in the condensed phase. Moreover, it has been reported[21] that the mechanism of action of POSS takes place in the condensed phase to form an insulative charring barrier. Thus it is particularly interesting to study the thermal degradation of PP/FQ-POSS before investigating the flammability of the fabrics.

The TG curve of FQ-POSS (Figure 7) reveals a slight degradation (3 wt% of weight loss) starting at 350°C and a main step of degradation occurring at 800°C. The final residue at 1000°C, 5 wt%. FQ-POSS, is thus thermally stable in the processing conditions and this may explain why no degradation is detected by ^{29}Si NMR.

PP and PP/FQ-POSS start to degrade 250°C but the degradation rate of PP/FQ-POSS is slower (Figure 7). The residue of PP is close to zero (0.2 wt%) at 500°C. PP/FQ-POSS exhibits a transient residue from 450 to 800°C corresponding to the stabilization by FQ-POSS. The final residue is 3 wt%.

To investigate the eventual interactions between PP and FQ-POSS during degradation, the curve of weight difference was plotted in Figure 8, which shows that interactions occur from 200 up to 320°C between PP and FQ-POSS. PP is stabilized in this temperature range. After 450°C, a slight destabilization of PP/FQ-POSS is observed.

Reaction to fire of PP and PP/FQ-POSS knitted fabrics is evaluated using cone calorimetry by oxygen consumption. The rate of heat release (RHR) curves of the two fabrics at an external heat flux of 35 kW m^{-2} are shown Figure 9.

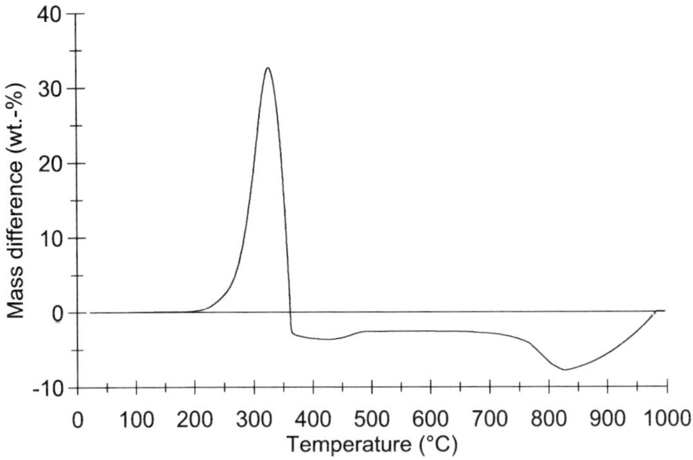

Figure 8 *Curve of mass difference of PP/FQ-POSS (10 °C min^{-1} air flow)*

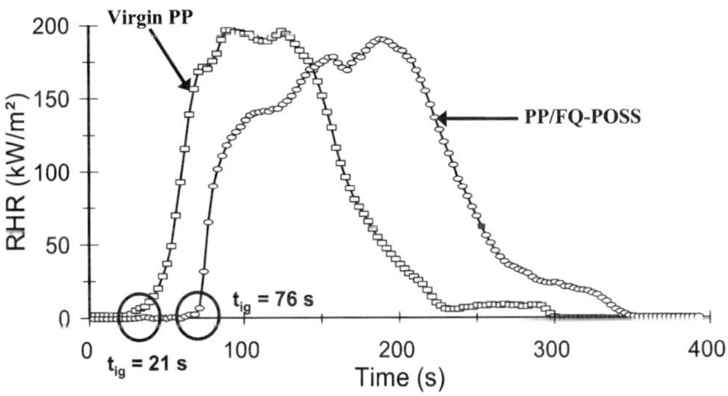

Figure 9 *Rate of heat release (RHR) curves of PP and PP/FQ-POSS knitted fabrics at 35 kW m^{-2}*

The time to ignition of virgin PP occurs at 21 s and is much shorter than that of PP/FQ-POSS. The better thermal stability of the system PP/FQ-POSS can explain this result. However, the peaks of the maximum of RHR of the two fabrics are similar, showing that the flammability properties of PP are not improved using FQ-POSS. The RHR curve of PP/FQ-POSS is only translated compared to PP because of the stabilization of PP by FQ-POSS. Total heat evolved (THE) is also not modified (about 200 kJ) whatever the fabrics, suggesting that FQ-POSS does not act as a flame retardant but only as heat stabilizer (decrease of the ignitability).

14.3.2 TPU-POSS Coating

TG curves of the coatings (TPU, TPU/DP-POSS and TPU/FQ-POSS) are shown in Figure 10. The main degradation of the three TPUs starts at the same

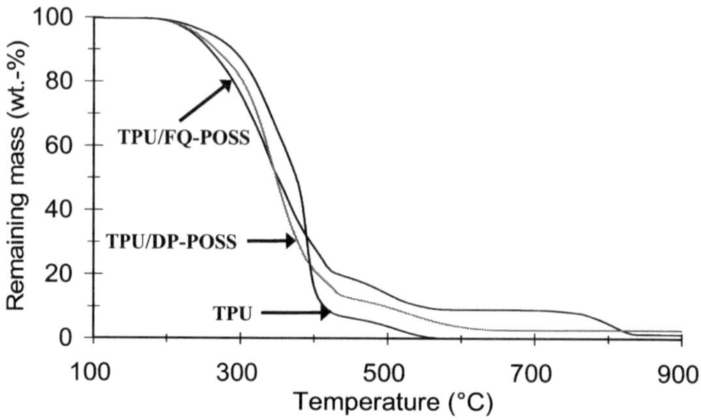

Figure 10 *TG curves of TPU, TPU/DP-POSS and TPU/FQ-POSS coatings (10 °C min^{-1} air flow)*

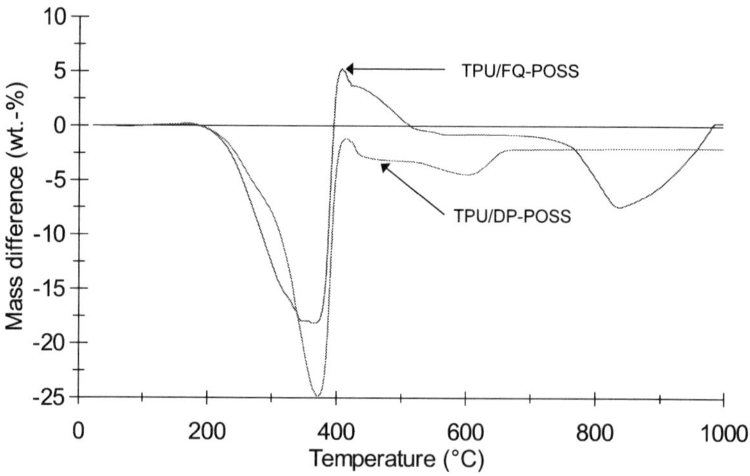

Figure 11 *Curve of mass difference of TPU/DP-POSS and TPU/FQ-POSS (10 °C min^{-1} air flow)*

temperature (220°C), but the degradation rate of TPUs containing POSS is faster than pure TPU until 400°C. A stabilization zone between 400 and 550°C is then observed. The final residues of TPU, TPU/DP-POSS and TPU/FQ-POSS at 900°C are 0, 5 and 3 wt% respectively.

Interactions between TPU and TPU-POSS during degradation are shown on the curve of weight difference on Figure 11. The two polymers TPU/DP-POSS and TPU/FQ-POSS are strongly destabilized from 200 and until 400°C. This effect is higher in the case of TPU/DP-POSS (–25% compared to –17%). Above 400°C, TPU/DP-POSS is still slightly destabilized, but TPU/FQ-POSS is stabilized by FQ-POSS between 400 and 550°C followed by a slight destabilization up to 1000°C.

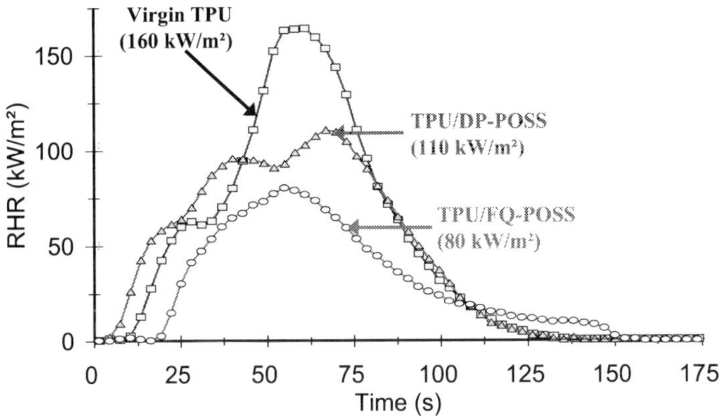

Figure 12 *Rate of heat release (RHR) curves of TPU, TPU/DP-POSS and TPU/FQ-POSS as coating of woven PET fabrics at an external heat flux of 35 kW m*$^{-2}$

RHR curves of TPU containing POSS coating on woven polyester (PET) fabric show a significant reduction in flammability (in terms of peak of RHR) with regard to virgin TPU (Figure 12). The two POSSs reveal different behaviors in TPU. The time to ignition of TPU/DP-POSS (7 s) is shorter than neat TPU (10 s) while that of TPU/FQ-POSS is longer (22 s). The RHR peaks of TPU/DP-POSS and TPU/FQ-POSS are decreased by 31% and by 50% respectively. During combustion, a char layer is formed at the surface of the materials. With TPU/FQ-POSS, the char is more uniform and only small cracks are observed at the surface (cracks can be seen with TPU/DP-POSS, which can explain its "hill and valley" curve). This char is more resistant and can smother the flame. It is also noteworthy that THEs of TPU/POSS's (60 kJ for TPU/DP-POSS and 40 kJ for TPU/FQ-POSS) are significantly lower than that of neat TPU (70 kJ), showing that DP-POSS and FQ-POSS act as real flame retardants (reaction to fire decreasing RHR and THE).

14.4 Conclusion

This work has investigated the use of POSS as flame retardant in textiles incorporated in yarns and in coating. PP containing POSS as multifilament yarns reveals that POSS permits the stabilization of PP; however, the flammability is not enhanced in terms of RHR, and only the time to ignition is much longer. It offers, therefore, the opportunity to make relatively heat resistance fabrics with low ignition. Concurrently, TPU-POSS coatings have been synthesized and the action of POSS as flame retardant has been demonstrated. In addition, the use of FQ-POSS permits both the increase of time of ignition and the decrease of peak of RHR. These results thus offer a promising route for flame retarding textile using POSS.

14.5 Acknowledgements

The authors are indebted to Mr. Dubusse from CREPIM for skilful experimental assistance in cone calorimetery. NMR experiments were made at the common research centre of the University of Lille and Mr. Bertrand Revel is acknowledged for helpful discussion and experimental assistance.

This work was partially supported by the European project FLAMERET ("New Surface Modified Flame Retarded Polymeric Systems to Improve Safety in Transportation and Other Areas" registered under the number No. G5RD-CT-1999-00120).

14.6 References

1. Q. Liu, J.J. Schwab, J.D. Lichtenhan, D. Mason and A. Lee, *Polym. Mater. Sci. Eng.*, 2002, **87**, 97.
2. J.E. Mark, C.Y.C. Lee and P.A. Bianconi, in *Hybrid Organic–Inorganic Composites*, American Chemical Society, Washington, DC, 1995.
3. Y. Chujo, *Solid State Mater. Sci.*, 1996, **1**, 806.
4. Y. Kojima, A. Usuki, M. Kawasumi, A. Okada, Y. Fukushima, T. Kurauchi and O. Kamigaito, *J. Mater. Res.*, 1993, **8(5)**, 1185.
5. P.B. Messersmith and E.P. Giannelis, *Chem. Mater.*, 1994, **6**, 1719–1725.
6. P.B. Messersmith and E.P. Giannelis, *J. Polym. Sci. A., Polym. Chem.*, 1995, **33(7)**, 1047.
7. J.W. Gilman, T. Kashiwagi and J.D. Lichtenhan, *SAMPE J.*, 1997, **33**, 40.
8. J.D. Lichtenhan, N.Q. Vu, Jason A.Carter, J.W. Gilman and F.J. Feher, *Macromolecules*, 1993, **26(8)**, 2141.
9. J.D. Lichtenhan, Y.A. Otonari and M.J. Carr, *Macromolecules*, 1995, **28**, 8435.
10. T.S. Haddad and J.D. Lichtenhan, *Macromolecules*, 1996, **29**, 7302.
11. J.D. Lichtenhan, J.W. Gilman and F.J. Feher, *US Patent 6,362,279*, 1996, assigned to University of California.
12. J. Choi, J. Harcup, A.F. Yee, Q. Zhu and R.M. Laine, *J. Am. Chem. Soc.*, 2001, **123**, 11420.
13. D.W. Scott, *J. Am. Chem. Soc.*, 1946, **68**, 356.
14. L. Zheng, R.M. Kasi, R.J. Farris and E.B. Coughlin, *J. Polym. Sci., Part A: Polym. Chem.*, 2002, **40**, 885.
15. L. Zheng, R.J. Farris and E.B. Coughlin, *Macromolecules*, 2001, **34**, 8034.
16. H. Xu, S.W. Kuo, J.S. Lee and F.C. Chang, *Macromolecules*, 2002, **35**, 8788.
17. http://www.hybridplastics.com
18. http://www.firesafetyinfo.org/Environment/eco-shemes.htm
19. J.D. Lichtenhan and J.W. Gilman, *US Patent 6,362,279,* 2002, assigned to US Air Force.

20. R.A. Mantz, P.F. Jones, K.P. Chaffee, J.D. Lichtenhan, J.W. Gilman, I.M.K. Ismail and M.J. Burmeister, *Chem. Mater.*, 1996, **8**, 1250.
21. S.K. Gupta, J.J. Schwab, A. Lee, B.X. Fu, and B.S. Hsiao, in *Affordable Materials Technology – Plateform to Global Value and Performance*, B.M. Rasmussen, L.A. Pilato and H.S. Kliger (eds.), SAMPE pub., Long Beach, CA, 2002, **47(2)**, p. 1517.
22. S. Bourbigot, E. Devaux and X. Flambard, *Polym. Degrad. Stab.*, 2002, **75(2)**, 397.
23. V. Babrauskas, *Development of Cone Calorimeter – A bench scale rate of heat release based on oxygen consumption*, NBS-IR 82-2611, US Nat. Bur. Stand., Gaithersburg, 1982.
24. V. Babrauskas, *Fire Mater.*, 1984, **8(2)**, 81.
25. E. Lippmaa, M. Maegi, A. Samoson, G. Engelhardt and A. R. Grimmer, *J. Am. Chem. Soc.*, 1980, **102(15)**, 4889.
26. E.T. Lippmaa, M. A. Alla, T.J. Pehk and G. Engelhardt, *J. Am. Chem. Soc.*, 1978, **100(6)**, 1929.

CHAPTER 15

Octaisobutyl POSS Thermal Degradation

ALBERTO FINA,[1] DANIELA TABUANI,[1] ALBERTO FRACHE,[2] ENRICO BOCCALERI[2] AND GIOVANNI CAMINO[1]

[1]Centro di Cultura per l'Ingegneria delle Materie Plastiche – Politecnico di Torino, V.le T. Michel, 5 - 15100 Alessandria – and INSTM member (alberto.fina@polial.polito.it)
[2]Università del Piemonte Orientale – Dip. Scienze e Tecn. Avanzate, P.za Ambrosoli, 5 - 15100 Alessandria – and INSTM member

15.1 Introduction

Silsesquioxanes are a family of compounds with general formula $(RSiO_{1.5})_n$ where R is hydrogen or an organic group, such as alkyl, aryl or any of their derivatives. Different structures are possible for these compounds,[1–3] depending on the position of R group, which can be bonded to a single silicon atom or between two silicon atoms to afford bridged polysilsesquioxanes.[4]

Silsesquioxanes can have various geometrical structure orders, including random structures, ladder structures and cage structures, better known as Polyhedral Oligomeric SilSesquioxanes (POSS). POSS are usually produced by hydrolytic condensation of trifunctional monomers $RSiX_3$, where X is a highly reactive substituent, such as Cl or alkoxy.[3,5] The first step is the hydrolysis of the monosilane to give the corresponding trisilanol. The second step is the condensation of the trisilanol to yield different silsesquioxane species. Many factors influence the hydrolytic condensation, and thus determine the silsesquioxane cage structures, such as the nature of R and X groups, the solvent, the concentration of monosilane $RSiX_3$, temperature, pH and reaction time.[6]

Even though oligomeric silsesquioxanes were first synthesized in 1946[7] and some research groups have worked on the subject since the 1960s,[8–12] recent years have seen a great research interest in POSS thanks to their organic-inorganic hybrid structure that gives remarkable properties in numerous applications, both as nanofiller for polymer-based nanocomposites and hybrids and as neat POSS.

In the field of hybrid and nanocomposite materials, POSS were successfully used to improve polymer properties, increasing use temperature, oxidation resistance and improving mechanical properties, as well as reducing viscosity during processing and polymer flammability.[13–15] Neat POSS have been used as low dielectric constant materials, new resists for electron beam lithography materials, high temperature lubricants or catalysts.[13,16]

Thermal degradation studies were developed on silsesquioxanes as ceramic precursors, obtaining silicon oxicarbides *via* high temperature pyrolysis in an inert atmosphere.[1]

Thermal properties of the homologous series of 8-fold alkyl substituted (from ethyl to decyl) silsesquioxanes have been studied by Bolln *et al.*[17] by thermogravimetric analysis, showing that, increasing the alkyl chains length, the weight loss onset shifts to higher temperatures. In nitrogen atmosphere at low heating rate (1°C min^{-1}) the onset temperature ranged from 166°C for octapropyl POSS to 355°C for octadecyl POSS. Residual weights at the end of analysis were in the range 5–15% of the initial weight; the authors related the residue to the formation of a siloxane structure. In air, residual weights at the end of analysis were higher (30 to 50%) than in inert atmosphere and the weight loss was higher with increasing alkyl chains length.

POSS degradation studies developed by Mantz *et al.*,[18] by the analysis of pyrolisis gases and chars, showed that fully condensed cyclohexyl POSS macromers [$R_6Si_6O_9$ and $R_8Si_8O_{12}$] had a propensity to sublimation on heating in an inert atmosphere. The incompletely condensed POSS ($R_8Si_8O_{11}(OH)_2$) underwent a more complicated degradation process, which resulted in a two-stage weight loss, with a consistent final residue of about 40% of the initial weight. In the first stage, sublimation was observed, together with a degradation process that resulted in the presence of trace amounts of water and carbon dioxide in the degradation gases. These phenomena produced an insoluble residue, due to macromer homopolymerisation, which occurred in competition with sublimation. During the second weight loss stage, cyclohexane and cyclohexene were released by degradation of the solid phase. Char analysis *via* solid-state ^{29}Si NMR and X-ray diffraction showed a progressive loss of cage structure order and crystalline structure with increasing thermal treatment temperature.

The thermal behaviour of octahydrogen POSS was also studied in order to produce low dielectric constant thin layers,[19–22] showing how POSS solution can produce a cross-linked solid thin film, after thermal treatment around 400°C in an inert atmosphere or even at lower temperatures in air. The mechanism proposed for cross-linking in an inert atmosphere[21] involves the release of SiH_4 and H_2, producing a silica-like phase.

FTIR analysis of octamethyl POSS and octahydrogen POSS based films were performed to monitor the structural behaviour at increasing temperatures by Liu *et al.*,[23] showing a transformation from cage into network structure.

On the basis of this background, in this chapter we investigate the thermal and thermooxidative degradation of a fully condensed alkyl substituted POSS, analysing both volatile pyrolysis products and solid residues.

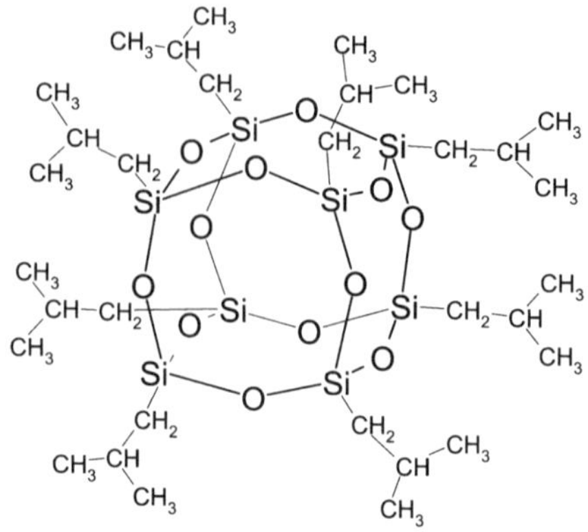

Figure 1 *Octaisobutyl POSS*

15.2 Experimental

Octaisobutyl POSS (Figure 1), in the following also named oib-POSS, was obtained as a fine powder from Hybrid Plastics Company and used as received.

Thermogravimetry (TGA) was performed on a TA Q 500 instrument, in platinum pans, with gas fluxes of 60 ml min^{-1} for sample gas (nitrogen or air) and 40 ml min^{-1} for balance protection gas (nitrogen). Analyses were carried out under isothermal conditions or at a fixed heating rate, between 50 and 800°C, with 10 mg samples (except where specified otherwise).

Differential scanning calorimetry (DSC) analyses were run using a TA Q1000 instrument; measurements were carried out in closed (but not hermetic) aluminium pans, under a nitrogen flow (50 ml min^{-1}). Heating rate was 10°C min^{-1}, with 2 mg samples.

Gas chromatography (GC) was performed on a Perkin-Elmer Autosystem XL, coupled with a Perkin-Elmer Turbomass Gold mass spectrometer (MS). A medium polarity silica capillary column was used, 30 m long and 0.25 mm diameter (VARIAN CP Sil 8 CB). Analysis conditions were set as follows: injector temperature 280°C, split flow 10 ml min^{-1}, 14 psi pressure controlled column fluxed with helium. oib-POSS was injected as a 3×10^{-4} M solution in chloroform (99% purity, Carlo Erba reagents) and a 7 min solvent delay time was set on the MS. The oven temperature for GC analysis was equilibrated at 50°C, held isothermal for 1 min, then raised with an heating ramp of 10°C min^{-1} to 300°C and held isothermally at the final temperature until analysis end.

Flash pyrolyses under inert conditions (He) were performed at 600 and 900°C in a SGE Pyrojector Mk2 pyroliser, assembled on the GC injector (Py GC-MS). Pyrolyses in air were performed in a quartz tube, heated in an oven purged with air flowing at 50 ml min^{-1}.

X-Ray powder diffraction (WAXD) patterns were obtained on a ARL XTRA48 diffractometer using Cu Kα radiation ($\lambda = 1.54062$ Å).

FTIR spectra (resolution 4 cm^{-1}) were recorded at room temperature on samples diluted in KBr pellets using a Bruker Equinox 55 spectrometer. Raman spectra (resolution 4 cm^{-1}) of powder samples were recorded on a RFS 100 Bruker FT Raman spectrometer using a 1064 nm wavelength excitation laser.

15.3 Results and Discussion

15.3.1 Thermal Degradation in Inert Conditions

DSC of oib-POSS shows on heating a sharp endothermic peak at 266°C and two other smaller endothermic peaks, at *ca.* 50°C. (Figure 2). The strongest peak at 266°C is attributed to the oib-POSS cage structure melting transition, as confirmed by visual observation of a sample heated in an oven.

Multiple melting transitions were previously observed by Abad *et al.* on glycidyleptaisobutyl POSS,[24,25] in which two melting peaks were found at 112 and 133°C. The authors attributed this thermal behaviour to different crystalline organization in crystal core or boundaries.

For oib-POSS discussed here, let us suppose a different explanation for the temperature difference between the two endothermic signals (*ca.* 50 and 266°C), one that is probably related to a crystalline organization of the oib-POSS organic fraction, which will be the subject of further studies.

Thermogravimetric curves obtained in nitrogen at 1, 10 and 100°C min^{-1} are shown in Figure 3. The weight loss takes place in two steps. The major one (80–90% weight loss) occurs with maximum weight loss rate at T_{max} = 237, 265,

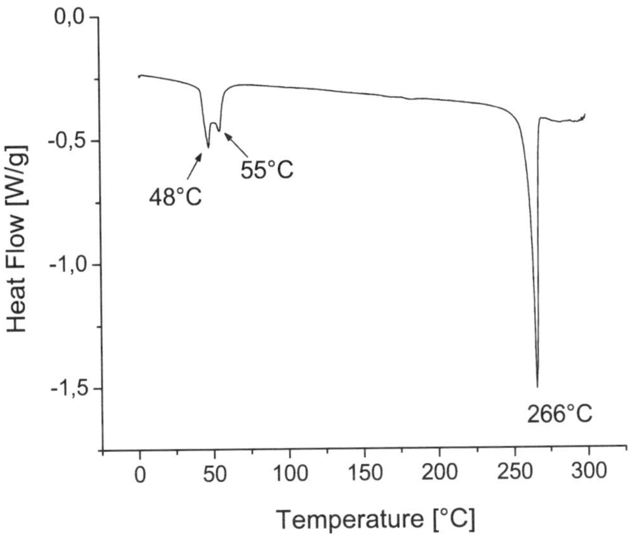

Figure 2 *Oib-POSS DSC analysis*

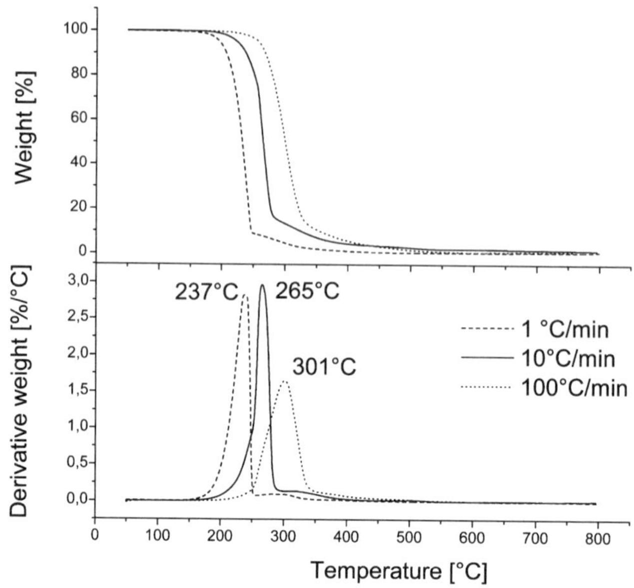

Figure 3 *TGA plots for oib-POSS in N_2 at different heating rates*

301°C depending on heating rate (1, 10, 100°C min^{-1}, respectively). The second, minor, weight loss occurs over a broader range of temperatures (T_{max} = 280 and 320°C on heating at 1 and 10°C min^{-1}, respectively) and is overlapped by the major volatilisation process on heating at 100°C min^{-1}. This second weight loss is probably related to degradation of a more stable phase formed during the previous heating throughout the first step. A negligible residue is left above 500°C.

Comparison of the TGA curves at 10°C min^{-1} of Figure 3 with the DSC curve of Figure 2 suggests that, on heating, oib-POSS mainly evaporates upon melting.

To confirm this interpretation, pyrolysis gas chromatograpy mass spectrometry (Py GC-MS) was carried out to analyse the volatilisation products evolved on heating oib-POSS.

The elution of a single peak at 34 min in GC-MS evidences oib-POSS purity (Figure 4). In the mass spectrum of oib-POSS (Figure 5) the molecular ion signal (*m/z* 874) is very low while electron impact gives rise to a series of fragments originating from oib-POSS by successive elimination of an isobutyl group as 2-methylpropene (*m/z* 56 in Figure 5), which leaves a series of POSS of general formula: $(iBu)_{8-n}H_n$POSS in which $n = 1$–7 (Scheme 1).

The most stable of these fragments is the *hepta* isobutyl POSS ion [$(iBu)_7H_1$POSS], producing the highest intensity peak at 816 *m/z*, followed by the signal at 422 *m/z*, corresponding to hydrogen POSS [H_8POSS], in which no organic groups are left on the molecule. Other macromer ion peaks having

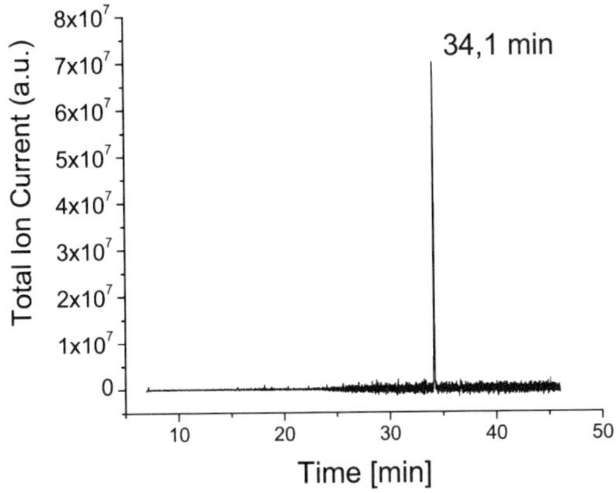

Figure 4 *Chromatogram for oib-POSS in* $CHCl_3$ *solution*

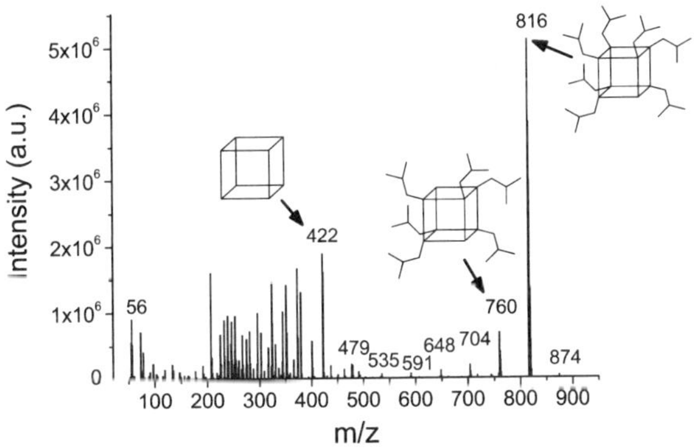

Figure 5 *Mass spectrum for oib-POSS in* $CHCl_3$ *solution*

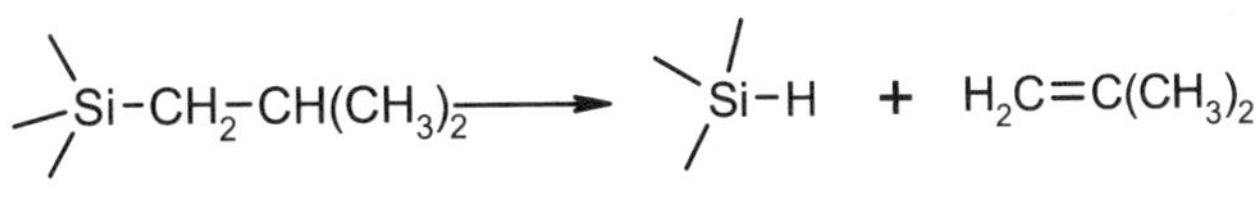

Scheme 1 *Elimination of 2-methylpropene from the POSS cage*

lower intensity correspond in the order, To *hexa*, *penta*, *tetra*, *tri*, *di* and *mono*isobutyl POSS (*m/z* 760, 704, 648, 591, 535 and 479 respectively).

Py GC-MS of oib-POSS at 600°C mainly leads to POSS evaporation, as shown in Figure 6 by the major elution peak (almost identical to that of Figure 4), whose mass spectrum is very close to that of volatilising oib-POSS reported in Figure 5 (see Figure 7 below).

The most evident degradation product of flash pyrolysis at 600°C is eluted at 32.6 min in GC (Figure 6). It can be identified using the mass spectrum reported in Figure 8 below; the molecular ion peak at 843 *m/z* indicates elimination of ethane (30 a.m.u., Scheme 2) from oib-POSS (874 a.m.u.) (see Scheme 2).

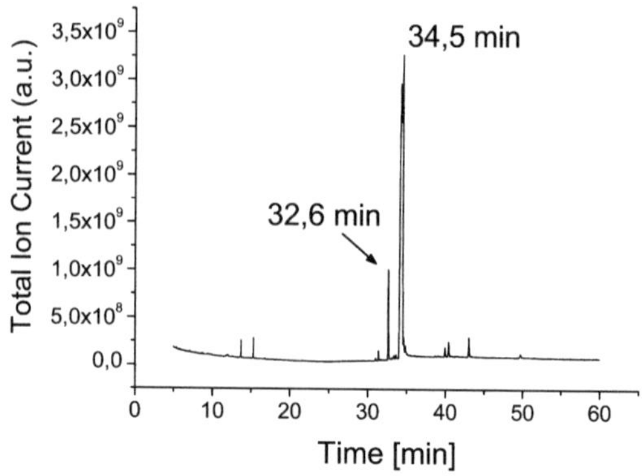

Figure 6 *Oib-POSS flash pyrolysis chromatogram at 600 °C in helium*

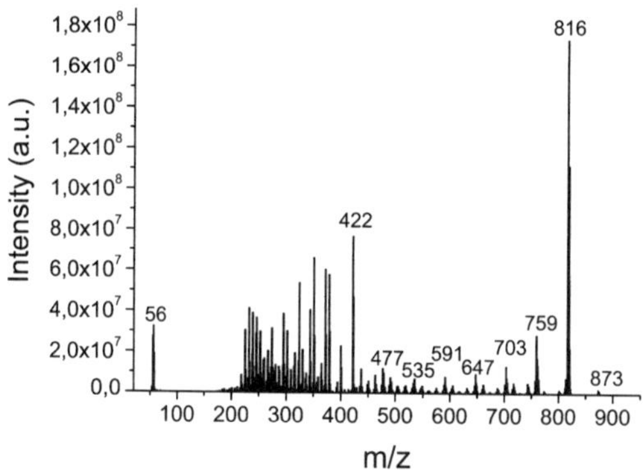

Figure 7 *Mass spectrum of 600 °C pyrolysis product eluted at 34.5 min in Figure 6*

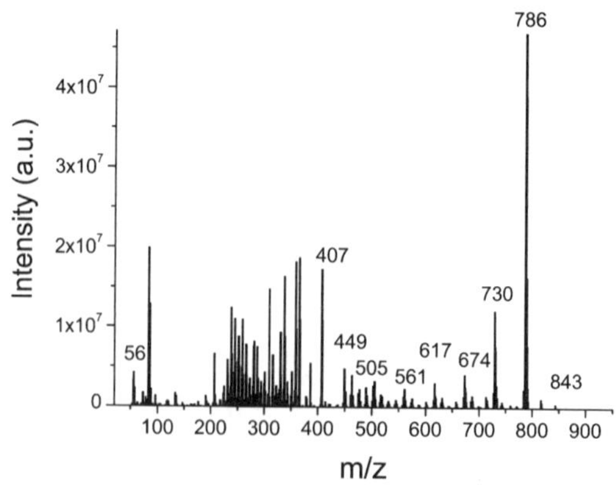

Figure 8 *Mass spectrum of 600 °C pyrolysis product eluted at 32.6 min in Figure 6*

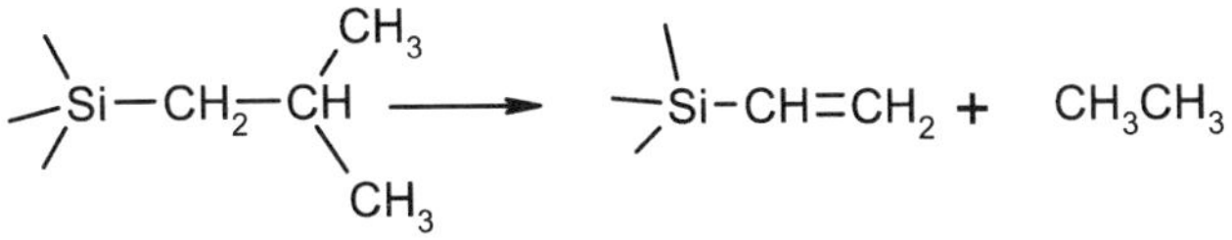

Scheme 2 *Elimination of a molecule of ethane from the oib-POSS cage after pyrolysis at 600 °C*

Electron impact fragmentation of the resulting heptaisobutylvynilPOSS occurs with successive elimination of 2-methylpropene (56 a.m.u.) as in Scheme 1, giving rise to the corresponding peaks at *m/z* 786, 730, 674, 617, 561, 505 and 449. A few other minor degradation products have been characterised, found both at shorter and longer elution times than original oib-POSS (Figure 6).

On the basis of these results, the *octa*-isobutyl POSS is essentially thermally stable when rapidly heated to 600°C, with minor partial thermolysis of organic isobutyl substituents.

On increasing the flash pyrolysis temperature to 900°C, completely different volatile products are obtained. The oib-POSS elution peak at 34 min disappears in the Py GC-MS chromatogram and low molecular weight compounds are formed (Figure 9), which have been identified, from their mass spectra, as unsatured organic molecules (Table 1). The first chromatogram peak around 1.5 min in Figure 9 is due to several unseparated light organic products not yet identified. Thus, at 900°C, degradation of the isobutyl groups overwhelms evaporation of oib-POSS, probably leaving silica as a residue.

The formation, at 900°C, of volatile compounds, which are so different from the isobutyl substituents of the oib-POSS, shows that a complex degradation process takes place at 900°C, involving molecular fragmentations and rearrangements of the substituents themselves, possibly catalysed by the silica-like structure of the degrading POSS.

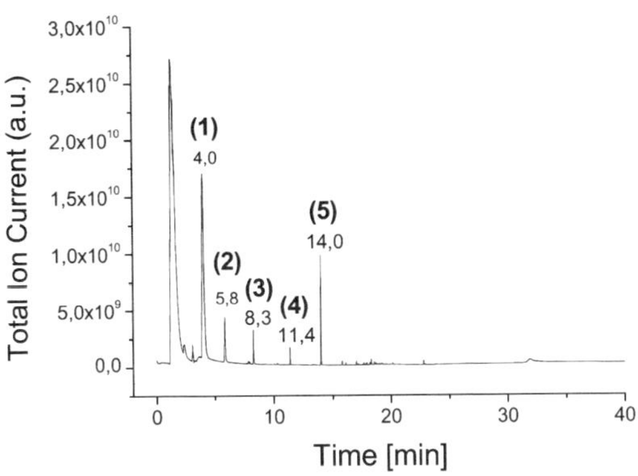

Figure 9 *Oib-POSS flash pyrolysis chromatogram at 900 °C in helium*

Table 1 *Volatile products of oib-POSS pyrolysis at 900 °C*

Elution peak	*Elution time* (min)	*Product*	*Structure*
1	4.0	1,5-Hexadiyne	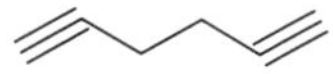
2	5.8	Cycloheptatriene	
3	8.3	Cyclooctatetraene	
4	11.4	Indene	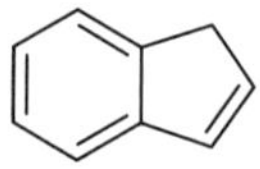
5	14.0	Azulene	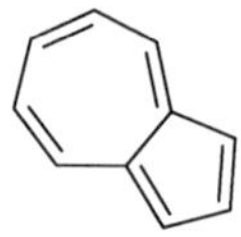

15.3.2 Thermal Degradation in Oxidative Conditions

The behaviour in air is very different from that observed in an inert atmosphere, as oxygen plays an active role in the degradation process. Indeed, a major weight loss step followed by a second broader and much smaller step is observed in air as in nitrogen. However, although the maximum rate of the major weight loss occurs at temperatures similar to those of the corresponding thermogravimetries performed in nitrogen (Figures 10 and 3), the residual

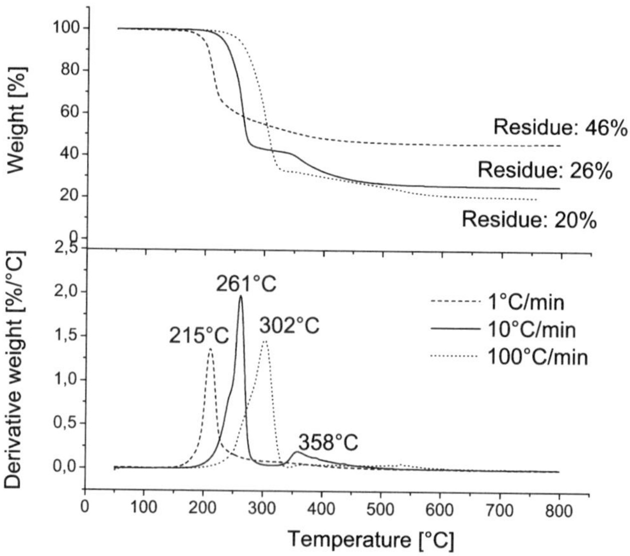

Figure 10 *TGA curves of oib-POSS in air; influence of the heating rates*

weight at 800°C in air is no longer negligible, but a significant amount of thermally stable residue is usually found. Furthermore, the residue amount depends on the heating rate: experiments run at at 1, 10, 100°C min^{-1} heating rates gave 46, 26 and 20 wt% of residue respectively.

The second minor step of weight loss is more evident in air on heating at 10°C min^{-1}, whereas in nitrogen its evidence decreases with increasing heating rate. Comparing the behaviour in air and in nitrogen (Figures 3 and 10) suggests that, under air, two concurrent phenomena are in competition: in fact, evaporation upon melting (typically found in inert atmosphere) is counterbalanced by the oxidation of oib-POSS.

Experiments at a fixed heating rate and different sample weights show an increase in the amount of residue at 800°C with an increase in sample weight (Figure 11). Differences in the amount of residual sample depending on the initial mass of oib-POSS can be explained by the formation of a glassy surface layer due to thermal oxidation of oib-POSS, which is protective toward mass and heat transfer, hindering volatilisation of the underlying material; a considerable amount of sample is thus available for oxidation to stable residue.

As the sample is in powder form, an increase in sample weight does not change significantly the exposed surface when placed in the TGA pan; conversely, the thickness of the powder lying in the sample cup changes. On this basis, it is reasonably clear that the weight of the sample protected is higher for larger samples, thus justifying the residual fraction increase shown in Figure 11.

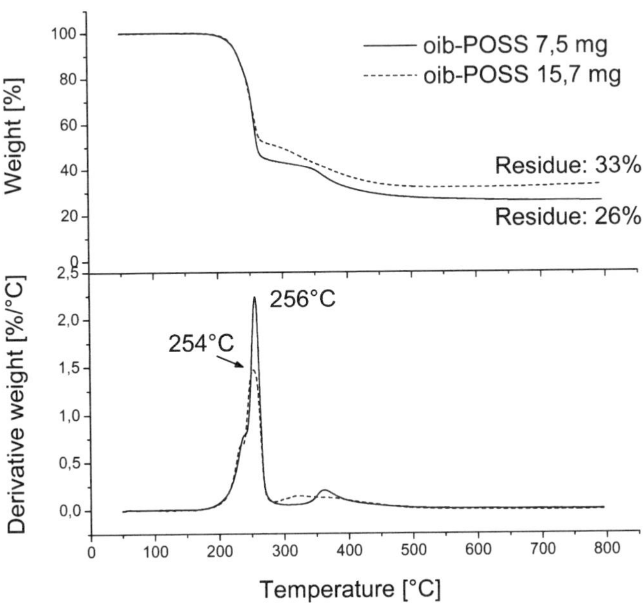

Figure 11 *Oib-POSS thermogravimetry; influence of sample weight*

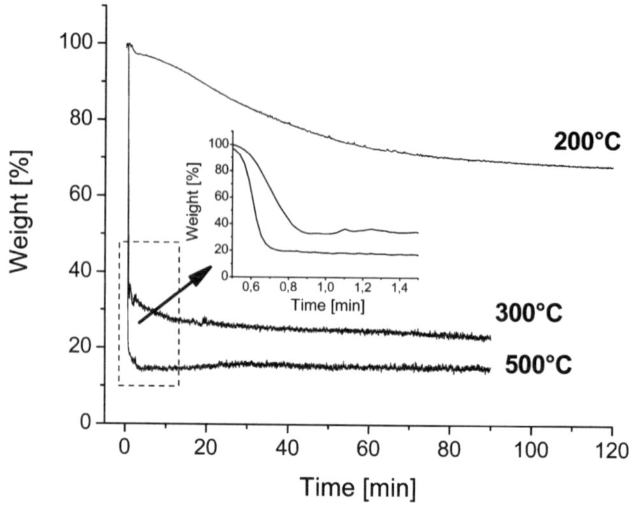

Figure 12 *TGA isothermal plots for oib-POSS, in air (inset shows magnification of initial weight loss)*

To understand better POSS oxidation phenomena, thermogravimetry measurements were performed in air in isothermal conditions, until the formation of stable residues was reached. Figure 12 shows that extensive oxidation of the organic structure of oib-POSS takes place at 200°C since 66% residue is left on heating, compared to 55% calculated for complete oxidation of POSS to SiO_2, which is the expected product left by oxidation of the organic substituents in POSS. Oxidation overcomes volatilisation at 200°C as the latter is relatively low at this temperature (Figure 3). On heating rapidly oib-POSS to 300 or 500°C, volatilisation increasingly overwhelms oxidation, with the formation of 25% and 17% residue respectively. The recovered residues show a glassy colourless surface when obtained at 200°C, whereas it turns to yellow at 300 or 500°C.

To further investigate the chemical and physical structure of the residues obtained by thermal treatments in air, studies by means of vibrational spectroscopy and X-ray diffraction techniques were performed.

The sensitivity of IR and Raman spectroscopy to inorganic and organic moieties, respectively, made these two techniques complementary in the characterisation of oib-POSS and of its degradation products. Figure 13 shows IR and Raman spectra for oib-POSS. Both Raman and IR spectra show a group of bands at *ca.* 2900 cm^{-1} due to the stretching deformation of the C–H bonds, which are more intense in the Raman spectrum. The bands in the region 1470–1200 cm^{-1} are due to bending deformation of $-CH_2$ groups of the aliphatic chains. Moreover, the IR spectrum presents an absorption at *ca.* 1100 cm^{-1}, relative to Si–O bonds stretching, which is not strongly evident in the Raman spectrum.

Spectra collected on the residue after treatment at 200°C for 120 min (Figure 14) show how the Si–O modes signal is still evident in the IR spectrum,

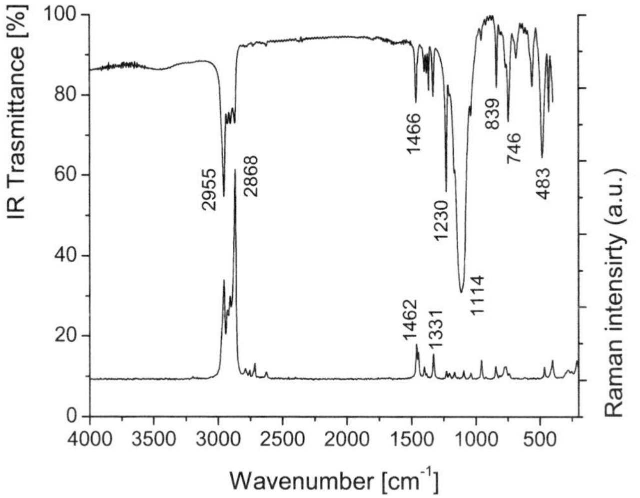

Figure 13 *IR and Raman spectra for neat oib-POSS*

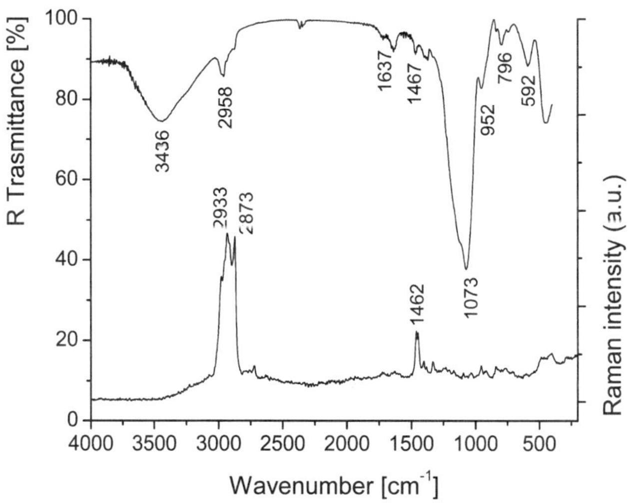

Figure 14 *IR and Raman spectra for isobutyl POSS after treatment at 200 °C for 120 min*

whereas the C–H bond signals are less intense than in the neat POSS but are still present, as is well shown by the Raman spectrum.

In the IR spectrum of Figure 14 a broad band centred at 3436 cm^{-1} is also observed, due to hydroxyl groups H-bonded to water molecules that are adsorbed on the surface, and an absorption at 1637 cm^{-1}, related to bending of water molecules. The presence of hydroxyl groups is probably due to partial cage fragmentation, as confirmed by WAXD analysis (see below).

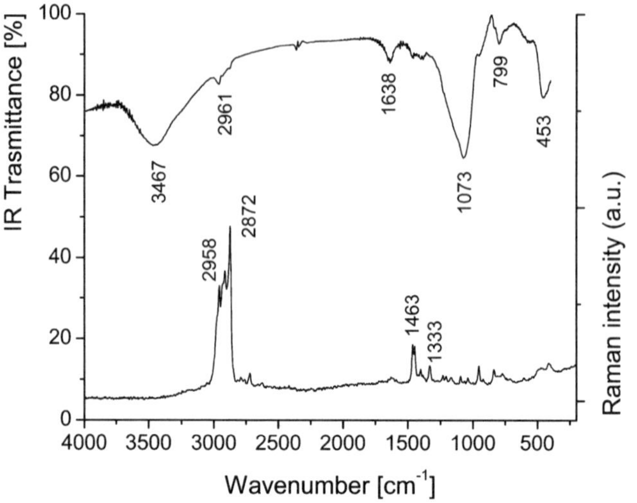

Figure 15 *IR and Raman spectra for oib-POSS after treatment at 300 °C, 60 min*

The POSS typical pattern of sharp peaks in the region between 1200 and 1550 cm^{-1} is attenuated in the IR spectrum of the residue, while the Si–O stretching band is broader than in original oib-POSS.

Spectra recorded after a 300°C (60 min) isothermal treatment (Figure 15) show a decrease in the intensities of the signals related to the organic fragments in the material, according to the proposed trend.

After treatment at 500°C (30 min) or higher temperatures, samples show fluorescence, which is typical of a silica phase, so that Raman spectroscopy is no longer helpful.

Finally, in the residue obtained after isothermal treatment of oib-POSS at 200°C followed by heating to 800°C no signal related to the organic fractions is detectable, the main vibrational features being very close to those of a typical silica (Aerosil) taken as reference (Figure 16).

The Si–O stretching band in IR spectra (centered at about 1110 cm^{-1} for neat oib-POSS) shifts towards 1070 cm^{-1} for residues treated at 200 and 300°C (Figure 17); we suppose this shift to be related to the formation, during decomposition, of a network structure in which Si–O bond energy is modified by the steric hindrance of residual organic groups, in agreement with a previous study.[23] Moreover, in the residue treated at 800°C, where no organic fraction is present in the material, the peak for Si–O stretching is aligned with that observed in silica.

As a result, heating oib-POSS in air leads to a partial loss of the POSS organic fraction, with the consequent formation of a silica-like phase, even at low temperature, *e.g.* 200°C. However, this phase obtained at low temperature is not fully stable; on heating the residue obtained at 200 up to 800°C a further weight

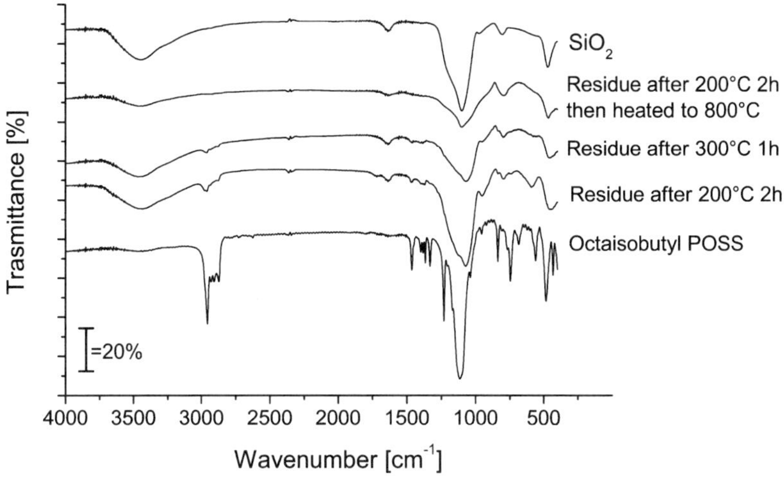

Figure 16 *IR spectra for POSS after different thermal treatments in air*

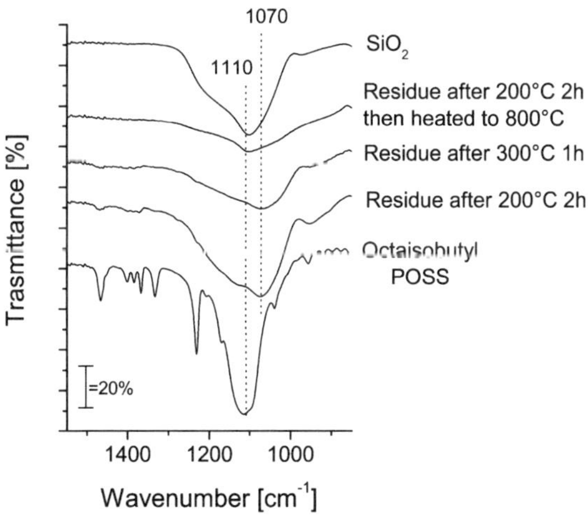

Figure 17 *IR signals for Si–O bonds (Dashed line at 1070 cm^{-1} highlights signal for network structure Si–O bonds, in comparison to signal for unstretched Si–O bonds at 1110 cm^{-1})*

loss occurs, confirming that an organic fraction is still present at the end of the first treatment (Figure 18).

Structural insights can be highlighted using X ray powder diffraction on the residues treated at different temperatures. The WAXD pattern for oib-POSS presents several sharp diffraction peaks due to product crystallinity; in particular, 2θ peaks at 8.0° and 8.8° are the most intense (Figure 19).

X-ray analysis on the residues from thermal treatments in air points out the formation of an amorphous phase. In particular, a very broad band between 15°

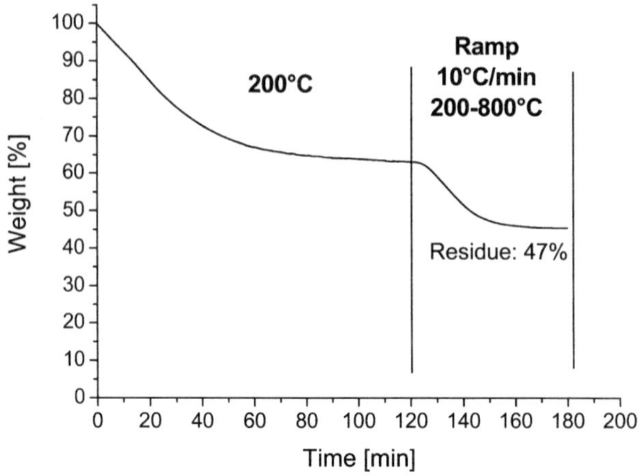

Figure 18 *Thermogravimetry for oib-POSS (Sample isothermally treatment at 200 °C and than heated at 10 °C min^{-1} to 800 °C)*

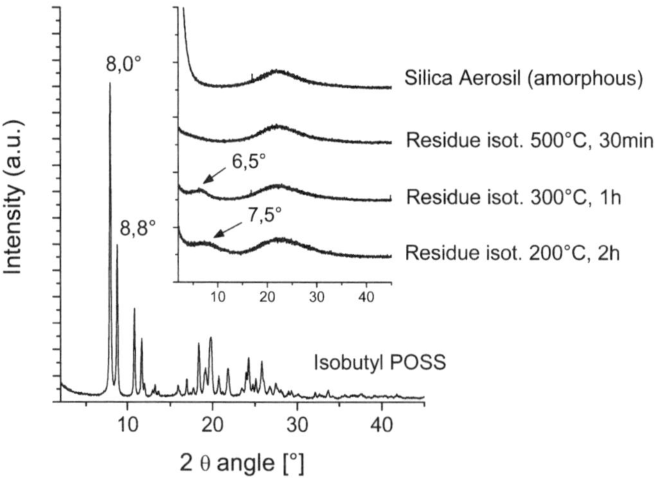

Figure 19 *WAXD patterns for POSS after different thermal treatments in air (residue patterns are shifted)*

and 30° 2θ angles is shown, which is typical of amorphous silica (Figure 19). Furthermore, in the X ray spectra of residues obtained at 200 and 300°C, a broad band appeared around $2\theta = 7°$, probably due to a rearrangement that partially retains the neat POSS order structure, as reported previously.[18] In the residue after treatment at 500°C no POSS-like residual order is observed so that the WAXD pattern is fully consistent with that of amorphous silica.

The chemical mechanism for conversion of oib-POSS into silica should involve thermal oxidation of the aliphatic isobutyl substituents; this process proceeds in organic structures through peroxidation, propagated by radicals created in the thermoxidative process, which occurs with higher probability on the tertiary carbon atom (Scheme 3).

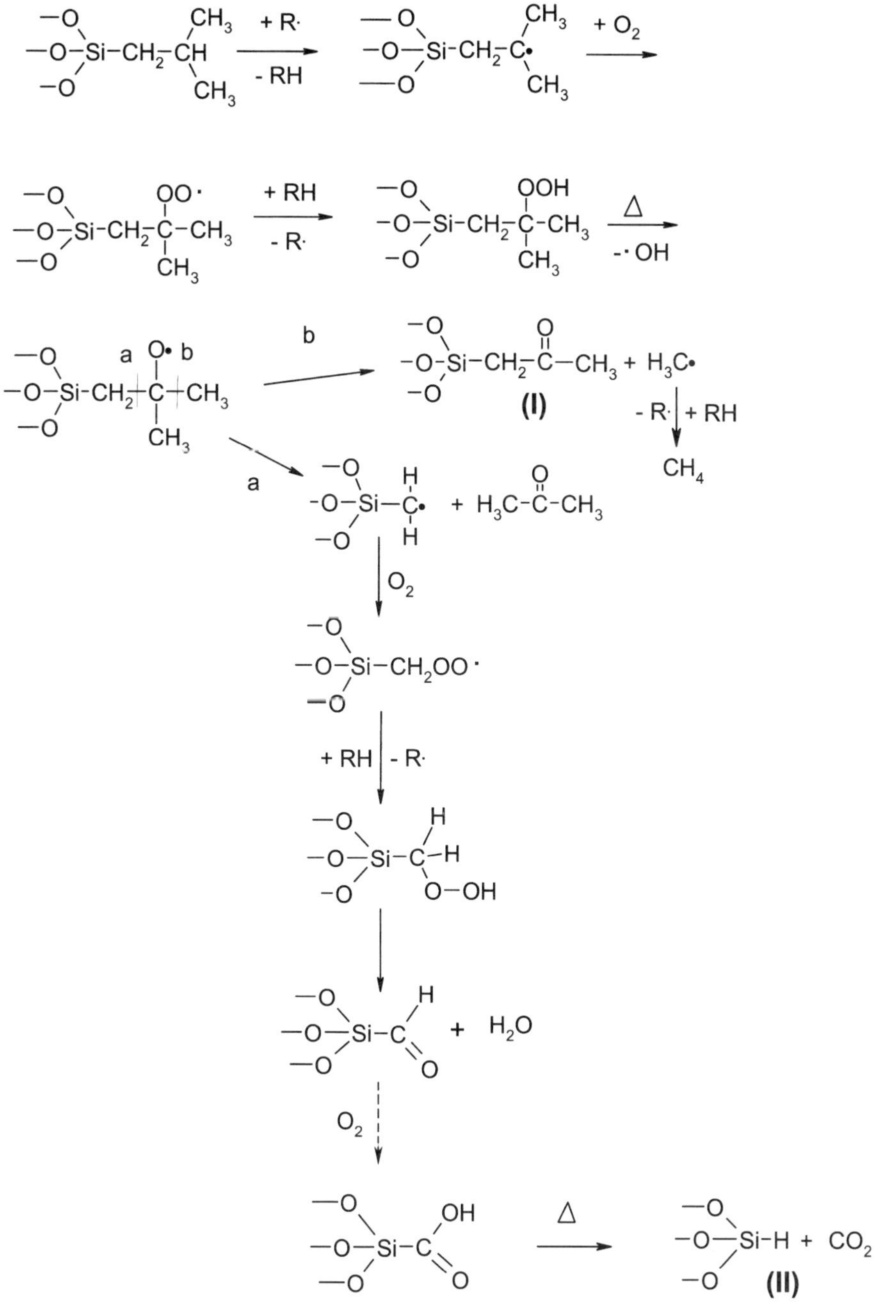

Scheme 3 *Peroxidation of the tertiary carbon atom of oib-POSS*

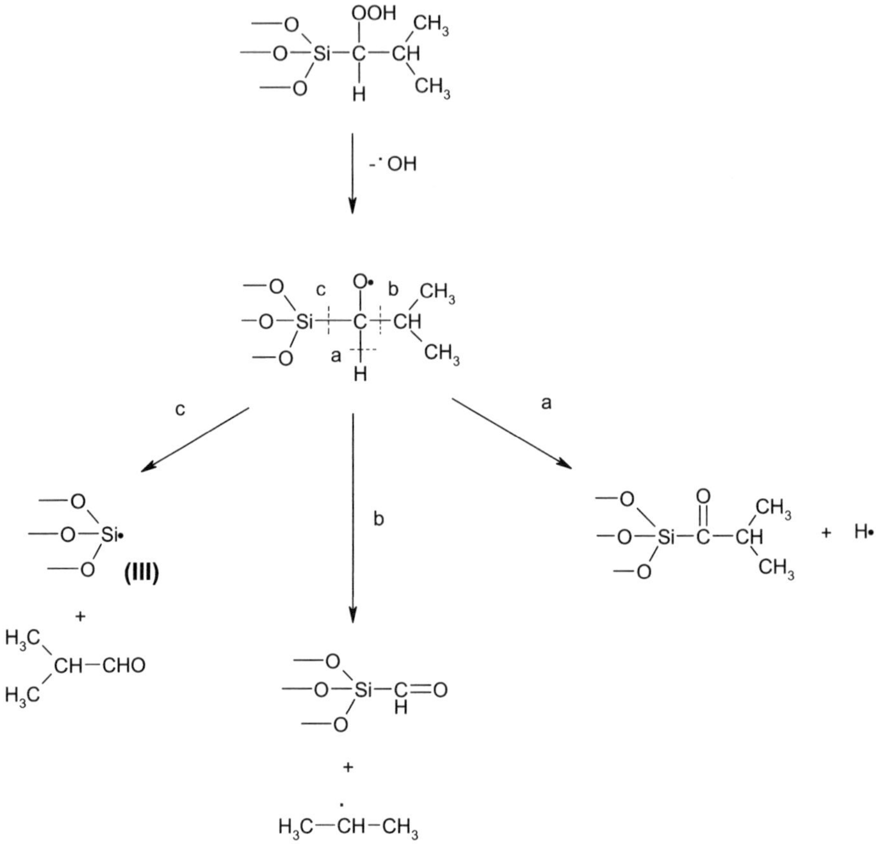

Scheme 4 *Peroxidation at the secondary carbon atom of oib-POSS*

Further peroxidation of (I) should also lead to the silane (II), whose mechanism of oxidation to silica through elimination of SiH_4 and H_2 has been published.[21] Peroxidation can also occur on the secondary carbon atom of oib-POSS (Scheme 4).

In this case, scissions a and b, by further oxidation shown in Scheme 3, lead to the silane (II), whereas route c gives a silicon-centered radical (III), which could react directly with oxygen to form a Si–O radical that is an effective precursor of the final structure of the silica residue obtained by thermal oxidation of oib-POSS.

15.4 Conclusions

Thermal programmed heat treatment of octaisobutyl POSS in inert conditions led to evaporation of the product, which is chemically stable until 600°C, at least. On flash heating at higher temperatures, the POSS organic fraction fragments, producing light organic compounds and a non-volatile silicon-containing phase.

In air, the degradation process is more complex; evaporation competes with an oxidation phenomenon that produced a silica-like thermally stable phase.

The structure of residual solid phase after thermoxidative treatments shows the coexistence of an inorganic siliceous part and a POSS-like organomodified phase. The amount of organic content in the residue depends on the thermal treatment conditions; the organic amount decreases with increasing treatment temperature, giving a silica-consistent structure for POSS treated in air up to 800°C.

The formation of such a ceramic phase can prefigure the use of POSS in polymeric materials fire retardancy.[15] In fact, during the burning process in the presence of oxygen, ceramisation could occur on the superficial layer thanks to the $(RSiO_{1,5})_n$ composition. The formation of this layer seems to produce a physical protection against combustion, by hampering diffusion of oxygen, evacuation of volatile and combustible products, and also limiting heat transfer, in analogy with the layered silicates mechanism.[26]

15.5 Acknowledgements

This project was carried out in the framework of an Italian interuniversity research program (COFIN 2002) funded by the Ministery of Education, University and Research (MIUR).

15.6 References

1. R.H. Baney, M. Itoh, A. Sakakibara and T. Suzuki, *Chem. Rev.*, 1995, **95**, 1409.
2. P.G. Harrison, *J. Organomet. Chem.*, 1997, **542**, 141.
3. G. Li, L. Wang, H. Ni and C.U. Pittman Jr., *J. Inorg. Organomet. Polym.*, 2001, **11**, 123.
4. K.J. Shea and D.A. Loy, *Chem. Mater.*, 2001, **13**, 3306
5. S. Lücke and K. Stoppek-Langner, *Appl. Surf. Sci.*, 1999, **144–145**, 713.
6. P.P. Pescarmona and T. Maschmeyer, *Aust. J. Chem.*, 2001, **54**, 583.
7. D.W. Scott, *J. Am. Chem. Soc.*, 1946, **68**, 356.
8. X. Brown, *J. Am. Chem. Soc.*, 1960, **82**, 6194.
9. F.J. Feher, D.A. Newman and J.F. Walzer, *J. Am. Chem. Soc.*, 1989, **111**, 1741.
10. R.L. Blanski, K.J. Weller and J.W. Ziller, *Organometallics*, 1991, **10**, 2526.
11. F.J. Feher and T.A. Budzichowski, *Polyhedron*, 1995, **14**, 3239.
12. P.A. Agaskar, *Inorg. Chem.*, 1990, **29**, 1603.
13. Proceedings of *"POSS Nanotechnology Conference"*, Huntington Beach, CA, September 2002.
14. S. Lu and I. Hamerton, *Prog.-Polym. Sci.*, 2002, **27**, 1661.
15. J.D. Lichtenhan and J.W. Gilman, *US Patent 6,362,279.*, 2002.
16. H.C.L. Abbenhuis, *Chem. Eur. J.*, 2000, **6**, 25.

17. C. Bolln, A. Tsuchida, H. Frey and R. Mülhaupt, *Chem. Mater.*, 1997, **9**, 1475.
18. R.A. Mantz, P.F. Jones, K.P. Chaffee, J.D. Lichtenhan and J.W. Gilman, *Chem. Mater.*, 1996, **8**, 1250.
19. M.J. Loboda and G.A. Toksey, *Solid State Technol.*, 1998, **41**, 99.
20. H.C. Liou and J. Pretzer, *Thin Solid Films*, 1998, **335**, 186.
21. Y.K. Siew, G. Sarkar, X. Hu, J. Hui, A. See and C.T. Chua, *J. Electrochem. Soc.*, 2000, **147**, 335.
22. C.C. Yang and W.C. Chen, *J. Mater. Chem.*, 2002, **12(4)**, 1138.
23. W.C. Liu, C.C. Yang, W.C. Chen, B.T. Dai and M.S. Tsai, *J. Non-Cryst. Solids*, 2002, **311**, 233.
24. C. Ramirez, M.J. Abad, L. Barral, J. Cano, F.J. Diez, J. Lopez, R. Montes and J. Polo, *J. Therm Anal. Calorim.*, 2003, **72**, 421.
25. M.J. Abad, L. Barral, D.P. Fasce and R.J.J. Williams, *Macromolecules*, 2003, **36**, 3128.
26. M. Zanetti, T. Kashiwagi, L. Falqui and G. Camino, *Chem. Mater.*, 2002, **14**, 881.

Mineral Fillers in Synergistic Systems

CHAPTER 16

Interactions between Nanoclays and Flame Retardant Additives in Polyamide 6 and Polyamide 6.6 Films

A. RICHARD HORROCKS, BALJINDER K. KANDOLA AND SUZANNE A. PADBURY

Centre for Materials Research and Innovation, Bolton Institute, Bolton, BL3 5AB, UK (arh1@bolton.ac.uk)

16.1 Introduction

We recently briefly reviewed the research published since 2000 regarding the consequences of introducing functionalised nanoclays into synthetic fibre-forming polymers such as polyamide 6, and 6.6 fibres.[1] This same review presents evidence, based on recent heat release studies of polyamide 6 composites of varying thickness at high heat fluxes such as 50 kW m^{-2},[2] that the ability of any nanodispersed clay particles present at 2 wt% will be marginal, and at higher concentrations of 5 wt% will produce less than a 10% reduction in peak heat release rate, PHRR. The only published work to date on the effect of nanoclays in polyamide 6 fibres is that by Bourbigot *et al.*[3,4] which shows a 33% reduction in PHRR when exposed to a heat flux of 35 kW m^{-2}. We have explained this larger effect at the lower heat flux as a consequence of the competition between nanoclay particle aggregation to form a surface barrier and volatilisation in a thermally thin condition, favouring the former, whereas at higher heat fluxes the latter is favoured.[1]

When condensed phase, sometimes intumescent, phosphorus-containing flame retardants are added to polyamide 6, and 6.6 polymers, the flame retardant property versus concentration relationship is not linear, and usually "S"-shaped as reviewed by Levchik, and Weil.[5] As indicated by ourselves,[1] significant flame retardancy is observed only at FR concentrations for ammonium polyphosphate, and similar species greater than 15 wt% in fibre-forming polymers such as polypropylene and polyamide 6. Usually, levels > 20% are required in bulk polyamide formulations to achieve V0 ratings in the UL94 test, and this is considered to correspond to an LOI > 25 or so. For potential fibre end-uses, such high levels are inappropriate because of the deleterious effects on desirable textile properties such as strength, lustre, and dyeability. Ideally, flame retardant levels < 10 wt% are more acceptable for fibre applications.

Recent research on a number of different polymeric substrates has led several workers to believe,[6] including ourselves, that the introduction of nanoclays, while enhancing char levels will only provide realistic fire resistant solutions if incorporated in polymers with other compatible and, hopefully, synergistic FR components.

While previous results have presented greater experimental detail,[7] this chapter concentrates on those nanoclay-flame retardant combinations that demonstrate at least additive, and at best synergistic, behaviour for several phosphorus-containing flame retardants in the presence of commercially available nanocomposite polyamides 6, and 6.6 cast into films. These are considered as possible models for fibres, and other thermally thin polymeric materials, and the results are interpreted in terms of our previously generated model.[1] The prime flame retardant parameter described here is the limiting oxygen index.

16.2 Experimental

16.2.1 Materials

A range of phosphorus-containing flame retardants was selected (Table 1). Proban CC polymer was prepared by introducing a commercial sample of the THPC-urea [tetrakis(hydroxymethyl)phosphonium chloride urea] condensate solution into a desiccator containing 0.91 g ml^{-1} ammonia solution. Once the solution had solidified it was then washed, and dried, prior to grinding to a fine powder. The ground polyphosphine-ammonia condensate was then oxidized in a 7 vol% hydrogen peroxide solution until the exothermic reaction had finished. Further washing, and drying was carried out, and the oxidised, stable Proban CC polymer was then reground to a finer powder that was then ready to use.

Polyamides 6 and 6.6, both with and without nanooclay (unspecified type present at a nominal 2 wt% level), were supplied as pellets by RTP Company (UK) Plastics Ltd.

Table 1 *Flame retardants*

FR	*Manufacturer*	*Constitution*
Antiblaze MCM	Rhodia	Ammonium polyphosphate
Antiblaze NH	Rhodia	Melamine phosphate
Antiblaze CU	Rhodia	Cyclic organophosphonate
Proban CC polymer	Rhodia	Poly (phosphine oxide)
Antiblaze MCM/pentaerythritol	Rhodia	APP/PER
MPC 1000	Rhodia	APP/PER/melamine
NH 1197	Great Lakes	Pentaerythritol phosphate
NH 1511	Great Lakes	PER phosphate/melamine

16.2.2 Film Preparation

Each polymer dope was made by dissolving a given weight of polymer chips (standard, or comprising a nanoclay, in this instance) in a calculated volume of 90% formic acid. For each polymer/additive solution, a total solid content of 33% w/v was maintained, irrespective of FR additive incorporated, as experimental work indicated this to be the most suitable viscosity to work with. Within each solution was a selected mass ratio of components–polymer, nanoclay if present, and FR such that approximate FR contents were 11, 15, 20, 23 or 27 wt%.

The film casting technique involved spreading the polymer dope on to a glass plate using a K-bar (selected on the basis of obtaining a film thickness of *ca.* 50 μm), and then leaving the film to stand in a fume cupboard for *ca.* 24 hours for the formic acid to fully evaporate off. The films were then peeled away from the glass plate. No attempt was made to analyse the size or distribution of the dispersed flame retardants, and it was evident that, except for films containing the liquid flame retardant, Antiblaze CU, all cast films had varying degrees of opacity compared with the translucent FR-free films. In fact, films containing APP were chalky in appearance, indicating their extremely heterogeneous character.

Phosphorus analysis was carried out on randomly selected films to ensure complete retention of the additive within the generated film. Thickness testing of each cast film was carried out, verifying overall uniformity of each cast film. Film thicknesses were typically in the range 40–50 μm.

16.2.3 Flammability Measurement

Limiting oxygen index measurements were carried out on a Stanton Redcroft FTA instrument for film samples according to ASTM D2863-77 (revised 1990). Since the area densities of cast films were 40–50 g m^{-2}, LOI measurements were undertaken on double-layered samples to give an area density of 80–100 g m^{-2}, which is similar to that of a lightweight textile fabric.

16.2.4 Thermal Analysis

Differential thermal analytical studies of 5 mg samples were undertaken in duplicate in a TA Instruments SDT 2960 under flowing nitrogen (100 ml min^{-1}) at a heating rate of 10 K min^{-1}.

16.3 Results and Discussion

16.3.1 Thermal Analytical Behaviour: Nanocomposite Character

The normal ways of ascertaining nanocomposite character in polymers are by use of transmission electron microscopy or X-ray diffraction.[8,9] However, the presence of a dispersed nanoclay in polyamide 6 favours the formation of the lower melting γ-form (melting point about 212°C) in unorientated samples.[10,11] While evidence of a similar favourable form in polyamide 6.6 has not been demonstrated, Figures 1 and 2 show DTA responses for both polyamide 6.6 and

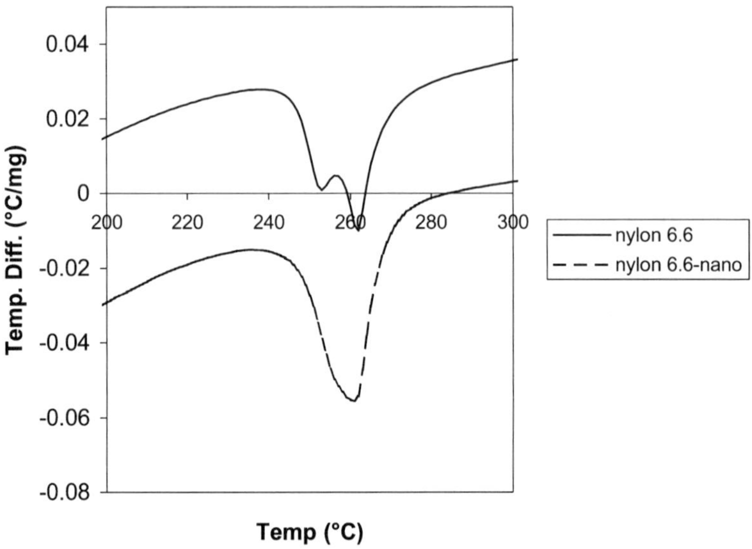

Figure 1 *DTA curves for the polyamide or nylon 6.6 polymer, with and without the inclusion of the nanoclay, and in the absence of FR additive*

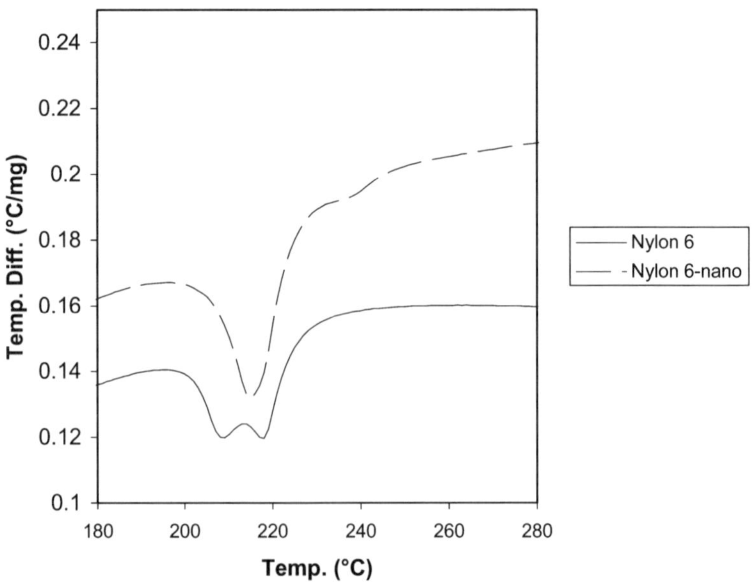

Figure 2 *DTA endotherms for polyamide or nylon 6 in combination in the presence and absence of the nanoclay, and with the exclusion of FR additive*

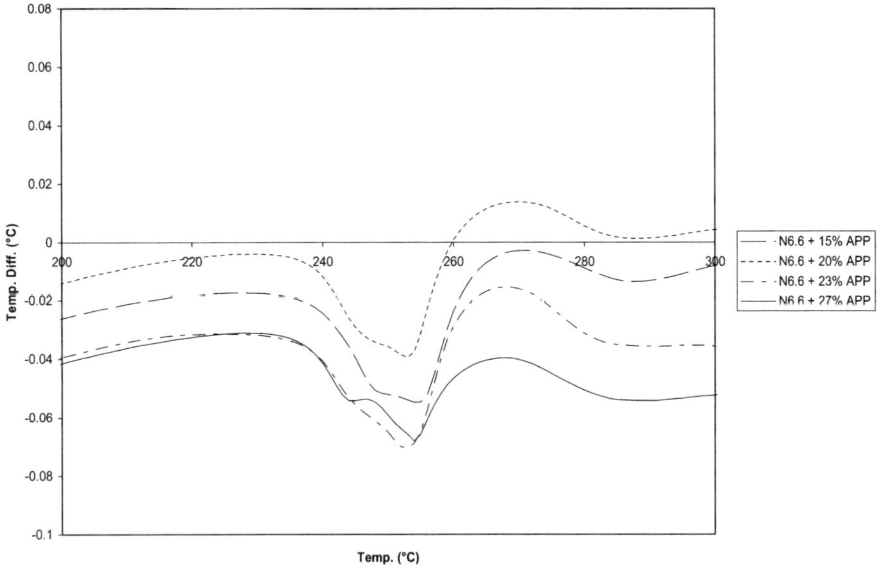

Figure 3 *Resultant DTA endotherms for polyamide 6.6 films with the inclusion of both nanoclay and APP at 15–27% FR additive concentration levels*

6 polymer samples, respectively, in the absence and presence of nanoclay for the commercial polymers selected in this work.

Figure 1 shows that the bimodal structure of the fusion endotherm for the pure polyamide 6.6 film, presumably the α-form, changes to a single peak when the nanoclay is present. The change in shape is indicative of a modification to the polycrystalline structure, and that this could be associated with the effect of the nanodispersed particles influencing the ordering of polyamide chains. Figure 3 typifies the effect of the addition of flame retardant, which for increasing concentrations of APP restore the bimodal character. Table 2 lists the fusion minima temperature for normal and nanocomposite polyamide 6.6 films, with and without flame retardants. Evidently, the positions of the single or major bimodal minima are similar for both unretarded films, and the addition of any of the flame retardants reduces minima temperatures slightly. However, increasing FR concentration has little further effect. That each of the flame retardants has a lowest endotherm, which may be lower than (*e.g.* 191°C for APP/PER as a consequence of PER melting) or greater than the polymer melting point (*e.g.* the volatilisation endotherm for Antiblaze CU at about 272°C or decomposition of APP at about 311°C), and that these appear to have little effect on polyamide 6.6 melting points, indicates that they are indeed micro-dispersed within films.

In Figure 2 the presence of nanoclay in polyamide 6 films not only causes a similar shift from a bimodal endotherm to a single minimum, but the temperature of the latter is less than that of the higher melting bimodal minimum. Table 3 shows this shift from 217.4 to 215.3°C to be indicative of the favouring of the γ-crystalline form, suggestive of a nanocomposite structure.[10,11] The addition of

Table 2 *Melting points for flame retardant polyamide 6.6 formulations, with and without the inclusion nanoclays, including decomposition/melting minimum for various pure additives*

	Standard film/FR (%)					*Film comprising nanoclay/FR* (%)				
Specimen	*11*	*15*	*20*	*23*	*27*	*11*	*15*	*20*	*23*	*27*
Nylon 6.6			262.6					262.1		
Pure APP						311.2				
N6.6 + APP	254.2	254.8	253.6	250.8	252.3	257.1	253.0	252.5	249.6	252.2
Pure NH						–				
N6.6 + NH	259.1	261.1	260.5	261.6	261.5	260.8	261.2	260.0	259.4	260.1
Pure CC						279.2				
N6.6 + CC	257.9	256.7	256.0	255.5	255.7	258.2	259.6	258.1	258.4	257.2
Pure 1197						–				
N6.6 + 1197	253.7	252.5	249.9	246.9	244.4	254	250.3	249.9	245.3	240.4
Pure 1511						–				
N6.6 + 1511	258.7	258.9	255.7	257.9	256.7	259.0	258.1	257.3	256.0	256.0
Pure CU						271.7				
N6.6 + CU	260.0	258.1	255.7	257.2	254.3	259.2	258.0	257.0	256.3	254.3
Pure MPC1000						–				
N6.6 + MPC1000	259.0	258.3	256.8	255.9	254.3	258.4	257.8	257.4	255.8	254.1
Pure APP/PER						191.1				
N6.6 + APP/PER	257.7	256.5	254.6	253.6	–	257.8	253.5	254.4	253.0	–

flame retardant has a similar effect to that seen in polyamide 6.6, where in the nanoclay-free film the bimodal shape is preserved, and the monomodal nanocomposite DTA fusion response gradually transforms to a bimodal one at higher FR concentrations. Table 3 shows that, for all nanoclay-containing, flame retarded films, fusion peak minima temperatures are less than the nanoclay-free analogue films, suggesting that the nanocomposite characters of the former films are preserved following addition of these retardant species. Interestingly, the lowest fusion endotherm temperature shifts occur when APP/PER and MPC retardant systems are introduced, and this may be a consequence of the proximity of the melting point of 191°C of the PER present in each being close to that of the polyamide 6.

Table 3 *Melting points for flame retardant polyamide 6 formulations in the absence, and presence of a nanoclay*

	Standard film/FR (%)					*Film comprising nanoclay/FR* (%)				
Specimen	*11*	*15*	*20*	*23*	*27*	*11*	*15*	*20*	*23*	*27*
Nylon 6			217.4					215.3		
N6 + APP	217.7	217.0	215.3	215.4	214.7	216.0	213.0	211.6	212.0	211.0
N6 + CC	220.1	220.4	215.7	215.6	215.0	216.1	216.7	216.8	216.5	216.3
N6 + MPC1000	218.5	207.9	211.3	213.3	212.2	217.1	212.5	213.4	214.2	212.7
N6 + APP/PER	212.3	214.5	210.4	210.2	–	213.2	212.5	208.2	205.0	–

16.3.2 Limiting Oxygen Index Measurements

16.3.2.1 Polyamide 6.6

To determine whether particular flame retardants acted positively or negatively in the presence of the nanoclay present, LOI results obtained for all the generated films were initially represented as the difference between respective film values, with and without the presence of a nanoclay, *i.e.* $\Delta LOI = LOI_{(nanoclay+FR)} - LOI_{(FR)}$. Tables 4 and 5 contain the individual sets of LOI data for these two respective flame retardant groups. These difference curves are illustrated in Figure 4, and suggest that the incorporation of a nanoclay, in conjunction with the FR additive, does not necessarily increase the LOI of the film sample containing FR only. Certain FR additives, in particular Antiblaze CU and melamine phosphate NH, behave in a negative manner. The remainder of the films examined indicate a positive effect at FR levels < 20%. Above this concentration level, however, ΔLOI values exhibit an overall decline.

LOI *versus* FR concentration data from Table 4 for the positive ΔLOI-generating polyamide 6.6/FR and polyamide 6.6/FR/nanoclay systems are illustrated in Figure 5. Trends for each formulation show general increases in LOI with FR concentration with APP, Proban CC, and MPC1000 retardants in the absence of nanoclays as "S"-shaped trends with significant LOI increases

Table 4 *Positive ΔLOI flame retardant data for polyamide 6.6 films*

Additive & actual level (wt%)	*%P*	*Standard films* LOI%	*Standard films* ΔLOI/P	*Nano films* LOI%	*Nano films* ΔLOI/P
No additive	–	21.0	–	21.8	–
11% APP	3.2	21.4	0.13	22.2	0.13
15% APP	4.8	21.4	0.08	23.0	0.25
20% APP	6.4	21.4	0.06	23.8	0.31
23% APP	7.0	23.4	0.34	25.0	0.46
27% APP	8.2	24.6	0.44	25.8	0.49
11% CC	1.8	21.6	0.33	22.2	0.22
15% CC	2.4	22.4	0.58	23.4	0.67
20% CC	3.2	22.8	0.56	24.2	0.75
23% CC	3.7	23.6	0.70	24.6	0.76
27% CC	4.3	24.4	0.79	24.6	0.65
11% MPC1000	2.0	21.8	0.40	21.8	0
15% MPC1000	2.7	21.8	0.30	22.6	0.30
20% MPC1000	3.6	23.8	0.78	24.2	0.67
23% MPC1000	4.1	24.2	0.78	24.6	0.68
27% MPC1000	4.9	25.4	0.90	24.6	0.57
11% APP/PER	2.2	21.8	0.36	22.2	0.18
15% APP/PER	3.0	22.6	0.53	22.6	0.27
20% APP/PER	4.0	23.0	0.50	23.0	0.30
23% APP/PER	4.6	24.2	0.70	23.4	0.35

Table 5 *Negative or zero ΔLOI flame retardant data for polyamide 6.6 films*

Additive & actual level (wt%)	*%P*	*Standard films*		*Nano films*	
		LOI	*ΔLOI/P* (%)	*LOI*	*ΔLOI/P* (%)
No additive	–	21.0	–	21.8	–
11% NH	1.5	22.0	0.67	22.2	0.27
15% NH	2.1	22.0	0.48	22.2	0.19
20% NH	2.8	22.0	0.36	22.6	0.29
23% NH	3.2	24.6	1.13	23.0	0.38
27% NH	3.8	26.2	1.37	23.4	0.42
11% 1197	1.9	21.4	0.21	21.8	0
15% 1197	2.6	21.8	0.31	21.8	0
20% 1197	3.4	21.8	0.24	22.6	0.24
23% 1197	4.0	22.2	0.30	23.0	0.30
27% 1197	4.7	22.6	0.34	23.0	0.22
11% 1511	1.7	21.4	0.24	21.8	0
15% 1511	2.3	21.4	0.17	21.8	0
20% 1511	3.0	21.8	0.27	21.8	0
23% 1511	3.5	22.2	0.34	22.2	0.11
27% 1511	4.1	22.6	0.39	22.6	0.20
11% CU	2.4	23.8	1.17	23.0	0.50
15% CU	3.3	24.2	0.97	23.0	0.36
20% CU	4.3	24.6	0.84	24.6	0.65
23% CU	4.9	25.0	0.82	24.6	0.57
27% CU	5.8	25.0	0.69	24.6	0.48

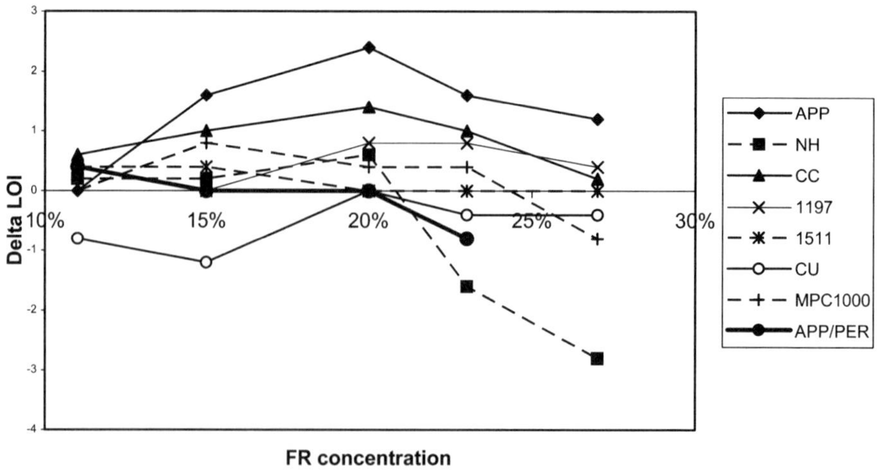

Figure 4 *ΔLOI for all additives examined in polyamide 6.6 films*

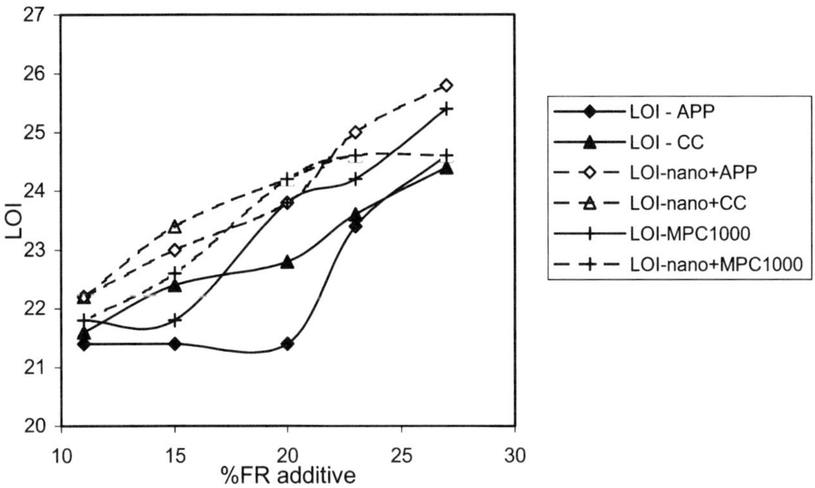

Figure 5 *LOI for FR/polyamide 6.6 and FR nanocomposite polyamide 6.6 films for FRs showing positive behaviour in the presence of nanoclay*

occurring only above 15 wt% presence. The presence of the nanoclay smoothes each of these trend shapes into a more uniform trend, and effectively shifts each to the left-hand side, demonstrating the origin of the positive ΔLOI trends in Figure 4.

Of specific interest is the behaviour of APP, which exhibits the lowest LOI at low concentration in the absence of nanoclay and yet yields the highest LOIs at 20% FR and above in the presence of nanoclay, as indicated also in Figure 4. The intumescent APP-containing MPC1000 films generally shows superior values in the absence of the nanoclay but, in contrast, exhibit the minimal increase when the nanoclay is present. However, all of the effective systems that include the nanoclay demonstrate possible synergistic behaviour, showing that lower FR addition levels produce higher LOI in comparison to the films without the clay.

While the melamine phosphate Antiblaze NH initially shows a good LOI increase with concentration, the inclusion of the nanoclay suggests that its presence is minor. However, during film casting, this sample produced a slightly grainy, surface-textured opaque film, suggesting that the nanoclay had aggregated, possibly because of the relative acidity (pH ~3.2) of this FR, which could have interacted with the positively charged, functionalised component of the nanophase. NH1197 and NH1511 additives, both possessing a pH of 4.5, produced clearer, translucent films, and still only achieve minimal activity when incorporated with the nanoclay. Interestingly, Antiblaze CU, although demonstrating much higher LOIs, behaves in a similar manner to NH1197, and NH1511.

Further reason for the positive nanoclay-FR activities of APP-containing and Proban CC films could be a consequence of their being highly activated at the melting temperature of polyamide 6.6 (~265°C). For example, APP starts to decompose at 251°C[12] with release of ammonia, water, and free P-OH acidic groups.[13] Melamine phosphate (Antiblaze NH) also starts to decompose at 256°C, but yields predominantly melamine pyrophosphate initially.[14] Both NH1511 and NH1197 pentaerythritol phosphate derivatives are far more stable with respective onsets of degradation at 280°C, and 288°C. Interestingly, Antiblaze CU is a liquid, and starts to lose mass at only 197°C,[13] although whether this involves major chemical decomposition is not known.

With regard to APP/PER polyamide 6.6 films, the results in Table 5 show that the presence of PER (as a char source, not as a specific flame retardant) increases the overall LOIs obtained but a negligible effect of nanoclay is evident, as discussed previously.[7]

Comparison of ΔLOI/P of each FR additive examined, both with and without the presence of the nanoclay, enables the effect of nanoclay on the potential effectiveness of a given phosphorus-containing FR to be assessed. For example, ΔLOI/Ps for nanoclay-containing APP and Proban CC formulations (Table 4) are greater than those not containing nanoclay, showing that the presence of the latter has effectively increased the respective FR efficiency. For MPC1000 and APP/PER, this increase with respect to phosphorus is not seen because the effect of nanoclay presence is only marginal up to 20% FR content, and ΔLOI values relate to slightly different respective zero% FR polymer values.

ΔLOI/P results presented in Table 5 show that the presence of the nanoclay significantly reduces, and even negates completely, the effect of the FR present. To determine the overall synergistic effectivity of each FR additive investigated the method of Lewin[15] has been employed in that $\Delta LOI_{(FR+synergist)}/\Delta LOI_{(FR)}$ values were calculated for each FR in the presence of the nanoclay (*i.e.* the synergist) and expressed per unit% of FR. Figure 6 shows plots of effectivity

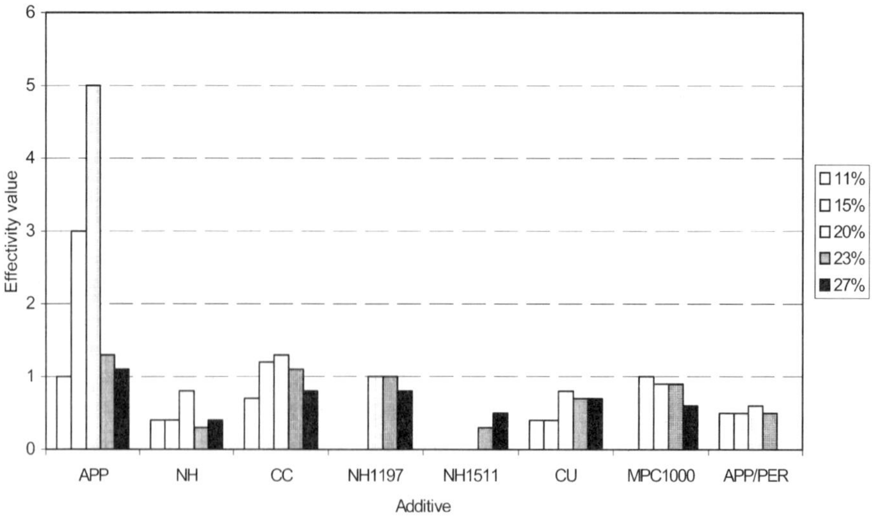

Figure 6 *Synergistic effectivity of nanoclay presence for all FR systems in polyamide 6.6 films*

represented as the increase in LOI for 1% of the FR element, and only those systems having effectivity values greater than unity are synergistic.

In conclusion, therefore, it is apparent from the results shown here that only APP and Proban CC demonstrate synergistic behaviour, while MPC1000 and APP/PER, previously categorised as effective systems in Figure 4, impart increased FR activity *via* the additive effect of nanoclay, and the respective FR.

16.3.2.2 Polyamide 6

Based on the above results, films were cast that contained only APP, Proban CC, MPC1000, and APP/PER as flame retardants. The LOIs obtained for the polyamide 6 films generated with and without the presence of the nanoclay versus concentration are shown in Table 6, and these results have been analysed in detail elsewhere.[7] When the results are plotted as $\Delta LOI = LOI_{(nanoclay+FR)} - LOI_{(FR)}$ *versus* FR concentration in Figure 7, it would appear that the presence of nanoclay acts in an antagonistic manner. However, the presence of nanoclay has depressed the LOI of polyamide 6 film alone from 22.6 to 18.8, and so the effect of added flame retardant should be referred to this reduced value. This may be undertaken by comparing the increase in LOI per unit phosphorus, ΔLOI/P in the absence, and presence of nanoclay as listed in Table 6.

Table 6 *LOI and ΔLOI flame retardant data for polyamide 6 films*

		Standard films		*Nano films*	
Additive	*%P*	*LOI*	*ΔLOI/P* (%)	*LOI*	*ΔLOI/P* (%)
No additive	–	22.6	–	18.8	–
11% APP	3.2	23.4	0.25	20.6	0.56
15% APP	4.8	23.4	0.17	22.4	0.75
20% APP	6.4	24.2	0.25	24.8	0.94
23% APP	7.0	26.0	0.49	26.4	1.02
27% APP	8.2	26.0	0.41	26.8	0.98
11% CC	1.8	23.4	0.44	21.4	1.44
15% CC	2.4	24.2	0.67	21.8	1.25
20% CC	3.2	25.0	0.75	22.2	1.01
23% CC	3.7	25.0	0.65	23.4	1.24
27% CC	4.3	25.0	0.28	23.8	1.16
11% MPC1000	2.0	23.8	0.60	21.0	1.10
15% MPC1000	2.7	23.8	0.44	21.8	1.11
20% MPC1000	3.6	25.4	0.78	22.2	0.94
23% MPC1000	4.1	25.4	0.68	23.4	1.12
27% MPC1000	4.9	25.8	0.65	25.4	1.35
11% APP/PER	2.2	23.8	0.55	21.0	1.00
15% APP/PER	3.0	24.2	0.53	21.8	1.00
20% APP/PER	4.0	24.6	0.50	22.6	0.95
23% APP/PER	4.6	24.6	0.43	23.8	1.09

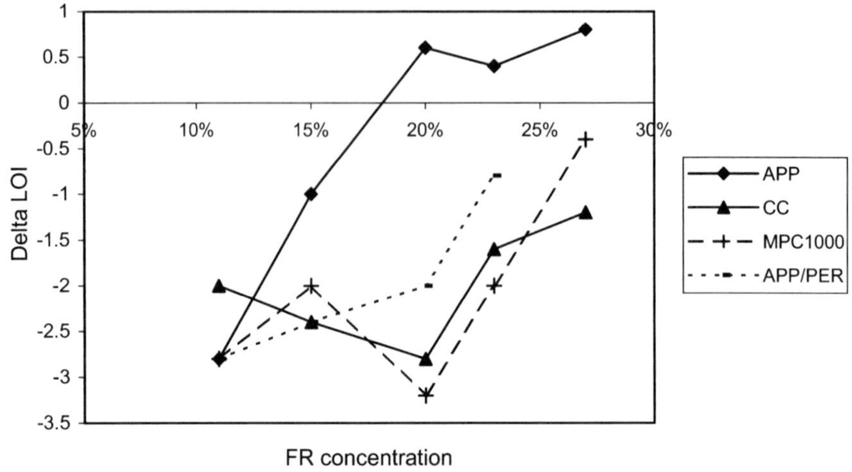

Figure 7 *ΔLOI for FR additives in polyamide 6 films*

These results show that ΔLOI/Ps are significantly greater for nanoclay-containing, flame retarded polyamide 6 films, as noted in Table 4 for polyamide 6.6 films. The lowering of the LOI of polyamide 6 to 18.8 prevents synergistic effectivity values being calculated as for polyamide 6. However, the potentially positive interaction between nanoclay and the selected FRs in polyamide 6, by defining the ratio $R = [\Delta LOI/P]_{nanoclay} / \Delta LOI/P$ for each flame retardant and by plotting R *versus* FR%, is shown in Figure 8.

These trends show clearly that the presence of nanoclay increases the FR efficiency by at least an average factor of 2, thereby indicating a positive nanoclay–FR interaction for each of these retardants. The effect of the nanoclay

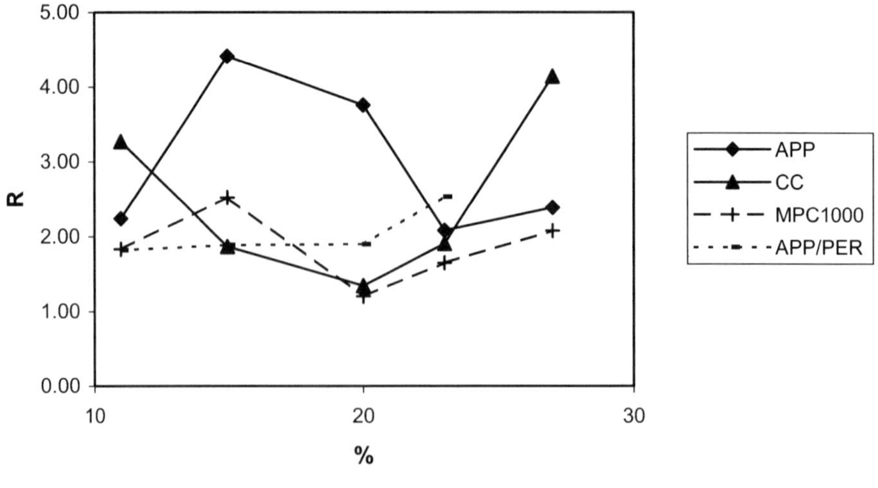

Figure 8 *Ratio, R, of increase in LOI per unit phosphorus in nanoclay-containing polyamide 6 film compared to the standard film*

alone on the burning behaviour of unretarded polyamide 6 films has been explained previously in terms of a change in the rheology of the melting film, and hence the value of the LOI measurement obtained.[7]

16.4 A Simple Model for Nanoclay–Fr Interation

In our previous publication[1] we suggested a simple way of quantitatively modelling the char-enhancing or "bridging" effect that the presence of nanodispersed functionalised clays may exert when present with microdispersed flame retardant particles. This original model involved calculating the mean distance between flame retardant particles having a certain size and dispersed in a polymer matrix, which for simplicity may be considered to be homogeneous by applying the simple concept of "mean free path" as an average distance, l, between dispersed species having a "collision diameter" or average diameter, σ. In a gas, it is assumed that individual gas molecules may be free to move as shown in Figure 9(a); in the situation of particles suspended in a polymer, where they are fixed, the "movement" is that of the reaction zone that spreads out from a given particle when heated above its reaction temperature. This is shown schematically in Figure 9(b). This may, for our purposes, also be influenced by the collisional frequency of reactive species in the molten polymer. Thus, the spread of this reaction zone in a polymer volume V may have a mean radius or "reaction length" before "colliding" or interacting with a second particle that may be defined for a single microdispersed flame retardant as follows shown in Equation (1).[16]

$$l = V/(\sqrt{2}\pi N\sigma^2) \tag{1}$$

for N dispersed spherical particles. If the retardant is present at a volume fraction v_{fa}, where $v_{fa} = (m_{fa}/\rho_a)/(m_{fa}/\rho_a + m_{fp}/\rho_p)$ and m_{fa}, and m_{fp}, and ρ_a, and ρ_p are respective additive and polymer mass fractions, and densities, then:

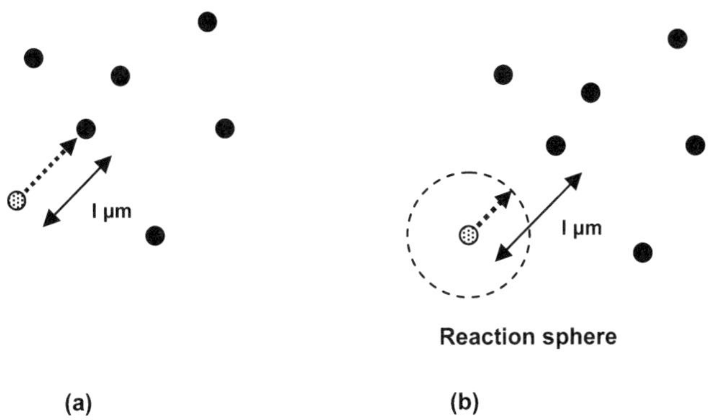

Figure 9 *Schematic representation of* (a) *mean free path of a gas molecule, and* (b) *of a randomly expanding reaction zone around a flame retardant particle*

$$N = (Vv_{fa})/[\frac{4}{3}\pi(\sigma/2)^3] \tag{2}$$

Combining Equations (1), and (2) yields:

$$l = \sigma/(\sqrt{2.6v_{fa}}) = \sigma/(8.5v_{fa}) \tag{3}$$

Thus for a given concentration, and hence volume fraction of flame retardant in a given volume V of polymer, as particle diameter increases so the mean "reaction zone" separation increases; conversely, as the diameter decreases so separation reduces.

The above model may be illustrated for a typical APP particle like Antiblaze MCM (Table 1) with reported density $\rho = 1.90$ g cm^{-3} and $\sigma_A = 25$ μm,[17] dispersed in a polyamide 6 or 6.6 matrix with $\rho = 1.14$ g cm^{-3}. If APP particles may be assumed to promote char at a lower temperature than the nanoclay particle, then we may apply Equation (3) for the range of APP concentrations used in Table 2. Table 7 lists these results. Similarly, assuming an average diameter for a nanoclay particle, $\sigma_B = 100$ nm, and generic clay density of 2 g cm^{-3} (reference 1) then the average reaction distance between any two nanoclay particles at 0.5–5.0 wt% (or pph) may also be calculated using Equation (3). The results are also listed in Table 7 which shows that for 2 wt% nanoclay $l = 0.95$ μm.

Our previous paper[1] also showed that the presence of nanoclay caused the shifts of LOI *versus* FR concentration plots in Figure 5, may be quantified as reductions in FR concentrations in Table 8 at LOIs of 23 and 24, which are seen to be about the inflexion points in each respective "S-shaped" curve. Thus,

Table 7 *Average interparticle distance, l μm, for Antiblaze MCM ammonium polyphosphate ($\sigma = 25$ μm, and $\rho = 1.90$ g cm^{-3}), and a nanoclay ($\sigma = 100$ nm and $\rho = 2$ g cm^{-3}) dispersed in polyamide 6 or 6.6 ($\rho = 1.14$ g cm^{-3})*

Additive, pph	*Mass fraction*	*Volume fraction*	l (μm)
APP			
11	0.11	0.07	42
15	0.15	0.10	30
20	0.2	0.13	22
23	0.23	0.16	19
27	0.27	0.19	16
28.5	0.285	0.20	15
Nanoclay			
0.5	0.005	0.0031	3.82
1	0.01	0.0062	1.90
2	0.02	0.0124	0.95
5	0.05	0.0315	0.37

Table 8 *Flame retardant concentrations required to achieve defined LOI values in polyamide 6.6 films*

	LOI=23		*LOI=24*	
FR	*PA6.6*	*PA6.6 nano*	*PA6.6*	*PA6.6 nano*
APP	23.8	15	28.5	20.1
MPC	16.3	14.5	20.5	18
CC	20.5	10.5	28.5	17.5

for ammonium polyphosphate, for example, the presence of nanoclay effectively reduces the APP concentration required for an LOI value of 24 from 28.5 to 20.1%. This LOI value approaches those required for many flame retardant thermoplastic polymer applications.

Table 7 shows that for APP present at 20%, and hence $v_{fb}=0.13$, yields $l=22$ μm as opposed to $l=15$ μm for 28.5 wt% APP. Based on the above simple model, it can be seen that the presence of 2 wt% nanoclay with particles on average separated by "reaction lengths" of about 1 nm provides an effective bridge between the more separated APP particles at 20.1 wt% concentration to yield the same flame retardant or char-generating property at 28.5 wt% presence in the absence of nanoclay.

Clearly, if this simple model is to be shown to provide even an approximate picture of the action of nanodispersed particles in the presence of microdispersed conventional flame retardant species, further more quantifiable research is necessary on such mixed systems. However, at the very least, the model suggests that reduction in particle size of the microdispersed component, and more effective char-promoting functionalisation of the nanoparticles, should enable lower overall quantities of flame retardant to be used to effect acceptable levels of flame retardancy.

16.5 References

1. A.R. Horrocks, B.K. Kandola and S.A. Padbury, in *Flame Retardants 2004*, Interscience Communications Ltd., London, 2004.
2. T. Kashiwagi, J.R. Shields, R.H. Harris Jr and W.A. Awad, *Proceedings of 14th conference 'Advances in Flame Retardant Polymers'*, Business Communications Inc., Norwalk, CT, 2003.
3. S. Bourbigot, E. Devaux, M. Rochery and X. Flambard, *Nanocomposite textiles: New Routes for flame retardancy, in 47th International SAMPE Symposium*, May 12–16, 2000, Volume 47, pp. 1108–1118.
4. S. Bourbigot, E. Devaux and X. Flambard, *Polym. Degrad. Stab.*, 2002, **75**, 397–402.
5. S.V. Levchik and E.D. Weil, *Polym. Int.*, 2000, **49**, 1033–1076.
6. C. Wilkie, in *Proceedings of 13th Conference on Recent Advances in Flame Retardancy of Polymeric Materials*, M. Lewin (ed.), Business Communications Inc, Norwalk, CT, 2002.

7. S.A. Padbury, A.R. Horrocks and B.K. Kandola, in *Proceedings of 14th Conference 'Advances in Flame Retardant Polymers*, M. Lewin (ed.), Business Communications Inc., Norwalk, CT, 2003.
8. J.W. Gilman, *Appl. Clay Sci.*, 1997, **15(1–2)**, 31–49.
9. J.W. Gilman and T. Kashiwagi, in *Polymer–clay Nanocomposites*, T.J. Pinnavaia, and G.W. Beall (eds.), John Wiley and Sons, New York, 2000. pp. 193–206.
10. S. Bourbigot, E. Devaux, M. Rochery and X. Flambard, in *Nanocomposite textiles: New Routes for flame retardancy, proceedings 47th International SAMPE Symposium*, May 12–16, 2000, Volume 47, pp. 1108–1118.
11. T.X. Liu, Z.H. Liu, K.X. Ma, L. Shen, K.Y. Zeng and C.B. He, *Compos. Sci. Technol.*, 2003, **63**, 331–337.
12. A.R. Horrocks, M.Y. Wang, M.E. Hall, F. Sunmomu and J.S. Pearson, *Polym. Int.*, 2000, **49**, 1079–1091.
13. G. Camino, L. Costa and L. Trossatelli, *Polym. Degrad, Stab.*, 1985, **12**, 203–211.
14. L. Costa, G. Camino and M.P. Luda, *Proc. Am. Chem. Soc.*, 1990, p. 211.
15. M Lewin and E.D. Weil, in *Fire Retardant Materials*, A.R. Horrocks and D. Price (eds.), Woodhead Publishing, Cambridge, UK, 2001, p. 39.
16. S. Glasstone, in *Textbook of Physical Chemistry*, Macmillan, London, 1960, pp. 274–277.
17. Anon, *Antiblaze MCM (formerly Amgard MCM) Data Sheets*, Rhodia Consumer Specialities (formerly Albright and Wilson Ltd), Oldbury, UK, 1989.

CHAPTER 17

Use of Clay–Nanocomposite Matrixes in Fire Retardant Polyolefin-Based Intumescent Systems

SOPHIE DUQUESNE, SERGE BOURBIGOT, MICHEL LE BRAS, CHARAFEDDINE JAMA AND RENÉ DELOBEL

Laboratoire des Procédés d'Elaboration de Revêtements Fonctionnels, Ecole Nationale Supérieure de Chimie de Lille, BP 108, F-59652 Villeneuve d'Ascq, France (sophie.duquesne@ensc-lille.fr)

17.1 Introduction

Polymer–clay nanocomposites are hybrid organic polymer/inorganic layered materials with unique properties when compared to conventional filled polymers. The fire retardant performance of clay nanocomposite polymers, with low clay mass fraction, show excellent improvement.[1–2] The mode of action is generally attributed to a "barrier effect" created by the dispersion of clay layers in the degraded matrix, which leads to a decrease in the feeding rate of the combustion products to the flame and, as a consequence, to a decrease in the rate of heat release.[3] Such a mechanism is similar to the mode of action of intumescent systems. Intumescent materials form, when heated, a foamed cellular charred layer on their surface, which limits the fuel transfer to the flame and the heat transfer to the polymer.[4–5] Hence, intumescent systems also act via a barrier mechanism. The association of the two concepts, consequently, appears interesting.

Generally, intumescent formulations contain three active ingredients: an acid source, a carbonization agent and a blowing agent.[6] The carbonization agents commonly used in intumescent formulations are polyhydric compounds such as pentaerythritol or sorbitol.[7–8] The use of such polyols involves problems such as migration, reactivity during processing, compatibility, *etc*. Therefore, new solutions are needed to avoid these problems.[9] Previous studies[10–12] demonstrated the efficiency of char-forming polymers (polyamide or polyurethane) as carbonization agents in intumescent formulations. Recent works of our laboratory,[13–14] have demonstrated that the use of nanocomposite polyamide as carbonization agent enables the improvement of fire performance of intumescent systems in a copolymer ethylene vinyl acetate (EVA). The clay allowed the thermal stabilization of a phosphorocarbonaceous structure in the intumescent char, which increased the efficiency of the shield, and, in addition, the formation of a ceramic that can act as a protective barrier.

The aim of this study is to investigate the effect of nanocomposite polymers on the fire retardant performance of polyolefin-based intumescent systems. In a first part, the influence of the use of nanocomposite polymer as matrix, as carbonization agent or as both on the fire retardancy, was investigated in intumescent EVA. Then, the fire retardant performance of intumescent polypropylene (PP) using polyamide-6 (PA-6) as carbonization agent was compared to those of nanocomposite intumescent polypropylene using a polyamide-6 clay nanocomposite hybrid (PA-6-nano).

17.2 Experimental

17.2.1 Materials

17.2.1.1 EVA Nanocomposite

Raw materials were EVA with 19 wt% vinyl acetate [Exxon's Escorene UL00119, MFI = 0.65 g/10 min] and Cloisite 30B for which the negative charges of its layers are compensated with methyl tallow bis(2-hydroxyethyl)-ammonium ions (Southern Clay Products Inc). The study has been carried out at constant EVA/30B ratio 95:5 wt/wt hereinafter called EVAnano. The materials were obtained by mixing the filler with the melted EVA in a Brabender Laboratory Mixer measuring head (type 350/EH, roller blades, checking the mixing conditions using the data processing torque rheometer system Brabender Plasticorder PL2000) at constant shear rate (50 rpm) and at constant temperature (160°C). The morphology of the materials has been previously investigated.[15] EVA copolymer appears to easily form a nanocomposite with Cloisite 30B even if the totally exfoliated structure is not been achieved but a mixed exfoliated/intercalated structure obtained.

17.2.1.2 PP Nanocomposite

Raw materials were PP [polypropylene supplied by Atofina – PPH7060 MFI = 12 g/10 min], PP-g-MA [maleic anhydride grafted polypropylene supplied by Crompton – Polybond 3200–2% MA, MFI = 110 g/10 min] and organically modified montmorillonite (Cloisite 20A, Southern Clay Product, organic modifier = dimethyl dihydrogenatedtallow quaternary ammonium). The study was carried out using the ratio PP/PP-g-MA/20A = 90:5:5 (wt/wt) hereinafter called PPnano. Mixtures were prepared using a Brabender mixer measuring head (described above) at constant shear rate (50 rpm) and constant temperature (190°C). Using those parameters, an intercalated nanocomposite is obtained.[16]

17.2.1.3 Intumescent Systems

The polyolefin matrixes used were EVA, EVAnano, PP and PPnano. The intumescent system was composed of ammonium polyphosphate (APP supplied by Clariant – Exolit AP422) and Polyamide 6 supplied by Nyltech (PA-6) or Polyamide 6 nanocomposite [PA-6nano supplied by UBE Industries (3 wt% clay content)]. PA6 nanocomposite is synthesized by ring-opening polymerization of caprolactame in the presence of cation-exchanged montmorillonite clay. An exfoliated structure is obtained.[17] Mixtures were prepared using a Brabender mixer measuring head (described above) at shear rate of 50 rpm at 230°C. Table 1 reports the formulations prepared. Sheets [3 (or 1.6) × 100 × 100 mm^3] were then obtained using a Darragon press at 230°C with a pressure of 10^6 Pa.

17.2.2 Fire Testing

17.2.2.1 Cone Calorimeter

A Stanton Redcroft Cone Calorimeter was used to carry out measurements on samples following the procedure defined in ASTM 1354–90. The method is based on oxygen consumption calorimetry.[18] The standard procedure used involves exposing specimens measuring 100 × 100 × 3 mm in horizontal orientation. An external heat flux of 50 kW m^{-2} has been used for running the experiments. This flux has been chosen because it is the common heat flux in mild fire scenario.[19] When measured, HRR (heat release rate) values are reproducible to within ±10%. The cone data reported here are the average of three replicated experiments.

17.2.2.2 Limiting Oxygen Index

LOI was measured using a Stanton Redcroft instrument on sheets (100 × 10 × 3 mm^3) according to a standard 'oxygen index' test (ASTM D2863/77).

Table 1 *Composition of the formulations*

Reference	*Matrix*	*Carbonization agent (CarbAgent)*	*Formulation Matrix/APP/CarbAgent*
EAP-60	EVA	PA6	60/33.3/6.7
EAPn-60	EVA	PA6nano	60/33.3/6.7
EnAP-60	EVAnano	PA6	60/33.3/6.7
EnAPn-60	EVAnano	PA6nano	60/33.3/6.7
PAP-80	PP	PA6	80/16.7/3.3
PAP-70	PP	PA6	70/25/5
PAP-60	PP	PA6	60/33.3/6.7
PnAPn-80	PPnano	PA6nano	80/16.7/3.3
PnAPn-70	PPnano	PA6nano	70/25/5
PnAPn-60	PPnano	PA6nano	60/33.3/6.7

Table 2 *LOI and UL-94 rating of intumescent EVA nanocomposites including polymer as carbonization agent*

Formulation	*LOI* (vol%)	*UL 94 rating*
EAP-60	30	V0
EnAP-60	34	V0
EAPn-60	29	V0
EnAPn-60	33	V0

17.2.2.3 UL-94

UL-94 tests were carried out on $100 \times 13 \times 1.6$ mm^3 specimens according to the American National Standard UL-94 (Test for flammability of plastics materials for part in devices and appliance, Underwriter laboratories, Northbook, ANSI/ASTM D-635/77).

17.3 Results and Discussion

17.3.1 Fire Retardant Performance of EVA Based Systems

Table 2 reports the LOI and UL-94 rating of the four systems studied. Interestingly, whatever the compound, V0 rating is achieved. Moreover, the LOI increases by 4 points when EVAnano is used as a matrix.

Figure 1 presents the heat release rate (HRR) curves *versus* time of the intumescent EVA-based formulations. Whatever the formulation, the curves present two peaks and a fire retardant system is obtained [PHRR (peak of heat release rate) = 1600 kW m^{-2} for neat EVA[15]]. The first peak corresponds to the formation of the protective layer, *i.e.* to the development of the intumescence, whereas the second one corresponds to its destruction or failure. When the protective layer

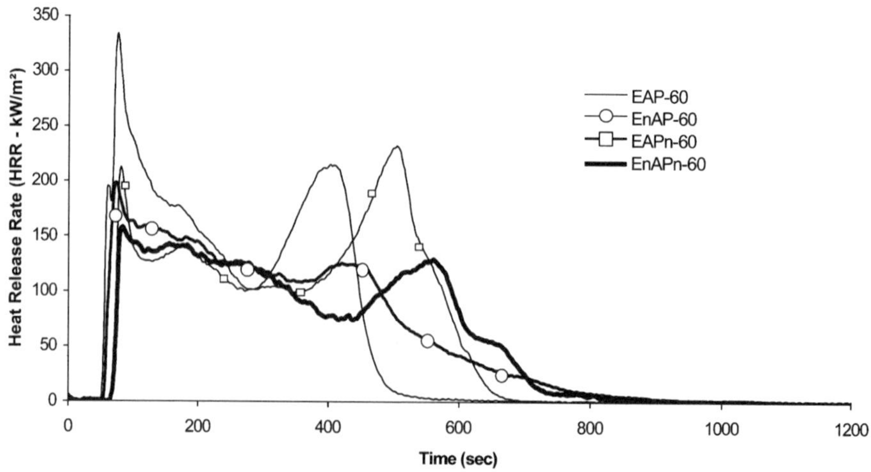

Figure 1 *Heat rate release vs. time for intumescent EVA-based formulations*

breaks, a high quantity of combustible gaseous products is evolved, leading to a sharp increase in the HRR. Clearly, when a nanocomposite is used in the formulation (as matrix, as carbonization agent or both), the first peak of heat release rate (PHRR) is reduced (from about 340 to 200 kW m^{-2}). However, the second peak decreases only when the nanocomposite is used as a matrix. It may be proposed that, in the last case, the clay dispersion allows the integrity of the intumescent shield to be maintained throughout the experiment.

Whatever the formulation, the total heat release (THR) is similar (around 67 MJ m^{-2}). Hence, the effect of the nanocomposite may be attributed to a barrier effect. The clay layers reinforce the protective intumescent shield. The mechanism of action could be attributed either to the improvement of the mechanical properties of the char by the filler and/or to a ceramisation effect.[20] For EVA/APP/PA6nano, it may be proposed that the quantity of clay (0.2 wt% of the total) is not high enough to maintain the integrity of the char, the quality and the homogeneity of the structure, which is linked with its properties throughout the experiment.

Figures 2 and 3 are in good agreement with the previously proposed assumption. In fact, when nanocomposites are used as component of the intumescent formulation, the mass loss rate decreases at the beginning of the experiment. So, it may be proposed that the barrier effect due to the presence of clay is efficient at the beginning of the experiment. Concerning the second peak, this remark is only true for the EVA nanocomposite-based formulation.

17.3.2 Fire Retardant Performance of PP Based Systems

Table 3 reports the fire retardant properties of PAP and PnAPn formulations according to cone calorimeter experiments. It appears that the intumescent

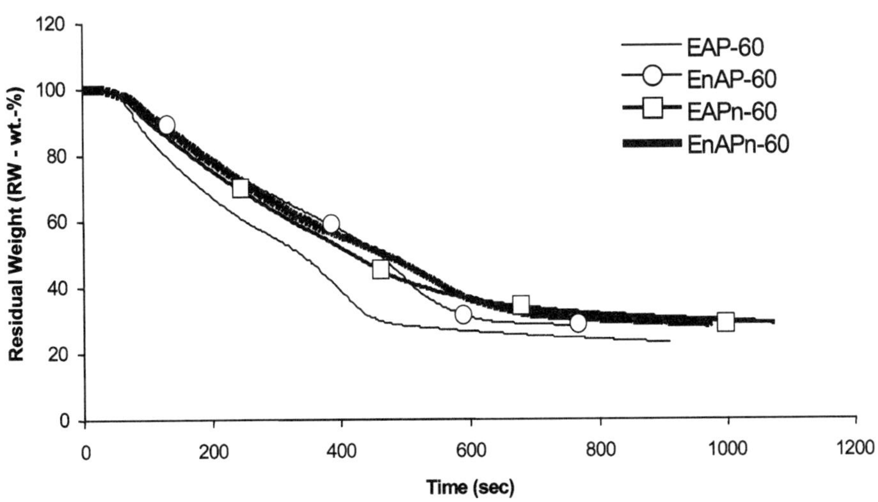

Figure 2 *Residual weight vs. time for intumescent EVA-based formulations*

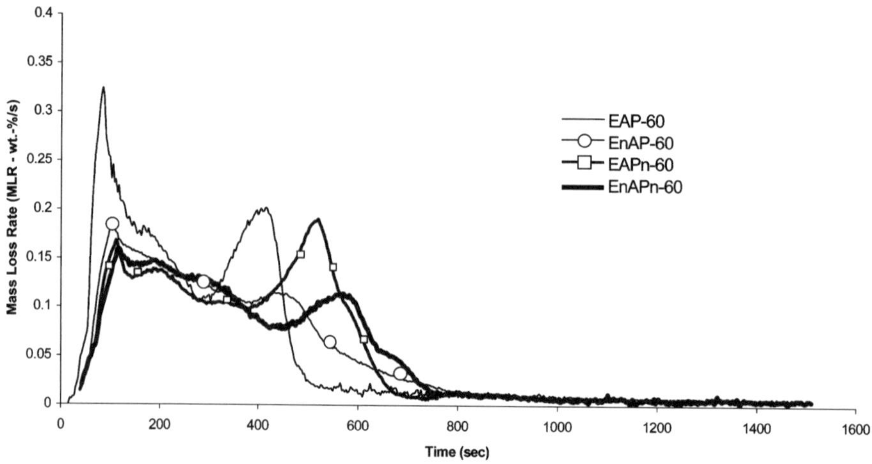

Figure 3 *Mass loss rate vs. time for intumescent EVA-based formulations*

Table 3 *Cone calorimeter data of intumescent PP nanocomposites including polymer as carbonization agent*

Formulation	*PHRR1*[a] (kW m^{-2})	*PHRR2*[b] (kW m^{-2})
PAP-80	405 ± 41	312 ± 32
PAP-70	278 ± 28	216 ± 22
PAP-60	170 ± 17	152 ± 16
PnAPn-80	354 ± 35	–
PnAPn-70	213 ± 21	177 ± 18
PnAPn-60	112 ± 11	113 ± 12

[a]First peak heat release rate. [b]Second peak heat release rate.

system is efficient (PHRR = 1000 – 1200 kW m^{-2} for neat PP) and that a small quantity of clay sharply decreases the first peak of heat release rate (up to 35% for PnAPn-60), as observed for EVA. Notably, using PPnano as a matrix, the decrease in the second peak of HRR is observed whatever the formulation, as reported for EVA.

Figures 4 and 5 report the HRR curves of PAP and PnAPn formulations respectively.

All the curves (expect for PnAPn-80) present two peaks, which is characteristic of intumescent systems, as described above. It appears that whereas the first peak occurs at a similar time for PP- and for PPnano-based systems, the second peak occurs at a longer time when clay is added into the formulation. This confirms that the addition of clay in intumescent systems enables the reinforcement of the protective charred layer created at the surface of the materials. Hence, the appearance of the second peak, which corresponds to the destruction of the intumescent shield, is delayed.

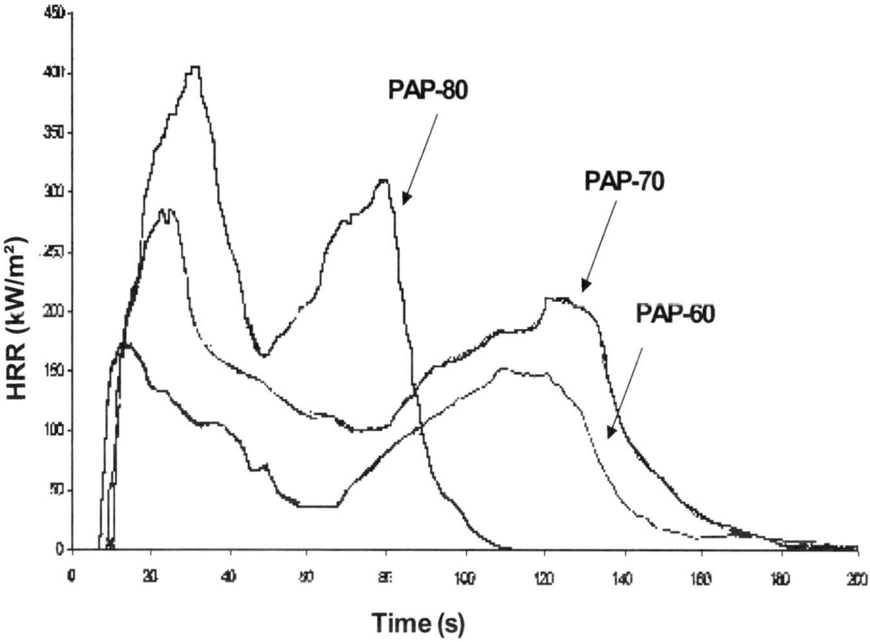

Figure 4 *Heat rate release vs. time for intumescent PP-based formulations*

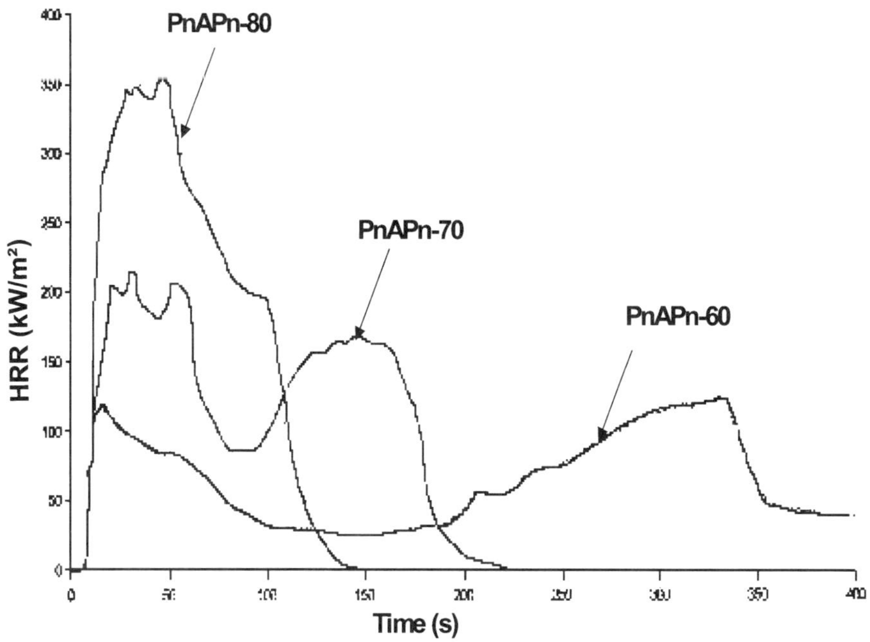

Figure 5 *Heat rate release vs. time for intumescent nanocomposite PP-based formulations*

17.4 Conclusion

This work demonstrates that the addition of clay in intumescent formulations improves the fire performance of polyolefin-based systems. Better improvement is achieved when clay is incorporated into the matrix to make a nanocomposite. This may be partially explained by the clay content, which is higher when a nanocomposite is used as matrix. However, it may be highlighted that even if a small quantity of clay (0.2 wt% of the total) is used the first peak of heat release is significantly reduced, which is particularly important since in a scenario of fire the beginning of the fire development is very important. It is assumed that the addition of clay reinforces the protective charred layer developed at the surface of the material.

17.5 Acknowledgement

The authors thank Nicolas Belverge for his helpful technical assistance.

17.6 References

1. J.W. Gilman, T. Kashiwagi, S. Lomakin, J.D. Lichtenhen and P Jones, in *Fire Retardancy of Polymers: The Use of Intumescence*, M. Le Bras, G. Camino, S. Bourbigot and R. Delobel (eds.) The Royal Chemical Society, Cambridge, UK, 1998, PP. 266–279.
2. C.A. Wilkie, in *Recent Advances in FR of Polymeric Materials*, M. Lewin (ed.) Business Communications Co Inc. Pub., Norwall, USA, 2002, Volume 13, pp. 155–159.
3. J.W. Gilman, C.L. Jackson, A.B. Morgan, R. Harris Jr., E. Manias, E.P. Giannelis, M. Wuthenow, D. Hilton and S.H. Phillips, *Chem. Mater.*, 2000, **12(7)**, 1866–1873.
4. H.L. Vandersall, *J. Fire Flammability*, 1971, **2**, 97–140.
5. S. Bourbigot, M. Le Bras, R. Delobel, R. Decressain and J.-P. Amoureux, *J. Chem. Soc., Faraday Trans.*, 1996, **92(1)**, 149–158.
6. M. Le Bras and S. Bourbigot, in reference 1, pp. 64–75.
7. M. Le Bras, S. Bourbigot, C. Delporte, C. Siat and Y. Le Tallec, *Fire Mater.*, 1996, **20**, 191–203.
8. M. Le Bras, S. Bourbigot, Y. Le Tallec and J. Laureyns, *Polym. Degrad. Stab.*, 1997, **56**, 11–21.
9. M. Le Bras and S. Bourbigot, in *Polypropylene: an A-Z Reference*, J. Karger-Kocsis (ed.), Chapman & Hall, London, 1998, p. 357.
10. S. Bourbigot, M. Le Bras and C. Siat, in *Recent Advances in FR of Polymeric Materials*, M. Lewin (ed.), Business Communications Co Inc. Pub., Norwall, USA, 1998, Volume 8, 146–160.
11. M. Le Bras, S. Bourbigot, C. Siat and R. Delobel, in reference 1, pp. 266–279.
12. M. Bugajny, M. Le Bras, S. Bourbigot and R. Delobel, *Polym. Degrad. Stab.*, 1999, **64(1)**, 157–163.

13. S. Bourbigot, M. Le Bras, F. Dabrowski, J. Gilman and T. Kashiwagi, *Fire Mater.*, 2000, **24**, 201–208.
14. F. Dabrowski, M. Le Bras, L. Cartier and S. Bourbigot, *J. Fire Sci.*, 2001, **19**, 219–241.
15. S. Duquesne, C. Jama, M. Le Bras, R. Delobel, P. Recourt and J.-M. Gloaguen, *Composites Sci. Technol.*, 2003, **63(8)**, 1141–1148.
16. A. Bendaoudi, S. Duquesne, C. Jama, M. Le Bras, R. Delobel, P. Recourt, J.-M. Gloaguen, J.-M. Lefebvre and A. Addad, this book, Chapter 8.
17. A. Usuki, Y. Kojima, M. Kawasumi, A. Okada, Y. Fukushima, T. Kurauchi and O. Kamigaito, *J. Mater. Res.*, 1993, **8(5)**, 1179–1184.
18. C. Huggett, *Fire Mater.*, 1980, **4(2)**, 61–65.
19. V. Babrauskas, *Fire Mater.*, 1984, **8(2)**, 81–95.
20. S. Duquesne, J. Lefebvre, S. Bourbigot, M. Le Bras, R. Delobel, P. Recourt, Oral communication, ACS 228th Fall meeting, 22–26 August 2004, Philadelphia, PA, USA, Abstr. Paper PMSE 247, ACS (Pub.) Washington, 2004.

CHAPTER 18

Effect of Hydroxides on Fire Retardance Mechanism of Intumescent EVA composition

GIOVANNI CAMINO,[1] ALESSANDRO RIVA, D. VIZZINI,[1] ANDRÉA CASTROVINCI,[2] PASCAL AMIGOUËT[3] AND PHILIPPE BRAS PEREIRA[3]

[1]Centro di Cultura per l'Ingegneria delle Materie Plastiche, V. T. Michel 5, 15100 Alessandria, Italy (giovanni.camino@polial.polito.it)
[2]Politecnico di Torino, Sede di Alessandria, V. T. Michel 5, 15100 Alessandria, Italy
[3]NEXANS-NRC, 170, Avenue Jean Jaurès, 69353 Lyon Cedex 7, France

18.1 Introduction

Polyethylenic polymers and co-polymers are widely used in many fields, particularly in electrical engineering. Due to their chemical composition these polymers are easily flammable and this is why the flame retardancy of these materials is a deeply studied matter. The main approach used to date to impart flame retardant properties to this class of polymeric materials has been the incorporation of additives, specifically halogen compounds. The combustion products coming from these materials have, however, many negative characteristics (corrosiveness, toxicity, *etc.*) that have pushed the industry and legislation to promote some new approaches to flame retardance.[1–3]

One of these developing approaches is that of intumescence. The intumescence mechanism consists in creating on the polymer surface a multicellular thermally stable expanded shield, able to reduce both the heat flux from the flame to the polymer matrix, responsible for the fuel production, and the transfer of the fuel to the flame, thus limiting the spread of fire.[4,5]

Generally, intumescent formulations consist of three basic components: an acid source (phosphates, borates *etc.*), a carbonising compound (polyols, polyamides, polyurethanes *etc.*) and a blowing agent (melamine and melamine compounds *etc.*). On heating, the acid source gives a mineral acid that takes part in the dehydration of the carbonising compound that forms a cellular structure when the blowing agent decomposes.[4–7] In the absence of the blowing agent, the gases evolved during carbonisation (*e.g.* water) perform the blowing action.

The association of a polyamide (*e.g.* PA6) or other char-forming polymers and APP as flame retardants for EVA and other thermoplastic polymers has already been reported.[8–12] Here we have studied the effect of combining the intumescent system APP-PA11 with magnesium hydroxide and aluminium hydroxide, which are widely used fire retardants in electrical cables sheeting materials, to explore possible synergies between the two fire retardants approaches that act with complementary physical actions. Indeed, the intumescent system provides a barrier to gas and heat transfer between flame and burning polymer whereas MH and ATH act by cooling the surface of the polymer through endothermal dehydration and water evaporation and by cooling the flame through dilution with water vapour.

18.2 Experimental

18.2.1 Materials

The following products where used: ethylene – vinyl acetate 28% copolymer (EVATANE 28-03, ATOFINA, EVA), Polyamide 11 (Rilsan ATOCHEM, PA11), magnesium hydroxide (Magnifin H10, Martinswerk, MH), ammonium polyphosphate (Exolit AP 422, Clariant, APP) and aluminium hydroxide (Apyral MD40, Nabeltec, ATH).

APP and hydroxides mixtures were prepared by grinding in a mortar. The polymers were mixed with inorganic components using a Brabender Mixer PLE, with roller blades and a 55 or 370 cm^3 mixing chamber for ATH and MH respectively, with a rotation speed of 60 rpm, mixing for 5–8 min at 180–190°C. Also, a Leistritz ZSE27 co-rotating twin screw extruder was used with the following temperature profile: 100/170/170/170/175/175/180/180/180. The screw rotation speed was set to 100 rpm, and the resulting throughput was 9.6 kg h^{-1}. The composition of the mixtures is reported in Table 1. Sample tests were prepared at 190°C by compression moulding using a Scamia type PC4 hydraulic press at 200 bar.

18.2.2 Combined Thermogravimetry–Infrared–Evolved Gas Analysis (TGA-FTIR-EGA)

TGA-FTIR-EGA analyses were performed on heating from 50 to 800°C 20 mg samples at 20°C min^{-1} (if not otherwise specified in figures legend) under a nitrogen flow (30 ml min^{-1}) using a Perkin-Elmer Pyris 1 TGA coupled by a Perkin-Elmer TG-IR interfaced with a Perkin-Elmer Spectrum GX Infrared Spectrometer equipped with IR gas cell (TGA-FTIR). The TGA-FTIR transfer line was heated at 220°C, while the IR gas cell was heated at 230°C to avoid condensation of degradation products inside the transfer line and the gas

Table 1 *Composition and combustion behaviour of the compounds*

	Composition (%)								
Sample n°.	*EVA*	*APP*	*MH*	*PA11*	*Ratio MH/AP*	*pkHRR* (kW m^{-2})	*TSR*	*TTI* (s)	*CO* (kg kg^{-1})
0	100	–	–	–	–	2660	–	–	–
1	42.5	40	15	2.5	0.38	225	605	54	0.29
2	68.5	10	15	6.5	1.50	487	1031	61	0.09
3	37.5	10	50	2.5	5.00	151	428	70	0.56
4	53	10	32.5	4.5	3.25	266	635	64	0.53
5	55.5	25	15	4.5	0.60	272	845	51	0.3
6	40	25	32.5	2.5	1.30	171	450	52	0.58
7	49	20	27	4	1.33	224	637	60	0.48
8	46	30	21	3	0.69	224	626	54	0.26
9	51	15	31	3	2.55	195	446	65	0.48
10	59	15	21	5	1.39	242	634	52	0.21

cell. The nitrogen flow was switched on at room temperature 10 min before the beginning of the analysis to have a stable IR background. An infrared spectrum of the evolved gases was sampled at 1°C (3 s) intervals.

An Amel Instruments ammonia probe and a Testo Instruments water probe were located in series on-line with the IR gas cell so that outcoming gases can be continuously analysed for ammonia and water content (EGA). The signal of the ammonia probe was recorded by means of an home-made software, while the signal from the water probe was collected by means of commercial dedicated software.

18.2.3 Expansion Measurements

Expansion measurements were performed by means of a laboratory made apparatus (Figure 1) consisting of a furnace (OCRAS Zambelli) hosting a quartz tube (diameter 30 mm, height 200 mm) at the bottom of which a sample disc (diameter 28 mm, thickness 3 mm) was placed. A probe resting on the sample upper surface is connected with a position transducer (DSEurope). The sample was heated with the same heating rate used during TGA-FTIR-EGA experiments (20°C min^{-1}) and expansion data were collected by means of an home-made software.

18.2.4 Oxygen Consumption Calorimetry (Cone Calorimeter)

The cone calorimeter tests were performed according to the ISO 5660-1 standard using a Fire Testing Technology Standard Cone Calorimeter; the samples (100 × 100 × 3 mm) were irradiated with a 50 kW m^{-2} heat flux and the ignition of the flame was obtained by a spark. Combustion behaviour was evaluated by: peak of heat release rate (pkHRR), total smoke release (TSR), time to ignition (TTI) and evolution of CO (CO).

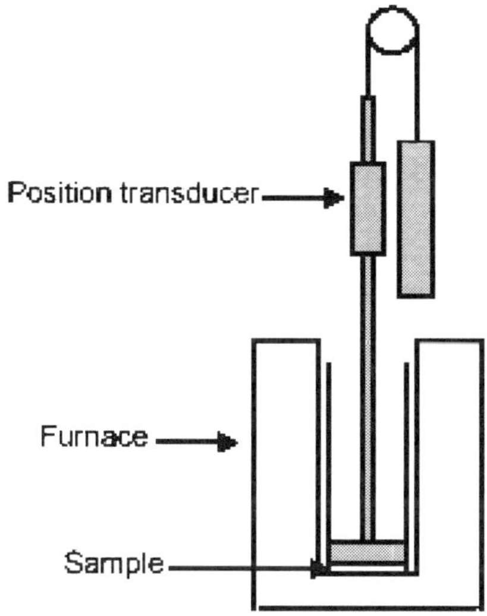

Figure 1 *Apparatus for expansion measurements*

18.3 Results and Discussion

18.3.1 Flammability Behaviour

Cone calorimeter tests were performed on a number of EVA/APP/MH/PA11 mixtures selected for a statistical approach to fire retardancy formulations (Table 1). Figure 2 shows the results obtained from the best performing sample (n 3), while the results for all the samples are summarised in Table 1.

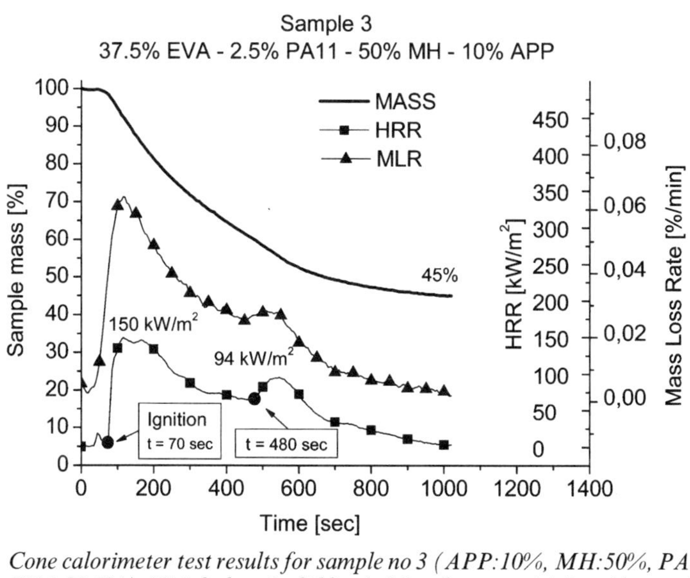

Figure 2 *Cone calorimeter test results for sample no 3 (APP:10%, MH:50%, PA11:2.5%, EVA:37,5%). Weight loss (solid line), Mass loss rate (▲) and heat release rate (■) curves*

Figure 2 shows that the intumescent shield is formed in a relatively short time after ignition, as evidenced by levelling off followed by a decrease of the HRR value 45 s after ignition ($t = 70$ s). When the intumescent shield begins to expand, HRR is progressively reduced, until the protective expanded charred layer begins to degrade. At this point, HRR may increase if a sufficient amount of organic material is still present under the protective layer, as with the sample in Figure 2 ($t = 480$ s). In other cases the polymeric material is completely consumed before the intumescent shield brakes down, so that no increase in HRR is observed after the development of intumescence.

Data of Table 1 show that the flame retardant performances of the samples appear to be mainly dependent on the total amount of fire retardant additives (30–60 wt%) in the intumescent compositions, especially if focusing on the pkHRR values. A relevant effect for the ratio between the components of the combined fire retardant system on combustion performances (*i.e.* MH/APP) is, however, noticeable.

For example, samples 1 and 6 contain a similar total loading of fire retardant additives (*ca.* 60%) with a very different ratio MH : APP, 0.38 and 1.30 respectively. The pkHRR is much larger in sample 1 (225 kW m^{-2}) as compered to sample 6 (171 kW m^{-2}). The same evidence is found comparing samples 8 and 9 (MH : APP 0.69 and 2.55, total filler loading: 54 and 49% respectively). Also in this case the larger pkHRR (224 kW m^{-2}) corresponds to the sample (8) with the smaller MH : APP ratio although it is the sample with larger fire retardant additives (54% compared to 49%).

Regardless of the amount of polymers, high MH/APP ratios not only reduce the pkHRR but also correspond to higher TTIs. See, for example, samples 3, 4 and 9 in which MH : APP ≥ 2.5 and TTI lies in the range 64 to 70 s, whereas it is from 51 to 61 s in all other cases.

The fire retardant system probably reduces smoke formation by reduction of EVA content of the compound since the lowest (428) and highest (1031) TSR values correspond to samples 3 and 2 in which EVA content is the lowest and highest, respectively, whereas CO formation shows an opposite trend, with lowest value (0.09) for the sample with highest EVA content (sample 2).

18.3.2 Thermal Degradation of APP in the Presence of MH or ATH

Chemical reactions occurring between MH and APP on heating that could be relevant to the fire retardance of the overall fire retardant system were studied by TGA-FTIR-EGA.

To confirm conclusions drawn from this study, ATH-APP reactions were also examined. ATH is a fire retardant hydroxide considered to act with the same mechanism as MH. TGA-FTIR-EGA analyses were also carried out on the pure components of the hydroxide-APP mixtures to identify the signals that could be used to monitor their degradation in the fire retardant complex mixtures heated to decomposition temperature.

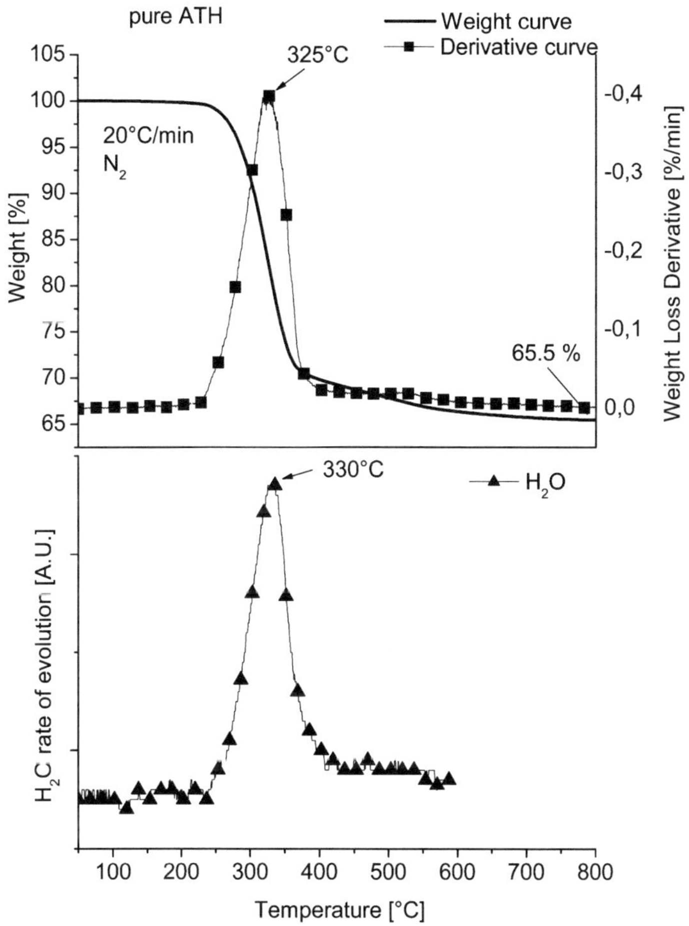

Figure 3 *TGA-coupled with water probe of pure ATH. TG (solid line), derivative TG (DTG, ■) and rate of water evolution (▲) curves*

18.3.2.1 ATH and MH

TGA curves of ATH (Figure 3) and MH (Figure 4) heated to 800°C show one main weight loss step with a weight loss rate maximum at about 330 and 430°C, respectively, which is due to water release, indicated by the signal from the water probe and responsible for the FR mechanism of inorganic hydroxides.[13]

18.3.2.2 APP

In agreement with published data,[6] TGA-FTIR-EGA analysis carried out on pure APP (Figure 5) shows elimination of ammonia and water between 300 and 450°C (maximum rate of weight loss at 380°C) with transformation of linear APP into cross-linked ultraphosphate (Scheme 1, reactions 1.1 and 1.2) which

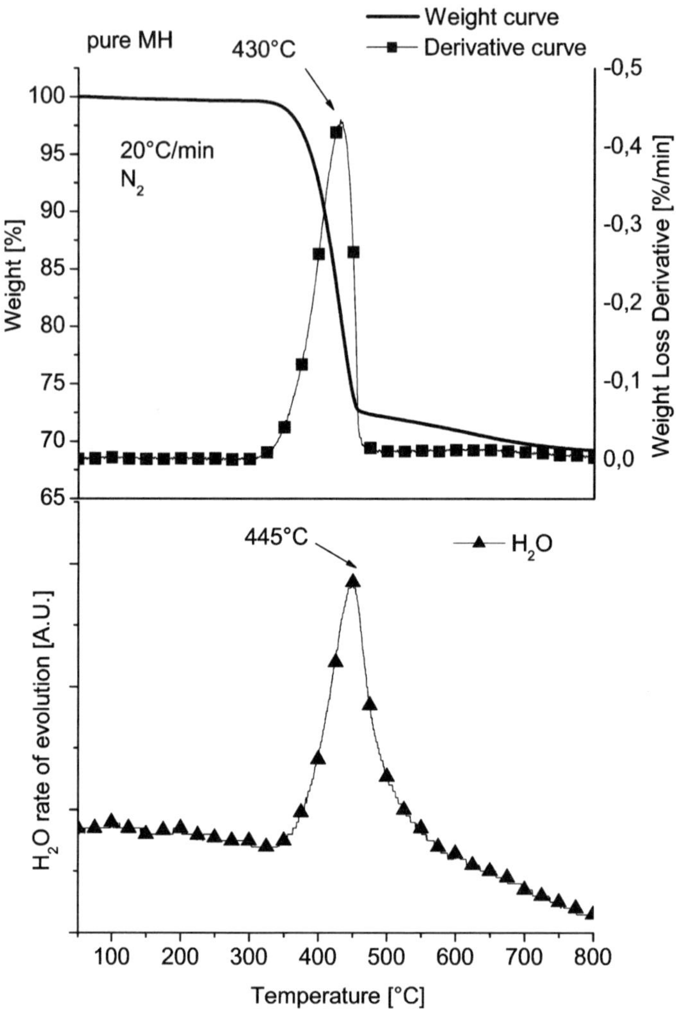

Figure 4 *TGA-coupled with water probe of pure MH. TG (solid line), derivative TG (DTG, ■) and rate of water evolution (▲) curves*

undergoes fragmentation to volatile P_2O_5-like moieties above 550°C. Ammonia evolution from APP is related to acidic site formation involved in the intumescence phenomenon, as already reported.[14] Since MH and ATH are bases, MH being the strongest, there is an interest in analysing the interaction between these components and APP, to investigate whether it could modify the FR behaviour of the intumescent mixture.

18.3.2.3 APP–MH Mixtures

In the upper part of Figure 6 the TG and DTG of a 50 wt%. APP/MH mixture are reported, together with the corresponding calculated curves obtained from

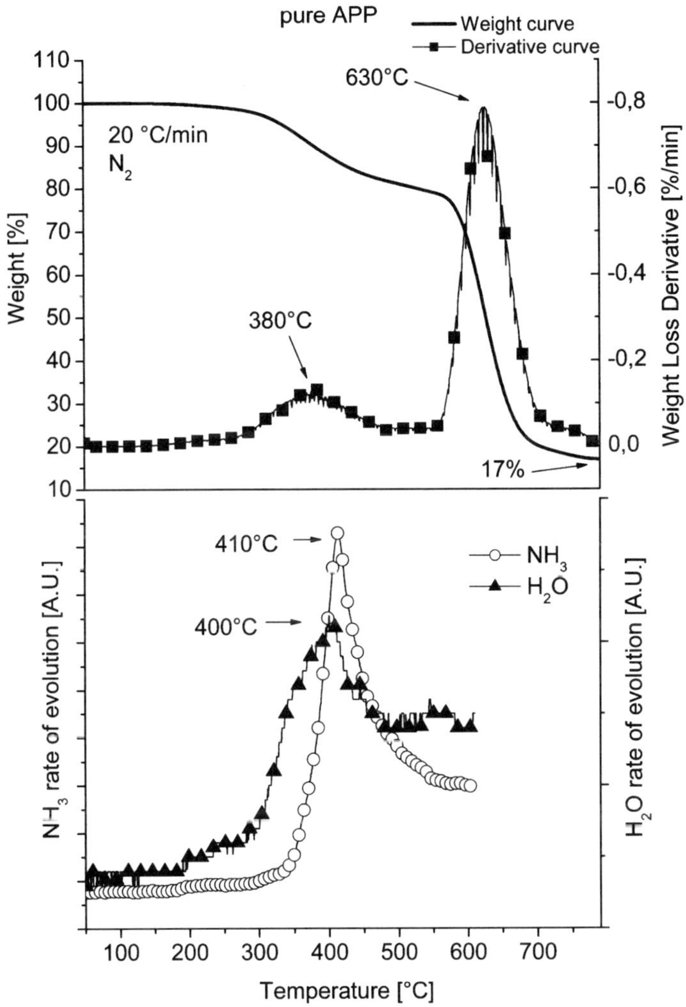

Figure 5 *TGA-coupled with water and ammonia probes of pure APP. TG (solid line) and DTG (■), rate of water (▲) and ammonia (○) evolution curves*

the analysis carried out on pure MH and APP heated separately. Whereas ammonia from APP and water from MH are expected to evolve in the same range of temperature (300–500°C) with a maximum rate at 396°C, Figure 6 shows that, instead, three steps of weight loss take place, with maximum rates at 346, 431 and 455°C.

The lower part of Figure 6 gives the water and ammonia evolution curves, showing that the first experimental weight loss step (DTG, T_{max} 346°C) is due to ammonia and water evolution, while the two overlapping weight loss steps (DTG, T_{max} 431 and 455°C) involve water evolution.

Evolution of ammonia at a lower temperature than expected (341°C, Figure 6, instead of 410°C Figure 5) is due to the basicity of MH, which shifts the

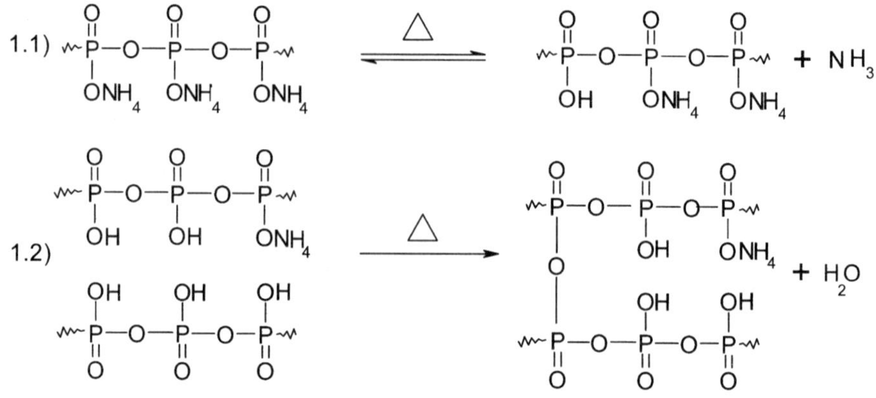

Scheme 1 *Reactions occurring during thermal degradation of APP*

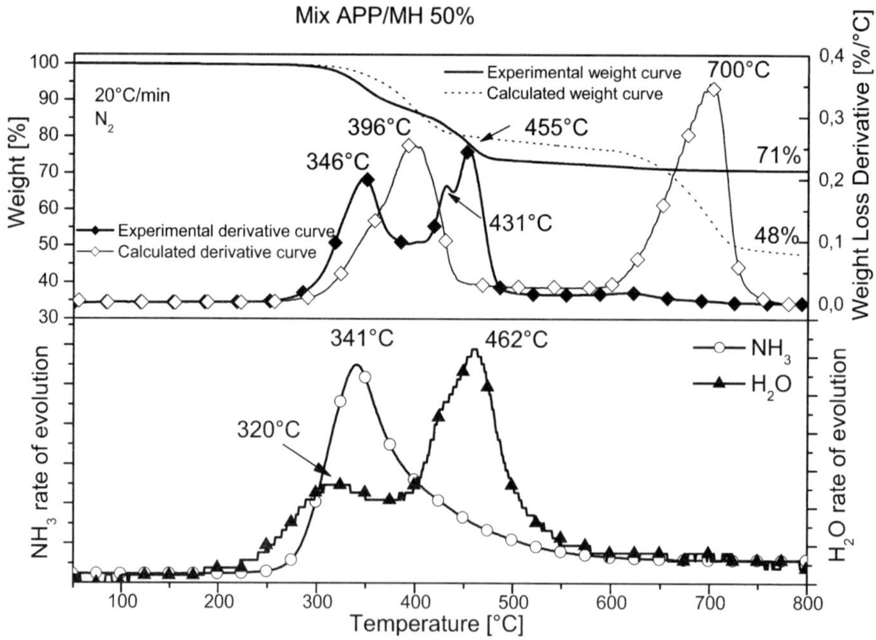

Figure 6 *Calculated and experimental TGA curves for a 50 wt% mixture of APP and MH. TG (solid line: experimental; dotted line: calculated) and DTG (♦, experimental – solid line; ◊, calculated) curves. Lower part water (▲) and ammonia (○) evolution curves*

ammonia evolution equilibrium (Scheme 1, reaction 1.1) to a temperature lower than that at which ammonia overcomes electrostatic attraction by the proton of polyphosphoric acid in pure APP. Indeed, the chemical reaction of MH hydroxyl groups with protons of polyphosphoric acid frees ammonia from the ammonium salt with formation of magnesium phosphate bonds (Scheme 2, reaction 2.1). In

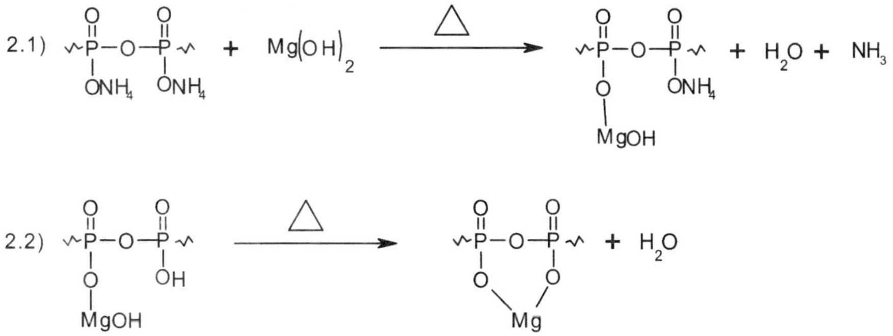

Scheme 2 *Reactions during thermal degradation of a APP/MH 50 wt% mixture. APP and MH interact to form magnesium phosphate bonds*

these conditions, thermal dehydration of MH (430°C, Figure 4) is partially replaced by a chemical reaction with APP that leads to elimination of water at 320°C (Figure 6).

The weight loss step at T_{max} = 431°C involves water elimination, as shown by a corresponding shoulder in the water evolution curve, which could be attributed to dehydration of unreacted MH occurring in the same range of temperature as on heating MH alone (max. weight loss 430°C, Figure 4). Finally, substantial water evolution is observed at 462°C (max. weight loss, 455°C), which could be explained by water elimination from the basic magnesium phosphate moieties formed by the partial neutralization of MH by APP (Scheme 2, reaction 2.2).

This interpretation is in agreement with decreasing importance of water elimination from unreacted MH at 385°C (Figures 7 and 8) and 431 (Figure 6) with increasing content of APP (40, 50 and 70%, Figures 6, 7 and 8 respectively). Indeed, whereas ammonia evolution is seen to occur in the first step of weight loss of the three mixtures, evolution of water from unreacted MH decreases progressively from Figure 7 to 6 and 8 with increasing APP : MH ratio, *i.e.* with increasing occurrence of reaction 2.1 of Scheme 2. Thermal degradation processes take place at about 40°C higher in Figure 6 than in Figures 7 and 8 because of a higher heating rate (20°C min^{-1} instead of 10).

Figure 6 shows that up to 650°C the experimental weight loss is higher than expected. Below 400°C this is due to acceleration of NH_3 and H_2O evolution. Above 400°C the reaction with MH makes dehydration of polyphosphoric acid more effective. Above 650°C the weight loss is lower than expected because, in the presence of MH, magnesium phosphate is formed, which prevents volatilization of phosphorous moieties deriving from ultraphosphate thermal decomposition above 650°C.

18.3.2.4 APP–ATH Mixtures

With ATH, thermal degradation of the hydroxide occurs in a lower temperature range (230–400°C) than that of ammonia and water elimination in pure APP (300–450°C). Thus, on heating APP-ATH mixtures a competition takes place

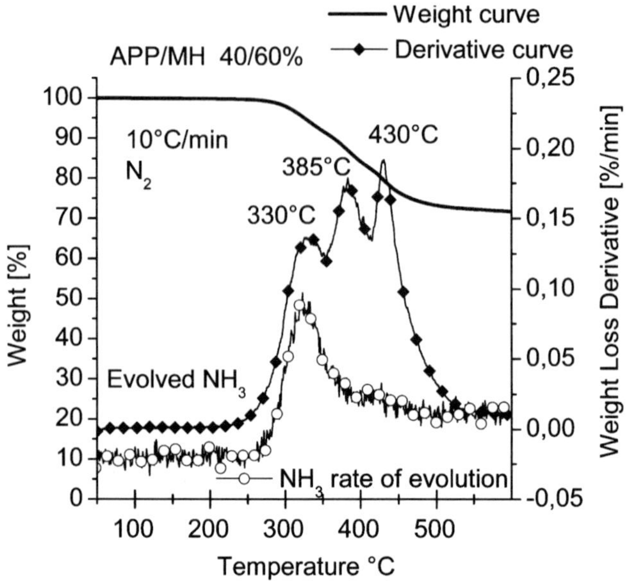

Figure 7 *TGA (at 10°C min^{-1}) coupled with ammonia probe of a 40/60% mixture of APP and MH respectively. TG (solid line), DTG (♦) and rate of ammonia evolution (○) curves*

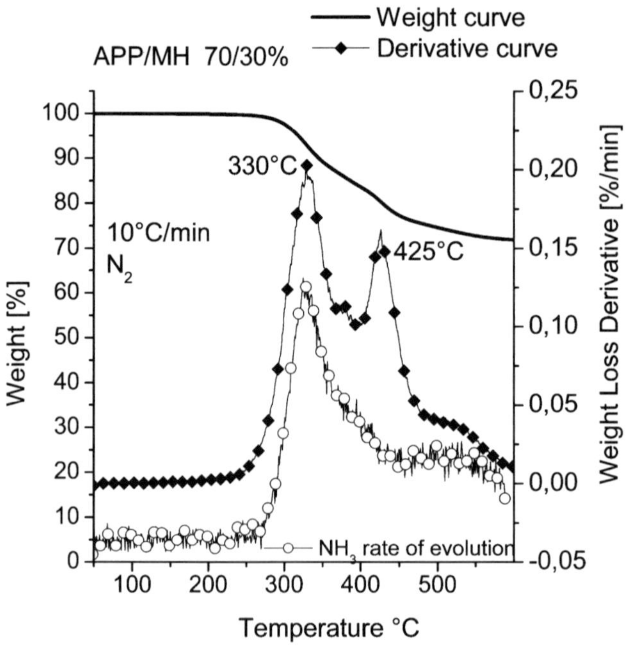

Figure 8 *TGA (at 10°C min^{-1}) coupled with ammonia probe of a 70/30% mixture of APP and MH respectively. TG (solid line), DTG (♦) and rate of ammonia evolution (○) curves*

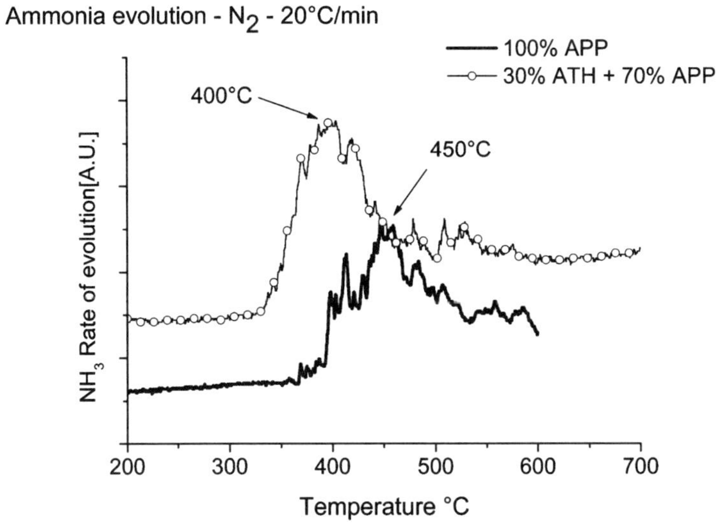

Figure 9 *Ammonia evolution curves for ATH/APP mixtures obtained by grinding the two compounds. Pure APP (solid line) and 70% APP + 30% ATH (○)*

between thermal and chemical degradation of ATH, which in the case of MH is shifted towards the chemical process owing to the higher temperature of MH thermal degradation compared to ATH.

As a consequence, acceleration of ammonia evolution from APP is less effective with ATH because dehydration to Al_2O_3 leads to loss of ATH basic properties on heating. Indeed, T_{max} for ammonia evolution from APP decreases when a high APP : ATH ratio (>1) shifts the competition in favour of chemical reaction between APP and ATH in comparison to ATH thermal dehydration, as shown, for example, in Figure 9 for a 70 : 30 APP : ATH mixture (T_{max} 400°C) compared to pure APP (T_{max} 450°C). Similarly, water evolution, which occurs with T_{max} at 336°C in ATH (Figure 10), is not affected in the presence of 50% APP whereas it takes place in two steps with T_{max} 315 and 490°C when APP is increased to 70%. In the stage at lower temperature, water is eliminated by chemical reaction between APP and ATH at temperatures similar to those observed in APP-MH mixtures. Dehydration at high temperature involves thermal degradation of the basic aluminium polyphosphate, which occurs with a mechanism similar to that of MH (Scheme 2, reaction 2.2), occurring, however, at a higher temperature.

18.3.3 Expansion Behaviour of Intumescent Mixtures Containing MH

Typical expansion behaviour taking place on heating the intumescent materials is shown for sample 10 in Figure 11, in which TGA-FTIR-EGA data are compared to the expansion behaviour of the sample, so that it is possible to understand which is the role of each component of the mixture (EVA, MH, APP,

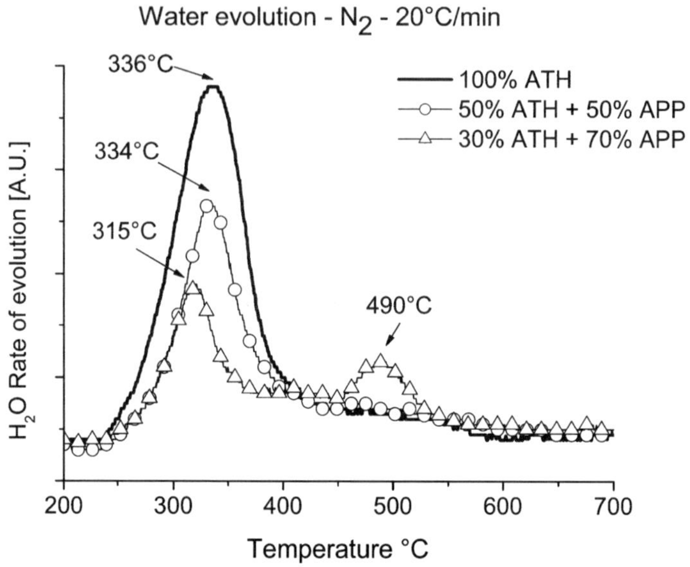

Figure 10 *Water evolution curves for ATH/APP mixtures obtained by grinding the two compounds in a mortar. Pure ATH (solid line), 50% ATH + 50% APP (○) and 30% ATH + 70% APP (△)*

PA11) in imparting the intumescent behaviour. Figure 11a shows the TG and DTG curves, Figure 11b shows the evolving gas (EG) profiles (acetic acid, ammonia and water), while in Figure 11c the expansion behaviour of the sample is reported.

On comparison of EG with expansion curves it can be seen that the sample expands in two steps. The first begins when ammonia is being released from APP by reaction with MH (T_{max} 337°C, Figure 11b), while reaction of APP with PA11 gives rise to the char.[15] The second blowing effect takes place when acetic acid is eliminated from EVA (T_{max} 394°C). In the same temperature range water also evolves (T_{max} 370°C), which contributes to expanding the char. However, instead of the three evolution steps expected from reaction between MH and APP (Figures 6–8), water evolves in a single step with a maximum rate at 370°C. Suppression of water evolution at high temperature (430–455°C, Figures 6–8) may be due to reaction of basic magnesium phosphate with evolving acetic acid (Scheme 3). This suppresses reaction 2.2 of Scheme 2. Water evolved at low temperature by the reaction between APP and MH (T_{max} 300–330°C, Figures 6–8), which is shown to occur by ammonia evolution at 337°C in Figure 11b, might be consumed by the hydrolysis reaction with PA11 of Scheme 4. Indeed, evolution of ammonia with a maximum rate at 510°C in Figure 11b could be explained by pyrolysis of amine groups, reacting from hydrolysis, which are included in the blowing charred structure.

The second ammonia evolution was not observed when PA6 was used instead of PA11 in a previous work,[16] thus confirming the suggested reaction mechanism. Indeed, PA6 would give depolymerisation to caprolactam which competes favourably with the hydrolysis of Scheme 4. The final expansion of

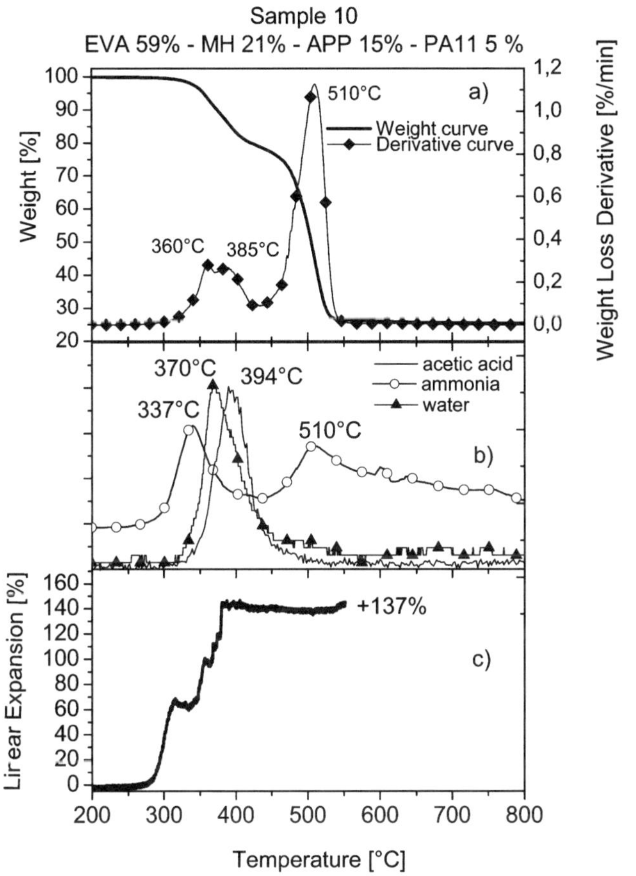

Figure 11 *TGA-FTIR-EGA-expansion measurements for sample no. 10 (EVA 59%, APP 15%, MH 21%, PA11 5%);* (a) *TG (solid line) and DTG (♦) curves;* (b) *ammonia (○), water (▲) and acetic acid (solid line) evolution curves;* (c) *relative expansion of the sample*

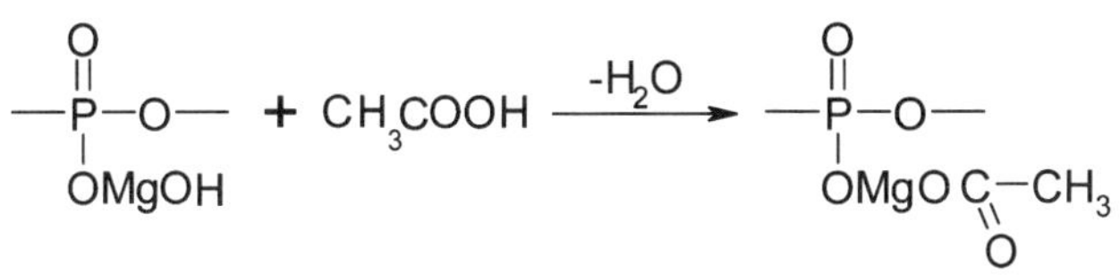

Scheme 3 *Reaction occuring between acetic acid and the basic magnesium phosphate*

the original samples on charring is about 140% and is not affected by the thermal decomposition of the polyene resulting from deacetylation of EVA (T_{max} 510°C).

Mixing of magnesium phosphate, resulting from the APP/MH reaction, with the carbonised residue from degradation of PA11 seems to give strength to the

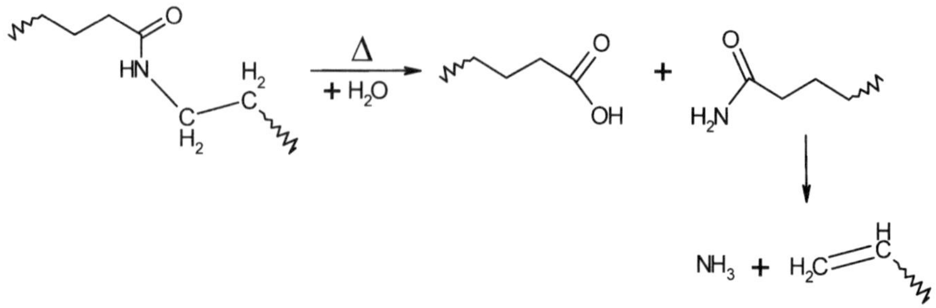

Scheme 4 *Reaction of PA11 with H_2O*

charred material, which does not collapse on EVA volatilisation. The expansion behaviour of the intumescent composition is mainly related to MH amount, as the expansion can be reduced to 100% or less if more than 30% of inorganic hydroxide is added, independently of the content of APP.

18.4 Conclusions

This study has shown that the interaction between the fire retardant additives APP and two basic metal hydroxides (ATH and MH) in their mixtures changes their degradation process, which is related to the basic strength of the hydoxide. MH, which is the strongest base, shifts ammonia from APP to a lower temperature than that of thermal evolution from pure APP and undergoes chemical dehydration by reaction with acidic OH groups of the resulting polyphosphoric acid (PPA). A basic magnesium phosphate-polyphosphate is formed that transforms into magnesium phosphate at high temperature with further elimination of water. Unreacted MH, which depends on APP to MH ratio, dehydrates as when heated alone.

In the case of ATH, chemical dehydration by reaction with APP unfavourably competes with thermal dehydration as compared to MH, because ATH eliminates water at a temperature much lower than MH, giving Al_2O_3, which does not react with APP. ATH reacts with APP when APP is in large excess.

Combination of the intumescent additive APP-PA11 with MH in the EVA matrix, gives, on heating evolution of NH_3 at a lower temperature than from APP heated alone owing to the APP–MH interaction discussed above. The PPA thus made available reacts with PA11, giving the char that is blown by the evolving NH_3. This early charring and blowing actions promoted by MH may explain the improvement of fire retardant performance of the intumescent system with larger MH : APP ratios, which enhance the rate of NH_3 evolution and PPA formation by right shifting the equilibrium of reaction 1.1.

Further blowing is due to evolution of water from thermal dehydration of unreacted MH, combined with evolution of acetic acid from thermal deacetylation of EVA, which together may add further protection to the polymer matrix.

Formation of magnesium phosphate-polyphosphate may increase the fire retardant efficiency of the intumescent shield by also acting as a refractive layer on the polymer surface, reducing heat transfer from the flame to the polymer and also re-irradiating the heat coming from the environment. Furthermore, magnesium phosphate mixed with the organic blown char could increase its mechanical strength to withstand air drafts in a fire, thus preserving its protective action.

18.5 References

1. G.L. Nelson, in *Fire retardancy of Polymeric Materials*, A.F. Grand and C.A. Wilkie (eds.), *Marcel Dekker Inc.*, New York, 2000, pp. 1–24.
2. T.J. O'Niel, *Proceedings of the 10th International Flame Retardants 2002 Conference,* S.J. Grayson ed., Interscience Communications, London, February 2002, pp. 33–43.
3. J.H. Troitzsch, *Proceed. 10th Int. Flame Retardants 2002 Conf.*, S.J. Grayson ed., Interscience Communications, London UK (February 2002), pp. 249–257.
4. M. Lewin, in *Fire retardancy of polymers: the use of intumescence*, M. Le Bras, G. Camino, S. Bourbigot and R. Delobel (eds.), The Royal Society of Chemistry, Cambridge, UK, 1998, pp. 3–32.
5. M. Le Bras and S. Bourbigot, in *Fire retardancy of polymers: the use of intumescence*, M. Le Bras, G. Camino, S. Bourbigot and R. Delobel (eds.), RSC, Cambridge, UK, 1998, pp. 64–75.
6. G. Camino and M.P. Luda, in *Fire retardancy of polymers: the use of intumescence*, M. Le Bras, G. Camino, S. Bourbigot and R. Delobel (eds.), RSC, Cambridge, UK, 1998, pp. 48–63.
7. G. Camino and R. Delobel, in *Fire retardancy of Polymeric Materials*, A.F. Grand and C.A. Wilkie (eds.), *Marcel Dekker Inc.*, New York, 2000, pp. 217–243.
8. G. Camino and S. Lomakin, in *Fire retardant materials*, A.R. Horrocks and D. Price (eds.), Woodhead Publ. Ltd., Cambridge, UK, 2001, pp. 318–336.
9. M. Le Bras and S. Bourbigot, in *Fire and Polymers : materials and solutions for hazard prevention*, G.L. Nelson and C.A. Wilkie (eds.), Washington DC, 2001, pp. 136–149.
10. X. Almeras, F. Dabrowski, M. Le Bras, F. Poutch, S. Bourbigot, G. Marosi and P. Anna, *Polym. Degrad. Stab.* 2002, **77**, 305–313.
11. M. Le Bras, S. Bourbigot and B. Revel, *J. Mater. Sci.*, 1999, **34**, 5777–5782.
12. M. Bugajny, M. Le Bras, A. Noel and S. Bourbigot, *J. Fire Sci.*, 1999, **17**, 494.
13. W.E. Horn, in *Fire retardancy of polymeric materials*, A.F. Grand and C.A. Wilkie (eds.), Marcel Dekker Inc., New York, 2000, pp. 285–352.
14. S.V. Levchik, G. Camino and L. Costa, *Fire Mater.*, 1995, **19**, 1–10.
15. S.V. Levchik, L. Costa and G. Camino, *Polym. Degrad. Stab.*, 1992, **36**, 31–41.
16. A. Riva, G. Camino, L. Fompiere and P. Amigouet, *Polym. Degrad. Stab.*, 2003, **82**, 341–346.

CHATPER 19

Barrier Effects for the Fire Retardancy of Polymers

BERNHARD SCHARTEL, MATTHIAS BARTHOLMAI AND ULRIKE BRAUN

Federal Institute for Materials Research and Testing (BAM), Unter den Eichen 87, 12205 Berlin, Germany (bernhard.schartel@bam.de)

19.1 Introduction

A clear change has been noticed in the way ecological and fire risks are balanced in the field of fire retardancy due to the current demands of consumer protection. A major trend is the reduction and substitution of all environmentally problematic compounds. One general approach to fulfilling this demand is to restrict the fire protection of products in key positions. This approach could be transferred to polymeric materials, concentrating fire retardancy at the interface between pyrolysis zone and gas phase. Obviously, such a surface fire retardancy is different to common fire retardancy of polymers based on homogeneously distributed fire retardant additives. Furthermore, it enables a decreased amount of fire retardants, which may reveal economical advantages as well.

The idea can be divided in two concepts: advanced surfaces and smart surfaces. The first term pools all fire retardant coatings. Of course, some fire retardant paints and coatings have been used successfully for decades already, for instance to protect steel and wood,[1,2] for which an incorporation of fire retardants had never been possible. However, fire retardant coatings recently have been discussed increasingly with respect to the protection of all kinds of polymeric materials and with respect to the use of new coating techniques such as intumescent gelcoats[3] and plasma polymerisation.[4,5] The second group, smart surfaces, pools the systems for which the protection layer is built up in the case of fire. This group includes char forming and intumescent materials, but also systems that yield inorganic residues. Indeed, all these three kinds of materials are addressed in this chapter, which aims to illuminate the cause-and-effect chain of such smart materials, especially the role of barrier properties. The chapter addresses the success and the problems of the concept as a valuable contribution to optimised materials development. Different examples are discussed, such as char-forming PA-66/red phosphorus compounds, the action of magnesia residue in HIPS, layered-silicate polymer nanocomposites and intumescent paints for steel constructions.

19.2 Experimental

The materials investigated were 25 wt% glass fibre reinforced polyamide 66 (PA-66/G5) materials without and with red phosphorus (P_r), Ultramid® (BASF AG, Germany). High impact polystyrene (HIPS) and HIPS containing 15 wt% magnesium hydroxide ($Mg(OH)_2$) were investigated. All systems were provided by BASF AG (Germany). They were compounded *via* extrusion and plates (10×10 cm^2, 2.8 mm thick) were prepared *via* injection moulding for the cone calorimeter investigations. Compounding and the injection moulding were performed in the technical scale laboratory at BASF AG in order to comply with industrial preparation standards. Polymer-layered silicate nanocomposites were prepared using polypropylene graft maleic anhydride (PP-g-MA) with a mass fraction of 0.6 wt% maleic anhydride (Aldrich Chemical Company, USA) and 5 wt% modified montmorillonite Cloisite® 20A (C20A) (Southern Clay Products, USA), which is based on dimethyl dehydrogenated tallow ammonium chloride as the organic modifier. The nanocomposites were compounded using a double screw extruder (ZSK 25, Werner and Pfeiderer, Germany) under vacuum with a rotational speed of 400 min^{-1}, with a throughput of 10 kg h^{-1} and with a temperature profile along the extruder of between 448 and 462 K from the feeder to the nozzle. Cone calorimeter plates (10×10 cm^2, 5 mm thick) were prepared by injection moulding with a temperature profile of between 428 and 448 K and an injection pressure of 900 bar. An intumescent coating was sprayed on square steel plates (100×100 mm^2 and 5 mm thick) of Euronorm S235JR. Layers of 0.3, 0.6, 1.0 and 1.5 mm were applied. A water-based product was used, which has been commercialised to increase the fire resistance of steel structures in buildings. The key components are polyacrylates as binding agents, polyhydric alcohols as carbonising substances, melamine as a foam-producing compound, ammonium polyphosphate as a dehydrating agent and an inorganic filler. Uncoated steel plates were investigated as a control.

The fire behaviour under forced flaming conditions was characterized using a cone calorimeter (Fire Testing Technology, UK) in accordance with ISO 5660, applying different external heat fluxes between 30 and 90 kW m^{-2}. The polymeric materials were measured in the horizontal position using the retainer frame. Data was evaluated using the decreased surface area of the sample (88.36 cm^2). The fire risks, heat release rate (HRR) and total heat release (THR) were monitored as well as the mass loss as a function of time. The residues were characterized with X-ray photoelectron spectroscopy (XPS) using a SAGE 100 (SPECS, Germany). Mg K_α radiation was used at an X-ray power of 250 W (12.5 kV). The flammability of the polymeric materials was characterized using the limiting oxygen index (LOI) according to ISO 4589. The coated steel plates were bedded on a thin layer of ceramic fibre, embedded in a 50 mm thick vermiculite block $15 \times 15\ cm^2$ in size, and investigated under the cone heater of the cone calorimeter. The temperature on the back of the steel plates was measured using thermocouples.

19.3 Results and Discussion

19.3.1 Role of Barrier Effects and Residue in Char Forming Systems

The fact that increasing char formation improves fire properties has been noted for decades.[6] In the following the role of barrier effects and of thermally stable residue is evaluated in detail for char-forming polymers by comparing PA-66/G5 and PA-66/G5/P_r. The thermal and thermo-oxidative decomposition of PA-66/G5 and PA-66/G5/P_r were investigated with thermogravimetry coupled with an evolved gas analysis. The results have been described in detail previously.[7,8] For both materials the polymer scission shows simultaneous formation of carbon dioxide, cyclopentanone, ammonia, methane and amines. Based on these reported results, a radical decomposition process and an amide hydrolysis are proposed as the main decomposition pathways. The radical decomposition process is not significantly influenced whereas the amide hydrolysis starts at lower temperatures for PA-66/G5/P_r than for PA-66/G5. Therefore, the decomposition temperature range was broadened. The lower onset temperature of decomposition indicates a catalysed amide hydrolysis.

The proposed decomposition pathways are summarized in Schemes 1 and 2 for PA-66/G5 and PA-66/G5/P_r, respectively. The evolution of cyclopentanone

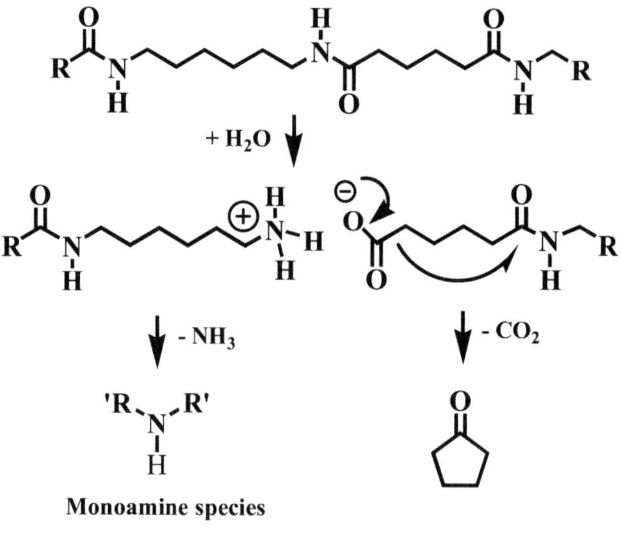

Scheme 1 *Amide hydrolysis decomposition pathway of PA-66/G5 in inert atmosphere. (Monoamine species: 2-methylpiperidine, hexamethylenimine, 1-methylcyclopentylamine, etc.)*

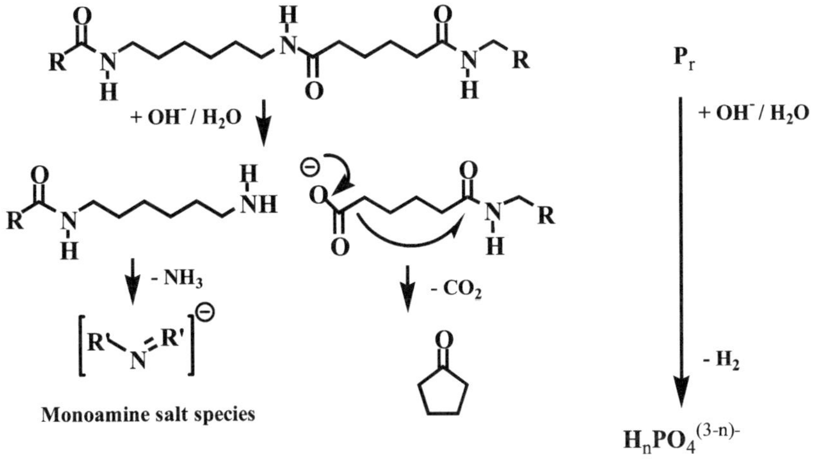

Scheme 2 *Amide hydrolysis decomposition pathway of PA-66/G5/P_r in inert atmosphere. (Monoamine species: 2-methylpiperidine, hexamethylenimine, 1-methylcyclopentylamine, etc.)*

indicates alkaline-catalysed scission of the polymer. It must be considered that, in the presence of water and a basic media, red phosphorus can be oxidized to phosphorus oxide species. In a fire scenario, the interaction of PA-66, H_2O, OH^- and red phosphorus results in a char-forming process. Based on results of thermal analysis it is presumed that the char is generated on carbon of the amine species.

A comprehensive fire behaviour characterization of PA-66/G5 and PA-66/G5/P_r has been given in recent literature.[7–9] The materials showed decreased flame zones and formed black residues. P_r triggers a condensed phase fire retardancy mechanism in PA-66. Indeed, 6–8 wt% P_r in glass fibre filled PA-66 samples were reported to be an outstanding combination that fulfilled UL 94 V-0 classification.[10] XPS investigation on the residue of PA-66/G5/P_r revealed carbon (identified peak for the binding energy C1s), nitrogen (N1s), phosphorus (P2s and P2p) and oxygen (O1s) besides the elements of the inorganic filler (Si2s, Si2p, Ca2p, a.o.). Evaluating the binding energy peak C1s at between 282 and 293 eV revealed carbon–carbon bonds typical for char structures, and also carbon–nitrogen and carbon–oxygen bonds. Based on the structure of the P2p (132–138 eV) and O1s (530–537 eV) peak, phosphorus–oxygen bonds were inferred. The less intensive peak of N1s (397–405 eV) exhibits small amounts of Monoamine salt species. Therefore the XPS data supported the charring of the polymer and phosphorus remaining in the condensed phase, mainly as phosphorus–oxides. The heat release and mass loss corresponded with respect to their rates and total values for PA-66/G5 and PA-66/G5/P_r. The effective heat of combustion was not changed by P_r, establishing the absence of important gas phase mechanisms.[8] The time to ignition was not improved. After ignition the strong initial increase of HRR was rather abruptly stopped for PA-66/G5/P_r when an effective residue layer was formed yielding a smaller peak of HRR.

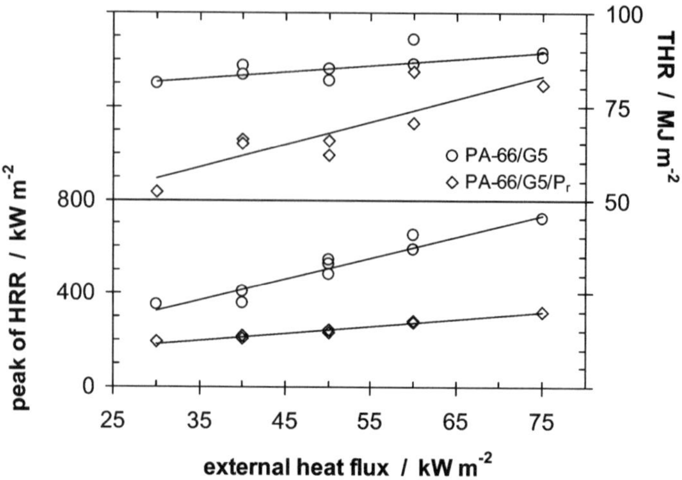

Figure 1 *Peak of HRR and THR vs. external heat flux for PA-66/G5 and PA-66/G5/P_r*

The further burning behaviour was different for samples with and without P_r. Samples with P_r were characterized by lower heat release rates (visually distinguished by a smaller flame zone), and longer burning times. Furthermore, the total mass loss was decreased for the increasing charring of red phosphorus containing samples and resulted in decreased total heat release.

Investigation of the fire behaviour for different external heat fluxes revealed a deeper understanding of the mechanisms. The residue due to the non-glass material increased from 5% up to 25% of the initial mass for decreasing external heat fluxes from 75 to 30 kW m^{-2}. Consequently, THR was strongly reduced for low external heat fluxes, whereas the effect vanished for higher external heat fluxes (Figure 1). Especially for low external heat fluxes, the amount was clearly greater than the effect given by replacing a combustible polymer with P_r. PA-66 was clearly transformed into a char-forming material. The peak of HRR of PA-66/G5 showed the polymer-typical strong increase with increasing external heat fluxes, whereas the samples with red phosphorus were less dependent on external heat flux (Figure 1). The barrier effect became active with progressive formation of a char at the surface, and depended on both the amount of char and its barrier properties rather than on the external energy impact.[9] Therefore, the fire retardancy effects showed self-enhancing characteristics through decreased thermal feedback from the flame zone on pyrolysis. What is more, with increasing external heat flux fire retardancy increased in terms of maximum heat release rate, but fire retardancy decreased in terms of total heat release at the same time (Figure 1). The char acted *via* two mechanisms: First, the barrier layer restricted the heat and mass transfer between the flame and the pyrolysis zone. Second, the thermally stable char was equivalent to reduced total fuel support of the flame zone. The barrier effect and the reduction of combustible volatiles were clearly identified as two separate fire retardancy mechanisms.

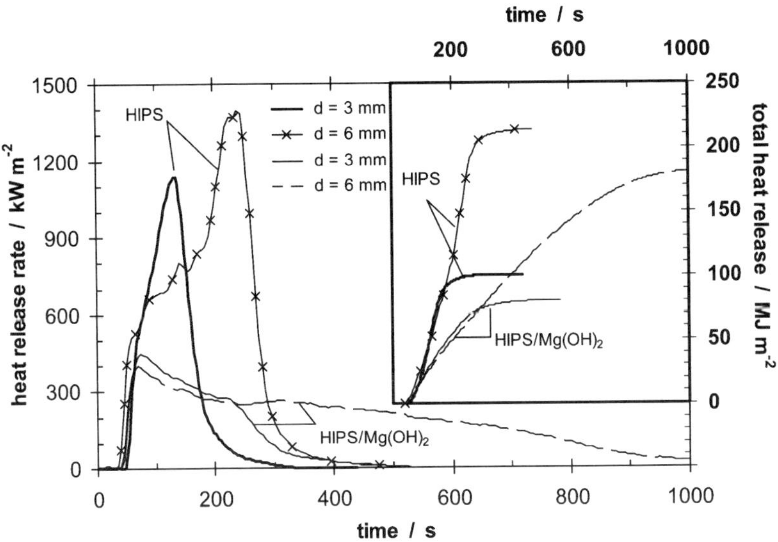

Figure 2 *HRR and THR vs. time for HIPS and HIPS/Mg(OH)$_2$ with different plate thickness (external heat flux = 50 kW m^{-2})*

The materials' surface was identified as a promising key position for fire retardancy.

19.3.2 The Effect of Inorganic Residue in Contrast to Char

In contrast to the induced char formation of the polymer in cases where P_r is added to PA-66/G5 materials, adding inorganic inert fillers resulted in a less pronounced effect with respect to thermal decomposition and THR. Obviously, there is an difference between inorganic residue and char, which is discussed in the following. Indeed, adding 15 wt% $Mg(OH)_2$ in HIPS, for instance, resulted in a 15% reduction of the total heat release under forced flaming conditions (Figure 2) due to the ordinary replacement of combustible material.

The weight and XPS data showed that the residue was just MgO with some $Mg(CO_3)$ caused by impurities, as has been reported before.[11] No trace of carbon was found to remain as char in the condensed phase. Furthermore, THR remained nearly unchanged for different external heat fluxes, since the total amount of released fuel remained unchanged even for very different burning times. However, $Mg(OH)_2$ influences burning behaviour by means of several effects.[11] Obviously, the endothermic reaction to MgO associated with the formation and release of H_2O is an efficient heat-sink mechanism. Furthermore, the released H_2O acts as a very effective cooling agent. Additionally, the MgO residue works as a barrier at the surface, suppressing the fuel support rate. This effect was clearly indicated by the changed HRR curve *versus* time for HIPS/$Mg(OH)_2$ in comparison to HIPS (Figure 2). After a similar time to ignition and a similar initial increase of HRR, the peak of heat release was significantly

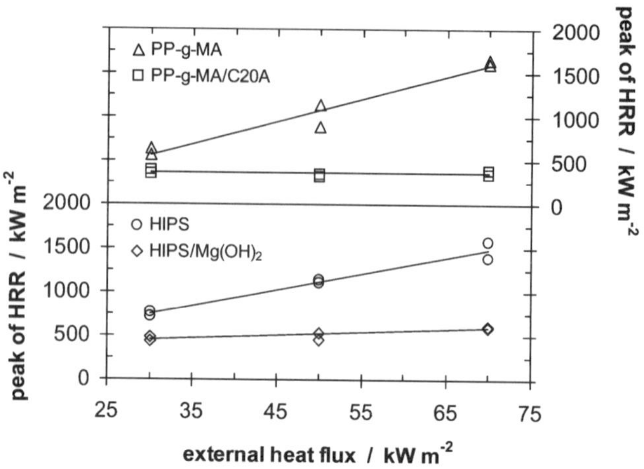

Figure 3 *Peak of vs. external heat flux for HIPS, HIPS/Mg(OH)$_2$, PP-g-MA and PP-g-MA/C20A*

reduced for HIPS/Mg$(OH)_2$. The MgO residue layer acted as an efficient barrier, resulting in a continuous HRR decrease when the barrier layer increased. This description is supported convincingly when samples with different thicknesses are compared. The thicker HIPS/Mg$(OH)_2$ samples showed the same initial increase, peak of HRR and subsequent decrease, which was merely elongated in terms of time (Figure 2). It became obvious that the main characteristics were dominated by barrier formation. The fire retardancy effect became strongly dependent on the formation of the barrier layer and on the barrier properties. Since both the MgO residue and the layered silicate residue of the example discussed below were very stable against heat treatment, the peak of HRR became nearly independent of the external heat flux (Figure 3). Consequently, the peak of HRR was strongly decreased for higher external heat fluxes as compared to the polymer.

In the last ten years, layered-silicate polymer nanocomposites have been heavily promoted as suitable materials to exploit barrier effects for fire retardancy.[12–14] Indeed, their performance was impressive in respect to peak of HRR (Figure 3). The performance of 5 wt% C20A, which forms good nanocomposites in PP-g-MA,[15] was up to the performance of 15 wt% Mg$(OH)_2$ in HIPS. The effect was reached even though no charring of the polymer was observed for the system of PP-g-MA/C20A either and although C20A did not provide the additional heat-sink mechanisms of Mg$(OH)_2$. It was pointed out that a distribution on the nanoscale enabled the exploration of the impressive aspect ratio of the silicate plates, forming an effective barrier layer. The increasing reduction of the peak of HRR for higher external heat fluxes yielded significantly decreased fire growth rates under forced flaming conditions.

Such performance is evaluated in modern test procedures, including those for building (SBI) and transportation (cone calorimeter test for applications in railway vehicles and aircrafts). However, analogous to the results reported

Table 1 *LOI for the investigated materials*

	LOI (%)		*LOI* (%)		*LOI* (%)
PA-66/G5	22.4 ± 0.5	HIPS	17.2 ± 0.5	PP-g-MA	19.2 ± 0.5
PA-66/G5/P_r	26.0 ± 0.5	HIPS/$Mg(OH)_2$	19.5 ± 0.5	PP-g-MA/C20A	19.3 ± 0.5

for PA-66/G5/P_r and HIPS/$Mg(OH)_2$, the time to ignition was not improved. Furthermore, for lower external heat flux the reduction became smaller, so that the fire retardancy effect vanished (Figure 3). Unfortunately, flammability tests such as LOI and UL 94 are connected with the performance for external heat fluxes towards zero and are the main criteria for polymeric materials. Therefore, PP-g-MA/C20A and HIPS/$Mg(OH)_2$ were to some extent not up to the demands of fire retarded polymer materials as opposed to the PA-66/G5/P_r example. Table 1 shows the LOI for the systems described and demonstrates the problem. Only the char-forming system PA-66/G5/P_r showed a relevant improvement of LOI in comparison to the corresponding polymer, whereas the nanocomposite did not show even a significant change. A moderate improvement was found for the HIPS/$Mg(OH)_2$ system, caused probably by the additional mechanisms beyond the barrier formation.

19.3.3 The role of Insulation Properties in Contrast to Mass Transfer Barrier

It is reasonable to assume that any surface layer works in principle as a barrier for heat transport and mass transport. Effectiveness in both respects depends on the specific properties of the layer. In practice the two effects often cannot be distinguished. Conversely, reduced fuel support in the flame zone due to a mass-barrier will result in a consequently decreased thermal feedback. A thermal barrier, however, will result in a reduced decomposition rate and therefore in reduced fuel support. Indeed, the implication of a mass-barrier and a heat-barrier, respectively, may be the same in terms of fire behaviour. Furthermore, to some extent the interdependence of both effects becomes self-enhancing with respect to fire retardancy.

The charring system discussed, PA-66/G5/P_r, showed a significant insulation contribution, resulting in reduced effective pyrolysis temperatures, whereas the characteristics of the residue-forming HIPS/$Mg(OH)_2$ and PP-g-MA/C20A may be dominated by mass-barrier effects. In contrast to this, intumescent systems were developed to optimise heat-barrier properties. Intumescent systems are smart systems that build up a multicellular char structure under heat treatment. Figure 4 shows a typical sequence of photographs taken during the intumescence of a fire protection layer on a steel plate under the cone heater. The thickness of the layer increased twenty, indeed up to one hundred, times. Only coactions of different decomposition processes–such as char formation, the release of non-flammable gases and heat-sink mechanisms–result in an effective intumescence. These have been described in detail in a previous paper on the investigated

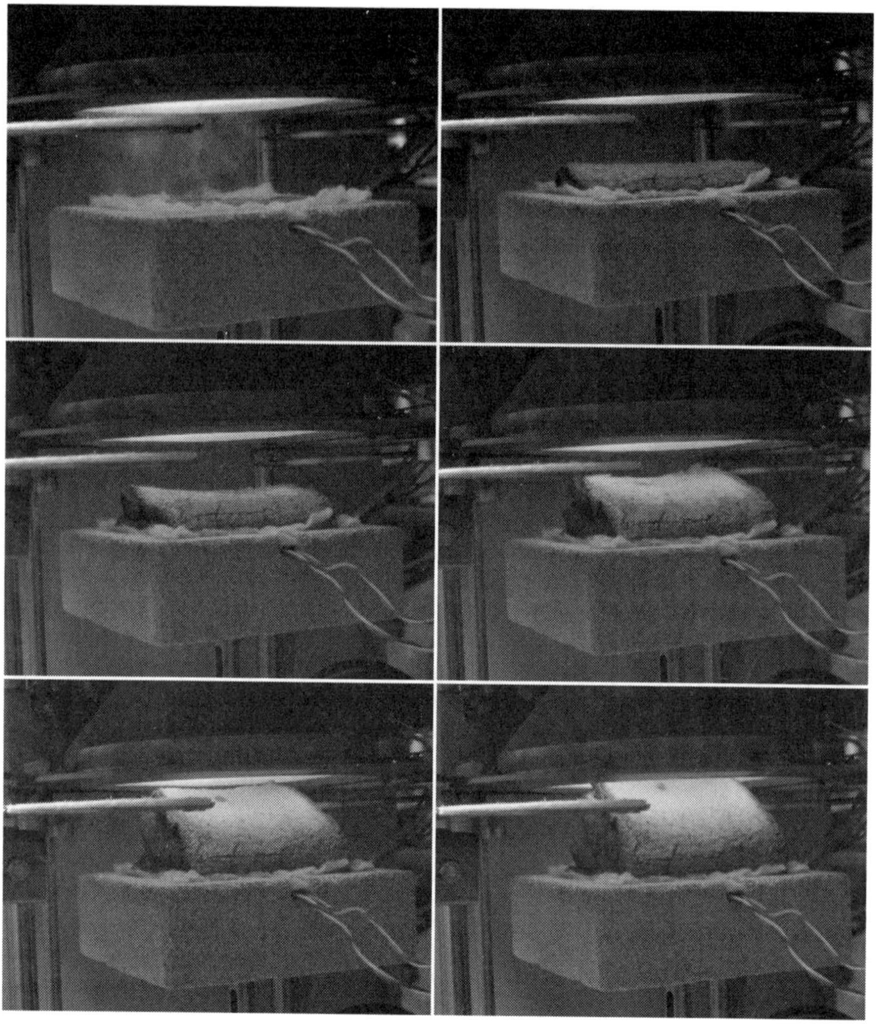

Figure 4 *Photographs taken during the intumescence of the 1.5 mm thick layer coated on a steel plate at an external heat flux of 60 kW m^{-2}*

material based on a thermogravimetric study.[16] Notably, the temperature ranges of intumescence under the cone heater and thermogravimetric results corresponded to each other when it was taken into account that a constant external heat flux was applied under the cone heater, whereas a constant heating rate was used for thermogravimetry.

The heat-insulating effects of the intumescent systems are illustrated in comparison to the temperature development of uncoated steel plates in Figure 5. Furthermore, the influences of varying the coating thickness and the external heat fluxes were investigated and are also summarized in Figure 5. The initial temperature increase was very similar for both the coated and the uncoated steel

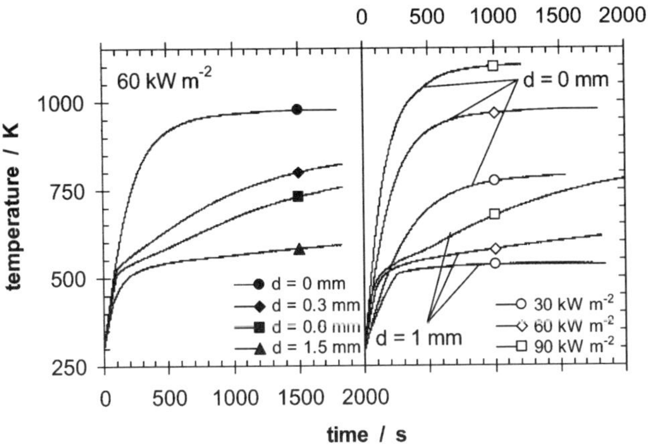

Figure 5 *Temperature at the back of uncoated and coated steel plates plotted against time for different coating thicknesses (d) and external heat fluxes*

plates. Multicellular char formation occurred at between 450 and 570 K and resulted in a significant slow-down of the temperature increase. The results were discussed recently, whereby the thermal conductivity was determined by means of numerical simulation.[17] The layer acted as a heat-barrier, resulting in an efficient thermal insulation. Hence, the steel temperature was dependent on the time, the layer thickness and the applied heat flux.

Decomposition onset temperatures of 750 K were reported for the multicellular char.[16] Depending on the external heat flux, corresponding temperatures were reached, especially at the top of the layer. This decomposition starting at the surface is clearly apparent in Figure 4, for instance. After the black char was decomposed, the white inorganic component of the material became visible. Different temperature curves were observed for longer times depending on the external heat flux, char decomposition depth and the remaining intact multicellular char (Figure 5). For smaller external heat fluxes and for larger initial layer thicknesses the insulation properties remained sufficient and the temperature of the steel plate was nearly constant, so that even for times over 30 min the steel temperature was below 600 K. Hence, the insulation properties of the intumescent coating built up a temperature difference larger than 150 K, and even up to 400 K, compared to those of uncoated steel plates. For higher external heat fluxes and thinner initial thicknesses, the insulation properties decreased significantly during the experiment. With polymeric materials, intumescence reduced the energy supporting the pyrolysis zone. Consequently, the flame spread can be significantly reduced, or self-extinguishing behaviour may be reached even when the effective temperature is reduced below the decomposition temperature.

19.4 Conclusion

Four different materials were investigated that exhibit smart surfaces in terms of fire retardancy. The two mechanisms for char-forming materials–barrier and reduction of total fuel–were discussed separately by means of an investigation of PA-66/G5 and PA-66/G5/P_r. The influence of barrier properties was discussed considering HIPS/$Mg(OH)_2$ and a PP-g-MA/C20A nanocomposite. An optimised heat-barrier forming system was presented considering an intumescent coating.

The surface was identified as a promising key position for fire retardancy. The different systems improved fire behaviour due to barrier properties. Some fire properties, especially the peak of the heat release, are rather easily improved by barrier layers, whereas others, such as time to ignition, flammability and the total heat release, may be not influenced significantly. Hence, only a comprehensive fire behaviour assessment illustrates whether barrier effects yield sufficient fire retarded materials. The results discussed illustrate the opportunistics and limits presented by the concept of smart surfaces.

19.5 Acknowledgements

The authors thank U. Knoll and Dr. Schulz for their measuring support. Parts of the work presented received financial support from the BASF AG (Germany) and the Volkswagen Foundation (I/77 974), respectively.

19.6 References

1. H.L. Vandersall, *J. Fire Flammability*, 1971, **2**, 97–140.
2. R. Kozlowski and M. Wladyka-Przybylak, "Natural polymers, wood and lignocellulosic materials", in *Fire Retardant Materials*, A.R. Horrocks and D. Price (eds.), Woodhead Publishing, Cambridge, UK, 2001, Chapter 9, pp. 293–317.
3. S. Hörhold and R. Walz, *Kunstst.-Plast. Eur.*, 1999, **89(8)**, A102.
4. S. Bourbigot, C. Jama, M. Le Bras, R. Delobel, O. Dessaux and P. Goudmand, *Polym. Degrad. Stab.*, 1999, **66(1)**, 153–155.
5. B. Schartel, G. Kühn, R. Mix and J. Friedrich, *Macromol. Mater. Eng.*, 2002, **287(9)**, 579–582.
6. D.W. van Krevelen, in *Properties of Polymers*, 2nd Edn., Elsevier, Amsterdam, 1976, Chapter 26B, pp. 525–536.
7. B. Schartel, R. Kunze and D. Neubert, *J. Appl. Polym. Sci.*, 2002, **83(10)**, 2060–2071.
8. B. Schartel, R. Kunze, D. Neubert and U. Braun, in *Recent Advances in Flame Retardancy of Polymers*, M. Lewin (ed.), Business Communications Co Inc, Norwalk, USA, 2002, Volume 13, pp. 93–103.
9. B. Schartel and U. Braun, *e-Polymers*, 2003, 13.
10. J. Davis and M. Huggard, *J. Vinyl Additive Technol.*, 1996, **2(1)**, 69–75.
11. W.E. Jr. Horn, "Inorganic hydroxides and hydroxycarbonates: their function and use as flame-retardant additives.", in *Fire Retardancy of Polymeric Materials*, A.F. Grand and C.A. Wilkie (eds.), Marcel Dekker, New York, 2000, Chapter 9, pp. 285–352.

12. M. Zanetti, T. Kashiwagi, L. Falqui and G. Camino, *Chem. Mater.*, 2002, **14(2)**, 881–887.
13. J.W. Gilman, T. Kashiwagi, E.P. Giannelis, E. Manias, S. Lomakin, J.D. Lichtenham and P. Jones, "Nanocomposites: radiative gasification and vinyl polymer flammability", in *Fire Retardancy of Polymers: The use of Intumescence*, M. Le Bras, G. Camino, S. Bourbigot and R. Delobel (eds.), Royal Society of Chemistry, Cambridge, 1998, pp. 203–221.
14. J.W. Gilman and T. Kashiwagi, "Polymer-layered silicate nanocomposites with conventional flame retardants.", in *Polymer-Clay Nanocomposites*, T.J. Pinnavaia and G.W. Beall (eds.), John Wiley & Sons, Chichester, 2000, Chapter 10, pp. 193–206.
15. A. Tidjani, O. Wald, M.-M. Pohl, M.P. Hentschel and B. Schartel, *Polym. Degrad. Stab.*, 2003, **82(1)**, 133–140.
16. R. Kunze, B. Schartel, M. Bartholmai, D. Neubert and R. Schriever, *J. Therm. Anal. Calorim.*, 2002, **70(3)**, 897–909.
17. M. Bartholmai, R. Schriever and B. Schartel, *Fire Mater.*, 2003, **27(4)**, 151–162.

CHAPTER 20

Plasma Assisted Process for Fire Properties Improvement of Polyamide and Clay Nanocomposite Reinforced Polyamide: A Scale-up Study

ANGÉLIQUE QUÉDÉ,[1] BRIGITTE MUTEL,[1] PHILIPPE SUPIOT,[1] ODILE DESSAUX,[1] CHARAFEDDINE JAMA,[2] MICHEL LE BRAS[2] AND RENÉ DELOBEL[2]

[1]Laboratoire de Génie des Procédés d'Interactions Fluides Réactifs - Matériaux, Université des Sciences et Technologies de Lille, F-59655 Villeneuve d'Ascq Cedex, France (Brigitte.Mutel@univ-lille1.fr)
[2]PERF, Ecole Nationale Supérieure de Chimie de Lille, U.S.T.L., F-59652 Villeneuve d'Ascq Cedex, France (Charafeddine.Jama@ensc-lille.fr)

20.1 Introduction

Polymers are used in many fields, but they generally require a modification before use. One of their inconveniences is their high inflammability. Fire risks can be reduced in several ways. The incorporation of flame retardant additives to the polymer[1] is a quite simple and cheap technique, but such additives (commonly hydroxides, halogenated or phosphated components), which have rather high loading rates (50–70 wt%), can reduce the mechanical properties of the polymer and lead to ecological problems.[2,3] Another way is a chemical modification of the macromolecule.[4] In this case, the thermal and mechanical properties of the polymer are preserved (the modified part of the macromolecule is lower than 10%) or enhanced. But, this rather expensive technique is not much used. Either physical[5] or chemical[6] surface modification processes seem attractive as they allow one to concentrate fireproof properties at the polymer surface, where the inflammability occurs, and do not modify the bulk properties of the material. However, literature data about these processes are rather poor. Polysiloxane-based films have good thermal stability and interesting flame retardant properties.[7] In previous work,[8–10] we gave evidence that it was possible to improve polyamide-6 (PA-6) and polyamide-6 clay nanocomposite (n-PA-6) flame retardancy thanks to the deposition of such films elaborated from cold remote nitrogen plasma (CRNP) assisted polymerization of 1,1,3,3-tetramethyldisiloxane (TMDS) pre-mixed with oxygen. Results were promising and the coating seems to be an efficient fire retardant for polyamide substrates.

The present work compares properties of samples treated in the same experimental conditions, either with the previous reactor[8–10] or with a larger one allowing the coating of standard size samples for limiting oxygen index tests and cone calorimeter measurements. The chemical composition of the films, their morphologies, their specific gravities and their growing rates are also compared. The influence of sample size on the evaluation of fire retardant properties is also presented.

20.2 Experimental

20.2.1 Reactor

The experimental set-up is schematically shown Figure 1. A nitrogen flow (purity: 99.995%) was excited in an electrodeless discharge by a microwave

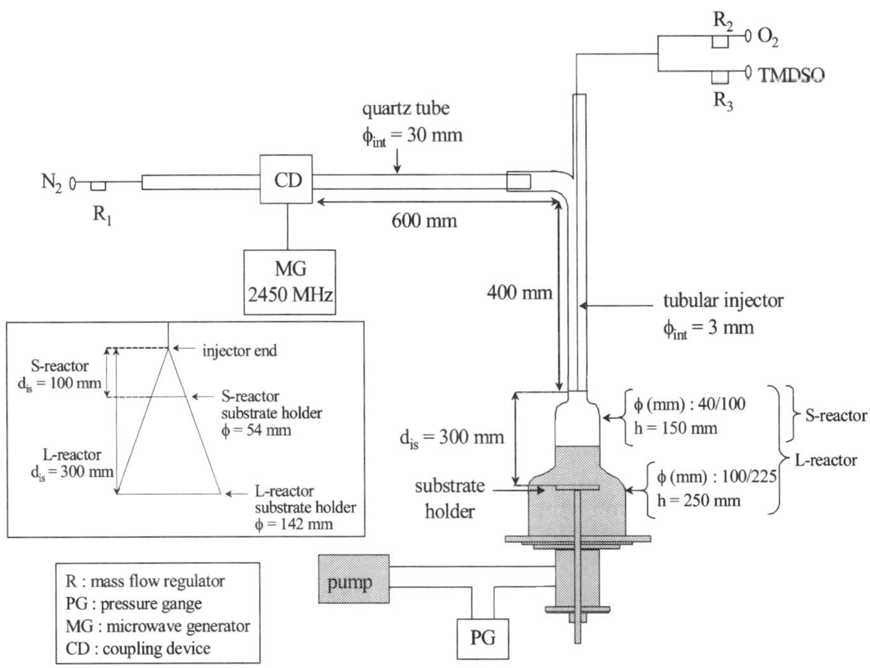

Figure 1 *Experimental set-up. Differences between S and L reactors are specified in grey*

generator. The discharge was produced in a fused silica tube and the gas was evacuated by continuous pumping. Excited species were led to the reactor chamber where the CRNP appears like a yellow afterglow. The CRNP is free of charged particles (and so substrate damage is avoided) and is characterized by a strong thermodynamic non-equilibrium. The monomer (TMDS – purity 97%, supplied from Aldrich Chemical Co), pre-mixed with oxygen (purity ≥99.5%) was injected (at a distance d_{is} from the sample) in the reaction chamber, through a coaxial injector. Only two elements (grey in Figure 1) make the difference between the two set-up used in this work: the Pyrex reactor size and the primary pump. For the small one (denoted by S-reactor), the Pyrex reactor was 150 mm high and the pump flow rate was 58×10^3 Nml min^{-1}. For the large one (denoted by L-reactor), the Pyrex reactor was constituted by the previous one added with a second part 250 mm high and the pump flow rate was 2×10^6 Nml min^{-1}. The deposition process was carried out in two steps with the same experimental parameters as the ones used in previous work.[8–10] At first, samples were treated for 5 min by the CRNP (N_2 flow rate (Q): 1800 Nml min^{-1}, transmitted microwave power (P): 560 W) to increase adhesion quality of the polymer.

Secondly, the deposition step (TMDS injection, duration denoted by t) was performed without air exposure. The polymerization process leads to a previously studied white glow.[11,12] ($Q_{N2} = 1800$ Nml min^{-1}, $Q_{O2} = 50$ Nml min^{-1}, $Q_{TMDS} = 5$ Nml min^{-1}, $P = 560$ W). The injector and substrate (d_{is}) were 100 mm apart in the S-reactor. In the L-reactor, the influence of this parameter is studied (section 20.3.1) to obtain a radial thickness homogeneity of the deposited film at least as good as that obtained in the small reactor. Samples were coated successively on each face with an open air exposure between the two steps for fireproof evaluation.

20.2.2 Characterization Techniques

Chemical characterization of the films was performed by FTIR spectrometry (Perkin-Elmer). Their morphologies were observed by scanning electron microscopy (Leo 982 Zeiss microscope) operating under 10^{-4} Pa (voltage 1 kV). For both studies, films were deposited on silicon 100.

The deposition rate determined in Å s^{-1} at a position x mm from the reactor center is denoted by V_x. It was evaluated from surface profilometry (Alphastep piezoelectric stylus, accuracy: $\pm 10^{-2}$ μm) on a coating performed on a glossy aluminum disk sample (diameter: 54 mm). The deposition rate determined in mg m^{-2} s at a position x mm from the reactor center is denoted by V'_x. It was evaluated from the mass deposited on a Si square sample (10×10 mm^2), the center of which was x mm from the reactor center. The mass was measured by weighing (accuracy: ±0.05 mg). The radial homogeneity of the coating was estimated from $H'_x = V'_x / V'_0$.

Adhesion quality was estimated by the cross-hatch cutter test: the deposited film was cut to the substrate as a cross-hatch pattern. The adhesion quality was quoted from the percentage of squares remaining stuck after the action of a normalized adhesive tape.

Fire properties of samples were evaluated from limiting oxygen index (LOI) tests and cone calorimeter (CC) measurements. LOI tests were performed using a Stranton Redcroft Instrument according to the ASTM D 2863/77 norm.[13] This test allows to determine the minimal oxygen rate, in an oxygen–nitrogen mixture, assuring the combustion of a sample vertically settled (standard size : $100 \times 10 \times 3$ mm^3). CC measurements were obtained with a Stranton Redcroft cone calorimeter according to the ASTM E 1354-90a norm.[14] Samples were exposed to a 35 kW/m^2 external heat flux, which represents the heat flux found in the vicinity of solid-fuel ignition source (standard size: $100 \times 100 \times 3$ mm^3). Conventional data can then be obtained, such as rate of heat release (RHR), ignition time (IT), total heat evolved (THE), volume of smoke production (VSP), CO and CO_2 rates of combustion gases and residual weights (RW).

20.2.3 Samples

Polyamide-6 and Polyamide-6 clay nanocomposite (clay mass fraction: 2wt%) were supplied by UBE as pellets. Sheets ($100 \times 100 \times 3$ mm^3) are obtained using a Darragon press at 255°C with a pressure of 10^6 Pa. LOI tests were performed with small samples ($50 \times 10 \times 3$ mm) denoted LOI-S and with normalized samples ($100 \times 10 \times 3$ mm^3) denoted LOI-N. CC studies were performed with small samples ($20 \times 20 \times 3$ mm^3) denoted CC-S and with normalized samples ($100 \times 100 \times 3$ mm^3) denoted CC-N. Coated PA-6 and n-PA-6 samples were respectively denoted by c-PA-6 and c.n-PA-6.

20.3 Results

20.3.1 Influence of d_{is} on Both Deposition Rate and Radial Thickness Homogeneity of Films Deposited in the L-reactor

As we aimed to deposit a film with a thickness that is as homogeneous as possible on (100×100 mm^2) samples, the influence of this parameter *versus* d_{is} was studied. Results are shown in Table 1. Taking into account accuracy values,

Table 1 *Influence of the distance between the injector and the substrate (d_{is}) on the deposition rates and on the radial thickness homogeneity of the film deposited in the L-reactor*

d_{is} (mm)	$V'_0 \pm 1.1$ (mg m^{-2} s^{-1})	$V'_{44} \pm 1.1$ (mg m^{-2} s^{-1})	$H'_{44} \pm \Delta H'_{44}$
295	8.9	3.3	0.37 ± 0.17
300	8.9	6.7	0.75 ± 0.24
305	7.8	3.3	0.42 ± 0.20
320	8.9	4.4	0.49 ± 0.18
350	7.8	4.4	0.56 ± 0.22
380	4.4	3.3	0.75 ± 0.44

H'_{44} is approximately constant whatever d_{is}. Except for $d_{is} = 380$ mm, the deposition rate at the center of the sample (V'_0) is also independant of d_{is}. Notably V'_0 is always larger than V'_{44}. At the sample edge, the deposition rate (V'_{44}) is maximum for $d_{is} = 300$ mm. This value was selected for the work reported here.

20.3.2 Comparison of Deposition Rate, Radial Homogeneity and Specific Gravity of the Films Obtained with the Two Reactors

Results are shown in Table 2. Taking into account H'_x accuracy values, the radial homogeneity of the deposited film seems to be better in the L-reactor. Besides, the deposition rate is strongly decreased (within a factor of 9). This factor is of the same order as the ratio between the crossing surface of reaction zone and substrate holder plan (Figure 1). In a first approximation, this can be explained supposing that all polymerization products are located inside this reaction zone. But, to confirm this hypothesis it would be necessary to take into account the kinetics of the mixture. According to accuracy, it appears that the film specific gravity is not dependant on the reactor size; it is approximately equal to 1.9 ± 0.3 g cm^{-3}.

20.3.3 FTIR Study: Comparison of the Chemical Structure of Films Obtained with the Two Reactors

Figure 2 shows FTIR spectra of 1 μm thick films elaborated in the two reactors. Whatever the reactor size, the main groups are $Si(CH_3)_x$ and Si–O–Si. Asymmetric and symmetric $\nu(CH_3)$ bands are at 2960 and 2910 cm^{-1} respectively,[15] the $\delta(CH_3)$ ones appear respectively at 1410 and 1250 cm^{-1}.[16,17] $\rho(CH_3)$ and ν(Si–C) bands are in the range 900–700 cm^{-1} [16–19]. The asymmetric ν(Si–O–Si) band, located towards 1200–1000 cm^{-1} [17,20] is the strongest. This chemical group is provided by the monomer; the deposited film has a polysiloxane-like structure. The ν(OH) band located between 3600 and 3000 cm^{-1} is also present. The only difference between the two films spectra is the relative decrease of OH and CH_3 bands when the reactor size increases.

20.3.4 SEM Study: Comparison of the Morphology of Films Obtained with the Two Reactors

Figure 3 shows SEM pictures of films elaborated in the two reactors. The 1 μm thick film elaborated in the L-reactor is uniform, smooth and without any defect (Figure 3a) while it shows a fine texture when it is elaborated in the S-reactor (Figure 3b). For the thicker film (≈ 4.5 μm), grains appear. Their size, about 0.1–0.3 μm for film elaborated in the L-reactor (Figure 3c–d), is about 1.2 μm for the one elaborated in the S-reactor (Figure 3e–f). Moreover, in this last case, a linear arrangement of these grains appears (Figure 3e–f). The film texture, finer with the L-reactor than with the S-reactor, may be correlated with the lowest deposition rate obtained when the reactor size increases.

Table 2 *Comparison of deposition rates, radial homogeneities and specific gravities of the films obtained with the two reactors*

S-reactor					*L-reactor*			
x (mm)	$V_x \pm 0.1$ (Å s^{-1})	$V'_x \pm 1.1$ (mg m^{-2} s^{-1})	H'_x	ρ (g cm^{-3})	$V_x \pm 0.1$ (Å s^{-1})	$V'_x \pm 1.1$ (mg m^{-2} s^{-1})	H'_x	ρ (g cm^{-3})
0	432.4	77.8	–	1.80 ± 0.03	46.5	8.9	–	1.91 ± 0.24
11	392.3	75.0	0.96 ± 0.03	1.91 ± 0.03	–	–	–	–
22	229.6	44.4	0.57 ± 0.02	1.93 ± 0.05	41.6	7.8	0.88 ± 0.23	1.88 ± 0.27
44	–	–	–	–	36.0	6.7	0.75 ± 0.22	1.86 ± 0.31

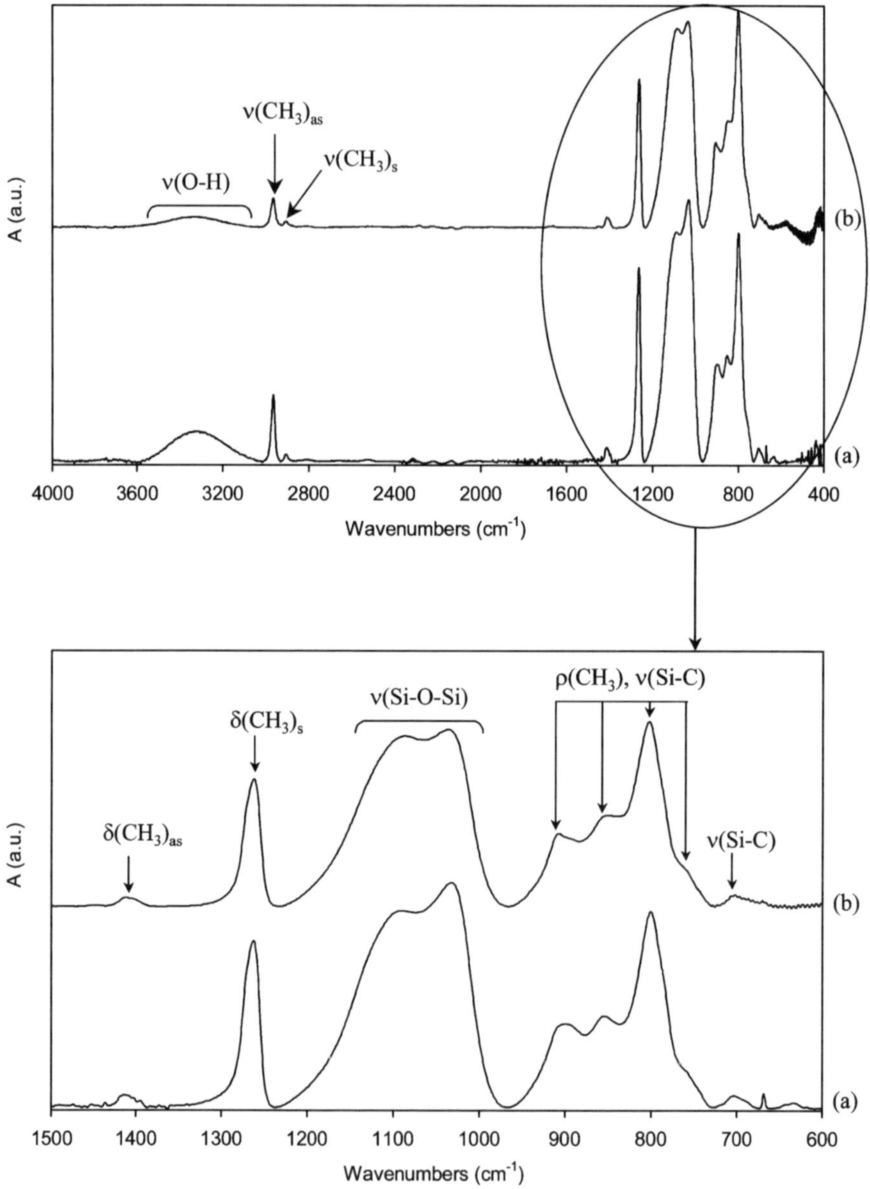

Figure 2 *FTIR spectra of films elaborated in the S-*(a) *and L-*(b) *reactors*

20.3.5 Flame Retardant Properties

20.3.5.1 LOI Tests

A comparison of results obtained with two samples sizes and with the two reactors is shown table 3. For virgin samples, LOI values are not dependant on sample size. Moreover, it appears that the clay incorporation (2 wt%) to PA-6

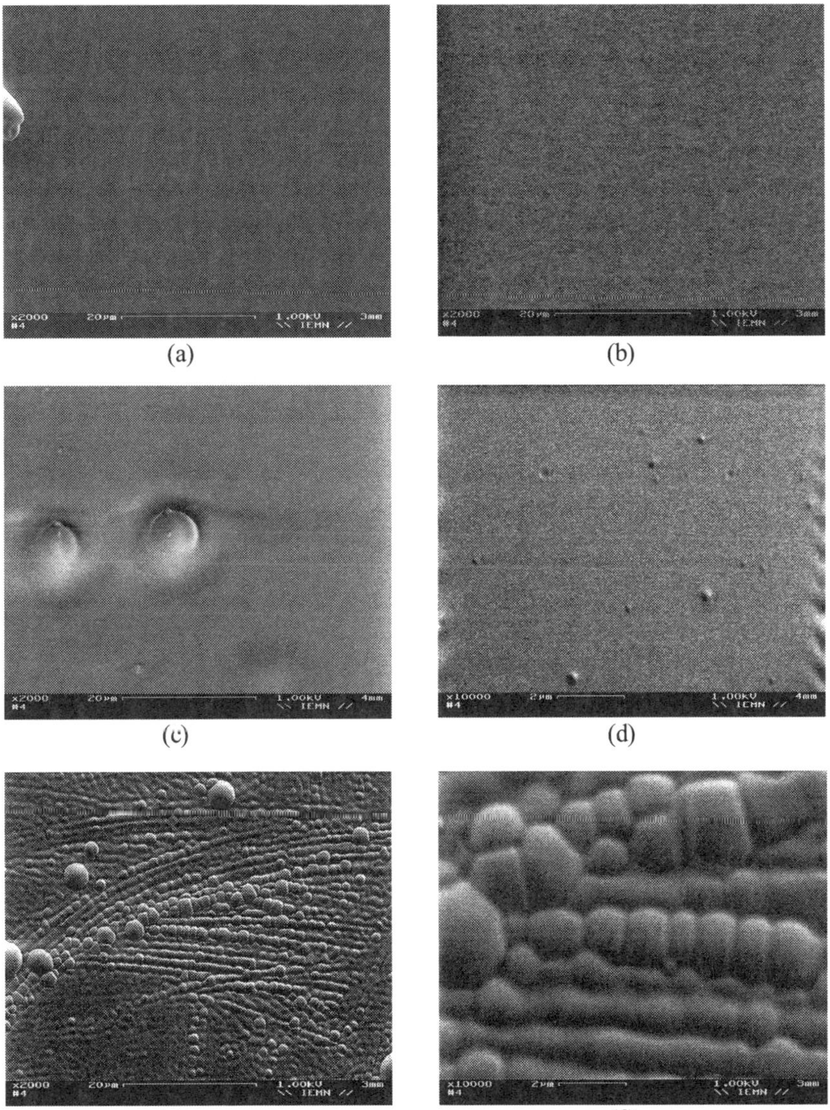

Figure 3 *SEM pictures of films elaborated in the two reactors*
Film thickness = 1 μm, × 2000: (a) *L-reactor,* (b) *S-reactor*
Film thickness = 4.5 μm, L-reactor: (c) *×2000,* (d) *×10000*
Film thickness = 4.5 μm, S-reactor: (e) *×2000,* (f) *×10000*

does not improve the LOI, which remains equal to 21 ± 1%. For coated samples, the film thickness at the sample edge (where ignition occurs during the test) was different. For the S-reactor, it was evaluated from V_{22} (Table 2) by supposing that the deposition rate was similar both for aluminum and polymer substrates.

Table 3 *LOI (vol%) vs. sample and reactor sizes*

	LOI-S	*LOI-N*	*Nature of LOI test residues*
PA6	21 ± 1	21 ± 1	–
n-PA6	22 ± 1	22 ± 1	–
Reactor	S-reactor	L-reactor	
c-PA6	25 ± 1	22 ± 1	Polysiloxane
c.n-PA6	46 ± 1	48 ± 1	Silica-like
Film thickness (μm)	28	1.5	

Table 4 *LOI (vol%; ±1%) vs. thickness of a film deposited at the edge of PA-6 and n-PA-6 samples (LOI-N) in the L-reactor*

Film thickness at the edge of the sample (μm)		*0*	*0.6*	*1.1*	*1.5*	*2.1*	*3.2*	*5.3*	*9.6*	*18.1*
LOI-N	PA-6	21	22	22	22	22	22	22	22	22
	n-PA-6	22	45	47	48	47	43	43	42	42

For the L-reactor, it was calculated from V_{44} (Table 2). For sample coated in the S-reactor, LOI is slightly improved when a film (28 mm thick) is deposited on PA-6 (LOI increases from 21 to 25%), but when it is deposited on n-PA-6, a strong improvement can be observed as LOI increases from 22 to 46%. These results can be explained by a better adhesion quality of the film deposited on n-PA-6 than on PA-6: after the cross-hatch cutter test, 84% of squares remain on c-n-PA, while only 60% remain on c-PA.

For the L-reactor, the influence of the film thickness at the edge of the sample was studied first. Results are shown table 4. Whatever the films thickness, ranging from 0.6 to 18.1 μm, LOI test results are not improved, neither by the coating nor by the clay addition to PA-6: PA-6, c-PA-6 and n-PA-6 have the same LOI (21 ± 1%). But the LOI of the c.n-PA-6 is strongly improved: it sharply increases as soon as a 0.6 μm thick film is deposited. A maximum value equal to 48% is obtained for 1.5 μm, then it decreases and remains quite stable (42%) for thickness ranging from 3 to 18 μm.

So, whatever the reactor size and film thickness, the combined clay addition and film deposition on PA-6 leads to a strong increase of LOI. This increase seems to be independent of film thickness (Table 3) and it is worth noting that a thin coating is sufficient. The effect of using either a clay addition or a coating is very slight or not noticeable.

FTIR spectra of LOI-S and LOI-L test residues were also recorded. c-PA-6 has a polysiloxane structure similar to the one of the deposited film, while the c.n-PA-6 residue is mainly silica-like. These results give evidence that, for PA-6, the use of both a nano-composite additive and a coating leads to a component that improved thermal stability.

Table 5 *Cone calorimeter measurements performed on virgin or coated in the S-reactor S-samples*

S-samples	*PA-6*	*c-PA6*	*n-PA6*	*c.n-PA6*
IT (s)	81 ± 18	71 ± 16	75 ± 6	83 ± 8
RHR (kW m^{-2})	1972 ± 80	1784 ± 199	1102 ± 112	807 ± 95
THE (kJ)	32 ± 2	32 ± 2	35 ± 1	29 ± 2
RW (%)	0.9 ± 0.2	1.2 ± 0.1	1.4 ± 0.2	1.3 ± 0.1

20.3.5.2 Cone calorimeter measurements:

CC measurements performed on small samples (S-samples), virgin or coated in the S-reactor are shown table 5. Film thickness at the sample edge, calculated from V_{11} (Table 2), was 47 µm. Taking into account the accuracy of measurement, the coating deposited on PA-6 does not improve results.

Likewise, no significant modification of IT, THE and RW parameters can be noticed between the four samples. Besides, a noticeable decrease (44 relative%) of the RHR peak can be observed after clay incorporation to PA-6; this decrease is reinforced (59%) after deposition of a film on the n-PA-6. To validate these promising results, a more complete study was performed on normalized size samples (N-samples) and with the L-reactor (summarized in table 6). For PA-6 (Table 6, column a), IT, RHR peak and RW results are in good agreement with literature data, respectively equal to 60 ± 3 s,[20] 1150 ± 115 kW/m^2 [20] or 1010 ± 101 kW/m^2 [21,22] and 1 ± 0.5%.[20] The RHR peak obtained in this work, larger than the one obtained in other work, leads to a higher THE (1346 ± 70 kJ instead of 1000 ± 100 kJ [20]). For n-PA-6 (Table 6, column c), RHR peak and RW are also in good agreement with literature data, respectively equal to 686 ± 69 kW/m^2 and 3 ± 0.5%.[21,22] Other parameters were not compared as no data were found. For coated samples, the film thickness at the edge was 1.5 µm (this value being chosen from LOI results (Section 20.3.5.1)). The coating deposited on PA-6 (Table 6, column b) does not allow a reduction of RHR or CO_2 peaks, but it leads to a decrease of THE (38%), CO (50%) and VSP (43%) peaks and of total quantities of CO (45%), of CO_2 (36%) and of VSP (41%). Combustion is not delayed, but is accelerated within ~100 s (Figure 4). This could be due to delamination of the coating under heating promoting faster degradation and/or to the presence of new chemical functions grafted at the PA-6 surface after the pre-treatment. These functions can lead to the formation of non-inflammable products, explaining the decrease of THE and of gaseous emissions. Clay incorporation to PA-6 (Table 6, column c) leads to a decrease of all peaks. (RHR: 34%; CO: 63%; CO_2: 39%; VSP: 32%) and of total quantities of energy (29%), of CO (24%) and of CO_2 (36%). The n-PA-6 combustion is delayed for 50 s in comparison to that of PA-6 (Figure 4), is slightly slowed and leads to a RW of 4% (1% for PA-6).

These results show that, thanks to clay incorporation, a protective coating is formed at the polymer surface during the combustion, reducing mass and heat transfers between the flame and the polymer. c.n-PA-6 (Table 6, column d) leads

Table 6 *Cone calorimeters measurements for virgin and coated PA-6 and n-PA-6 samples. Influence of film thickness and of samples size. Samples were coated in the L-reactor. Each result is the average value of three measurements*

Sample size	*Standard (100 × 100 × 3 mm³)*					*Small (20 × 20 × 3 mm³)*	
Film thickness	*1.5*				*10*	*10*	
Samples	*PA-6*	*c-P.A-6*	*n-PA-6*	*c.n-PA-6*	*c.n-PA-6*	*n-PA-6*	*c.n-PA-6*
Ignition time (s)	66 ± 3	67 ± 11	98 ± 2	96 ± 2	110 ± 2	75 ± 6	107 ± 12
RHR peak (kWm^{-2})	1053 ± 30	967 ± 70	699 ± 34	623 ± 10	534 ± 28	1102 ± 112	712 ± 112
THE (kJ)	1346 ± 70	829 ± 39	949 ± 45	900 ± 23	858 ± 10	35 ± 1	37 ± 2
CO peak (ppm)	253 ± 1	127 ± 15	94 ± 5	82 ± 5	71 ± 9	7.0 ± 1.3	–
Total CO emission(ppm s)	14564 ± 370	7961 ± 939	11011 ± 1398	10944 ± 880	10924 ± 677	157 ± 39	–
CO_2 peak (vol%)	2.46 ± 0.07	2.29 ± 0.29	1.50 ± 0.10	1.20 ± 0.10	1.1 ± 0.1	0.084 ± 0.002	0.056 ± 0.007
Total CO_2 (vol% s)	304.8 ± 4.1	195.3 ± 15.0	194.8 ± 7.1	170.9 ± 4.3	177.9 ± 6.2	5.85 ± 0.21	3.85 ± 0.39
VSP peak (10^3 m^3 s^{-1})	4.4 ± 0.4	2.5 ± 0.1	3.0 ± 0.2	2.7 ± 0.5	2.6 ± 0.4	0.50 ± 0.03	0.28 ± 0.03
Total VSP emission (m^3)	0.228 ± 0.017	0.134 ± 0.013	0.312 ± 0.034	0.304 ± 0.013	0.441 ± 0.47	0.022 ± 0.001	0.014 ± 0.002
Residual weight (%)	1.0 ± 0.2	1.9 ± 0.2	4.0 ± 0.3	4.2 ± 0.2	3.7 ± 0.3	1.4 ± 0.2	2.4 ± 0.3
Column number	a	b	c	d	e	f	g

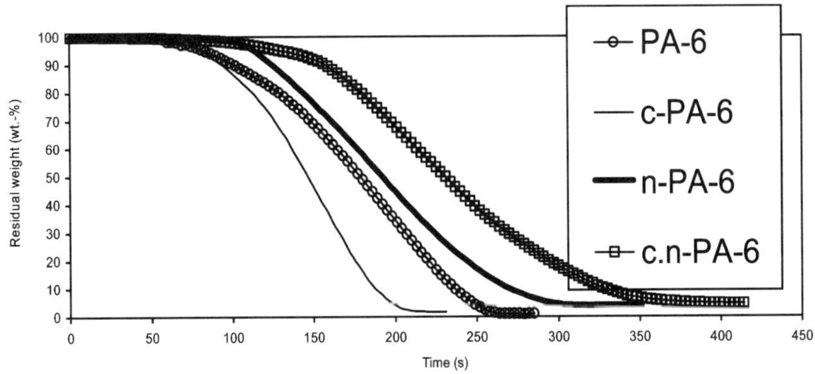

Figure 4 *Evolution of the residual weight during the combustion of virgin and coated (in the L-reactor) PA-6 and n-PA-6*

to very good flame retardant properties. In comparison to PA-6, the ignition time is increased by ≈ 30 s. RHR, CO, CO_2 and VSP peaks are decreased (41, 68, 51 and 39 respectively) as well as total CO_2 and CO quantities (44 and 25%). The combustion is delayed for 50 s and slowed and the corresponding RW is 4%.

From this study, it seems that results obtained in the S-reactor are in good agreement with those obtained with the L-reactor, but a comparison is difficult as the sample thickness was different. Thus, the film thickness influence on CC measurements was firstly studied. Results obtained for c.n-PA-6 (N-samples) in the L-reactor with a 10 μm thick film are compared with the 1.5 μm thick one (Table 6, b and c, d and e columns). It appears that the CC results are improved when the coating thickness increases : IT increases (12%) and the RHR peak decreases (14%) as well as THE (5%). Then, the influence of the sample size was studied from a comparison of results obtained with S- and N-n-PA-6 samples, either virgin or coated with a 10 μm thick coating elaborated in the L-reactor. Except for VSP results, evolutions of CC parameters are similar for the two sets of samples, but the effect is stronger for S-samples (Table 6, c and e, f and g columns).

To determine the nature of the layer formed during the combustion, residues were collected before ignition (≈ 105 s), after ignition (≈ 114 s), at RHR peak (≈ 204 s), and at the end of the cone calorimeter measurements (≈ 400 s) performed with c.n-PA-6 sample, and analyzed by FTIR. Figure 5 shows that the formed layer is mainly silica-like.

20.4 Conclusions

This study aimed to improve the fire retardant properties of PA-6 and to validate the process by performing fire tests in standard conditions. The technique used involves both incorporation of clay nano-composite (2 wt%) in the PA-6

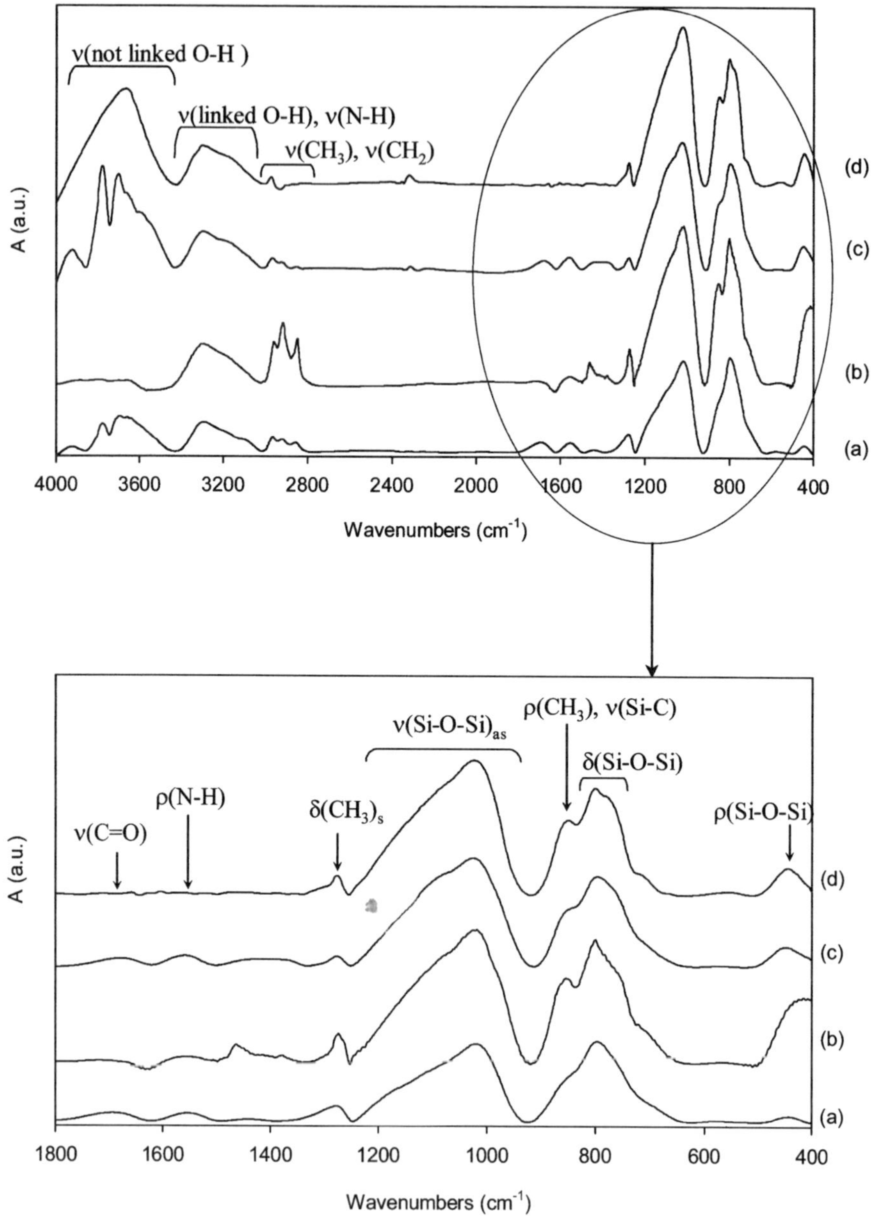

Figure 5 *FTIR spectra of the residue of coated n-PA-6 (film thickness = 10 μm) obtained during CC experiment.* (a) *Before ignition,* (b) *after ignition,* (c) *at RHR peak,* (d) *at the end of combustion*

and a coating obtained from cold remote nitrogen plasma assisted polymerization of 1,1,3,3-tetramethyldisiloxane monomer pre-mixed with oxygen. In comparison with virgin PA-6, the fire retardant performances of the c.n.PA-6 performed in standard conditions are characterized by an increase of LOI (130%) and a decrease of both the RHR peak (41%) and the THE (33%). During combustion, the nanocomposite structure of the polymer leads to the formation of a surface protective layer, the action of which is reinforced by the coating. This carbonaceous and silica-like layer acts as a barrier, limiting mass and heat transfers between the flame and the polymer and slows down toxic gases emission produced by polymer combustion. The advantage of our process, in comparison with other techniques used for PA-6, is a resulting simultaneous improvement of the three main parameters (LOI, RHR, THE). For comparison with literature, some other processes results can be considered. The incorporation of ethylene acetate vinyl (EVA 10.3%) and ammonium polyphosphate (APP 28%) to PA-6[20] leads to an increase of LOI (39%) and to a decrease of RHR peak (55%), but the THE is not modified. A coating obtained from plasma assisted polymerization of hexamethyldisiloxane[23] leads to a decrease of RHR peak (30%), but the THE is not modified. LOI tests were not performed.

The influence of sample size on cone calorimeter studies performed on virgin and coated n-PA-6 shows that the use of small samples instead of standard size ones leads to a correct prediction of the evolution of cone calorimeter parameters, but with an enhanced effect. The coating efficiency increases with its thickness. The use of two set-ups also shows that the radial homogeneity of the thickness film is preserved when the reactor size increases, but the growth rate is decreased, leading to a smoother structure. The chemical nature of the deposited film, mainly polysiloxane-like, and its specific density (≈ 1.9 mg/cm^3) are similar whatever the reactor size.

From this work[24] it may be concluded that a polysiloxane-like coating deposited on n-PA-6 is an efficient way to improve PA-6 fire retardancy. However, the mechanism occurring between the nanocomposite and the coating has to be specified.

20.5 Acknowledgments

The authors thank C. Boyaval from the Institut d'Electronique et de Microélectronique du Nord (Villeneuve d'Ascq, France) for his technical assistance in Scanning Electron Microscopy.

20.6 References

1. M. Lewin, in *Fire Retardancy of Polymers – The Use of Intumescence*, M. Le Bras, G. Camino, S. Bourbigot and R. Delobel (eds.), The Royal Society of Chemistry, Cambridge, U.K., 1998, p. 3.
2. L.W.D. Weber and H. Greim, *J. Toxicol Environ. Health*, 1997, **50(3)**, 195–215.
3. G. Camino, M.P. Luda, and L. Costa, In *Chemical Industry and Environment*, Volume. I, *General Analysis-Risk Analysis*, J. Casal (eds.), 1993, p. 221.
4. S.Y. Lu and I. Hamerton, *Prog. Polym. Sci.*, 2002, **27(8)**, 1661–1712.

5. A.I. Balabanovich, G.F. Levchik, S.V. Levchik and W. Schnabel, *Fire Mater.*, 2001, **25**, 179–184.
6. T. Jana, B.C. Roy and S. Maiti, *Polym. Degrad. Stab.*, 2000, **69**, 79–82.
7. G. Camino, S.M. Lomakin and M. Lazzari, *Polymer*, 2001, **42(6)**, 2395–2402.
8. C. Jama, A. Quédé, P. Goudmand, O. Dessaux, M. Le Bras, R. Delobel, S. Bourbigot, J.W. Gilman and T. Kashiwagi, in *Fire and Polymers, Materials and Solutions for Hazard Prevention*, G.L. Nelson and C.A. Wilkie (eds.), American Chemical Society Publication, ACS Symposium Series 797, Washington, DC, 2001, Chapter 16, p. 200.
9. C. Jama, A. Quédé, H. Sadiki, O. Dessaux, P. Goudmand, R. Delobel and M. Le Bras, in *Recent Advances in Flame Retardancy of Polymeric Materials*, M. Lewin (ed.), Business Communications Co. Inc., Norwalk, USA, 2001, Volume 12, p. 127.
10. A. Quédé, C. Jama, P. Supiot, M. Le Bras, R. Delobel, O. Dessaux and P. Goudmand, *Surf. Coat. Technol.*, 2002, **151–152**, 424–428.
11. P. Supiot, F. Callebert, O. Dessaux and P. Goudmand, *Plasma Chem. Plasma Proc.*, 1993, **13**, 539–554.
12. F. Callebert, P. Supiot, P. Goudmand and O. Dessaux, *11th International Symposium on Plasma Chemistry*, D.E. Harry (ed.), Loughborough University Pub., UK, 1993, p. 1493.
13. "Standard test method for measuring the minimum oxygen concentration to support candle-like combustion of plastics (Oxygen Index)", ASTM D 2863/77, Philadelphia, 1977.
14. "Standard test method for heat and visible smoke release for materials and products using an oxygen depletion calorimeter", ASTM E 1354–90a, Philadelphia, 1990.
15. L.L. Tedder, G. Lu and J.E. Crowel, *J. Appl. Phys.*, 1991, **69**, 7073.
16. D.R. Anderson (ed.), in *Analysis of silicones, chapter 10: Infrared, Raman and Ultraviolet spectroscopy*. eds. Wiley John and Sons, New- York, 1974. 247.
17. C. Rau and W. Kulisch, *Thin Solid Films*, 1997, **249**, 28.
18. F. Callebert, P. Supiot, K. Asfardjani, O. Dessaux, P. Goudmand, P. Dhamelincourt and J. Laureyns, *J. Appl. Polym. Sci.*, 1994, **52**, 1595–1606.
19. P.G. Pai, S.S. Chao, Y. Takagi, G. Lucovsky, *J. Vac. Sci. Technol.*, 1986, **A4**, 689.
20. C. Siat, *Thesis n° 2000ART00403*, Lens, France, 2000.
21. J.W. Gilman, *Appl. Clay Sci.*, 1999, **15**, 31–49.
22. J.W. Gilman, C.L. Jackson, A.B. Morgan, R. Harris, E. Manias, E.P. Giannelis, M. Wuthernow, D. Hilton and S.H. Philips, *Chem. Mater.*, 2000, **12**, 1866–1873.
23. B. Schartel, G. Kühn, R. Mix, J. Friedrich, *Macromol. Mater. Eng.*, 2002, **287**, 579–582.
24. A. Quédé, *Thesis n°3301*, Lille, France 2003.

CHAPTER 21

Fire Retardant Polypropylene / Flax Blends: Use of Hydroxides

MAGALI FOIS,[1] MICHEL GRISEL,[1] MICHEL LE BRAS,[2] SOPHIE DUQUESNE[2] AND FRANCK POUTCH[3]

[1]URCOM, EA 3221, Université du Havre, 25 rue P. Lebon, BP 540, F-76058 Le Havre,France.(michel.grisel@univ-lehavre.fr)
[2]PERF, ENSCLille/USTL, UPRES EA 1040, BP108, F-59652 Villeneuve d'Ascq Cedex, France
[3]CREPIM, Bruay-la-Bussière, France

21.1 Introduction

Over the last few years, several studies have investigated the exploitation of cellulosic fibres as load bearing constituents in composite materials that are easily moulded for automotive applications. The use of these materials in composites has increased due to their unlimited availability, their relative cheapness compared to conventional materials such as glass and aramid fibers, their low abrasion, their multi-functionality, their ability to recycle, and because they compete well in terms of strength per weight of material.[1] Many varieties of plant fibres exist, such as "hairs" (cotton, kapok), fibre-sheafs of dicoltylic plants or vessel-sheafs of monocotylic plants (flax, hemp, jute and ramie), and hard fibres (sisal, henequen and coir).[2]

The flax plant (*Linum usitatissimum*) is a member of the family Linaceae that is important for the production of low-density fibre. The seed of the flax plant is known as linseed, and from it are obtained the linseed oil for commerce, alkyd resins for paints, printing inks and some varnishes. Different varieties of the plant, which are mainly grown in The Netherlands, Belgium and France for linseed and for fibres, are well known (reference 3 and references therein). In this study plant grown for fibre in Normandy (France) was used, the weight ratio of cellulose in flax being about 71%.

Plant derivatives have been previously investigated as flame-retardants additives (FR) for isotactic polypropylene (PP). To this end they can be used alone (lignin,[4] flax fibres in 12.5 and a 40 wt% amounts[5,6]) as char formers or in synergy with conventional FR additives [lignin in association with triglycidylisocyanurate, monoammonium phosphate and melamine,[7,8] flax fibres (fibber in this text) in association with ammonium polyphosphate (APP) or expandable graphite[8]]. These previous results seem quite successful and lead one to assume that ecological friendly fire retardancy is a technological breakthrough for PP/flax composites.[9] Moreover, in a recent paper, we showed that flax fibres act as synergist agent in a conventional PP-based intumescent formulation.[6]

Hornsby and Rothon discussed the economic importance of mineral fire retardants fillers such as aluminium and/or magnesium hydroxides in Chapter 2. It is now known that efficiency of these additives is related to multiple activities:

- Dilution of the polymer in the condensed phase;
- decrease of the amount of available fuel, thus increasing the amount of thermal energy needed to raise the temperature of the composition to the pyrolysis level, due to the high heat capacity of the fillers;
- increase of the enthalpy of decomposition
- emission of water vapour, involving a dilution of gaseous phase by water vapour
- decrease of amount of fuel and oxygen in the flame;
- possible endothermic interactions between the water and decomposition products in the flame (reactions (1) and (2)):

$$2Al(OH)_3 \rightarrow Al_2O_3 + 3H_2O \qquad \Delta H = 298\ \text{kJ mol}^{-1} \tag{1}$$
$$Mg(OH)_2 \rightarrow MgO + H_2O \qquad \Delta H = 380\ \text{kJ mol}^{-1} \tag{2}$$

- a decrease of feedback energy to the pyrolysing polymer;
- finally, an insulative effect of the oxides remaining in the char and the eventual charring of the materials.[10–16]

The use of these flame-retardants acting through physical effects require relatively large amounts of additives: 50–65 wt% in the case of aluminium hydroxide [ATH; $Al(OH)_3$] or magnesium hydroxide [MH; $Mg(OH)_2$]. If in these composites the impact and tensile strengths are reduced, their stiffness can be improved (increase of the flexural modulus from 700–1500 up to 2000–5000 MPa for, respectively, standard PP grades to mineral filled systems[17]). Stiffness of PP/flax composites is high too (flexural modulus close to 3500 MPa with a 30 wt% fibres level) and addition of natural flax fibres to a standard PP increases the tensile strength from *ca.* 700 to *ca.* 1500 MPa.[18] So, it may be presumed that PP/flax/hydroxide composites show high stiffness with preserved tensile performances and then may be suitable polymeric material for transportation.

The present chapter presents preliminary results concerning fire, thermal and mechanical performances of typical PP/flax fibres/ATH and PP/flax fibres/MH composites. The discussion will be carried out considering oxygen consumption calorimetry (cone calorimeter), thermogravimetry and conventional tensile and flexural tests.

Table 1 *Composite composition*

	PP (wt%)	*Fibber* (wt%)	*Additives* (wt%)
PP	100		
PP/Fibre	60	40	
PP/Fibre/ATH	40	26.5	33.5
PP/Fibre/MH	40	26.5	33.5

21.2 Experimental

21.2.1 Materials

Raw materials were PP [Finapro grade, as pellets supplied by Fina (France); MFI: 12 g (10 min)$^{-1}$ (230°C/2.16 kg)]; ATH and MH were commercial grade additives (supplied, respectively, by Alcan and Dead Sea Bromine).

Flax "Fibres" are short (mean length about 20 mm), non-textile residue from flax tow obtained from natural tangling after harvesting and natural weathering retting [Fibber, technilin grade supplied by La Centrale Linière Cauchoise (France), generally used in paper or composites manufacturing].

Polymer and additives mixtures (Table 1) were mixed in a Brabender Laboratory Mixer with roller blades monitored by a Brabender Plasticorder PL 3200 data processing torque rheometer system operating at 180°C and 20 rpm. The polymer was first introduced in the heating chamber and, after 3 min, the additives were introduced into the chamber and mixed for 6 min, the total mixing time being 9 min. The obtained mixture was then pressed in a Daragon press at 185°C and 70 bar on a 10×10 cm^2 plateau to obtain 3 mm thick sheets, from which all tested specimens were produced.

21.2.2 Fire Testings

Samples ($100 \times 100 \times 3$ mm^3) were exposed, in a Stanton Redcroft Cone Calorimeter, to a heat flux of 50 kW m^{-2}. This external flux has been chosen because it corresponds to the heat evolved during a well-developed fire.[19] Three tests were carried out on each material. Mean values were extracted from these tests to limit measurements uncertainties.

UL-94 testing was carried out on $127 \times 12.7 \times 3$ mm^3 sheets according to UL-94 testing.[20]

21.2.3 Thermogravimetric Analyses

TG analyses were performed using a Setaram Setsys 12 thermo balance at a 5°C min^{-1} heating rate from 25 to 800°C under airflow. Samples (about 20 mg) were placed in a platinum pan. The precision of temperature measurements was ±1.5°C.

21.2.4 Mechanical Characterisations

Three-point bending tests, according to ASTM D-790M standard, and tensile tests were carried out using a universal mechanical testing machine, Instron 4204H0610. For each test and composite a minimum of five samples were tested, at a crosshead speed of 2 mm min^{-1}. Specimens were cut with dimensions $120 \times 12.5 \times 3$ mm^3 for tensile tests and $60 \times 25 \times 3$ mm^3 for flexural tests.

21.3 Results and Discussion

21.3.1 Fire Performances

Composites contribution to a fire was first studied considering oxygen consumption calorimetry data and hazards resulting from the combustion products of composites. Figure 1 shows that PP/Fibber, PP//ATH and PP//MH systems lead to a decrease from 45 to respectively about 30, 40 and 40 s of the time necessary for inflammation (t_i) when compared to PP. This result is important because it shows that the introduction of flax fibres in the polyolefin decreases the apparent stability of the material and increases the ease of ignition. This may be explained either by formation of defects during the mixing process and/or by a conductibility of the composite that is comparatively higher than that of the virgin polymer. Moreover, this addition forbids the existence of a significant protective endothermic effect of the hydroxide additives: the increase of t_i (10 s) is low and the hydroxide-based composites contribute to a flame at a time lower than that of the virgin polymer.

Nevertheless, comparison of the rate of heat release (RHR) curves shows that addition of flax fibres decreases the contribution of the resin to a fire [drop from *ca.* 1800 to 640 kW m^{-2} in the peak of rate of heat release (RHRpk)]. Addition of ATH and MH in PP/Fibber increases this effect (respectively to 360 and

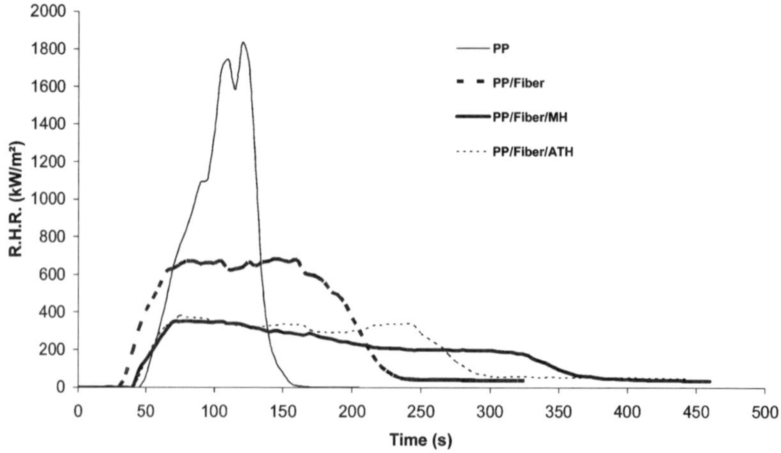

Figure 1 *Rate of heat release curves of PP and of the PP/flax composites vs. time (external heat flux: 50 kW m^{-2})*

365 kW m^{-2} RHRpk values). The decrease of the RHR peak values may be explained by the formation of a protective coating, which may restrict the diffusion of heat to the underlying resin and/or the diffusion of fuels (products of the degradation of the polymer) to the flame.

TG curves under 50 kW m^{-2} irradiance (Figure 2) show that addition of flax fibber leads to a decrease of weight loss *versus* time after a 40 wt% loss. Optical observations (Figure 3a) show a protective surface charring of the material that decreases the weight loss rate. After flame extinguishing (*i.e.* 200 s) the residue weight, lower than 5 wt%, corresponds to the formation of a light ash-like material (Figure 3b).

The behaviours of PP//ATH and PP//MH are quite similar: at *ca.* 70 s, charred ceramics form. Both are thermally stable at, respectively, around 250 and 325 s. It appears that the oxides resulting from the degradation of the

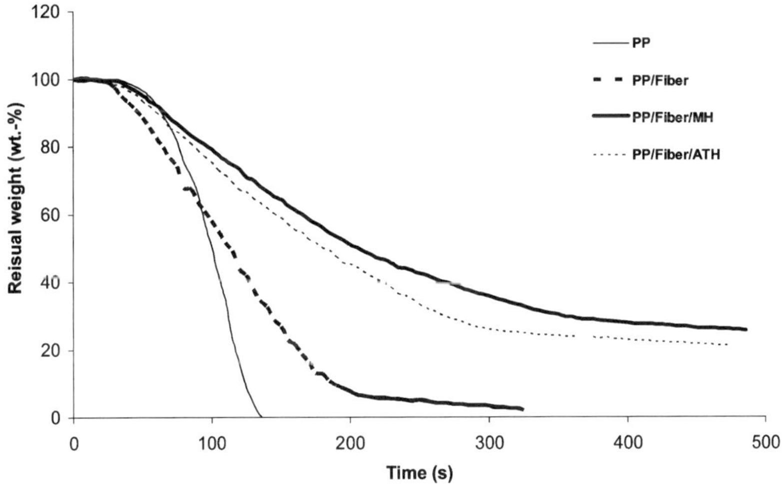

Figure 2 *Weight loss curves of PP and of the PP/flax composites vs. time (external heat flux: 50 kW m^{-2})*

(a)

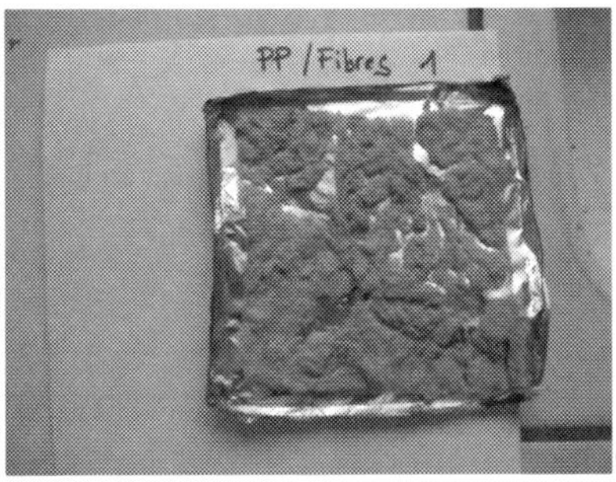

(b)

Figure 3 *Optical observation of the PP/Fibber after a 40 wt% 100 s;* (a) *and a 97 wt% loss 300 s,* (b) *under an external heat flux: 50 kW m^{-2}*

hydroxides and char resulting from the flax degradation form the protective coating, the presence of the oxides increasing the stability of the char.

The residual weight of the samples when flame extinguishing occurs (respectively 28 and 32 wt%) is higher than the weight computed assuming the residue is composed of the oxide (product of the degradation of the hydroxide) and ashes (product of the decomposition of flax), respectively 23 and 27 wt%. This implies a participation of PP, or the products resulting from its degradation, to the charring process and allows one to presume a catalytic part played by the oxides. Optical observation of the residues confirms this additional char [grey colour of the oxide-based residue and residual surface char (Figure 4)].

TG curves (Figure 5) confirm the charring process, showing that char from flax degrades at *ca.* 410°C. This implies that PP takes a part in the charring

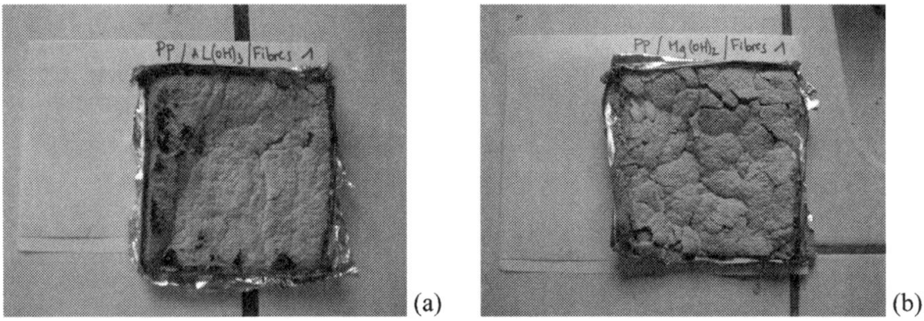

Figure 4 *Optical observation of PP/ATH [after a 60 wt% loss (250 s)] and of PP/MH [after a 60 wt% loss (285 s)] under an external heat flux: 50 kW m^{-2}*

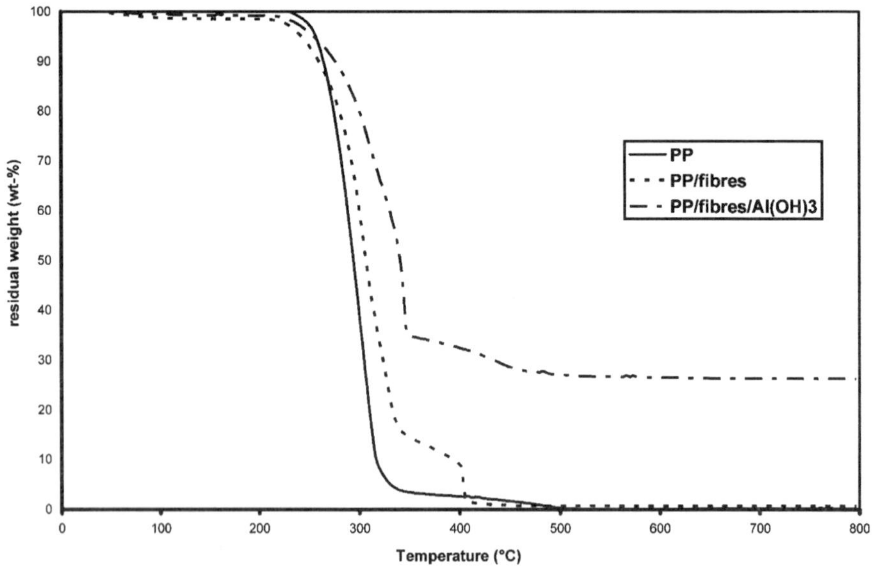

Figure 5 *Thermogravimetric curves of PP and of PP/flax composites (under air heating rate: 5°C min^{-1})*

process of PP/Fibber and that the resulting char is comparatively thermally stable. Moreover, the residual weight of PP//ATH at 800°C (28 wt% including 18 wt% of Al_2O_3) confirms the formation of a comparatively stable char during the thermo-oxidative degradation of the composite.

The total heat release curves (Figure 6) show that this char degrades slowly after the flame extinguishing. It may take a part in post glowing and carbon oxides evolving during this step. Nevertheless, Table 2 shows that addition of flax to PP leads to an appreciable decrease of the amounts of CO and CO_2 evolved and that addition of hydroxides plays only an additional part in reducing the CO evolution.

Figures 7 and 8 show the precise effect of each filler. Addition of flax leads to a decrease of the CO and CO_2 evolution peaks that are related to the corresponding decrease of the RHRpk values, explained by the formation of the charred surface coating. An eventual additional catalytic effect from ATH and MH may be presumed, which should increase the selectivity for the formation of polyaromatic species in the condensed phase to the detriment of CO formation.

No classification in class 94 V using the vertical flame test method is observed with the composite materials. This is explained by the fall of burning surface

Table 2 *Total carbon oxides evolution (external heat flux: 50 kW m⁻²)*

	PP	*PP/Fiber*	*PP/Fiber/ATH*	*PP/Fiber/MH*
CO (g)	0.011	0.007	0.007	0.006
CO_2 (g)	0.48	0.40	0.33	0.33

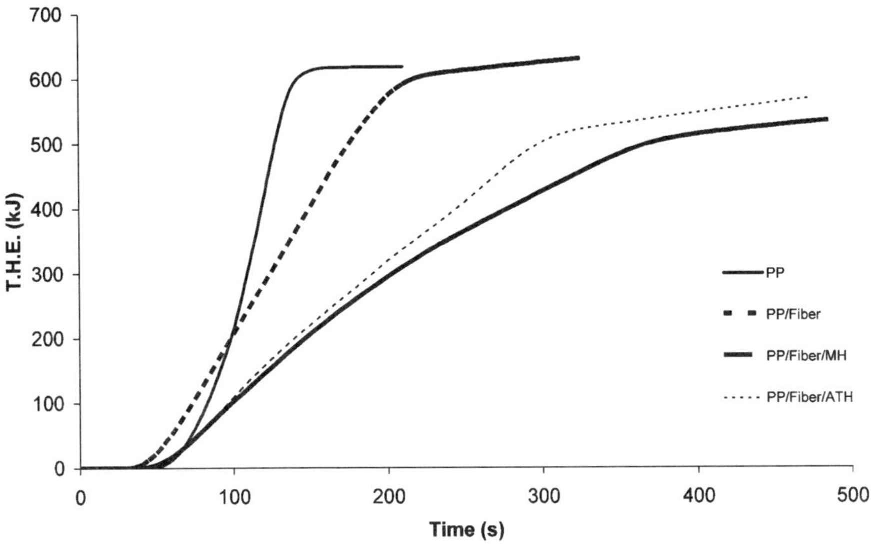

Figure 6 *Total heat release curves of PP and of PP/flax composites vs. time (external heat flux: 50 kW m^{-2})*

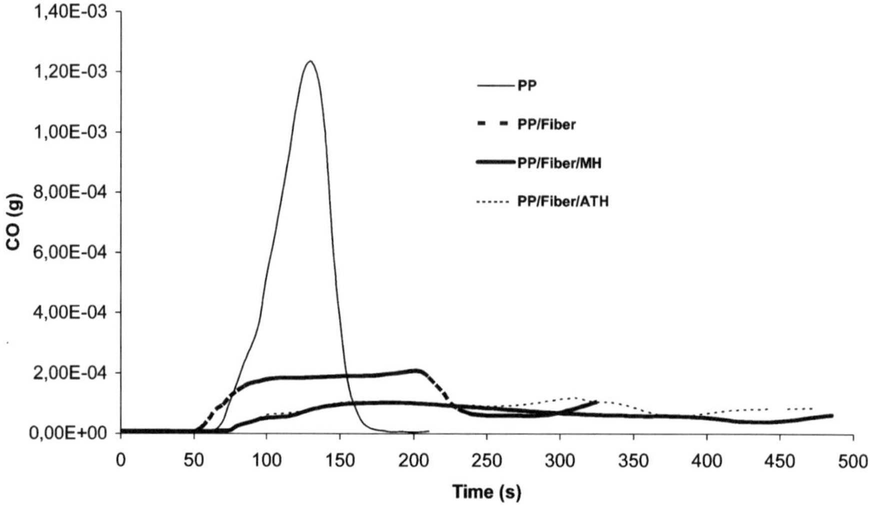

Figure 7 *Carbon monoxide evolution from PP and from PP/flax composites vs. time (external heat flux: 50 kW m^{-2})*

material with PP/Fibber and PP//ATH. Conversely, the surface material formed from PP//MH reveals its insulative property *via* avoiding melting of the polymer and subsequent dripping. Nevertheless, this insulative character is not efficient because of the formation of cracks in the coating (Figure 4b).

21.3.2 *Mechanical Properties*

A low value of the elongation at break associated with an increase of the tensile strength is a classic characteristic of natural fibre/PP composites containing FR hydroxides.[21] It may be noticed that fibrous filler can usually improve the tensile strength.[22]

Results of the mechanical tests are reported in Table 3. Addition of fibres gives such an effect, a very low value of the elongation and high tensile modulus resulting from the addition of both fibres and hydroxides.

Fibber addition leads to a slight increase of the flexural modulus of composites. Relatively high values of flexural modulus standard deviation is a probe of the low homogeneity of materials, *i.e.* local accumulation of fibre in the composite as experimentally evidenced by us.

Table 3 *Mechanical properties of polypropylene and composites*

	Mixes			
Property	*PP*	*PP/Fibber*	*PP//ATH*	*PP//MH*
Tensile modulus, (GPa)	1.5	3.2	4.1	5.4
Elongation at break, (%)	5	0.6	0.35	0.23
Flexural modulus, *E* (GPa)	2.0	2.6	4.8	4.9
E standard deviation, (%)	3.3	16	12	9.2

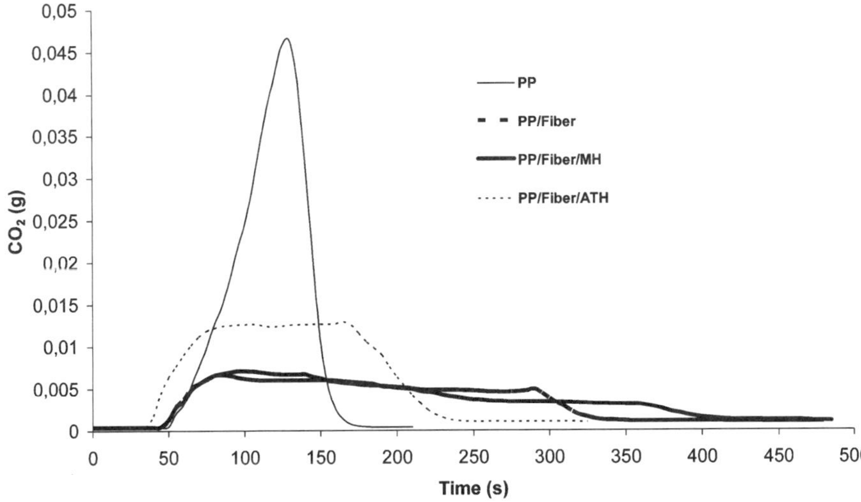

Figure 8 *Carbon dioxide evolution from PP and from the PP/flax composites vs. time (External heat flux: 50 kW m^{-2})*

Addition of hydroxides in the PP/Flax composite increases significantly the flexural modulus and decreases its standard deviation. So, a best dispersion of the fillers may be presumed, which may decrease the number of stress-concentration points through the composites.

Mechanical properties of these materials are significantly influenced by the interfacial interactions, which depend on the size of the interface and the strength of the interaction.[23] As a consequence, optimized mechanical properties should be obtained *via* the optimization of particle sizes[18] the use of coupling agents[24] or chemical treatments (such as maleic anhydride or vinyl trimethoxy-silane treatments,[18] which may affect the hydrophilic nature the fibres[25]) of the natural filler.

21.4 Conclusion

The effects of hydroxides flame retardant fillers on the flammability and mechanical properties of flax-filled polypropylene composites were compared. 33.5 wt% of aluminum or magnesium hydroxide can effectively reduce the flammability of the composites. Moreover, these hydroxides have been proved to be reinforcing fillers for polypropylene on the basis of increases in the tensile and flexural strength.

21.5 References

1. D. Robson, J. Hague, G. Newman, G Jeronomidis and M. Ansell, in *Survey of Natural Materials for Use in Structural Composites as Reinforcement and*

Matrices, D. Robson *et al.* (eds.), Woodland Pub. Ltd., Abingdon, UK, 1996.
2. S.J. Eichhorn, C.A. Baillie, N. Zafeiropoulos, L.Y. Mwaicambo, M.P. Ansell, A. Dufresne, K.M. Entwistle, P.J. Herrada-Franco, G.C. Escamilla, L. Groom, M. Hughes, C. Hill, T.G. Rials and P.M. Wild, *J. Mater. Sci.*, 2001, **36**, 2107–2131.
3. J.E.G. Van Dam, G.E.T. Van Vilsteren, F.H.A. Zomers, W.B. Shannon and I.T. Hamilton, in *Increased Application of Domestically Produced Plant fibres in Textiles, Pulp, and Paper Production, and Composite Materials*, European Commission Directorate-General XII, Sci. Res. Development, EUR 16101, EN 1994, 58–78.
4. A. De Chirico, M. Armanini, P. Chini, G. Cioccolo, F. Provasoli and G. Audisio, *Polym. Degrad. Stab.*, 2003, **79**, 139–145.
5. M. Helwig and D. Paukszta, *Mol. Cryst. Liq. Cryst.*, 2000, **354**, 961–968.
6. M. Le Bras, S. Duquesne, M. Fois, M. Grisel and F. Poutch, *Polym. Degrad. Stab.*, 2005, in press.
7. G. Audisio, A. De Chirico, B. Focher and G. Gallina, in *Fourth European Workshop on Lignocellulosics and Pulp, Extended Abstracts*, Stresa (Italy), 8–11 September 1996, p. 528.
8. G. Gallina, E. Bravin, C. Badalucco, G. Audisio, M. Armanini, A. De Chirico and F. Provasoli, *Fire Mater.*, 1998, **22**, 15–18.
9. B. Schartel, U. Braun, U. Schwarz, S. Reinemann, *Polymer*, 2003, **44**, 6241–6250.
10. P. R. Hornsby, *Fire Mater.*, 1994, **18**, 269.
11. P. R. Hornsby and C. L. Watson, *Polym. Degrad. Stab.*, 1990, **30**, 73.
12. F. Molesky, in *Recent Advances in Flame Retardancy of Polymeric Materials, Volume 1*, M. Lewin and G. Kirshenbaum (eds.), BCC Inc. Pub., Norwalk, USA, 1990, p. 92.
13. J. Levesque, in reference 12, p. 102.
14. O. Kalisky, R. J. Mureinik, A. Weismann and E. Reznik, in *Recent Advances in Flame Retardancy of Polymeric Materials, Volume 4*, M. Lewin and G. Kirshenbaum (eds.), BCC Inc. Pub., Norwalk, USA, 1993, p. 140.
15. S. Bourbigot, M. LeBras, R. Leeuwendal, K.K. Shen and D. Schubert, *Polym. Degrad. Stab.*, 1999, **64**, 419–425.
16. S. Bourbigot and M. Le Bras, in "Flame retardant plastics (Chapter 5)" *Plastics Flammability Handbook – Principles, Regulations, Testing, and Approval*, J. Troitzsch (eds.), Hanser Pub., Munich, 2004, pp. 145–148.
17. K. Bernreitner and H. Hammerschmid, in *Polypropylene: An A-Z Reference*, J. Karger-Kocsis (ed.), Kluwer Publishers, Dordrecht, 1999, pp. 148–158.
18. G. Cantero, A. Arbelaiz, R. Llano-Ponte and I. Mondragon, *Comput Sci. Technol.*, 2003, **63**, 1247–1254.
19. V. Babraukas, *Fire Mater.*, 1984, **8(2)**, 81.
20. "Tests for flammability of plastics materials for part devices and appliances", Underwriters Laboratories, Northbrook, ANSI//ASTM D-635/77 (1977).

21. M. Sain, S.H. Park, F. Suhara and S. Law, *Polym. Degrad. Stab.*, 2004, **83**, 363–367.
22. Y.W. Leong, M.B. Abu Bakar, Z.A.M. Ishak and B. Pukansky, *J. Appl. Polym. Sci.*, 2004, **91**, 3315–3326.
23. M. Bugjany, M. Fois and M. Grisel, unpublished results (Private communication, 2003).
24. Z. Demjen, B. Pukansky and J. Nagy, *J. Compos. A*, 1998, **29**, 323.
25. Y.S Thio, A.S. Argon, R.E. Cohen and M. Weinberg, *Polymer*, 2002, **43**, 3661.
26. B.V. Kokta, R.G. Raj and C.Daneault, *Polym.-Plast. Technol. Eng.*, 1989, **28**, 247–259.

CHAPTER 22

Intumescence in Ethylene-vinyl Acetate Copolymer Filled with Magnesium Hydroxide and Organoclays

LAURENT FERRY, PIERRE GAUDON, ERIC LEROY AND JOSÉ-MARIE LOPEZ CUESTA

Ecole des Mines d'Alès, Centre des Matériaux de Grande Diffusion, 6, avenue de Clavières 30319 Alès Cedex, France (Eric.Leroy@ema.fr)

22.1 Introduction

Ethylene vinyl acetate (EVA) copolymers are commonly used in the cable industry due to their flexibility and processing characteristics. In these applications, fire retardancy (FR) can be achieved using hydrated mineral fillers such as alumina trihydrate (ATH) or magnesium hydroxide (MH). However, high filler contents are required to obtain satisfying fire properties.[1,2] This high mineral loading decreases the mechanical performance of the materials. Therefore, to obtain a set of competitive properties, it becomes interesting to enhance the efficiency of the hydrated minerals by partially substituting them with synergistic additives, allowing a reduction of the global filler content.

During the last decade, several works have shown that the fire behaviour of EVA can be improved by associating hydrated fillers with other mineral fillers. Beside the classical endothermic and dilution effects of hydrated fillers, authors have tried to develop other interesting effects such as the constitution of protective layers or the promotion of charring. Shen[3] highlighted a synergism between ATH and zinc borate for a total filler loading higher than 70%. He obtained a limiting oxygen index (LOI) higher than 50%. Bourbigot *et al.* underlined that synergism can also exist between MH and zinc borate. This association leads to an increase in LOI, zinc borate promoting a protective vitreous layer during combustion.[4] To enhance the efficiency of MH, Cross *et al.* found that zinc stannate or zinc hydroxystannate can be applied as a coating on the filler surface.[5] This treatment permits a decrease of the effective particle size in the matrix, leading to an improvement of fire properties as well as mechanical properties.

In a previous paper, we mentioned that a synergism can also be observed when EVA is flame retarded by MH combined with talc.[6] Due to its lamellar structure, talc plays a positive role in the constitution of a mineral barrier. More recently, great interest has been found in the use of nanofillers. Camino *et al.* used fluorohectorite and montmorillonite as FR in EVA.[7] They showed that the silicates accelerate deacetylation of the polymer but slow down the thermal degradation of the deacetylated polymer due to the formation of a barrier at the surface of the materials. Beyer reported that modified nanoclays combined with ATH allow an improvement of the fire behaviour by charring effect.[8] Char formation reduces the amount of fuel available for combustion and, therefore, the amount of heat released. Apart from nanoclays, char promotion can also be achieved by other mineral fillers exhibiting a high specific surface area. Several papers of Gilman and Kashiwagi highlighted that silica or silica gel promotes char, which reduces the peak of heat release rate.[9,10]

In the present chapter, EVA has been flame retarded using MH in association with montmorillonite and silica or talc. The strategy consists in combining the effects of water release, barrier formation and char promotion of these different fillers.

22.2. Experimental

22.2.1 Materials

The EVA used was thermoplastic-elastomeric grade ELVAX® 260 (Du Pont), with a melt flow index of 6 g $(10\ \text{min})^{-1}$ and a vinyl acetate content of 28 wt%. The organoclay used in EVA was Cloisite 15A (Southern Clays), which is a natural montmorillonite (MMT) modified with dimethyldihydrogenated tallow ammonium salt. A non-modified MMT, Cloisite Na+, was also used, as a reference, for structural characterisation. The magnesium hydroxide (MH) was Magnifin® H10 (median particle size: $d_{50} = 0.85\ \mu m$ and BET surface area: $S_{BET} = 10\ m^2\ g^{-1}$) provided by Martinswerk. The silica was an amorphous precipitated silica Tixosil® 73 ($d_{50} = 6\ \mu m$, $S_{BET} = 79\ m^2\ g^{-1}$) provided by Rhodia. The talc was a non-commercial product ($d_{50} = 0.5\ \mu m$, $S_{BET} = 34\ m^2\ g^{-1}$). Concerning these two high specific surface area fillers, one has to keep in mind that silica is composed of micronic particles that are nanoporous and talc consists of submicronic platelets.

22.2.2 Processing

Processing was performed by mixing the fillers with the molten EVA pellets in a Haake internal mixer at 170°C at 60 rpm for 10 min. Sheets of 4 mm thickness were then compression moulded at 140°C under a pressure of 100 bar for 5 min.

Table 1 *Epiradiateur test characteristics and LOI values of the various formulations*

Formulations	*TTI* (s)	*IP* (s)	*N*	*LOI* (vol%)
EVA 60 MH	116	9.0	15	49
EVA-57MH-3MMT	123	8.6		42
EVA-55MH-5MMT	133	7.0	19	39
EVA-50MH-10MMT	139	7.8	17	34
EVA-50MH-5MMT-5Si	122	7.4	18	38
EVA-45MH-7.5MMT-7.5Si	117	7.1	14	35
EVA-50MH-5MMT-5T	138	6.9	17	39

These were cut to the requisite size, depending on the experiment to be performed. In all cases, the total filler content was 60 wt%, but the various relative proportions of organoclay, talc and silica were varied (see Table 1 below for the various formulations studied).

22.2.3 Experimental Techniques

X-ray diffraction (XRD) analysis was performed using a Philips PW 1710 diffractometer with a monochromatized Cu K_α radiation ($\lambda = 0.154$ nm). The scattering angle (2θ) domain studied ranged from 3° to 63°, with a rotation step scanning of 0.05° with a count time of 1.2 s. For the pure MMT (reference MMT Cloisite Na^+ and Cloisite 15A) measurements were carried out on powders, whereas for EVA composites, measurements were done on compression moulded sheets.

Thermo-oxidative degradation of the various polymer/filler materials was studied using standard techniques of thermogravimetric analysis (TGA) and differential thermal analysis (DTA), Setaram TGDTA92, at a temperature scanning rate of 5°C min^{-1} from room temperature to 700°C under air flow. A typical sample mass for those experiments was 30 mg.

"Epiradiateur test" (AFNOR NF P 92-505) was carried out on sheets ($70 \times 70 \times 4$ mm^3) to determine the flammability and the self-extinguishability of the various formulations. In this test, a radiator (500 W) is placed at 30 mm over the upper face of a sheet of plastic. When the sample starts burning, the time to (first) ignition (TTI) is noted. Afterwards, the radiator is removed and replaced as soon as extinction occurs. Application of this device is repeated by steps until successive extinctions for 5 min, the number of steps (N), the mean inflammation period (IP) are noted. For each filled composition, not less than four specimens were tested.

The limiting oxygen index (LOI) was been measured using a Stanton Redcroft instrument on barrels ($80 \times 10 \times 4$ mm^3) according to ISO 4589 specifications.

The cone calorimeter, manufactured by Fire Testing Technology (FTT), is a standard apparatus used for fire retardancy tests (ISO 5660), but a brief description follows. The polymer sample ($100 \times 100 \times 4$ mm^3) is placed horizontally on a balance and irradiated from above by a truncated conical heater supplying a heat flux of 50 kW m^{-2}. Combustion is initiated by spark emission in the proximity of the upper sample surface. Smoke and gas emissions can be monitored from an evacuation pipe connected to the cone apex. In this technique, measured data provide detailed information about sample ignition, heat release rate (HRR), smoke release rate (SRS) and weight loss. Among other things, the following characteristic parameters may be obtained: time to ignition, TTI (s): determined visually and taken to be the period required for the entire surface of the sample to burn with a sustained luminous flame, and peak of heat release rate, (Peak HRR) (kW m^{-2}), taken as the peak value of the heat release rate *vs.* time curve, and considered to be the variable that best expresses the maximum intensity of a fire, indicating the rate and extent of fire spread.

22.3 Results and Discussion

In the following, the composition EVA-60MH has been considered as reference. The other FR systems consist of a partial substitution of MH by other mineral fillers, *i.e.* MMT, silica and talc.

22.3.1 Structural Characterization

Figure 1 shows a comparison between natural montmorillonite (Closite Na+) and organically modified montmorillonite XRD patterns. Natural montmorillonite exhibits a peak at $2\theta = 7.5°$, corresponding to a 12 Å interlayer spacing. The curve of modified MMT shows two new peaks at $2\theta = 4.5$ and 10°. These peaks correspond to an interlayer spacing of 19 Å (with $n = 1$ and $n = 2$ in Bragg's law). Therefore, we can conclude that the chemical modification leads

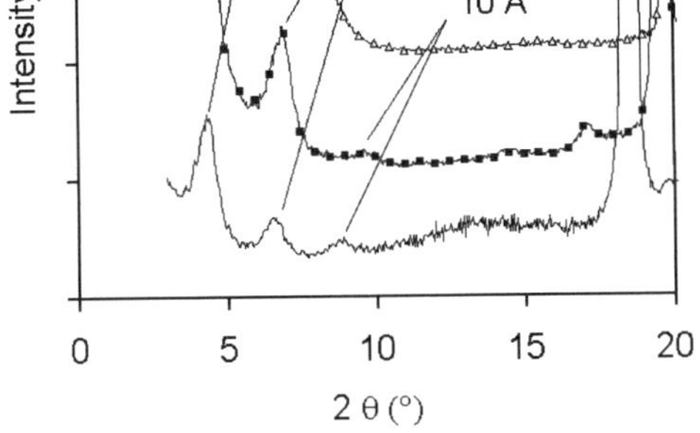

Figure 1 *XRD patterns, (Δ) natural montmorillonite, (■) Cloisite 15A (MMT), (—) EVA-55MH-5 MMT*

to an intercalated structure. However, the 12 Å peak is still observed, probably caused by a certain amount of not-exchanged MMT. When modified MMT is introduced in EVA in association with MH (EVA-55MH-5MMT), we noted that the XRD pattern remains quasi-unchanged in the small-angle region. Three peaks are observed at 20, 13 and 10 Å, the latter being an harmonic of the first peak.

These results seems to indicate that processing with an internal mixer did not induce exfoliation of a significative part of the phyllosilicate; nevertheless it is possible that a partial exfoliation occurred. Some TEM observation should be performed in the future to clarify this point.

22.3.2 Thermal Analysis

Figures 2 and 3 show, respectively, the mass loss and heat flow curves obtained for the different samples tested. The thermo-oxydative degradation mechanism of pure EVA copolymers has been described in literature[11] and consists of two clearly identified steps: first a deacetylation reaction occurs (first mass loss between 300 and 400°C on Figure 2), with production of acetic acid (a non-combustible gas) and leading to acetylene-ethylene copolymers, with a significant amount of crosslinking.[11] These chains are then degraded in a second step (second mass loss between 400 and 450°C on Figure 2), leading to a decrease of molecular masses and to the production of combustible gases and a small amount of char that is finally degraded at higher temperatures [final mass loss above 450°C on Figure 2]. Figure 3 shows that these two degradation steps are accompanied by exothermic peaks, the second peak corresponding to the production of combustible gases, showing a shoulder around 480°C, which can be partially attributed to char decomposition.

When EVA is filled with MH, the deacetylation reaction is shifted to higher temperatures (Figure 2) due to the endothermic decomposition of MH. The second step of the decomposition is also modified. Effectively, Figure 3 clearly

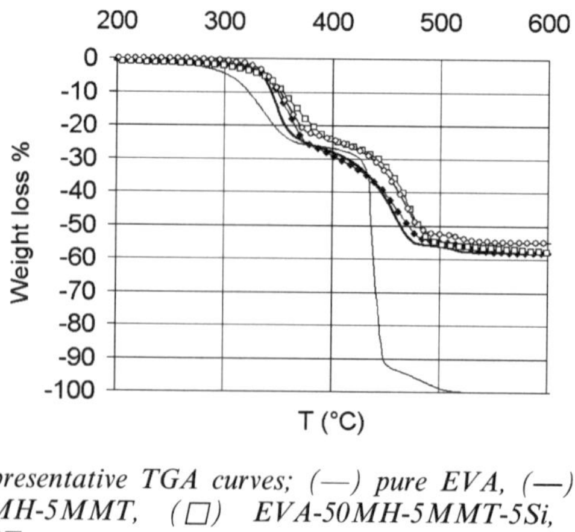

Figure 2 *Some representative TGA curves; (—) pure EVA, (—) EVA-60MH, (◆) EVA-55MH-5MMT, (□) EVA-50MH-5MMT-5Si, (○) EVA-45MH-5MMT-5T*

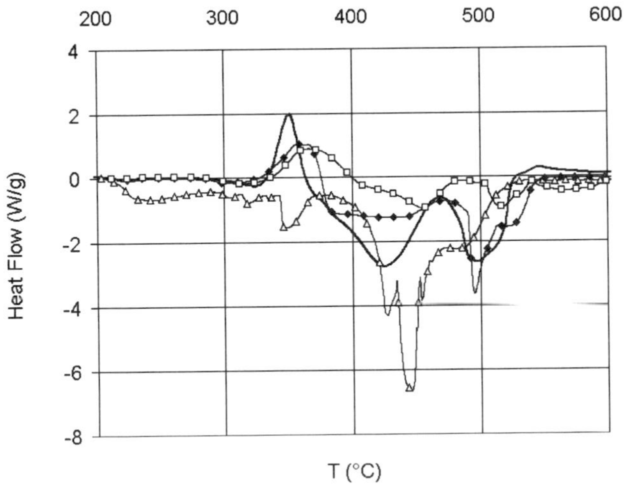

Figure 3 *Some representative DTA curves; (Δ) pure EVA, (—) EVA-60MH, (◆) EVA-55MH-5MMT, (□) EVA-50MH-5MMT-5Si*

shows the presence of two exothermic peaks: one below 450°C and another around 500°C. If MH in EVA is partially substituted by MMT (Figure 3), the peak around 450°C is flattened and that around 500°C is amplified, with a shoulder towards higher temperatures, showing an even better thermal stability. This behaviour, which has been already observed by Camino *et al.* for EVA/MMT formulations[11] and by Beyer for EVA/ATH/MMT formulations,[8] is due to the promotion of polymer charring, the char formed being degraded at higher temperatures (second peak around 500°C) than EVA chains (first peak below 450°C).

These features seem to be even more enhanced when MH is partially substituted by both MMT and talc or silica. The exothermic decomposition peaks are smaller, particularly in the presence of silica. (Figure 3) and are shifted to higher temperatures. Regarding the first stages of degradation, DTA curves indicate that thermal events observed in pure EVA and EVA 60 MH compositions are strongly modified (Figure 3). The endothermic loss of water is shifted towards high temperatures for compositions containing talc or silica. In addition, the peak width is enlarged, possibly due to a limitation of heat transfer from the sample core to the outside.

22.3.3 Fire Properties

22.3.3.1 Epiradiateur Test

The results of the "Epiradiateur test" show that the partial substitution of MH by MMT leads to an improvement of the resistance to ignition, as well as the self-extinguishability of EVA-filler composites. We can see in Table 1 that the time to ignition (TTI) increases from 116 to 139 s when MMT increases from

0 to 10 wt%. Besides, we note that self-extinguishability seems to exhibit an optimum for intermediary MMT content since the mean inflammation period (IP) is lower for 5 wt% of MMT than for 3 or 10%. This assumption is corroborated by the number of inflammations (*N*) which shows the highest value for the EVA-55MH-5MMT composition.

When MH is partially substituted by both MMT and silica, we remarked that, on the one hand, the resistance to ignition is better than that of the reference system but lower than that of EVA-MH-MMT compositions; on the other hand, we can notice that the self-extinguishability remains almost the same, whereas the number of inflammations tends to decrease. This result means that, in the presence of silica, the material is hard to re-ignite. This can be correlated to the thermal analysis results, showing the formation of a more stable char for this composition (second exothermal peak around 520°C on Figure 3).

When MH is partially substituted by both MMT and talc, we clearly see that resistance to ignition as well as self-extinguishability are improved since time to ignition is as high as 138 s whereas the mean inflammation period is as low as 6.9 s. This FR system gives the best compromise between both flammability and auto-extinction properties.

22.3.3.2 LOI Test

Table 1 indicates that the presence of MMT strongly affects the LOI of filled EVA composites. LOI decreases from 49 to 34 when MMT content goes from 0 to 10 wt%. No dripping was observed. The flame spreads very rapidly to the bottom of the sample, reaching the critical line located 50 mm below the top of sample which stops the test. We have estimated the flame spread rate by measuring the burned length during LOI test. Figure 4 shows that the spread rate is inversely proportional to MMT content at a given oxygen index. One can observe that the flame spread is slower for a lower oxygen index. For these

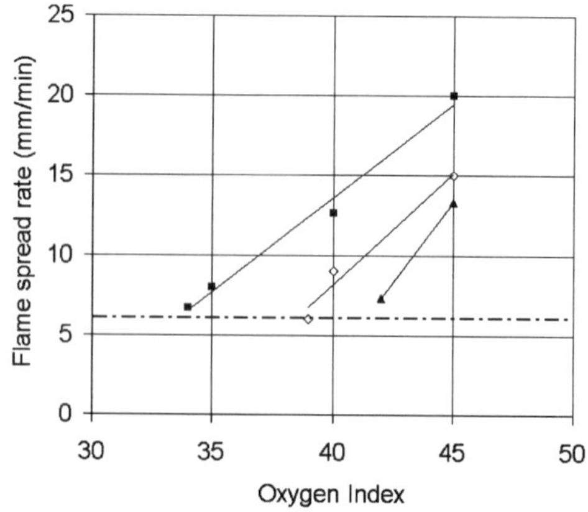

Figure 4 *Flame spread rate as a function of oxygen index, (■) 10wt% MMT, (◇) 5 wt% MMT, (▲) 3 wt% MMT*

systems, we remarked that the limiting oxygen index corresponds to the oxygen index where the flame spread rate becomes lower than *ca.* 6 mm min^{-1}. At the end of the test, we observed that the samples are still cohesive. They are charred at the surface but remain undegraded in the bulk, showing that the flame spread is only superficial. We assigned these observations to a wick effect induced by the presence of MMT. Due to their lamellar structure, organoclays ensure a high connexity between mineral fillers, which is likely to promote heat (and thus flame) propagation.

Partial substitution of MH by both MMT and silica, on the one hand, and both MMT and talc, on the other hand, leads to similar behaviours during the LOI test. The values obtained (Table 1) shows that this property seems to be governed either by the MH content or by the MMT content, the former tending to increase LOI, while the latter causing a decrease due to the wick effect previously mentioned.

22.3.3.3 Cone Calorimeter

The results of the cone calorimeter study are presented in Table 2, the most typical HRR *vs.* time curves being shown on Figure 5. Notably, the presence

Table 2 *Results of cone calorimeter tests for the various formulations*

Formulations	*TTI* (s)	*HRR max* (kW m^{-2})
EVA	35	657
EVA-60MH	72	370
EVA-57MH-3MMT	90	230
EVA-55MH-5MMT	97	220
EVA-50MH-10MMT	98	333
EVA-50MH-5MMT-5Si	79	190
EVA-45MH-7.5MMT-7.5Si	82	228
EVA-50MH-5MMT-5T	8288	204

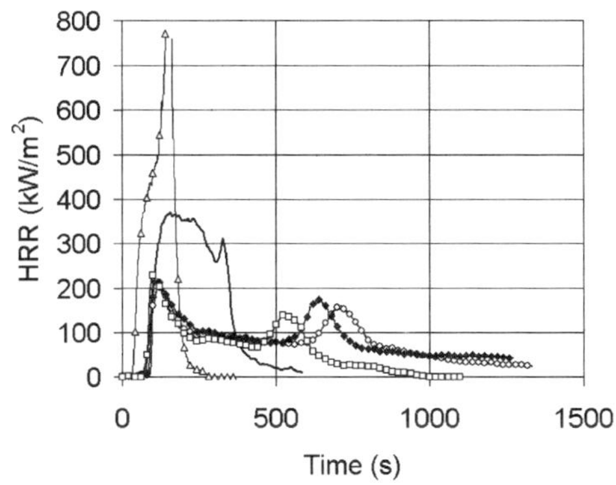

Figure 5 *Some representative HRR vs. time curves; (Δ) pure EVA, (—) EVA-60MH, (◆)EVA-55MH-5MMT, (□)EVA-45MH-7.5MMT-7.5Si, (O)-EVA-45MH-5MMT-5T*

of organoclays highly modifies the combustion behaviour of the filled EVA. Regarding the peak of HRR, which is a key parameter, we can observe that, as for self extinguishability, there seems to be an optimum reduction for 5% of MMT content, the peak value being higher for a 10% content.

As previously described for systems with ATH as hydrated filler,[8,12] a rigid residue is obtained after burning for all specimens. This residue, which looks like a foam, is mainly inorganic but contains about 12% of char (this carbon content was estimated by measuring the weight loss after 3 hours at 1050°C). This result confirms the charring effect of MMT observed in thermal analysis experiments.

One interesting point is the foam-like structure of the residue. Effectively, the sample thickness increases from 4 mm, at the beginning of the test, to more than 10 mm for the final residues (Figure 6). The creation of an expanded foam can be assimilated to a kind of intumescence. During cone calorimeter experiments, we observed that the expansion (which finally leads to such residues) starts before the ignition of the sample. As opposed to samples filled with MH only, the samples containing organoclays do not show any bubble bursting at the irradiated surface during the pre-ignition period, whereas, at the same time, the sample thickness grows. This suggests that the bubbles resulting from the degradation of MH (water), MMT organic part and EVA (acetic acid), are trapped inside the sample, which starts to foam. Effectively, when a cone calorimeter experiment is stopped just after ignition, one can observe that the whole sample has been converted into a foam, the upper part being charred and brittle, and the lower part remaining still flexible like undegraded filled polymer (Figure 7).

When a complete experiment is performed, the foamed structure created before ignition then slows down both heat transfer and diffusion of fuel and oxygen, thus promoting charring, and finally, a cohesive foam is obtained at the end of combustion (containing char and minerals). The complex mechanism leading to the formation of the foam during the ignition period involves probably many phenomena influenced by the presence of MMT, such as bubble

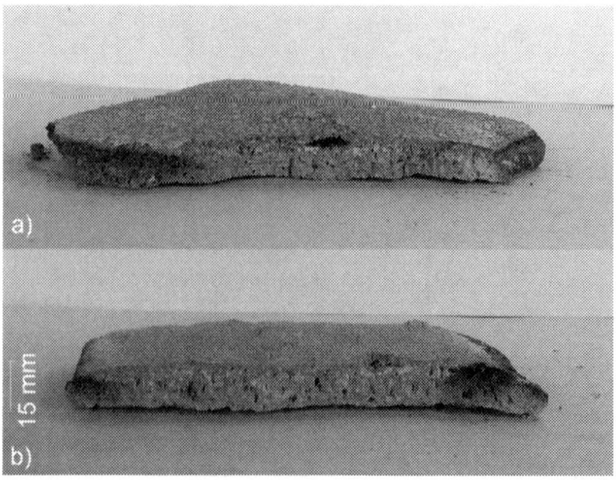

Figure 6 *Residues of cone calorimeter tests (a) EVA-55MH-5MMT and (b) EVA-50MH-5MMT-5T*

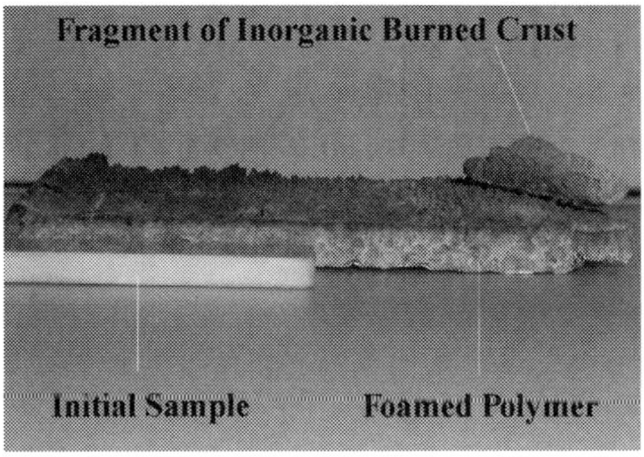

Figure 7 *EVA cone calorimeter specimen 57 wt% of MH and 3 wt% of Cloisite 15A, experiment was stopped just after ignition*

heterogeneous nucleation, increased viscosity, cross-linking and charring promotion.

The final thickness of the foam residues was 16 mm for 3% of MMT, 14 mm for 5% of MMT and only 10 mm for 10% of MMT. This may be partially related to a strong increase of melt viscosity with MMT content, which has already been described in the literature.[13] Concurrently, the foam obtained with only 3% of MMT showed cracks and was much more brittle than those of the two other compositions. Consequently, the optimum reduction of HRR peak for 5% MMT is probably due to a compromise between an important foaming and a good mechanical stability of the foam structure created.

All the compositions involving talc or silica in combination with MMT and MH lead to foam-like residues similar to those obtained with EVA/MH/MMT (Figure 6).

A positive effect on HRR peak reduction was observed but was counterbalanced by a decrease in ignition time, the resulting values being however higher than that of EVA 60 MH. The compositions containing EVA-50MH-5MMT- 5Silica and EVA-50MH-5MMT-5Talc exhibit lower peaks of HRR than EVA-55MH-5MMT (Table 2). This result can be related to the lowering of exothermicity and the enhancement of char thermal stability observed by thermal analysis, especially in the presence of silica (Figure 3). For EVA-50MH-5MMT-5Talc, the final thickness of the residue (15 mm) was also slightly increased compared to EVA-55MH-5MMT, suggesting an influence of other factors governing the foaming phenomena. Talc acts, probably, as a bubble nucleating agent similarly as it does in polymer foams.[14] Moreover, the lamellar morphology of talc enables it to form a mineral barrier similar to that of MMT.

An increase of the substitution rate of MH up to 15% using both MMT and silica does not lead to an improvement of HRR peak (Table 2). This seems to confirm the optimum reduction of HRR for 5% MMT content.

Other possible actions of talc and silica could concern either the control of molten polymer viscosity or the trapping by interfacial interactions of the bubbles made of water vapour and degradation products of EVA and MMT. The porous structure of silica could be able to trap chains fragments issued from the decomposition of EVA. Besides, both talc and silica present a better thermal stability than MMT (which loses around 30% of its initial weight at 700°C due to the decomposition of alkylammonium modifiers). Consequently, compared to MMT, these two high specific surface area fillers may have a more active effect on charring processes at high temperature, that is to say, in the last stages of EVA degradation.

22.4 Conclusions

The incorporation of organoclays in partial substitution of magnesium hydroxide leads to efficient flame retardant systems in EVA. The action of organoclays can be reinforced by a combination with delaminated talcs or silica. This leads to shifts of thermal analysis curves. Improvements of self-extinguishability and peaks of HRR are conditioned by the type of composition elaborated. The main mechanism of action of these fire retarded compositions is connected to a phenomenon of intumescence leading to the formation of a foam-like structure during the pre-ignition period. This foamed structure then burns slowly due to both limited heat transfer and limited diffusion of fuel and oxygen.

22.5 References

1. F. Montezin, J.M. Lopez-Cuesta, A. Crespy and P. Georlette, *Fire Mater.*, 1997, **21**, 245.
2. R.N. Rothon, in *Particulate-Filled Polymer Composites*, 1st Edn, Longman Scientific, Harlow, UK, 1995.
3. K. Shen, *Plastic Composites*, 1988, Nov.–Dec., 26.
4. S. Bourbigot, M. Le Bras, R. Leeuwendal, K.K. Shen and D. Schubert, *Polym. Degrad. Stab.*, 1999, **64**, 419.
5. M.S. Cross, P.A. Cusack and P.R. Hornsby, *Polym. Degrad. Stab.*, 2003, **79**, 309.
6. A. Durin-France, L. Ferry, J.M. Lopez-Cuesta and A. Crespy, *Polym. Inter.*, 2000, **49**, 1101.
7. M. Zanetti, G. Camino, R. Thomann and R. Mülhaupt, *Polymer*, 2001, **42**, 4501.
8. G. Beyer, *Fire Mater.*, 2001, **25**, 193.
9. T. Kashiwagi, A.B. Morgan, J.M. Antonucci, M.R. VanLandingham, R.H. Harris,W.H. Awad and J.R. Shields, *J. Appl. Polym. Sci.*, 2003, **89(8)**, 2072.
10. J.W. Gilman, S.J. Richie, T. Kashiwagi and S. Lomakin, *Fire Mater.*, 1997, **21**, 23.
11. G. Camino, R. Sgobbi, A. Zaopo, S. Colombier and C. Scelza, *Fire Mater.*, 2000, **24**, 85.
12. S.C. Brown, M.L. David, K.A. Evans and J.P. Garcia, *WO 00/66657*, 2000, assigned to Alcan International Ltd.
13. B. Hoffmann, C. Dietricha, R. Thomann, C. Friedrich and R. Mülhaupt *Macromol. Rapid Commun.*, 2000, **21**, 57.
14. C.B. Park, L.K. Cheung and S.-W. Song, *Cellular Polym.*, 1998, **17(4)**, 221.

CHAPTER 23

Spent Oil Refinery Catalyst: A Synergistic Agent in Intumescent Formulations for Polyethylenic Materials

LUCIANA R. DE MOURA ESTEVÃO,[1] REGINA SANDRA V. NASCIMENTO,[1] MICHEL LE BRAS[2] AND RENÉ DELOBEL[3]

[1]Instituto de Química – DQO, Universidade Federal do Rio de Janeiro, CT Bloco A, 6° andar, Cidade Universitária, Ilha do Fundão, Rio de Janeiro, RJ, CEP 21941-590, Brazil (lestevad@iq.ufrj.br)
[2]PERF, ENSCL, BP 108, F-59652 Villeneuve d'Ascq Cedex, France
[3]CREPIM, Parc de la Porte du Nord, F-65700, Bruay la Buissière, France

23.1 Introduction

In recent years, concern has been widely expressed about the toxicity of the versatile and highly efficient halogenated flame retardants.[1] The modern concept of flame retardancy implies that flame retardants should effectively reduce the probability of fire development and also its consequences, both on humans and on structures.[2] Following this concept, halogen-based flame retardants become somewhat unsatisfactory since, on burning, they give rise to dense smoke and to acidic corrosive fumes. In the quest for halogen-free flame retardants much research is being centred on intumescent additives. These systems have provided efficient means for enhancing fire safety performance while presenting an environmentally friendlier approach than the traditional halogen systems. The intumescence process results from a combination of charring and foaming of the surface of the burning polymer, forming a shield that protects the underlying material from the action of the heat flux or flame.[3]

However, regardless the efficiency of flame-retarded systems, the additional cost of the end products still limits them largely to institutional sales.[4] This chapter reviews the role played by spent petroleum refining catalyst from the fluid-bed catalytic cracking (FCC) unit in classical ammonium polyphosphate (APP) and pentaerythritol (PER) intumescent formulations. This zeolite-based waste material is discharged from the cracking units in large quantities, giving rise to a number of environmental hazards. Thus, its reutilization in intumescent formulations not only contributes in lowering environmental hazard due to waste discharge, but may also provide a way to produce materials that meet flame-retardancy standards at lower cost, allowing them to find their way into various residential and commercial markets.

The basic concepts of intumescence are first briefly reviewed, followed by an outline of the characteristics of the FCC catalyst. Once both intumescence and the spent catalyst have been presented, the synergistic action of the catalyst in the intumescent formulations can then be fully discussed. In this section, the effect of catalyst loading, particle size, and its components on flame retardancy is evaluated by traditional fire tests, such as the limiting oxygen index (LOI) and cone calorimetry. The final section covers the participation of the spent catalyst in the formation of the intumescent shield, as observed by heating microscopy and by scanning electron microscopy (SEM) of burnt surfaces.

23.2 Protection *Via* Intumescence

Intumescent systems interrupt the self-sustained combustion of a polymer matrix at its earliest stage, that is, the thermal degradation accompanied by the evolution of gaseous fuels.[3] On heating, these systems swell to form foamed cellular charred layers on the surface of the burning material. This layer acts as a physical barrier to heat and mass transfer, protecting the underlying material from the heat flux and limiting the diffusion of both the combustible gases generated by pyrolysis that feed the flame and of the oxygen that sustains the burning process[5–8] (Figure 1). Hence, intumescent systems interfere with the action of all the necessary components for the fire triangle, namely heat, oxygen, and fuel.

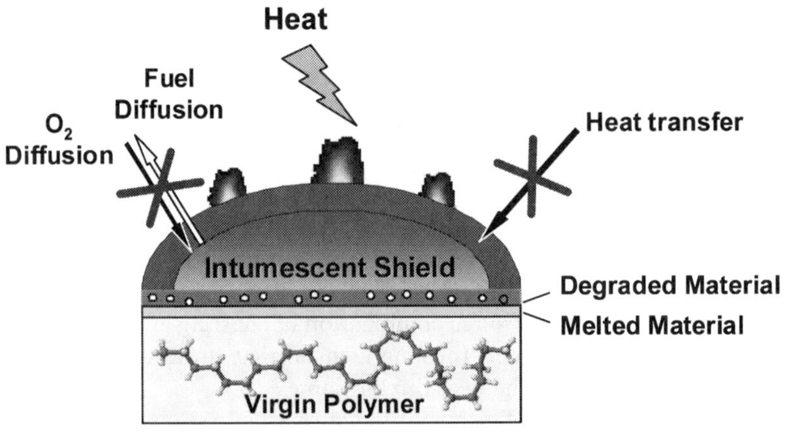

Figure 1 *Schematic action of an intumescent polymeric formulation (adapted from reference 9)*

23.2.1 Intumescent Formulations

An intumescent formulation generally contains three active components:[3,10]

- An acid source (precursor of acid species), such as ammonium phosphate, ammonium polyphosphate (APP), diammonium diphosphate or diammonium pentaborate.
- A carbonific compound, usually polyhydroxy compounds such as pentaerythritol (PER), xylitol, mannitol, sorbitol and polymers that naturally carbonise under heat or fire (polyamides, polycarbonates and polyurethane).
- A spumific (or blowing) compound that releases large quantities of non-combustible gases such as NH_3 and CO_2. Salts of phosphoric acid, melamine and guanidine have been used for this purpose.

Intumescent formulations should contain components that fulfil all three functions. There are compounds that may function in more than one way, *viz.* ammonium polyphosphate (APP), which acts both as an acid source and as a blowing agent by producing the corresponding acid and by releasing NH_3 on heating.

The first stage of the accepted mechanism of intumescence involves the decomposition of the acid source to give a mineral acid. The mineral acid then reacts with the carbonific agent to form a carbonaceous layer (char). In the final step, the spumific compound decomposes, generating gaseous products, which cause the char to swell, forming a foam-like insulating layer. Continued heating causes the decomposition of the intumescent material and loss of the foamed character.[3,11]

23.3 Synergistic Agents

The incorporation of active components into an additive system may lead to an additional effect,[12] an antagonistic effect[13] or a synergistic effect.[12–15] The use of synergistic agents has deserved increasing attention in flame retardancy. These agents produce more efficient systems while also making it possible to reduce the amount of flame retardants necessary to deliver efficient flame retardancy performance within the stringent regulations imposed.[16] Clays and zeolites[13,14] have been used as synergistic agents in intumescent ammonium polyphosphate (APP) and pentaerythritol (PER) formulations, and we now approach the use of spent oil refinery cracking catalyst for the same purpose.[17,18]

23.4 Oil Cracking Catalyst

After use, large amounts of waste catalyst are discharged from the reactors, giving rise to a number of environmental hazards. The worldwide annual demand for fluid bed catalytic cracking (FCC) catalyst in oil refineries is around

300000 tons[19] and, while many forms of reutilization of the material have been suggested, only its use in the cement industry has gained any wide range industrial status.[20]

23.4.1 The FCC Process and Catalyst – Basic Concepts

The FCC process aims at the conversion of heavy fractions from oil distillation (gas oils and resid fuels) into lighter fractions with greater economic value such as gasoline, diesel fuel and light gases, by the joint action of heat and a zeolite-based catalyst.[21,22] During the cracking process the catalyst is continuously being deactivated, and, to maintain catalytic activity and selectivity, part of the catalyst inventory is periodically discharged and new catalyst is introduced into the process. The removed catalyst is known as spent catalyst, exhaust catalyst or equilibrium catalyst.

The FCC catalyst is typically composed of a zeolite [a crystalline aluminosilicate with general formula $M_{x/n}(AlO_2)_x(SiO_2)_y{\cdot}wH_2O$, were $y/x \geq 1$ and M represents a cation with formal charge n][23] embedded in a silica-alumina matrix. The matrix consists mainly of a binder, such as silica hydrosols or alumina gels, an active matrix component, usually alumina, and a filler, typically kaolin clays.[22,24–25] Zeolite Y was first used as the active component in the cracking catalyst in the early 1960s. Shortly after, ultrastable Y zeolites (USY) and rare earth exchanged zeolites (REY and REUSY) were introduced into the catalysts,[26] both possessing greater thermal/hydrothermal stability and acidity than the Y type zeolite originally used.[27,28]

23.4.2 Chemical Composition and Physical Properties of the Spent FCC Catalyst

Throughout the following sections E.Cat. has been used to designate Exhaust or Equilibrium Catalyst, representing the spent FCC catalyst withdrawn from the FCC unit without further treatment. Due to its relatively large particles, E.Cat. was milled and wet sifted and the catalyst obtained will be represented by MEC. The numbers following MEC refer to the fraction being studied. The material that stayed on a given sieve is the oversize of the sieve, and is represented by the plus (+) sign. Analogously, the passing material is the undersize and represented by the minus (–) sign.[29] The morphology of E.Cat. and its –635# fraction is shown in Figure 2.

The chemical composition of E.Cat. and its fractions, obtained using X-ray fluorescence, are shown in Table 1, while the average particle size, zeolite content and textural properties of the materials are presented in Table 2. An increase in zeolite content was observed in all MEC fractions, indicating a partial dissolution of the amorphous components.

For simplicity, whenever the sieve size is not specified, MEC will refer to the undersize of the 635 mesh sieve (MEC = MEC – 635#).

Table 1 *Chemical composition of E.Cat and its fractions obtained by milling and sifting*

Component	*E.Cat.*	*MEC* +150#	+270#	+400#	+635#	−635#	*Units*
Aluminium oxide	31.9	31.1	31.0	31.4	30.8	31.5	%
Silicon oxide	66.2	67.1	67.3	66.9	67.4	66.7	%
Nickel	< 100	< 100	< 100	< 100	< 100	< 100	mg kg^{-1}
Iron(III) oxide	0.37	0.47	0.37	0.40	0.38	0.37	%
Vanadium	< 100	< 100	< 100	< 100	< 100	< 100	mg kg^{-1}
Titanium(IV) oxide	0.92	0.90	0.89	0.91	0.91	0.88	%
Antimony	< 100	< 100	< 100	< 100	< 100	< 100	mg kg^{-1}
Lanthanum(III) oxide	0.36	0.35	0.35	0.34	0.35	0.38	%
Cerium(III) oxide	0.03	0.02	0.02	0.01	0.01	0.03	%
Praseodymium(III) oxide	0.03	0.03	0.02	0.03	0.03	0.04	%
Neodymium(III) oxide	0.10	0.09	0.09	0.09	0.09	0.10	%
Sodium oxide	0.38	0.32	0.31	0.28	0.44	0.37	%
Phosphorous(V) oxide	0.09	0.08	0.08	0.09	0.08	0.09	%
Bismuth(III) oxide	96	104	102	107	115	96	mg kg^{-1}
Molar SAR[a]	3.52	3.67	3.69	3.62	3.73	3.60	

[a]SAR = molar silica:alumina ratio.

Table 2 *Physical properties and zeolite content of E.Cat and its fractions*

Fraction	*Average particle* size (μm)	*Zeolite content* (wt%)	*Textural properties* SA_{BET} (m^2 g^{-1})[a]	*MiPA* (m^2 g^{-1})[b]
E. Cat.	75.5	35.6	210.93	157.19
MEC +150#	125.3	37.6	237.22	176.78
MEC +270#	77.8	39.7	232.18	177.65
MEC +400#	50.8	39.6	231.86	170.42
MEC +635#	33.0	38.9	220.08	161.50
MEC −635#	8.77	37.6	224.58	166.19

[a]Surface area obtained by the BET method. [b]Micropore area.

23.5 Effect of the Catalyst on Fire Performance of Intumescent Formulations: Are the Additives in Synergy?

All intumescent systems studied were based on ammonium polyphosphate (APP, supplied by Clariant under the trade name Exolit 422) and pentaerythritol (PER, from Acros and Sigma Aldrich). The mixture, having an APP:PER weight ratio of 3, was added to the polymeric matrix to account for 30 wt% of the final mass. The materials studied were either processed in laboratory mixers (Brabender or Haake 90 at 160 and 150°C, respectively) or in a double screw extruder (Brabender) with temperatures varying from 120 to 160°C. Polyethylene copolymers and terpolymers were used as matrices.

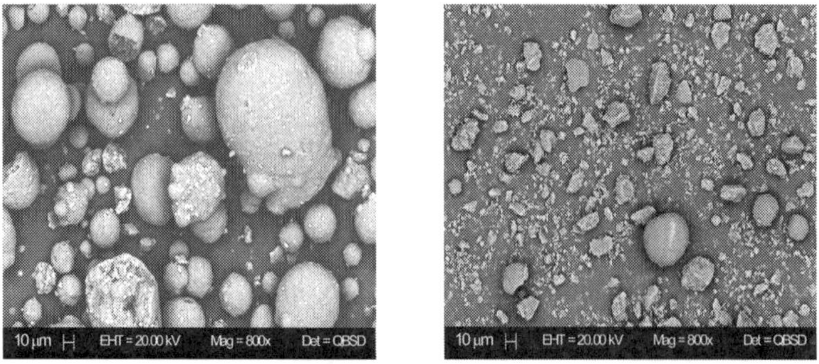

(a) (b)

Figure 2 *Scanning electron microscopy images of* (a) *E.Cat and* (b) *MEC–635#*

23.5.1 Effect of Catalyst Loading

The addition of various amounts of MEC to the polymer containing the intumescent formulation has shown that the catalyst increases the flame retardancy performance of these systems. However, no effect is observed when the catalyst is added to the pure polymer, indicating that the intumescent additives and the catalyst are in synergy. The LOIs shown in Table 3 illustrate well the phenomenon in an ethylene-butyl acrylate copolymer matrix. Maximum values are achieved with 5 wt% of MEC in this matrix, but even smaller amounts, like 1.3%, already increase the APP/PER oxygen index by 7 units, corresponding to a 33% gain. Thus, the addition of MEC has taken the intumescent APP/PER materials from "slow-burning" to "self-extinguishing"[30] composites.

Other polyethylene matrices gave similar results, although the optimum MEC content varied from 2.5 to 5% depending on the matrix.[31,32] Maximum values for the rate of heat release (RHR) from the systems, measured by cone calorimetry, confirm the observations made based on the LOI results. MEC addition to the APP/PER formulations significantly lowers RHR of the intumescent materials. In Figure 3, the maximum RHR of the APP/PER formulations without MEC is *ca.* 75% higher than for the intumescent system containing 5 wt% of the waste material. MEC incorporation into other polyethylenic matrices supports the proposition that a synergistic effect occurs between MEC and the APP/PER intumescent additives.

Table 3 *Influence of MEC addition on the LOI (accurate to ±1 unit) of the pure ethylene (70 wt%)-butyl acrylate (30 wt%) copolymer, PEBA, and of the copolymer containing the intumescent mixture*

	LOI (vol%)				
MEC content (wt%) added to	*0*	*1.3*	*2.5*	*5.0*	*10.0*
Pure polymer	18	18	18	18	19
Polymer + intucescent APP/PER mixture	21	28	30	31	29

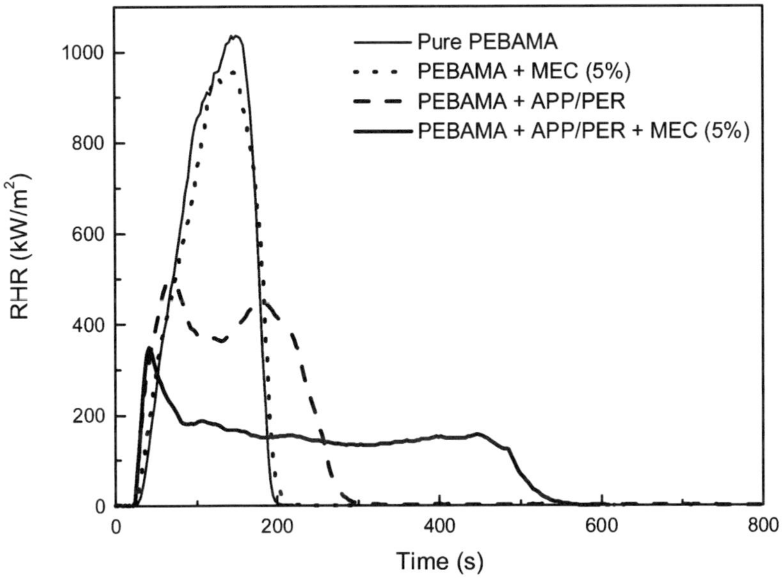

Figure 3 *Effect of MEC addition to PEBAMA [terpolymer of ethylene (91.5 wt%)-butyl acrylate (5 wt%)-maleic anhydride (3.5 wt%)] formulations on the rate of heat release*

23.5.2 Effect of the Catalyst's Particle Size

The thermal properties of many systems can be influenced by the particle size of the additives. Differences in particle size can determine the extent of fire retardancy performance of $Al(OH)_3$ and $Mg(OH)_2$ filled systems.[33]

Hence, it is not surprising that the extent of the synergy observed between the intumescent additives and the spent catalyst should depend on the average particle size of the catalyst samples. The effect is clearly illustrated in Figure 4. The higher LOI are attained with the finest fractions and decrease with increasing particle size. This tendency could possibly be attributed to the generation of higher tension points in the char by particles of greater dimensions, leading to a decrease in the shield's strength.[34] The limiting oxygen index of the E.Cat. system shows good accordance with the average particle size trend observed.

23.5.3 Effect of the Catalyst's Components on Flame Retardancy

Having established that the spent catalyst greatly enhances the fire performance of APP/PER intumescent formulations, each of one of the catalyst's components was studied individually to determine which of them contribute to the final synergistic effect. Fire testing revealed that zeolites, whatever the type, are the components that exhibit the highest synergy with the intumescent additives. The LOIs shown in Table 4 illustrate the effect of the various components on fire performance of the intumescent mixtures. Comparison between the zeolites has

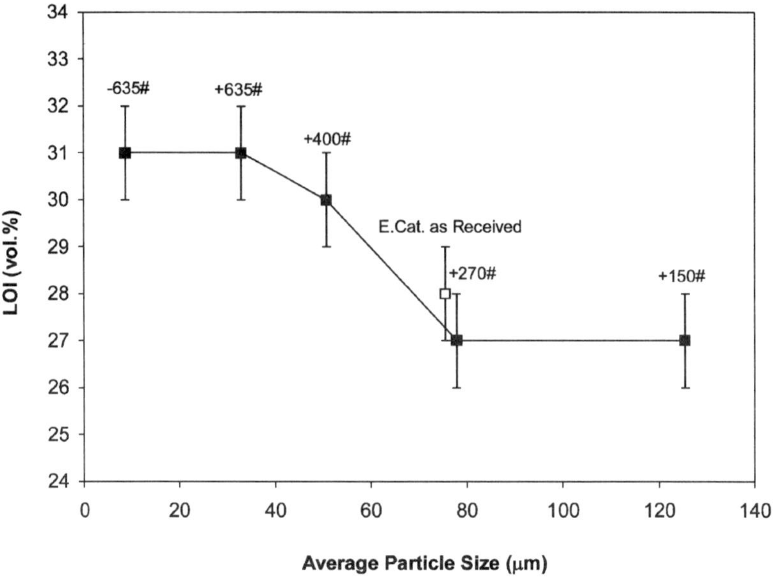

Figure 4 *Effect of particle size on LOI values of intumescent formulations in a PEBA [ethylene (70 wt%)-butyl acrylate (30 wt%)] matrix. (LOI accurate to ±1 unit)*

Table 4 *Influence of spent catalyst and its components on LOI of intumescent formulations in PEBA matrix (LOI accurate to ±1 unit)*

	LOI (vol%)						
					Zeolites		
Filler content (wt%)	*MEC* (−635#)	*Kaolin*	*Alumina*	*Silica*	*NaY*	*USY*	*REY*
0	21	21	21	21	21	21	21
1.3	28	22	21	21	31	31	31
2.5	30	23	21	22	32	31	31
5.0	31	26	22	22	31	31	32

shown that the nature of the counter-ion (NaY and REY) and the strength of the acid site (NaY and USY) do not significantly affect the fire performance of the materials. Maximum LOIs are achieved with smaller amounts of zeolite than with MEC, but the ultimate values attained are equivalent in both materials. Kaolin also increases flame retardancy performance of the APP/PER systems, though to a lesser extent than zeolites or MEC. Silica and alumina presented no synergistic effect detectable by the method.

The results seem to indicate that the increased efficiency of the APP/PER systems occurs by means of aluminosilicate species. The use of silica and alumina separately did not enhance the fire properties of APP/PER formulations, while the benefits of kaolin, zeolites and spent catalyst are obvious.

23.5.4 Spent Catalyst and the Intumescent Layer

The material resulting from the degradation of the intumescent additives is heterogeneous, composed of gaseous products trapped in a phosphocarbonaceous cellular material, also known as the condensed phase. The condensed phase is a mixture of solid and liquid phases.[3] The nature and proportion of these phases dictates the dynamic properties of the insulating layer and hence the fire performance of the material. Thus, an active participation of the catalyst in the generation of the intumescent shield may lead to modified dynamic properties and account for the increased fire performance of the systems. Heating microscopy and scanning electron microscopy (SEM) coupled with an X-ray energy dispersive system (EDS) both made clear that the catalyst effectively takes part in the protective layer and increases the thermal stability of the charred surface structure.

Heating microscopy has recently been proposed by our group as a useful tool for monitoring the intumescence phenomenon continuously *in situ.*[17,35–38] The degree of intumescence can be estimated by evaluating the sample's projected area at different temperatures throughout the experiment. The procedure involves heating 3 mm sided cubic samples from 30 to 700°C, under static air atmosphere, with a 10.5 A electric current during an average interval of 30 min. The projected areas are quantified by means of an image analysing software. Selected images, obtained at 30, 100, 350 and 700°C of the PEBAMA terpolymer formulations are presented in Figure 5. A quantitative analysis of the intumescent behaviour of the systems was carried out in which the sample's projected area was recorded and divided by the initial area, thus reducing the errors due deviations in the specimen's dimensions. The intumescence process can be easily followed by the association of Figures 5 and 6.

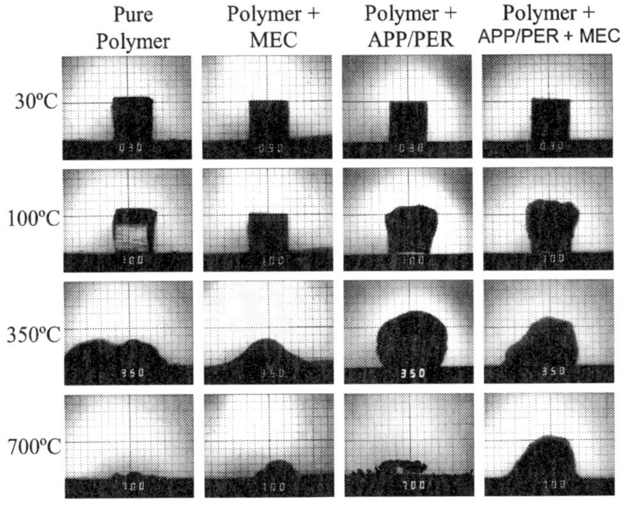

Figure 5 *Heating microscopy images of the PEBAMA systems*[17]

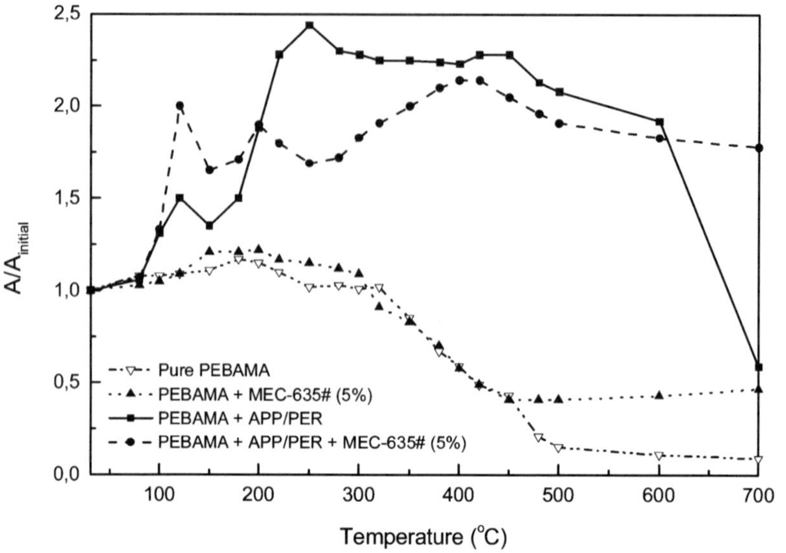

Figure 6 *Ratio between the sample's projected area at a determined temperature and its initial projected area, determined by heating microscopy followed by image analysis. Values are the average of 5 measurements, error bars omitted for clarity*[17]

During heating, pure polymer becomes completely transparent at around 100°C, where the melting process also becomes evident. No significant intumescence is observed for the pure polymer and the polymer containing MEC.

The intumescent process, as expected, became evident only by the addition of APP/PER to the mixture. The sample containing only APP/PER suffered greater swelling than that to which the catalyst had been added. However, the presence of the catalyst helped maintain the specimens' structure at temperatures up to 700°C, when the system containing only APP/PER had already collapsed. Noteably the final value of the area ratio for polymer + APP/PER + MEC (5%) system at 700°C involved a large standard deviation. Many of the samples tested presented an area ratio of practically zero, while a few appeared to partially maintain the structure. However, by keeping the temperature at 700°C the residue gradually disappeared, indicating that its lack of total degradation was time-dependent. This was not the case for the APP/PER + catalyst system, where the residue at 700°C was maintained for even at longer times.

The participation of the catalyst in the intumescent shield formed was observed by scanning electron microscopy (SEM) coupled with an X-ray energy dispersive system (EDS). The surfaces of the intumescent samples were burnt by applying a 4 cm flame for 25 s on a 20 × 4 mm area and the resulting surfaces are presented in Figures 7 and 8. Vast regions of the burnt surface of the sample containing only APP/PER have the aspect shown in Figure 7a. On the sample's

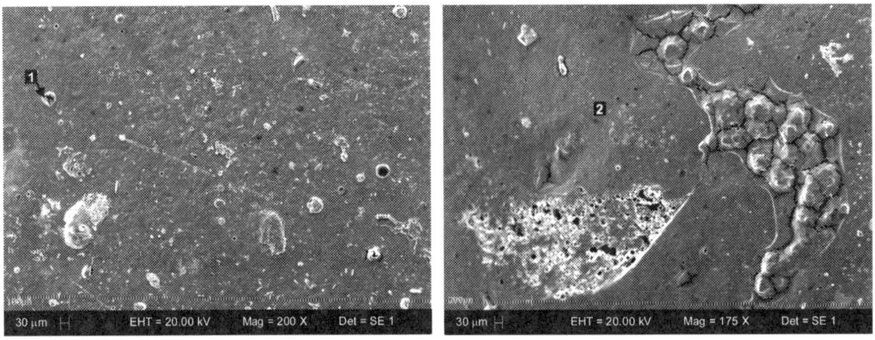

(a) (b)

Figure 7 (a) *and* (b) *Burnt surface of a PEBAMA + APP/PER sample. EDS analysis of points 1* (a) *and 2* (b) *revealed the elements P, O and C*

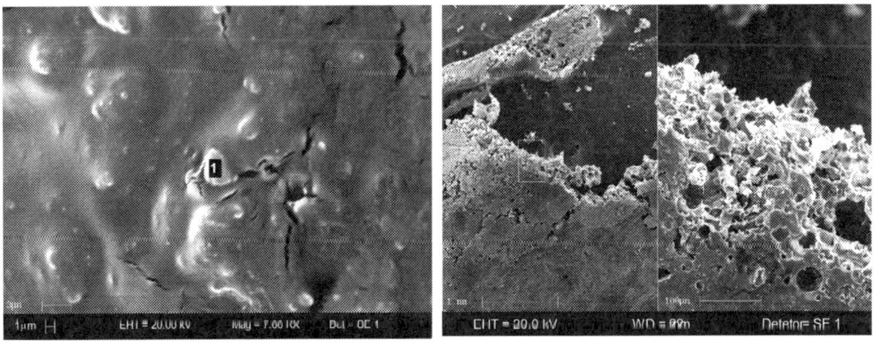

(a) (b)

Figure 8 (a) *and* (b) *Burnt surface of a PEBAMA + APP/PER + MEC (5%) sample. EDS analysis of point 1* (a) *revealed the elements Si, Al, P, O and C*

surface are small rods and particles that an EDS analysis revealed to contain phosphorous, carbon and oxygen (Figure 7a, point "1"). Similar results were obtained by analysing the matrix (point "2", Figure 7b).

The foamed structure that characterises an intumescent shield could only be detected in a few isolated areas, such as the one shown in Figure 7b, beside a region of high phosphate concentration. The low abundance of these foamed regions indicates that the intumescent shield lacked adequate dynamical/mechanical properties for its maintenance after continued flame application and cooling.

Contrasting with these observations, the APP/PER system to which MEC had been added kept its intumescent shield almost unimpaired during burning and cooling, except for the few cracks shown in Figure 8. An EDS analysis of point "1" in Figure 8a revealed the presence of Si and Al, indicating that MEC particles take part in the intumescent shield. A crack in the shield, caused by sample manipulation, revealed the porous foam-like structure responsible for limiting heat and mass transfer (Figure 8b).

Thus, the increased flame retardancy of the APP/PER + MEC systems can possibly be attributed to the higher thermal stability of the protective shield formed during heat exposure of the material.

23.6 Conclusion

Waste FCC catalyst strongly enhances the fire retardant properties of intumescent APP/PER formulations. However, the incorporation of the catalyst to polyethylenic matrices without the intumescent formulations does not modify its fire performance. This way, the waste material appears to be a powerful synergistic agent in APP/PER intumescent formulations. Fire performance increases with the catalyst content up to an optimum value that depends on the nature of the matrix (2.5 to 5.0 wt%). Moreover, the efficiency of the catalyst is strongly dependent on its particle size.

From the major catalyst components only the zeolite and kaolin enhance fire performance of the intumescent systems. Silica and alumina individually do not show synergistic effects with APP/PER, indicating that aluminosilicates are responsible for the enhanced effect.

The association of heating microscopy with image analysis makes it is possible to conclude that the use of the FCC waste catalyst contributes to the maintenance of the charred surface structure at temperatures higher than would be observed in APP/PER systems that do not make use of this component. SEM-EDS results made clear that the catalyst takes part in the protective intumescent shield formed. The presence of the catalyst possibly results in the modification of the dynamic-mechanical properties of the shield, accounting for its increased resistance.

23.7 Acknowledgements

The authors gratefully acknowledge CAPES and PRONEX for financial support, CETEM/Brazil for the SEM analysis, CENPES/Petrobras for the characterisation of the waste catalyst and F. Pouch, L. Pankewitch, N. Debusse and O. Dobosz (CREPIM) for technical assistance in fire testing.

23.8 References

1. M.S. Cross, P.A Cusack and P.R. Hornsby, *Polym. Degrad. Stab.*, 2003, **79**, 309.
2. G. Camino in *Fire Retardancy of Polymers: The Use of Intumescence*, M. Le Bras, G. Camino, S. Bourbigot and R. Delobel (eds.), The Royal Society of Chemistry, Cambridge, UK, 1998, p. v.
3. M. Le Bras and S. Bourbigot, in reference 1, p. 64.
4. M.S. Reisch, *Chem. Eng. News*, 1997, Feb. 24, 19.

5. G. Berttelli, G. Camino, E. Marchetti, L. Costa, E. Casorati and R. Locatelli, *Polym. Degrad. Stab.,* 1989, **25**, 277.
6. S. Bourbigot, M. Le Bras, P. Breant, J.M. Tremillon and R. Delobel, *Fire Mater.*, 1996, **20**, 145.
7. S.-Y. Lu and I. Hamerton, *Prog. Polym. Sci.,* 2002, **27**, 1661.
8. M. Le Bras, S. Bourbigot, E. Felix, F. Pouille, C. Siat and M. Traisnel, *Polymer*, 2000, **41**, 5283.
9. X. Almeras, F. Dabrowski, M. Le Bras, R. Delobel, S. Bourbigot, G. Marosi and P. Anna, *Polym. Degrad. Stab.*, 2002, **77**, 315.
10. M. Elomaa, L. Sarvaranta, E. Mokkola, R. Kallonen, A. Zitting, C.A.P. Zevenhoven and M. Hupa, *Crit. Rev. Biochem. Mol. Biol.*, 1997, **27**, 137.
11. D.B. Dahm, *Prog. Org. Coatings*, 1996, **29**, 61.
12. M. Le Bras, S. Bourbigot, Y. Le Tallec and J. Laureyns, *Polym. Deg. Stab.*, 1997, **56**, 11.
13. M. Le Bras and S. Bourbigot, *Fire Mater.,* 1996, **20**, 39.
14. S. Bourbigot, M. Le Bras, R. Delobel, P. Bréant and J.M. Tremillon, *Polym Degrad Stab.*, 1996, **54**, 275.
15. S. Bourbigot, M. Le Bras, P. Breant, J.M. Tremillon and R. Delobel, *Fire Mater*, 1996, **20**, 145.
16. J. Murphy, *Reinforced Plastics*, 2001, **45**, 42.
17. L.R.M. Estevão and R.S.V. Nascimento, *Polym. Degrad. Stab.*, 2002, **75**, 517.
18. L.R.M. Estevão, M. Le Bras, R. Delobel and R.S.V. Nascimento, *J. Fire Sci.*, 2004, **22(3)**, 211–227.
19. F.C. Barbosa and M.G. Tavares, *Petrobras Mag.*, 1999, **27**, 17.
20. R.T. Oliveira, *FCC em Revista*, 1997, **10**, 2.
21. R.A. Meyers, *Handbook of Petroleum Refining Process*, Mc Graw Hill Book Company, New York, 1986.
22. P.G. Smirniotis and L. Davydov, *Catal. Rev. – Sci. Eng.*, 1999, **41**, 43.
23. J. L.F. Monteiro, *2° Curso Iberoamericano sobre Peneiras Moleculares*, São Carlos, Brazil, 1995, 1.
24. W.C. Cheng, G. Kim, A.W. Peters, X. Zhao and K. Rajagopalan, *Catal. Rev. – Sci. Eng.*, 1998, **40**, 39.
25. E.F. Sousa-Aguiar, “O papel das zeólitas am catalisadores de craqueamento. Relatório” CENPES/PETROBRAS, 199x, 1.
26. R.H. Harding, A.W. Peters and J.R.D. Nee, *Appl. Catal. A: General*, 2001, **221**, 389.
27. P.B. Venuto, *Microporous Mater.*, 1994, **2**, 297.
28. G.L. Baugis, H.F. Brito, W. Oliveira, F.R. Castro and E.F. Sousa-Aguiar, *Microporous and Mesoporous Mater.*, 2001, **49**, 179.
29. J.H. Perry, *Chemical Engineers' Handbook*, McGraw-Hill Book Company, New York, 1950.
30. G. Camino and L. Costa, *Polym. Degrad. Stab.*, 1988, **20**, 271.
31. L.R.M. Estevão, M. Le Bras, R. Delobel and R.S.V. Nascimento, *Eur. Polym J.*, 2004, **40(7)**, 1503–1513.

32. L.R.M. Estevão, Doctoral Thesis, DQO – Instituto de Química, UFRJ, Rio de Janeiro, Brazil, 2002.
33. R.N. Rothon, in *Particulate-Filled Polymer Composites*, R.N. Rothon (ed.), Longman Scientific & Technical, Bath, UK, 1995, 207.
34. M. Le Bras, Thèse de Doctorat ès Sciences Physiques, Lille, France, 1997.
35. R.S.V. Nascimento, C.R. Perruso and P.R. Hornsby, *Extended Abstracts from Eurofillers 97*, Manchester, 1997, 457.
36. R.S.V. Nascimento, A.N. Pereira and C.R. Perruso, *Proceedings from VI Simposio Latinoamericano de Polimeros*, Viña del Mar, Chile, 1998, 126.
37. L.R.M. Estevao, M. Le Bras, L.C.S. Mendonça-Hagler and R.S.V. Nascimento, *Proceedings of the 17th World Petroleum Congress*, Rio de Janeiro, Brazil, 2002, 152.
38. L.R.M. Estevão, M. Le Bras, R. Delobel and R.S.V. Nascimento, *Proceedings of the 9th European Meeting on Fire Retardancy and Protection of Materials – FRPM'03*, M. Le Bras *et al.* (eds.), USTL Pub., Lille, France, 2003, p. 37.

CHAPTER 24

Zinc Borates as Synergists for Flame Retarded Polymers

SERGE BOURBIGOT, MICHEL LE BRAS AND SOPHIE DUQUESNE

Laboratoire des Procédés d'Élaboration des Revêtements Fonctionnels, UPRES EA 1040, École Nationale Supérieure de Chimie de Lille (ENSCL), Université des Sciences et Technologies de Lille, BP 108, 59652 Villeneuve d'Ascq Cedex, France (serge.bourbigot@ensc-lille.fr)

24.1 Introduction

Previous studies have demonstrated that there are major advantages (smoke suppressant, afterglow suppressant, corrosion inhibitor, anti-tracking agent and synergistic agent) of combining zinc borates, in particular in halogen-free systems, with other flame retardants in several kinds of polymers (EVA, PVC, polyamides, *etc.*).[1–5] This chapter surveys the use of zinc borates as synergist in halogen-free flame retarded (FR) polymers.

Several crystalline structures of zinc borates are known but a few find industrial use in significant tonnage. Zinc borates can be divided into two categories, hydrated and anhydrous.[6] So-called hydrated borates, which account for most known boron-containing minerals and synthetic borates consumed by industry, have structures containing B–OH groups (hydroxyl-hydrated borates) and may, also, contain interstitial water. Here, we only evaluate the performance of different synthetically made zinc borates commercialized under the brand name Firebrake®:[7] FBZB ($2ZnO{\cdot}3B_2O_3{\cdot}3{\cdot}5H_2O$), FB415 ($4ZnO{\cdot}B_2O_3{\cdot}H_2O$) and FB500 ($2ZnO{\cdot}3B_2O_3$). Notably the chemical formula and the structure of FB290 have been recently revisited by Schubert *et al.*[8] and his investigation led to a revision of the chemical formula as $2ZnO{\cdot}3B_2O_3{\cdot}3H_2O$. Each of these compounds can be prepared selectively by reactions of zinc oxide with boric acid in water. The specific product is then obtained by adjusting the ratio B_2O_3/ZnO and temperature.

The thermal behavior of those zinc borates is also important to process polymers containing them. Dehydration of FBZB commences upon heating above 290°C and complete dehydration requires 445 J g^{-1}. The dehydration onset temperature of FB415 is much higher than that of FB290 since it is 415°C. This property allows for processing in engineering thermoplastics and other relatively high temperature systems. The dehydration sequence of FB290 involves the loss of three molar equivalent of water through condensation of B–OH groups and yields FB500, a substantially amorphous material of composition $2ZnO{\cdot}3B_2O_3$. The selection of zinc borate depends, therefore, on the processing temperature of the polymer but its chemical composition might play a role in the FR mechanisms. As far as we know, no paper describes and compares the effect of the chemical composition of zinc borates as synergist in a FR polymer. Here, it is also one of our goals to point out this aspect.

Zinc borate itself is not really a flame retardant. An example is an ethylene-vinyl acetate copolymer [EVA containing 19% vinyl acetate (VA) and hereafter called EVA19] filled by 50 wt% zinc borate (FBZB and FB500). LOI values (Limiting Oxygen Index: Standard Test Method for Measuring the Minimum Oxygen Concentration to Support Candle-like Combustion of Plastics measured according to the standard ASTM D2863/77) of such formulations are as low as 21 wt% (samples drip and burn) and no classification is achieved at the UL-94 test [UL-94 test: the test was carried out on sheets (3 mm thick) according to ASTM D-635-77 standard]. Rate of heat release (RHR) curves (samples exposed to an external heat flux of 50 kW m^{-2} according to ASTM 1356-90 standard) of EVA19-FBZB and EVA19-FB500 are similar, suggesting that there is no effect of the chemical composition of zinc borate (Figure 1). Nevertheless, zinc borate plays here the role of flame retardant since the RHR peak of virgin EVA in the same conditions is about 1700 kW m^{-2}. This effect may be assigned to a "filler effect" (dilution of fuel, *i.e.*, of polymer) and to the formation of low viscosity vitreous protective coating due the decomposition of zinc borate.[4,5] It also explains the poor performance obtained at LOI and UL-94 tests because of the dripping of the samples when heated.

This short discussion shows the versatility of zinc borates and its potential as synergists in flame retarded polymers. Here, we will then focus on the performance of zinc borate as synergist and on the influence of the chemical composition of the three zinc borates mentioned above on the flammability properties of different thermoplastics [polypropylene (PP) and EVA] filled by metal hydroxides and by intumescent systems.

24.2 Zinc Borates In Eva-Metal Hydroxides Systems

Sectors of industry, notably those concerned with aerospace, microelectronics, cable and wire manufacture, are particularly interested in alternative halogen-free fire retardants such as aluminium or magnesium hydroxides [$Al(OH)_3$ or $Mg(OH)_2$], hereafter called ATH and MH.[10] These latter achieve their effects by decomposing endothermally with the release of water close to the temperatures at which polymers themselves decompose, and by forming a protective ceramic-like structure.[3,5]

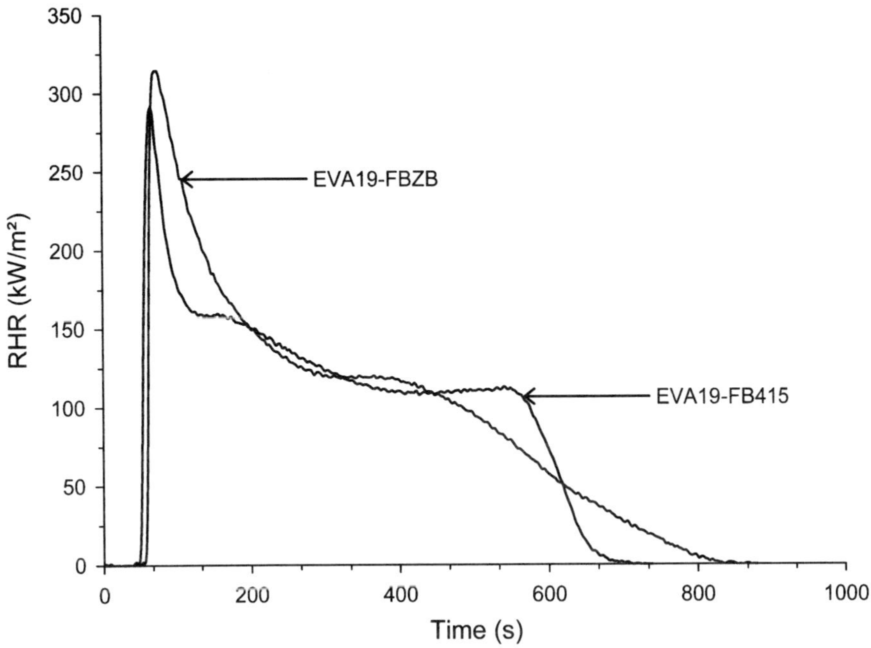

Figure 1 *RHR (external heat flux = 50 kW m^{-2}) versus time of the formulations EVA19-zinc borate (total loading 50 wt% remains constant, compounded in a roller mixer at 160°C at 50 rpm)*[9]

Zinc borates act as synergists when substituting partially metal hydroxides in thermoplastics. A typical example is given on Figure 2 for an EVA (EVA containing 24% VA and hereafter called EVA24) filled by 60 wt% of different ratios of MH/Zinc borate. LOI *versus* the substitution amount of MH in zinc borates reveal a synergistic effect for the two zinc borates (FBZB and FB415). Maximum LOI is reached at 3 wt% for FBZB (LOI = 41 vol%, *i.e.* 20% of relative increase) and at 5 wt% (LOI = 44 vol%, *i.e.* 10% of relative increase) for FB415. The difference between the two zinc borates is that the LOI maximum lies at different substitution level (3 wt% compared to 5 wt%) and that the relative increase of the performance of the formulation containing FB415 is twice as high as that of the formulation containing FBZB. Conversely, LOIs of the formulation containing FB415 decrease slower than those containing FBZB at high substitution level (> 7 wt%). These trends due to the two different zinc borates are not clear to us at this time and work is in progress to explain them. Comparison of the median particle size of the two zinc borates reveals that it is of 5 μm for FB415 and of 9 μm for FBZB. This suggests that this parameter might act upon the performance, modifying the viscosity of the polymer. As LOI is a vertical test, the protective material drips down when the viscosity becomes too low. It was previously reported by Carpentier *et al.*[4] who demonstrated the direct influence of the rheological behavior on LOI values and who explained the decrease of LOI when substituting MH by zinc borate at too high level.

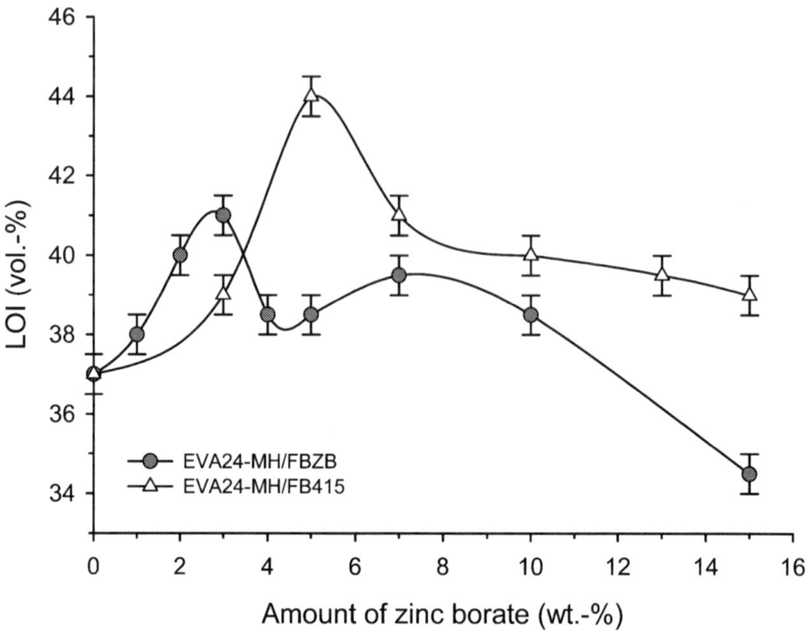

Figure 2 *LOI vs. the substitution amount of MH in zinc borate in the formulation EVA24-MH/zinc borate (total loading 60 wt% remains constant, compounded in a roller mixer at 160°C and 50 rpm)*[11]

The second explanation is the high dehydration temperature of FB415 (415°C) compared to that of FBZB (290°C). Hydrolysis of polymeric chains might take place at relatively low temperature with FBZB and, so, the viscosity would fall.

Cone calorimetry by oxygen consumption permits evaluation of materials in different fire scenarii conditions. It is then a tool of choice to investigate the effect of zinc borates in metal hydroxides filled polymers. RHR curves (samples exposed to an external heat flux of 50 kW m^{-2} according to ASTM 1356-90 standard) of different EVA19-based formulations show that the peak of RHR in time and in intensity (RHR peak = 185 kW m^{-2} at 80 s) is not modified upon incorporating zinc borates (Figure 3). Time to ignition is also unchanged. The effect of zinc borates is revealed at longer times, when the second peak of RHR appears. Except for FBZB, the second RHR peak is slightly decreased and is spread out over time for formulations containing zinc borates compared to that without. Notably FBZB-XF has the same chemical composition as FBZB but has a smaller particle size (median particle size is of 2 μm compared to 9 μm). RHR values of the formulation containing FBZB-XF are significantly lower than that containing FBZB and the same as that containing FB415 after the first RHR peak. This result suggests that the particle size of zinc borates plays a significant role in the flammability properties of the EVA/ATH systems. It makes sense, because the mechanism of action of ATH is to form a protective alumina ceramic and zinc borate reinforces its efficiency by acting as a binder

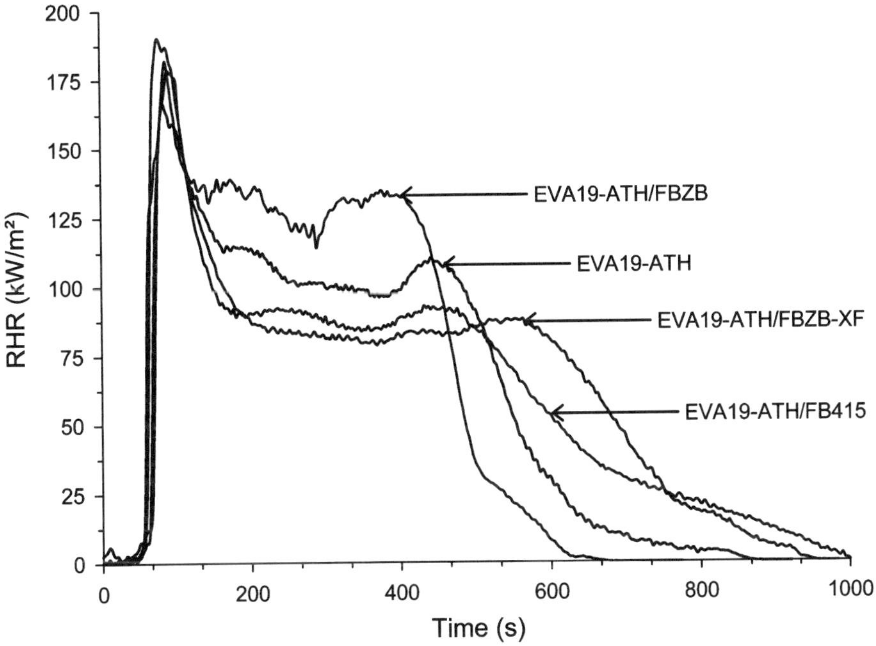

Figure 3 *RHR (external heat flux = 50 kW m^{-2}) vs. time of the formulations EVA19-ATH/zinc borate (total loading 65 wt% remains constant, formulations contains 35 wt% EVA19, 43.3 wt% ATH and 21.7 wt% zinc borate, compounded in a roller mixer at 160°C and 50 rpm)*[9]

(formation of B_2O_3/ZnO glass),[3,5] filling up pores of the ceramic. If we assume that the dispersion would be better using finer particles, then the B_2O_3/ZnO glass coming from the degradation of zinc borate would be homogenously dispersed in the polymeric matrix and would interact better with the alumina ceramic. However, there is no observed effect of the chemical composition of zinc borate on flammability properties. It seems, therefore, that temperature of dehydration of zinc borate does not influence flammability of the systems. According to our previous discussion, we can say that the viscosity of the melt FR polymer is the key factor to pass a vertical test.

To complete these results, a comparison of EVA24-based formulations containing 60 wt% fillers (combination MH with zinc borate) is shown on Figure 4. The figure shows that the incorporation of zinc borates (7 wt% substitution) reduces RHR values when added in EVA24-MH systems. It was not observed in the previous case because the ratio ATH/zinc borate was not adjusted to get the lowest RHR (as an example, the ratio ATH/zinc borate = 2 provides the best performance in EVA19-ATH/FBZB-XF formulations).[3,9] The chemical composition of zinc borate does not also play a role in the flammability properties (in terms of RHR) in the EVA24-MH/zinc borate systems, as observed above. Nevertheless, different ratios of substitution should be evaluated to draw a final conclusion.

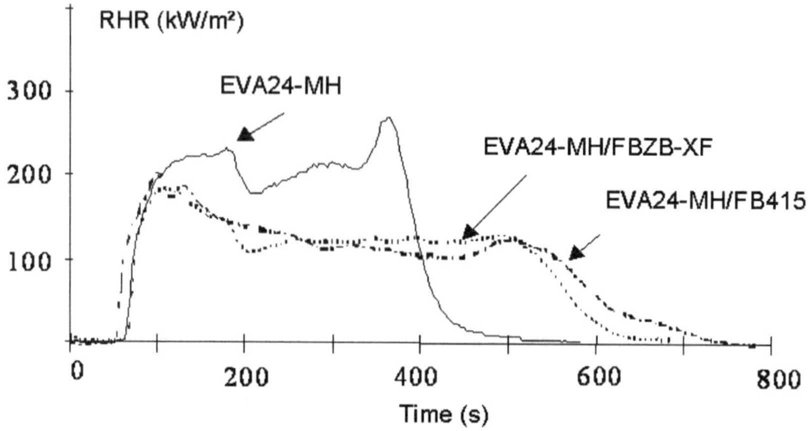

Figure 4 *RHR (external heat flux = 50 kW m^{-2}) vs. time of the formulations EVA24-MH/zinc borate (total loading 60 wt% remains constant, formulations contains 40 wt% EVA19, 53 wt% MH and 7 wt% zinc borate, compounded in a roller mixer at 160°C and 50 rpm)*[12]

24.3 Zinc Borates in PP-Based Intumescent Systems

Fire protection of flammable materials by an intumescence process has been known for several years. Flame retarding polymers or textiles by intumescence are essentially a special case of a condensed phase mechanism.[13–17] Intumescent systems interrupt the self-sustained combustion of the polymer at its earliest stage, *i.e.* the thermal degradation with evolution of the gaseous fuels. The intumescence process results from a combination of charring and foaming of the surface of the burning polymer. The resulting foamed cellular charred layer, the density of which decreases as a function of temperature, protects the underlying material from the action of the heat flux or of the flame. A recent paper reviews the latest development in intumescence.[18]

Intumescent molecules are synthesized using novel procedures to obtain derivatives of bis(2,6,7-trioxa-1-phosphabicyclo[2.2.2]octane-4-methanol)phosphate hereafter called melabis[19] (Figure 5). The new melabis is evaluated in

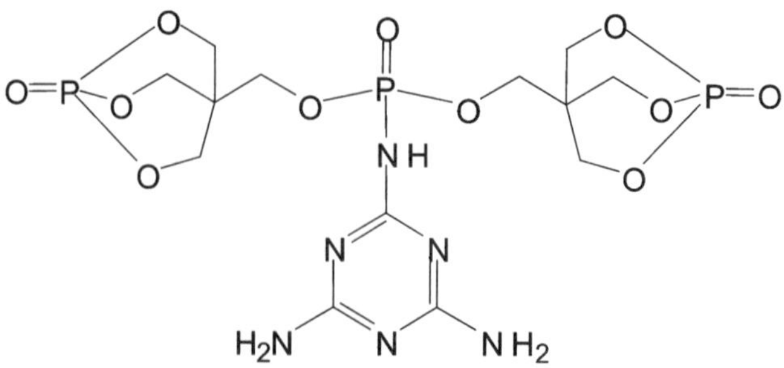

Figure 5 *bis(2,6,7-Trioxa-1-phosphabicyclo[2.2.2]octane-4-methanol)phosphate molecule (derivatives were synthesized in this work*[19]*)*

polypropylene (PP) filled at 30 wt%. The LOI is 31 vol% and jumps to 34 vol% upon adding 1 wt% melamine at 30 wt% total loading. All those formulations are V-0 rated at the UL-94 test [UL-94 test: carried out on sheets (3 mm thick) according to ASTM D-635-77 standard].

Previous papers reported that zeolites,[20] TiO_2,[21] talc,[22] or MnO_2,[23] could be used as synergists in intumescents. No work has been published on the potential use of zinc borates as synergistic agents in intumescent systems and it is the purpose of this section to investigate its efficiency in PP-melabis formulations. A large synergistic effect is observed on substituting a small amount of melabis by zinc borate (Figure 6). The LOI jumps from 30 vol% for the formulation without zinc borate to 35.5 vol% at 1 wt% substitution in FBZB, to 38 vol% at 1 wt% substitution in FB415, and to 39 vol% at 2 wt% substitution in FB500. This suggests that the chemical composition may play a role in the synergistic action of zinc borate in intumescence. Further work is in progress to understand the interaction of zinc borate with intumescent formulation. Preliminary results show that boronphosphate is formed by reaction of phosphate with boron oxides.[24] This permits the stabilization of phosphate at high temperature, and then to keep the integrity of the intumescent structure.

Cone calorimetry at an external heat flux of 50 kW m^{-2} of PP-melabis systems containing FB500 (2 wt% substitution) reveals that RHR curves (not shown) are similar to those of the system without FB500. The development of an intumescent structure is always observed and the peak of RHR is as low as

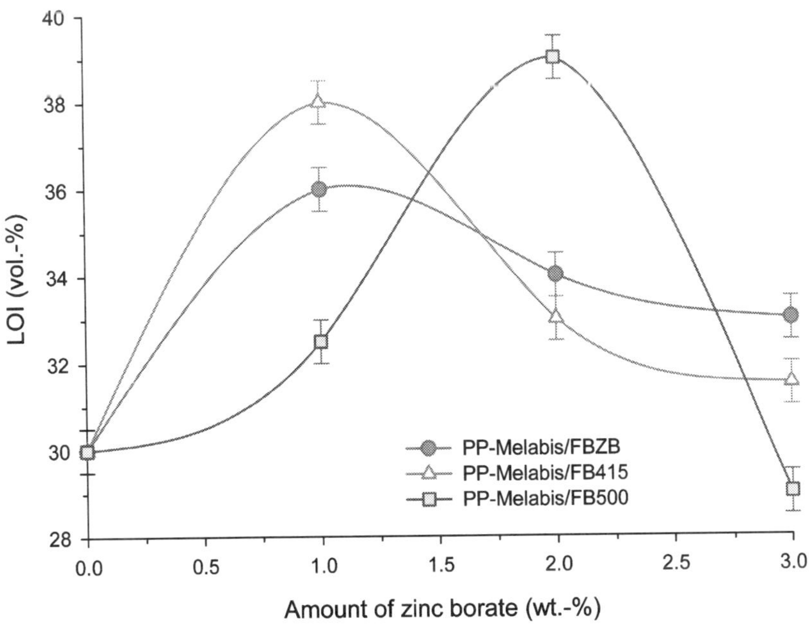

Figure 6 *LOI vs. the substitution in zinc borate in the formulation PP-melabis/zinc borate (total loading 30 wt% remains constant, compounded in a roller mixer at 200°C and 50 rpm)*[25]

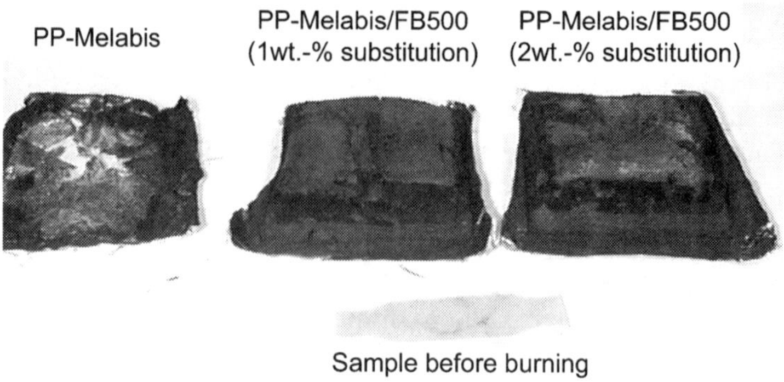

Figure 7 *Cone calorimeter residues of formulations PP-melabis and PP-melabis/zinc borate (total loading 30 wt% remains constant, 1 and 2 wt% zinc borate substitution compounded in a roller mixer at 200°C and 50 rpm)*[25]

150 kW m^{-2}. Notably, the residue of the system without zinc borate is not expanded. Only agglomerates of black char particles are observed (Figure 7). With FB500, the expanded shape is kept and the inside structure looks like a foamed cake (Figure 7). Even if the influence of zinc borate is not detectable on RHR curves, visual observation of the residues confirms that adding zinc borate to the intumescent permits the maintaince of the foamed expanded character of the intumescent structure.

24.4 Conclusions

Zinc borate is an efficient synergistic agent in different combinations of flame retarded polymeric materials. Two concepts were evaluated: intumescence and ceramic formation. In these two cases the LOI increases dramatically at 1–3 wt% substitution in zinc borate, V-0 classification is achieved, and RHR may be decreased.

The chemical composition of zinc borates may also play a part in the flammability properties of the materials. According to the presented results, we can propose some rules to design a flame retarded polymer containing zinc borate as synergist:

- Low particle size to achieve good dispersion.
- Between 1 and 5 wt% substitution of the main flame retardant by zinc borate to get the highest LOI.
- High dehydration/decomposition temperature to get the best efficiency and to pass vertical tests.

24.5 References

1. K.K. Shen, *Plastics Compounding*, 1988, Nov./Dec.
2. K.K. Shen and D.F. Ferm, *in Proceedings of Recent Advances in Flame Retardancy of Polymeric Materials*, M. Lewin (ed.), BCC Pub., Stamford, USA, 1997, Volume 8.

3. S. Bourbigot, M. Le Bras, R. Leeuwendal, K.K. Shen and D.M. Schubert, *Polym. Deg. Stab.*, 1999, **64**, 419.
4. F. Carpentier, S. Bourbigot, M. Le Bras and R. Delobel, *Polym. Int.*, 2000, **49**, 1216.
5. S. Bourbigot, F. Carpentier and M. Le Bras, *ACS Symposium Series N° 797*, C.A. Wilkie and G.L. Nelson (eds.), American Chemical Society Pub., Washington DC, 2001, Chapter 14, pp. 173–185.
6. "Boric oxide, boric acid and borates", in *Ulmann's Encyclopediaa of Industrial Chemistry*, VCH, Munich, Germany, 1972, Volume A4, pp. 263–280.
7. http://www.boraxfr.com
8. D.M. Schubert, F. Alam, M. Z. Visi and C.B. Knobler, *Chem. Mater.*, 2003, **15**, 866.
9. F. Carpentier and S. Duquesne, unpublished results.
10. P.R. Hornsby and C.L. Watson, *Plastic Rubber Process Applicat.*, 2003, **6**, 169.
11. F. Carpentier, PhD Thesis, University of Lille, France 2000.
12. M. Le Bras, S. Bourbigot, F. Carpentier, R. Leeuwendal and D.M. Schubert, *Gummi Fasern Kunststoffe*, 1998, **92**, 972.
13. G. Camino, L. Costa and L. Trossarelli, *Polym. Degrad. Stab.*, 1985, **12**, 213.
14. R. Delobel, M. Le Bras, N. Ouassou and F. Alistiqsa, *J. Fire Sci.*, 1990, **8(2)**, 85.
15. S. Bourbigot, M. Le Bras and R. Delobel, *J. Fire Sci.*, 1995, **13(1–2)**, 3.
16. A.R. Horrocks, *Polym. Deg. Stab.*, 1996, **54**, 143.
17. S. Zhang and A.R. Horrocks, *J. Appl. Polym. Sci.*, 2003, **90(12)**, 3165.
18. S. Bourbigot, M. Le Bras, S. Duquesne and M. Rochery, *Macromol. Eng. Mater.*, 2004, **289**, 499.
19. F. Chemin, Master Thesis, CNAM, Paris, France, 2001.
20. S. Bourbigot, M. Le Bras, J.M. Trémillon, P. Bréant and R. Delobel, *Fire Mater*, 1996, **20**, 145.
21. D. Scharf, R. Nalepa and T. Wusu, *Fire Safety J.*, 1992, **19(1)**, 103.
22. S.V. Levchik, G.F. Levchik, G. Camino and L. Costa, *J. Fire Sci.*, 1995, **13**, 43.
23. G.F. Levchik, S.V. Levchik, P.D. Sachok, A.F. Selevich, A.S. Lyakhov and A.I. Lesnikovich, *Thermochim. Acta*, 1995, **257(1–2)**, 117.
24. M. Jimenez, S. Duquesne and S. Bourbigot, unpublished results.
25. S. Bourbigot, E. Chemin and M. Le Bras, *Pli Soleau N° 38234* (5th March 1999).

CHAPTER 25

Fire Retardancy of Engineering Polymer Composites

PÉTER ANNA,[1] SZABOLCS MATKÓ,[1] GYÖRGY MAROSI,[1] GÁBOR NAGY,[2] XAVIER ALMÉRAS[3] AND MICHEL LE BRAS[3]

[1]Organic Chemical Technology Department, Budapest University of Technology and Economics, Müegyetem rkp.3, Budapest, H-1111, Hungary (gmarosi@mail.bme.hu)
[2]Polinvent Kft., Ady Endre út 59, H-1221 Budapest, Hungary
[3]Perf, Upres Ea 1040, Ensc-lille,Ustl, F-59650 Villeneuve d'Ascq Cédex, France

25.1 Introduction

Polypropylene (PP) in reinforced and flame retarded form is considered as an engineering thermoplastic material appropriate to use, for example, in transportation industry. It can be flame retarded with an ammonium polyphosphate (APP)–polyol intumescent flame retardant additive system efficiently;[1] however, these systems have some limitations, such as limit of processing temperature, sensitivity to humidity,[2] limited elongation.[3] To eliminate these limitations various trials were made to substitute the charring polyol compound with blocked derivatives of polyol,[4] or derivatives of N-containing heterocyclic compounds.[2] Recently, some heteroatom-containing oligomer[5] and polymer, actually polycaproamide (PA 6)[6,7] was suggested as charring component. PA 6 fulfils the charring function, but does not satisfy the melt rheological and mechanical requirements. This chapter demonstrates, through flammability, mechanical and rheological measurements, that the mentioned requirements can be influenced favourably further by incorporation of two-dimensional fillers, such as talc and layered silicate, and ethylene vinylacetate copolymer in an appropriate sequence.

A special inorganic–organic polymer hybrid resin, known as "3P resin", prepared basically from polyisocyanate and polysilicic acid[8] is already a commercial product. It has versatile applications, such as renovation of pipelines, especially underground sewers made of concrete or reinforced concrete,[9] making impermeable the substruction of engineering constructive works lying under ground[10] and special binding materials under wet conditions.[11] Its favourable characteristics are good adhesion, weather resistance, flame-resistancy, and low cost. Some engineering applications were limited by low flexural modulus and low strength, which could be improved by use of fiber reinforcement.[12] The incorporation of reinforcing fibres, however, reduces the flame retardancy. The influence of a surface-treated aluminium trihydrate on the flame ratardancy and the mechanical properties of this engineering thermosetting polymer was also studied to extend its application field to the transportation industry.

25.2 Experimental

25.2.1 Components of Polypropylene Compounds

Polypropylene (PP) homopolymer, Tipplen H 535 (TVK, Hungary); Talc (T) FINTALC M 15 (Suomen Takki, Finnland), average particle diameter: 15 μm, specific surface area: 7.2 $m^2\,g^{-1}$; specific gravity: 2.7 $g\,cm^{-3}$; Ethylene vinylacetate copolymer (EVA), IBUCELL K 100 (H.B. Fuller) vinyl acetate content 28%; Maleic anhydride grafted PP wax (PP-gMA), Licomont AR 504 (Clariant); Polybutene intercalated nanoclay (Nano.clay) was prepared by dispersing organophilic montmorillonite, Bentone SD-1 (Rheox Inc.), in toluene (10 g for 400 ml) at room temperature during a steady mixing, followed by addition of 5 g of polybutene (Hyvis 2000, Britsh Petrol) to the nano-clay dispersion and, finally, after 3 hours, removing the solvent by vacuum distillation.[13]

25.2.2 Components of 3P Composites

Methylene-diphenyl-diisocyanate oligomer mixture, MDI L 30 E1 Borsodchem Rt. Hungary); Na-water glass, Betol (Kemikal Építőanyagipari Vállalat, Hungary); Aluminium trihydrate (ATH), Alolt 8 (Ajka Timföld, Hungary); Basalt fibre (TOPLAN Tapolcai Bazaltgyapot Kft. Hungary); Aminosilane coupling agent, Silan GF 91 (Wacker-Chemie Hungary Kft.).

Surface treatment of ATH was performed as follows: a solution containing 20% amino silane coupling agent was prepared by use of ethanol solution containing 10% water. This solution was used for intensive wetting the ATH in a quantity that contained 1% coupling agent related to the ATH. After wetting, the system was agitated for a further 10 min, and then the solution was evaporated at 105°C.

25.2.3 Compounding of Thermoplastic Composites

Compounds were prepared by homogenising the components for 10 min in the mixing chamber 350 of a Brabender Plasti-Corder PL2000 with a rotor speed of 50 rpm at 235°C. Sheets ($100 \times 100 \times 3$ mm^3) were obtained by compression molding in a Collin P200E press at 170°C for PE and at 190°C for PP (pressure 3 MPa).

Rheological measurements were carried out by use of an AR 2000 Rheometer type (TA Instruments) thermal scanning rheometer in a parallel plate configuration. Samples of $25 \times 25 \times 2$ mm size were positioned between the plates, with a starting gap of 1.5 mm. The applied normal force was 0.6 N, the shear rate 0.008 s^{-1}. The heating rate was 15°C min^{-1}.

SEM images were produced by use of a 5500 LV type GEOL Instrument scanning electron microscope.

Flammability was characterised by LOI values, according to ASTMD 2863 standard, and by UL 94 tests.

25.3 Results and Discussion

25.3.1 Intumescent PP Compounds Containing PA 6 Charring Component and Talc as Melt Rheology Controller

In previous work the charring polyol component of intumescent PP (IPP) compounds was substituted by PA 6 polymer. Further improvement of flame retardancy, however, requires the application of a melt rheology-modifying component, such as talc. As well as the components, the way of incorporating talc also has a determining effect on the flammability and mechanical properties. Table 1 shows talc containing various formulations. The IPP composition is the reference compound prepared without talc. The talc-containing IPP/T composition was prepared in a single-step process in which all of the additives were fed into the PP melt simultaneously. The IPP/T* composition contains talc as well, but the preparation of the composition was performed in a double-step process, that is, in a first step the PA 6 and talc were added into the melted EVA. In a second step the other additives, the PP and APP were added to the components previously homogenised in the first step. The composition of IPP/T/M is equal to the IPP/T composition with the difference that it also contains PP-gMA. The relative values of the mechanical characteristics of compounds (measured absolute values related to the values of the compound containing no talc) are given in the Figure 1. The UL 94 ratings are given in the Table 1.

Table 1 *Composition of talc-reinforced intumescent PP compounds*

	IPP	*IPP/T*	*IPP/T*[a]	*IPP/T/M*
PP	60	47	47	46
APP	26.3	26.3	26.3	26.3
PA 6	8.7	8.7	8.7	8.7
EVA	5	5	5	5
Talc		13	13	13
PP-gMA				1
UL 94	HB	V0	V0	V0

Glossary: IPP = PP + APP + PA6 + EVA; PP/T = IPP + Talc; IPP/T/M = IPP-T + PP-gMA.
[a]Double-step preparation, 1. EVA + PA 6 + T, 2. Other components

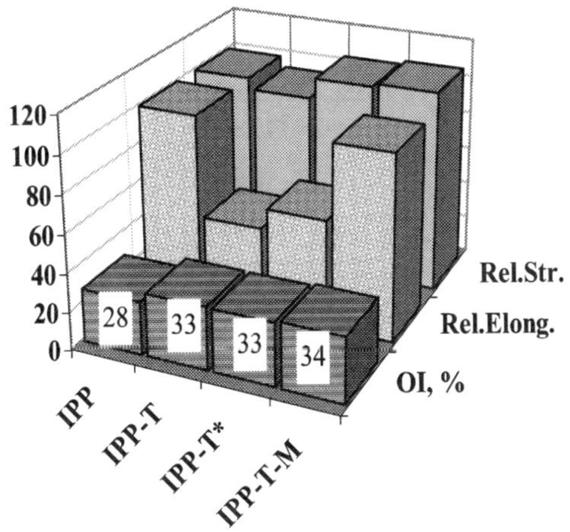

Figure 1 *Mechanical and flammability characteristics of intumescent PP compounds prepared with different compounding technologies*

The IPP compound containing no talc shows moderate flame-retardant character with 28% LOI value and a UL 94 rating of HB qualification. The compound has a high susceptibility for melt dripping. Incorporation of talc in the composition by the single-step method, in composition IPP/T, improves the flame retardancy, but the mechanical properties, mainly the elongation at break, fall considerably. Preparing the compound by the double-step method, in composition IPP/T*, the flammability is not influenced, the elongation at break increases and the strength improves, achieving a higher value than the compound without talc.

Figure 2 compares the apparent viscosity of the compounds, which means the melt viscosity at lower temperature, and the visco-elastic behaviour of the char at high temperature, as reported earlier.[14]

The viscosity of the melted compounds in the temperature range 180–270°C is strongly increased by the incorporation of the talc independently of the mode of compounding. This viscosity increase effect reduces the dripping susceptibility of the compounds during the ignition process. The presence of talc also increases the visco-elastic modulus of charred foam formed in the burning phase, improving the durability of the burning surface. Both effects manifest improved flame retardancy.

The improvement of mechanical properties of the compound prepared by the double-step process compared to the single-step one can be explained by different phase structures. It is more advantageous if a phase consisting of EVA, talc and PA 6 is dispersed in the PP + APP phase (double-step process) than the compounds prepared by the single-step mode, in which there is poor direct interaction between PP and talc.

The IPP/T/M compound containing talc and an adhesion promoter shows simultaneous improvement of flame-retardancy and mechanical properties. The

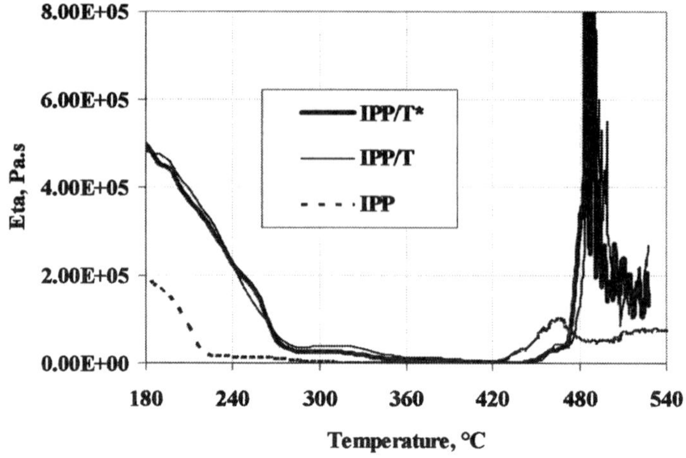

Figure 2 *Apparent viscosity of various IPP compounds measured by thermal scanning rheometer*

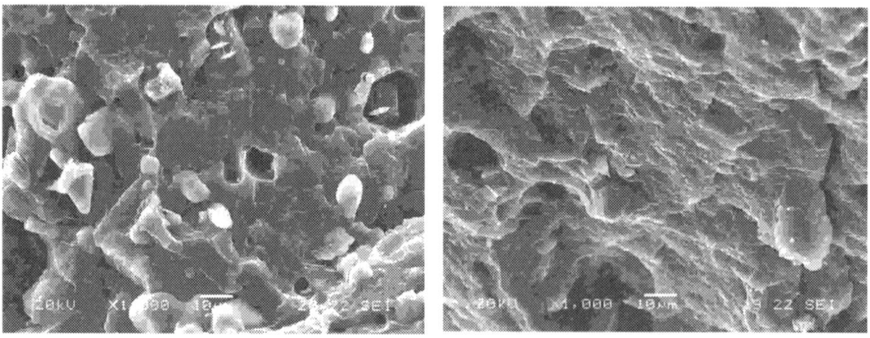

3/A IPP-T* **3/B IPP-T-M**

Figure 3 *SEM image of talc-reinforced intumescent PP compound prepared without and with interfacial additive*

favourable effect of talc on the flame-retardancy can be explained as discussed above. The improvement of the mechanical properties is a consequence of the improved interaction between the components, especially between the talc and PP matrix, as demonstrated in Figure 3.

The broken surface of the compound prepared without adhesion promoter (3/A in Figure 3) contains talc in the form of discrete particles, and holes remained in the place of the removed talc particles. In the presence of the additive (3/B in Figure 3), however, nearly perfect embedding of talc can be observed.

Table 2 *Nano-clay containing intumescent PP composites*

	IPP	*IPP+N*	*IPP+N*[a]*
PP	60	58	58
APP	26.25	26.25	26.25
PA 6	8.75	8.75	8.75
EVA24	5	5	5
Nano-clay		2	2
UL 94	HB	HB	V0

Glossary: IPP = PP + APP + PA 6; IPP-N = IPP + Nano clay. [a](Double-step preparation, 1. EVA + PA 6 + Nano clay, 2. Other components)

25.3.2 Intumescent PP Compounds Containing PA 6 Charring Component and Nano-Clay as Melt Rheology Controller

The influence of polymer intercalated nano-clay on the flammability characteristics of IPP was studied by comparing the characteristics of compositions given in Table 2. The composition of IPP is the same as given in Table 1. IPP + N consists of intumescent components and 2% polymer-intercalated nano-clay prepared by the single-step compounding mode, while the IPP + N* composition was prepared by the double-step mode. The LOI, relative strength and elongation are given in Figure 4. The UL 94 rating is given in Table 2.

Incorporation of the nano-clay in the IPP compound by the single mode increases the LOI from 28 to 32, but this is not enough to improve the UL 94 rating. Incorporation of nano-clay into the compound influences the relative strength to a negligible degree but the relative elongation is considerably altered. The elongation increase effect can be attributed to the elastomer content of the intercalated nano-clay.

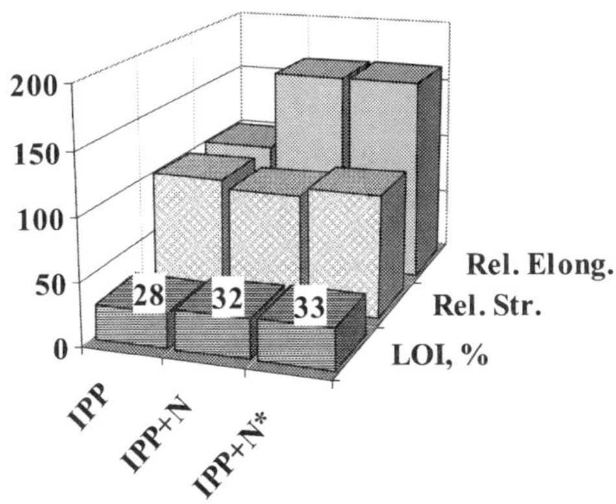

Figure 4 *Mechanical and flammability characteristics of nano-clay containing intumescent PP composites prepared by single- and double-step compounding*

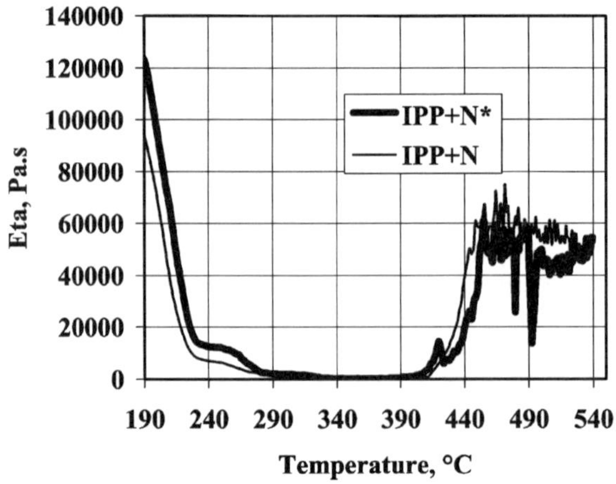

Figure 5 *Apparent viscosity of intumescent PP compounds containing nano-clay incorporated by single- and double-step modes*

Incorporation of the nano-clay by the double-step mode hardly affects the mechanical properties and the LOI, but a change can be observed in the UL 94 rating, as the compound achieves the V0 grade. This unexpected phenomenon can be explained on the basis of the rheological test of the compounds given in Figure 5.

Incorporation of the nano-clay by the double-step mode increases the melt viscosity of the compound in the temperature range 190–270°C considerably, reducing the melt dripping of the compound. This melt viscosity increase effect seems to be enough to improve the UL 94 rating. As the amount of applied nano-clay is small, the mechanical properties do not deteriorate, and the application of an adhesion promoter (PP-gMA) is not necessary.

One can conclude that good intumescent PP compounds can be formed from PP-PA 6 components if rheology controlling additives and double-step compounding technology are applied.

25.3.3 Flame Retarded and Basalt Fibre Reinforced Thermosetting Polymer (3P) Composites

Isocyanate-silicic acid based 3P resin is a fireproof material but the incorporation of reinforcing, such as basalt fibers, reduces the flame retardancy of the composite. The basalt fiber used in the experiments was a short fibre, prepared with the so-called Junkers technology (Figure 6). The melted basalt rock flows on the rapidly rotating spinning head, which spreads the thin, melt trips in the cooling air, where they solidify to short fibres.

The form of the primary fibres can be seen in Figure 7/A, and after some refining treatment in Figure 7/B. The refined fibre[15] was used for the experiments.

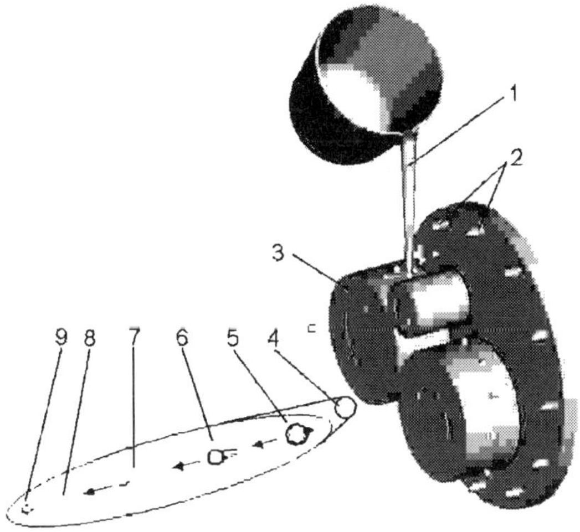

Figure 6 *Principle of preparation technology of basalt fibre*

Figure 7 *Primary basalt fibre* (A) *and refined basalt fibre* (B)

The 3P resin is a cross-linked system of methyldiphenyl isocyanate (MDI) oligomer and water-glass, with additives regulating the rate of reactions. The general reaction scheme of 3P resin formation is given in Figure 8. The 3P resin has many beneficial properties, such as good adhesion, weather fastness, abrasion proof character, low cost and excellent flame retardancy thanks to its very stable tri-isocyanurate main structural element.

The main drawback is its low flexural modulus at high temperature, even if it is applied with glass fibre reinforcement. It was an obvious idea to prepare 3P resin with basalt fibre reinforcement. Reduced flame resistancy, due to incorporation of basalt fibres, was attempted to improve using surface-treated aluminium trihydrate.

The composition of various 3P composites can be seen in Table 3. LOI values, the relative strength and elongation are given in Figure 9.

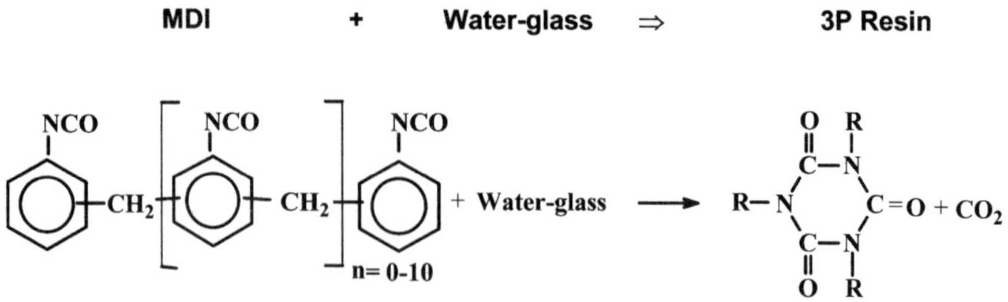

Figure 8 *Formation reaction of 3P resin*

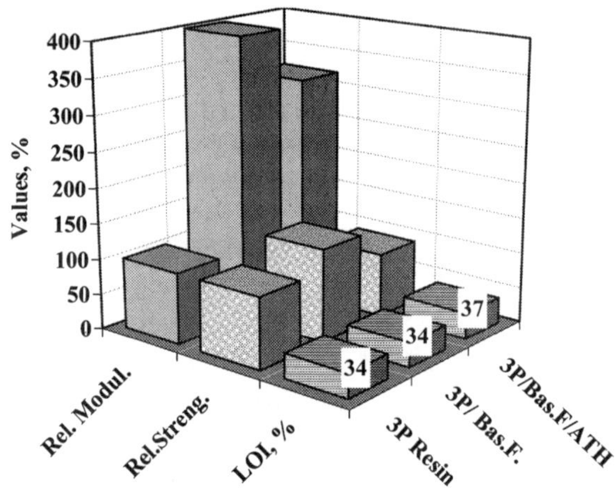

Figure 9 *Characteristics of modified 3P resins*

Table 3 *Composition of 3P resin compounds*

	3P resin	*3P/Bas.F*	*3P/Bas.F/ATH*
3P resin	100	70	40
Basalt fibre		30	30
ATH surf. tr.			30
UL 94	V0	V2	V0

The UL 94 rating is also given in Table 3. The 3P resin has excellent flame retardancy but poor mechanical properties. The addition of basalt fibre increased the flexural modulus radically and the strength moderately; however, the flame retardancy deteriorates (the composite burns too long after the second ignition in the UL 94 test). The incorporation of a moderate amount of surface-modified ATH restores the original good flame resistance, preserving the high modulus.

The appropriate combination of the known components of composite materials and efficient FR additives may result in contemporary new composite formulations meeting the requirements of the strict environment regulation, finding application as flame and sound protection covers in public transportation vehicles.

25.4 Conclusion

The influence of a micro- and a nano-sized fillers on the flammability and mechanical properties of polypropylene based thermoplastic and isocyanate-silicic acid based (3P) thermosetting polymer systems have been investigated. Intumescent flame retardant polypropylene compound can be prepared using ammonium polyphosphate as acid source, polyamide as char-forming component, ethylene vinylacetate copolymer and talc or thermoplastic polymer intercalated nano-clay as rheology controlling additive; however, the mechanical properties for some applications are not satisfactory in all respect. The application of a double-step compounding technology, that is, the homogenisation first of the ethylene vinylacetate, polyamide and the rheology modifying additives, followed by the dispersing of these blends is polypropylene-ammonium polyphosphate system, resulted in improved flame-retardancy and mechanical properties. Application of maleic anhydride grafted polypropylene wax in a talc-containing system brought improvement by the single-step compounding technology. The isocyanate-silicic acid system is a fireproof resin; however, the incorporated basalt fibrous reinforcing filler reduces the flame retardancy. Incorporation of an aminosilane surface treated aluminium trihydrate together with basalt fibre guarantees the flame retardancy and the improvement of mechanical properties simultaneously.

25.5 Acknowledgement

This work has been financially supported by the Ministry of Education Hungary through projects Széchenyi OM-00169/2001, 3A/0036/2002 and by the Hungarian Research Fund through project OTKA T026182. The scholarship from IKMA foundation is also acknowledged.

This work was also partially supported by the European project FLAMERET ("New Surface Modified Flame Retarded Polymeric Systems to Improve Safety in Transportation and Other Areas" registered under No. G5RD-CT-1999-00120).

25.6 References

1. G. Camino, L. Costa, L. Trossarelli and G. Landoni, *Polym. Degrad. Stab.*, 1984, **8**, 13.
2. I. Kouji and T. Ryoji, "Flame-retardant thermoplastic composition", *JP 154302.93; EP 0 627 460 A1*, 1994.

3. Gy. Bertalan, Gy. Marosi, P. Anna, L. Vigh, K. Szentirmai and G. Budai, "Flame-retardant, self-extinguishing polyolefine composition", *Hung. Pat. 209 135*, 1993.
4. M. Sicken and W. Wanczke, "Fire retardant polymer compositions with increased stability", *EP 0 584 567 A3* (02.03.1994).
5. P. Anna, Gy. Bertalan, Gy. Marosi, M. Márton and E. Zimonyi, "Flame retardant additive for polymer compounds with improved hydro-thermal stability", *German Patent Application, R4700/2003.*
6. M. Le Bras and S. Bourbigot, in "*Fire and Polymers, Materials and Solution for Hazard Prevention*", G.L. Nelson and C.A. Wilkie (eds.), ASC Symp. Ser. 797, Washington DC, 2001, pp. 136–149.
7. X. Almeras, M. Le Bras, P. Hornsby, S. Bourbigot, Gy. Marosi, S. Keszei and F. Poutch, *Polym. Degrad. Stab.*, 2003, **82**, 325–331.
8. G. Nagy, "Materials composites and foams made from polysilicic acid and polyisocyanate, and process for producing them", *WO 9221713.*
9. G. Nagy, Procedure for the repair and/or renovation of pipelines, especially underground and sewers made of concrete or reinforced concrete, *Aus. Pat. 2311900.*
10. G. Nagy, "Process for making impermeable the engineering constructive works laid under the ground by filling up method", *Hung. Pat. 71910.*
11. G. Nagy, "Polysilicic acid/polyisocyanate basic material materials, binding materials and foams and process for preparing them", *US Pat. 5622999.*
12. G. Nagy, "Structure for introducing a system of light fibres", *Pol. Pat. 329 435.*
13. A. Usuki, M. Kato, A. Okada and T. Kurauchi, *J. Appl. Polym. Sci.*, 1997, **63**, 137–139.
14. P. Anna, Gy. Marosi, I. Csontos, S. Bourbigot, M. Le Bras and R. Delobel, *Polym. Degrad. Stab.*, 2001, **74**, 423–426.
15. Sz. Matkó, P. Anna, Gy. Marosi, A. Szép, S. Keszei, T. Czigány and K. Pölöskei, *Macromol. Symp.*, 2003, **202**, 255–267.

CHAPTER 26

Flame Retardant Mechanisms Facilitating Safety in Transportation

GYÖRGY MAROSI,[1] SÁNDOR KESZEI,[1] ANDREA MÁRTON,[1] ANDREA SZÉP,[1] MICHEL LE BRAS,[2] RENÉ DELOBEL[2] AND PETER HORNSBY[3]

[1]Budapest University of Technology and Economics, Department of Organic Chemical Technology H-1111 Budapest, Mûegyetem rkp. 3, Hungary (gmarosi@mail.bme.hu)
[2]PERF, Ecole Nationale Superieure de Chimie Lille, USTL, F-59652 Villeneuve d'Ascq Cédex, France
[3]Brunel University Kingston Lane, Uxbridge, Middlesex, UB8 3PH United Kingdom

26.1 Introduction

Polymeric materials used in the field of transportation have increased by 120% during the last 20 years. The stock of buses and coaches in Europe is almost the double (~600,000) their number of the 1970s. This increase, however, has not been accompanied with a consideration of fire safety issues, because severe price-competition does not allow assigning additional budget for fire safety. Directive 95/28/EC of the European Parliament (24 October 1995) determined the requirement for the "Burning behavior of materials used in the interior construction of certain categories of motor vehicle" limiting the "Horizontal Burning Rate" to 60 mm min^{-1}, which is a very weak category and only slightly better than the 102 mm min^{-1} limit in the USA. The relevant standards are given in Table 1. The low level of fire safety requirements play a determining role in the increasing number of fatalities accompanying a fire.

Table 2 shows that in 2000 20% of all fires were vehicle fires in the USA, while more than 10% of fire death and property losses were in connection with vehicles. The increase compared to 1995 is also considerable.

Table 1 *Current standards regulating flammability of materials used in road transportation*

Standard No.	*Title*
ISO 3795	Road vehicles and tractors and machinery for agriculture and forestry – determination of burning behaviour of interior materials
DIN 75200	Bestimmung des Brennverhaltens von Werkstoffen der Kraftfahrzeuginnenausstattung
ASTM E-162	Surface Flammability of Materials Using a Radiant Heat Energy Source
FMVSS No.302	Flammability of Interior Materials – Passenger Cars, Multi-purpose Passenger Vehicles, Trucks and Buses
SAE J369	Flammability of Polymeric Interior Materials. Horizontal Test Method

Table 2 *Fire statistics*

	Fires (× 1000)	*Vehicle fires* (per 10^6 persons)	*Fire death* (× 1000)	*Fire-death in vehicles* (per 10^6 persons)	*Injuries* (× 1000)	*Property loss* (billion $)
World/year	6000–24000		166			400
USA-total/2000	1708		4		22	11
USA-vehicle fire/1995	300	1.1	0.37	1.4	1.83	0.61
USA-vehicle fire/2000	350	1.3	0.47	1.7	1.85	1.3
Germany-vehicle fires/year	40	0.5	0.04–0.08	0.5–1		
Switzerland-vehicle fires/year	7–9	1.1				
Austria-vehicle fires /year	3	0.4				

The interest of vehicle producers in advancing safety can be increased if other requirements of the used materials are improved together with the fire properties. Thus, parallel development is needed to solve the recyclability, acoustic, mechanical, economic, and flammability issues, which poses a complex challenge for material science. A wide scope is required as well when the various parts of vehicles are concerned. Fires originate mainly from the motor parts (40%) followed by mechanical, electrical and external reasons (26, 17 and 17% resp.). At present it takes only 5–10 min for the fire to spread from the engine to passenger's compartment.

Complex work, considering the above-mentioned aspect, has been carried out within a European fifth framework research program (FLAMERET) dedicated to developing FR materials for vehicles. The research work was performed in cooperation of the author's institutes, their industrial partners (MAL aluminum trihydroxide producing comp., PEMÜ plastic processing comp., RATI car accessories producing comp., RÁBA–Mór vehicle seat producing comp. in Hungary and Clariant in Germany), textile institutes (GEMTEX-France, Natural Fiber Institute of Poland), a quality control institute (EMI, Hungary) and user companies (IRISBUS and NABI bus producers in Hungary, Bombardier Transportation). The project focused on surface/interface modification considering an early patent of BUTE,[1] which applied the Buzágh–Ostwald principle[2] for the first time to polymer systems. (According to this principle the interphases in heterogeneous disperse systems should allow a continuous, harmonic transition between the phases.) However, in flame retarded (FR) polymers the phase structure to be modified and its temperature dependence is quite complex. Additionally to the modification of the interfaces around the dispersed particles, the surface (to be contacted with the fire), and in some cases sandwich layers, has to be altered as well. The aims are also more complex, including stabilization (fire, hydrolytic, or photooxidative), controlling the rheology, the charring, promoting the accumulation of FR additives on the surface at high temperature, formation of barrier layer and preserving the mechanical properties by reactive coupling of phases.

The polymers used in vehicles inside should not contain halogen, thus the phosphorus compounds and metal hydroxides are preferred as flame retardants. The formation of intumescent flame retardant systems combining acid source and charring components is one promising route.[3] Various such systems have been proposed but their positive effect is mostly accompanied with certain disadvantages, such as stability or processability problems.[4] Nanocomposite-type thermoplastic polymers belong to the most thoroughly studied materials in recent years. Mechanical properties and flame retardancy were improved due to introduction of nanoparticles at low concentration.[5] However, despite their promising effect in reducing the rate of heat release in most polymer systems, clay nanoparticles alone do not act as real fire retardant additives.[5,6] More significant advancement is expected when nanoparticles are combined with other flame retardant additives. Migration of the particles to the surface followed by formation of a protective layer is the potential mechanism that explains how the nanoparticles could contribute to the effect of other flame retardants.[7–9] Depending on the composition the combined systems show either a synergistic effect or less advantageous characteristics.[5,10] In intumescent systems, for example, the introduction of nanoparticles may cause a positive or negative change, depending on the composition, compounding technology and type of testing method. Understanding the reason behind these differences may contribute to knowledge about the mechanism of fire retardancy.

In this chapter materials developed for forming internal panels and noise insulating sheets for vehicles are discussed. These contain intumescent flame retardant (IFR) systems interacting with montmorillonite-type clay nanoparticles and $BaSO_4$ combined with metal hydroxides or IFR respectively. Based on structure–property relationship of nanocomposites two methods are proposed for promoting the formation of a protective layer on the surface. Further results of the above-mentioned project, including the multilayer injection molding, flame retarded engineering composites, and textiles, is discussed in other chapters. All the results are utilized for improving the performance of the materials used in vehicles.

26.2 Experimental

26.2.1 Materials

Polymer matrices were polypropylene (PP) of Tipplen H535 type, density 0.9 g cm^{-3}, melt index 4 g $(10\ min)^{-1}$ at 21.6 N, 230°C (product of TVK Co. Hungary) and a thermoplastic elastomer blend, called *PEMÜBEL*, consisting of EVA (vinyl-acetate content 27%), SBS (styrene content 48%), and PS, density 0.98 g cm^{-3}, melt index 29 g $(10\ min)^{-1}$ at 21.6 N, 190°C (product of PEMÜ Co. Hungary).

Two types of intumescent system were applied, one consisting of ammonium polyphosphate (APP, Exolit AP 422, Clariant, Germany) and polyol (POL, pentaerythritol, Aldrich) as char forming component, the other contained a reduced amount of APP and a phosphorylated polyol char-forming component. The method of phosphorylation has been published elsewhere.[11,12] The P content was the same in both systems. Montmorillonite nanoparticles (MMT) were applied using bentonite-based organoclay product, Bentone SD-1 (Rheox Inc.), which was intercalated[13] with the char-forming components of intumescent as follows: the char-forming components (either phosphorylated or not) were added into the diluted dispersion of Bentone SD-1 in toluene. Then the toluene was removed by distillation and drying. Ceramic precursors were aliphatic and aromatic polyboroxosiloxane (BSil) elastomers prepared in our laboratory according to earlier described process.[14,15] The siloxanes were compatibilized at their silanol end groups by grafting with on alkyl siloxane compatibilizing agent having a long hydrocarbon chain.

By varying the sequence of mixing and composition of the additives various types of additive system were formed as follows:

1. APP and polyol (pentaerythritol) applied in a ratio of 3 : 1 (IFR)
2. Mixture of APP and nanoparticles intercalated with polyol (APP + MMT/POL)
3. Components of intumescent system enclosed separately between the layers of nanoparticles (MMT/APP + MMT/POL)
4. Mixture of APP and nanoparticles intercalated with phosphorylated polyol (PPOL) component of intumescent system (APP + MMT/PPOL)
5. Ceramic precursor based on aliphatic BSil used as coating for APP and polyol
6. Ceramic precursor based on aromatic BSil used as coating for APP and polyol

The total amount of additives (30%), P content (21%), MMT and BSil contents (2%) were constant.

Aluminum trihydroxide (ATH) flame retardant filler was Alolt 60 FLS, average particle size 1.5 μm (product of MAL Co. Hungary). Barium sulphate ($BaSO_4$) filler white powder, average particle size 40 μm, molecular mass 233 g mol^{-1}, (product of KLÖCKNER GmbH, Austria).

26.2.2 Methods

Compounds were prepared by homogenisation of components in the mixing chamber WE350 of a Brabender Plasti-Corder PL2000 with rotor speed of 50 rpm, at 200°C. Sheets (120 × 120 × 1.8 mm) were obtained by compression moulding using a Collin P 200 E type laboratory press at 200°C and a pressure of 3 MPa. For micro-Raman measurements a Labram Raman Microscope system, produced by Jobin Yvon Horiba, was used at 632.81 nm excitation of a He/Ne laser. Thermogravimetry (TG) was performed using Setaram Labsys equipment, sample weight 10 mg, heating rate 7.5°C min^{-1} in a nitrogen atmosphere. Fire resistance was characterized with a Cone calorimeter (Stanton Redcroft, heat flux 50 kW m^{-2}), UL 94 test (according to ASTM 1356–90 and ANSI//ASTM D-635/77) and limiting oxygen index measurement (LOI, according to ASTMD 2863). X-ray photoelectron spectroscopic (XPS) characterisation of samples was performed with a Kratos XSAM 800 spectrometer using $MgK\alpha_{1,2}$ radiation. The spectra were referenced to the hydrocarbon-type carbon at binding energy BE = 285 eV. Bending loss factor measurements were performed on small, narrow bands with bending wave excitation. The samples were 300 mm long, 20 mm wide and 68 mm thick. Using this size the important acoustic frequency range can be covered. Samples were exited by an electrodynamic exciter. The sample was fixed to the base steel band of the exciter by a thin layer of adhesive. Based on measurement of eigenfrequencies of samples and the bending wave the loss factor results were calculated using the following equation:

$$\xi(f) = 1/(2\Pi f_i \tau)$$

where $\xi(f)$ is the frequency function of bending loss factor, f_i the i-th eigenfrequency of the sample and τ the time constant of vibration ring off in sample [time of (–8.7) dB drop of vibration amplitude].

26.3 Results and Discussion

26.3.1 Development of Nanocomposites for Forming Internal Panels

To meet the recyclability criteria polyolefin matrix was selected for all the internal vehicle elements to be developed. The flammability of internal panels of vehicles were to be decreased using an intumescent flame retardant (IFR) system. We decided to combine it, based on recent cone calorimeter results,[16] with nanoparticles to improve the efficiency and processability. Most papers

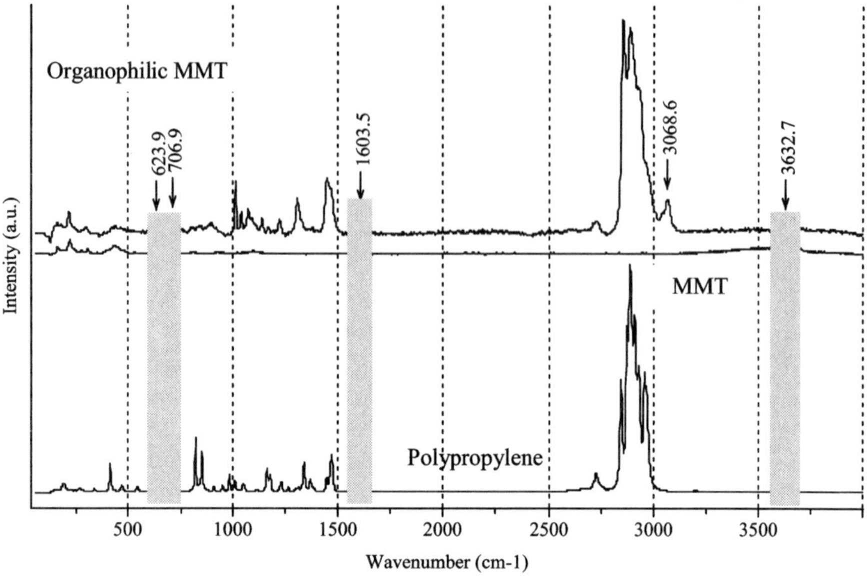

Figure 1 *Raman spectra of pure and organophilic montmorillonite compared with PP*

published on the reduction of heat release rate (RHR) explain this effect with the formation of a protecting layer of nanoparticles on the surface. The driving force for the migration of nanoparticles to the surface is, however, not clarified yet. To get a better insight into this mechanism, before applying nanocomposites in the vehicles, a series of model systems of different structure were formed (as given in the Experimental part).

Differences in the level of homogeneity of the samples may affect the comparison. A Raman microscope was used to check the homogeneity of dispersion of nanoparticles and other components in the system. Raman spectra of pure and organophilic MMT are given in Figure 1. As the strongest peak of organophilic MMT overlap with peaks of PP the characteristic bands at wavenumbers 700, 1600 and 3600 could be used for identification. The structure was not exfoliated, but the dispersion was equally homogeneous for all samples.

Screening of the flame retardant effect of systems with similar composition but with a wide variety of structures was made using LOI and UL94 tests. The results are given in Figure 2.

Based on the results one can conclude that the combination of nanoparticles with an intumescent flame retardant system affects the performance significantly. Separate modification of the acid source (APP) and the charring component (POL) with nanoparticles leads to a decline of flame retardant characteristics. In this system the nanoparticles probably separate the components, thereby hindering the charring process. The same nanoparticles, however, may cause considerable improvement, especially when the nanoparticles are intercalated with P-containing polyol (which is in fact a complete intumescent system). Two

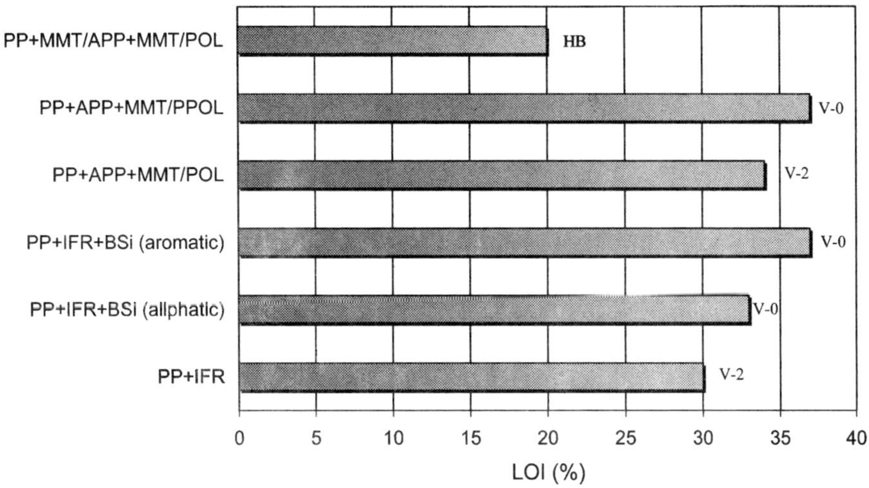

Figure 2 *Comparison of flame retardancy of various model systems (compositions of samples are given in the Experimental part)*

types of boroxosiloxanes (BSil) proved to be effective synergists as well. The results suggest new mechanisms for improving fire retardancy.

26.3.2 New Mechanisms for Delivering FR Components to the Surface

1. Comparing the TG curves of nanoparticles intercalated with phosphorylated and non-phosphorylated polyol in Figure 3, shows much earlier decomposition in the former case. This means that with phosphorylated polyol the action of flame retardant starts with gas formation between layers of MMT, thereby separating the layers, and the formed gas bubbles tend to move out from the bulk, driving the nanoparticles to the surface. The polymer matrix is protected this way in a very early phase. The intercalated system, due to similarities with the mechanism of expandable graphite, can be called an *expandable nanocomposite (ENC)*. This mechanism seems to result in the best flame retardant efficiency among the compared nanocomposites.
2. There are other two samples in Figure 2 that reached the V0 level. These contain BSil additive that is, according to recent papers, a ceramic precursor, forming a durable protective (barrier) layer on the surface of flame-damaged material.[17] It was pointed out earlier that such additives form an interlayer around APP particles and promote, in the case of fire, its migration to the surface. This migration may occur, however, even at lower temperature if the adhesion energy between the interlayer and polyolefin matrix is lower than the cohesion energy ($W_a<W_c$) causing phase separation. In contrast a compatible interlayer (when $W_a>W_c$)

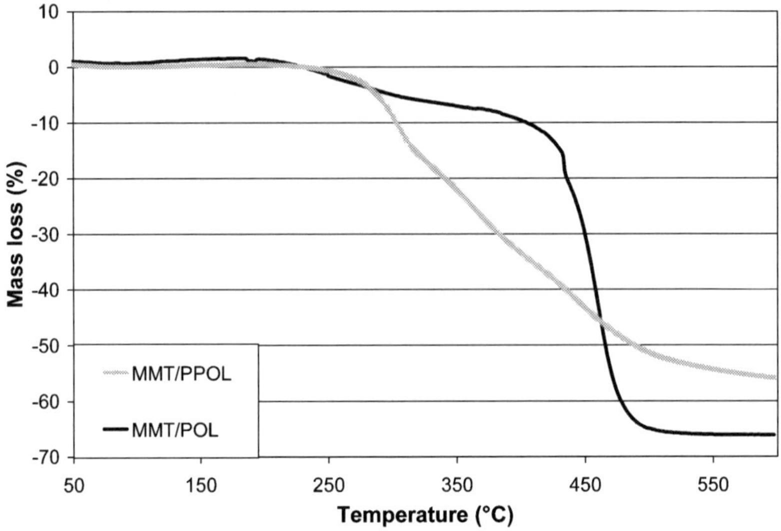

Figure 3 *Weight loss–temperature plots of nanoparticles intercalated with polyol and with phosphorylated polyol*

promotes the homogeneity, but presents no driving force for migration at high temperature. The solution was the grafting of a surfactant type molecule into the interlayer (as described in the Experimental part), which acts as a compatibilizing segment under common processing circumstances but decomposes at the temperature of fire, initiating the accumulation of FR additives on the surface. Such an interlayer is defined as an '*adaptive interphase*' *(AIP)* that is able to react to the change of its environment.

Investigation of the three V0 samples in a cone calorimeter gave a chance to compare the two ways for formation of barrier layers. The cone calorimeter plots are given in Figure 4, where the unmodified intumescent system is depicted as reference.

Nanoparticles and both types of the BSil caused a considerable decrease and delay in RHR peaks compared to the simple APP-polyol system. Interestingly, to see the ignition range of the curve of nanocomposite sample (APP + MMT/PPOL) starts with needle-like peak, after which the RHR value becomes zero for almost a one minute period. This supports the above-described mechanism about the formation of a barrier layer at the beginning of fire action. Comparing the three samples the run of the RHR curve of nanocomposite sample is similar to aliphatic BSil, while the aromatic ceramic precursor gives somewhat better results. The aromatic and aliphatic ceramic precursors differ due to the higher thermal and mechanical stability of the surface layer formed in the presence of the aromatic one.

The surface composition of the FR polymers in the presence of MMT and aromatic BSil was detected by XPS (Figure 5).

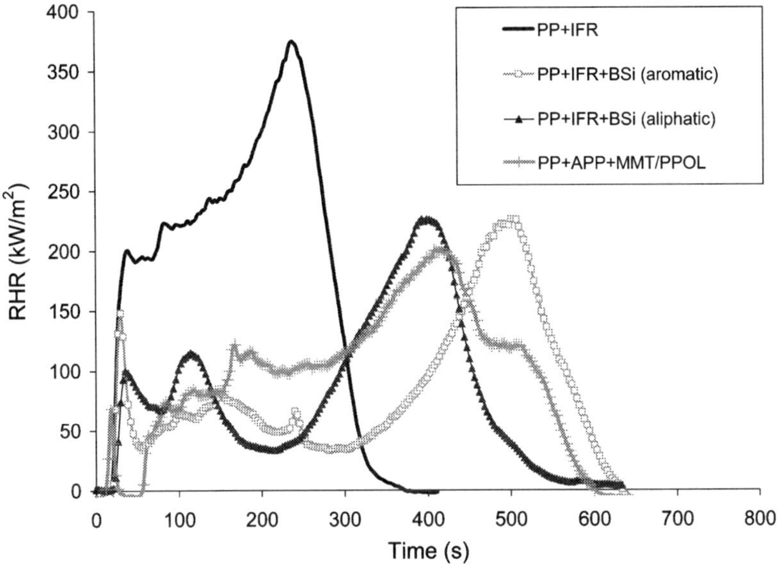

Figure 4 *RHR curves of IFR-PP systems containing nanoparticles intercalated with phosphorylated polyol and ceramic precursors (BSil) compared to unmodified IFR-PP*

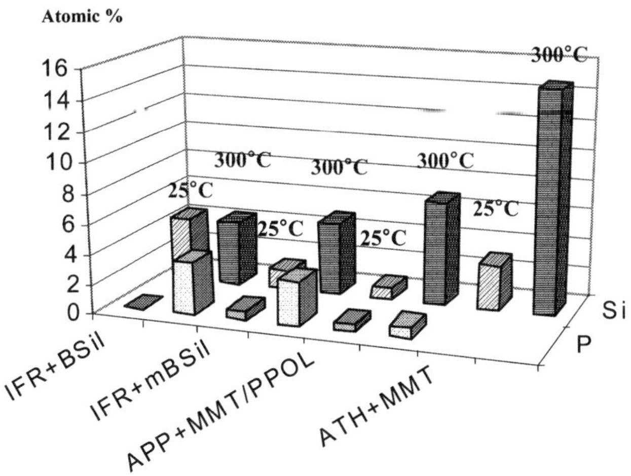

Figure 5 *Concentration of silicon and phosphorus atoms on the surface of various flame retarded samples at ambient and elevated temperatures, determined by XPS analysis (compositions of samples are given in the Experimental part)*

The data show that the P concentration on the surface of the BSil-containing IFR system at 25°C is about the detectable limit, but the amount of Si atoms is considerably higher, suggesting undesirable migration at low temperature. Compatibilized interlayer (mBSil) and phosphorylated polyol intercalated

Table 3 *Properties of vehicle panels produced on industrial scale*

Property	*Unit*	*Value*	*Method*
MFI (230°C/2.16 kg)	g (10 min)$^{-1}$	2.31	ISO 1133
MFI (190°C/5 kg)		3.75	
Density	g cm^{-3}	0.98	ISO 1183
E-modulus	MPa	2600	ISO 527
Izod impact strength notched 23°C	kJ m^{-2}	5.5	ISO 180
Izod impact strength notched −20°C	kJ m^{-2}	3.4	ISO 180
Flammability		V0	UL 94
Oxygen index (LOI)	%	34	ASTMD 2863

MMT promote the additive migration to the surface only at 300°C, which is optimal for FR polyolefins. Further worth is needed to explain the high increase of Si concentration with the ATH + MMT system at 300°C.

The properties of sheets produced on an industrial scale, based on an the described experiments, are given in Table 3.

26.3.3 Development of Flame Retarded Noise Insulating Sheets

Another challenging task was trying to meet both the acoustic and flame retardancy requirements of noise insulating sheets. Good insulation of sound is achieved generally by introduction of 60% $BaSO_4$ into the polymers. Another 60% additive is generally necessary to achieve good flame retardancy with metal hydroxides, which were altogether 120%. To resolve this paradox a series of samples were prepared with 60% filler content consisting of $BaSO_4$ and ATH of varied ratio in an EVA/SBS/PS blend (PEMÜBEL) matrix.

Figure 6 shows the change of FR performance and density against the ratio of the two fillers. Of course the flame resistance decreases and the density increases upon replacing ATH with $BaSO_4$ but there is a limit composition (36% ATH and 24% $BaSO_4$) up to which the V0 level is preserved. The mechanical properties are given in Figure 7. The tensile strength and elasticity modulus gradually decrease as the $BaSO_4$ content decreases on the expense of ATH, while the elongation at break changes in the opposite direction.

To predict the acoustic characteristics of the filler mixture containing polymers their dynamic mechanical behavior were measured. The magnitude of damping (loss factor) determined this way is proportional to the noise insulating capacity. Selecting the V0 composition containing the largest amount of $BaSO_4$ its loss factor (damping) was determined over a wide range of frequencies. The reference material was the 60% $BaSO_4$ containing non-flame retarded compound. A system containing 50–10% $BaSO_4$ and IFR was also examined as comparison. The results are given in Figure 8.

Surprisingly, the flame retardant additives assist $BaSO_4$ in absorbing the vibration energy very efficiently. The run of the loss factor curves of FR materials is similar to the reference curve or even better in some frequency

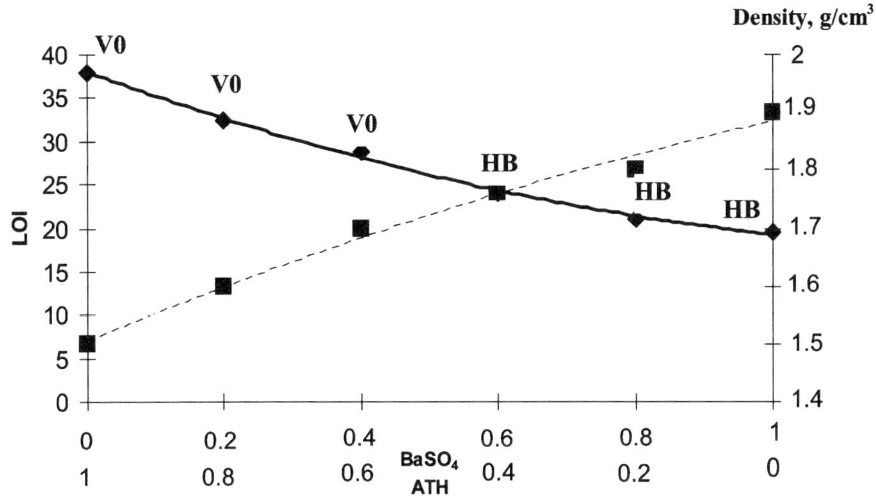

Figure 6 *Change in limiting oxygen index (LOI) and density vs. share of $BaSO_4$ and ATH within the 60% filler content of an elastomeric compound prepared for producing noise insulating sheets (—— LOI, ---- density)*

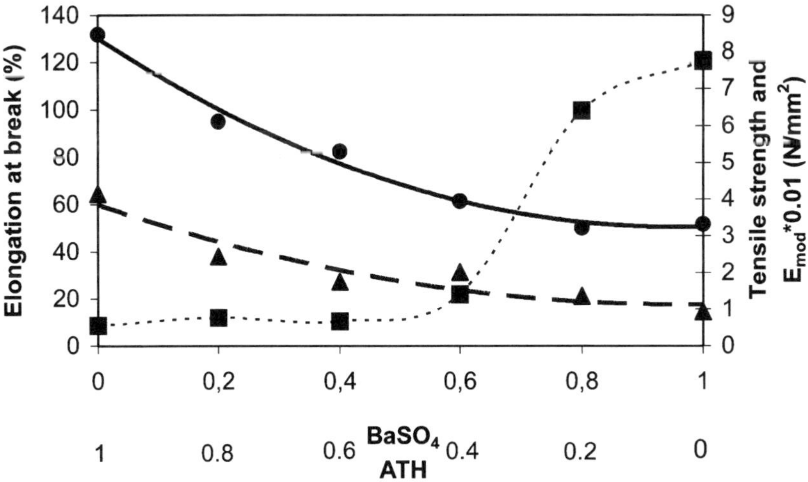

Figure 7 *Change of tensile strength (•), elasticity modulus (E_{mod}) (▲) and elongation (■) at break against the share of $BaSO_4$ and ATH within the 60% filler content of elastomeric compound prepared for producing noise insulating sheets (—— tensile strength, – – – E_{mod}, --- elongation at break)*

ranges. Especially, the ATH-containing material is interesting, showing a large maximum in the 200–800 Hz frequency range.

Table 4 gives the properties of noise insulating sheets produced on an industrial scale based on the described experiments.

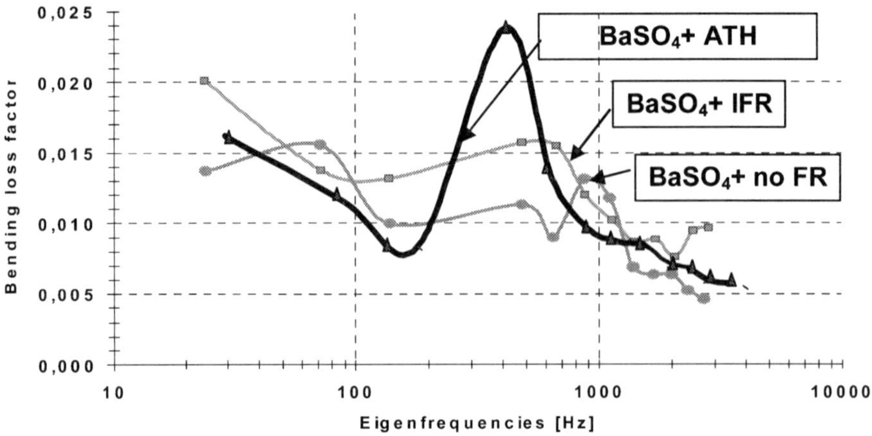

Figure 8 *Frequency function of loss factor of elastomeric compound (PEMUBEL, see in Experimental part) containing different filler mixtures (60% total amount)*

Table 4 *Properties of noise insulating sheets produced on an industrial scale*

Property	*Unit*	*Value*	*Method*
Oxygen index (LOI)	%	30	ASTMD 2863
Flammability		V0	UL 94
Horizontal burning rate	mm min^{-1}	0	ISO 3795
Density	g cm^{-3}	1.52	ISO 1183
Tensile strength	N mm^{-2}	8.48	ASTM 638
Elongation at break	%	8.53	ASTM 638
Tensile modulus	N mm^{-2}	413	ISO 527

26.4 Conclusions

Materials to be used in the vehicle industry should meet a complex set of requirements. Combined additive systems are used for this purpose. The efficiency of combined flame retardant/filler systems depends on the components, composition and structure. To produce internal panels on an industrial scale, intumescent flame retardants were combined with montmorillonite clay mineral. Considerable improvement was found when P and polyol (for example phosphorylated polyol) were intercalated together between the nanolayers of the clay, while isolation of the components of an intumescent additive system from each other by nanolayers resulted in decreased performance. Intercalated intumescent additives initiate an early formation of gases between the nanolayers, separating and driving them to the surface. Consequently, there is no need for exfoliation at the temperature of processing in the case of flame retarded nanocomposites – a quick flame initiated exfoliation can be efficient. The protective effect achieved this way is comparable with the effect of compatibilized boroxosiloxane ceramic precursors, which tend to accumulate on the surface upon heat treatment.

Aromatic boroxosiloxane elastomer proved to be a better ceramic precursor than the aliphatic one.

To develop noise insulating sheets of high flame retardancy level, $BaSO_4$ filler was combined with metal hydroxide and an intumescent-type fire retardant in an elastomeric matrix. Both combinations resulted in a good balance of acoustic and stability characteristics because, according to the dynamic mechanical analysis, the flame retardant additives participate in the damping process.

26.5 Acknowledgement

This work has been financially supported by the Ministry of Education Hungary through projects Széchényi OM-00169/2001, 3A/0036/2002 and by the Hungarian Research Fund through project OTKA T026182. The scholarship from the IKMA foundation is also acknowledged.

26.6 References

1. Gy. Bertalan, I. Rusznák, L. Trezl, A. Huszar and G. Szekely, *Hungarian Pat. 167063* (1975); *US Pat. 4116897* (1978); *German Pat. 2453491* (1986).
2. A. Buzágh, *Kolloid Z.*, 1952, **125**, 14–21.
3. R. Delobel, M. Le Bras, N. Ouassou and F. Alistiqsa, *J. Fire Sci.*, 1990, **8(2)**, 85.
4. Gy. Marosi, I. Ravadits, Gy. Bertalan, P. Anna, M.A. Maatoug, A. Tóth and M.D. Tran, in *Fire Retardancy of Polymers: The Use of Intumescence*, M. Le Bras, G. Camino, S. Bourbigot and R. Delobel (eds.), The Royal Society of Chemistry, Cambridge, UK, 1998, pp. 325.
5. C.A. Wilkie, "Fire retardancy in polymer-clay nanocomposites", in *Recent Advances in Flame Retardancy of Polymeric Materials, XIII*, M. Lewin (ed.), BCC Publ., Norwalk, USA, 2002.
6. J. Zhu, F. Uhl and C.A. Wilkie, in *Fire and Polymers; Materials and Solutions for Hazard Prevention*, ACS Symp. Ser. 797. G.L. Nelson and C.A. Wilkie (eds.), ACS, Washington DC, USA, 2001, 24–33.
7. J.W. Gilman, T. Kashiwagi and J.D. Lichtenhan, *SAMPE J.*, 1997, **33(4)**, 40.
8. J.W. Gilman, T. Kashiwagi, S. Lomakin, J.D. Lichtenhan, P. Jones, E.P. Giannelis and E. Manias, in *Fire Retardancy of Polymers: The Use of Intumescence*, M. Le Bras, G. Camino, S. Bourbigot and R. Delobel (eds.), The Royal Chemical Society Pub., Cambridge, UK, 1998, pp. 203–221.
9. J.W. Gilman, A. Morgan, E.P. Giannelis, M. Wuthenov and E. Manias, "Flammability and thermal stability studies of polymer layered-silicate nanocomposites II", *Conference on Recent Advances in Flame Retardancy of Polymeric Materials*, M. Lewin (ed.), BCC, Stamford, USA, 2000.

10. M. Le Bras and S. Bourbigot, in *Fire and Polymers; Materials and Solutions for Hazard Prevention*, ACS Symp. Ser. 797, G.L. Nelson and C.A. Wilkie (eds.), ACS, Washington DC, 2001, pp. 136–149,.
11. Gy. Marosi, A. Toldy, Gy. Parlagh, Z. Nagy, K. Ludányi, P. Anna and Gy. Keglevich, *Heteroatom Chem.*, 2002, **13(2)**, 126–130.
12. J.E. Telschow, *Phosphorus, Sulfur and Silicon*, 1999, **144–146**, 33–36.
13. Gy. Marosi, P. Anna, A. Márton, Gy. Bertalan, A. Bóta, A. Tóth, M. Mohai and I. Rácz, *Polym. Adv. Technol.*, 2002, **13**, 1–9.
14. Gy. Bertalan, Gy. Marosi, P. Anna and L. Víg, "Process for production of flame retardant polyolefin compounds", *Hungarian patent 209 135/93* (1993).
15. Gy. Marosi, P. Anna, I. Csontos, I. Ravadits and A. Márton, *Macromol. Symp.*, 2001, **176**, 198.
16. S. Bourbigot, M. Le Bras, F. Dabrowski, J.W. Gilman and T. Kashiwagi, *Fire Mater.*, 2000, **24(4)**, 201–208
17. Gy. Marosi, A. Márton, P. Anna, Gy. Bertalan, B. Marosfõi and A. Szép, *Polym. Degrad. Stab.*, 2002, **77**, 259–265.

Effect of the Addition of Mineral Fillers and Additives on the Toxicity of Fire Effluents from Polymers

CHAPTER 27

Comparison of the Degradation Products of Polyurethane and Polyurethane–Organophilic Clay Nanocomposite–A Toxicological Approach

GENNADY E. ZAIKOV,[1] SERGEI M. LOMAKIN[1] AND ROMAN A. SHEPTALIN[2]

[1]Institute of Biochemical Physics of Russian Academy of Sciences, 4 Kosygin Street, Moscow, 119991, Russia, (chembio@sky.chph.ras.ru)
[2]D.I. Mendeleyev Russian Chemical-Technological University, Moscow, Russia

In Chapter 10 we showed that addition of a low level of organic clay in polyurethane (PU)- based nanocomposite reduces both the mass loss rate under an external heat flux and the flammability of PU. This chapter will evaluate the effect of this addition on the evolution of the gaseous products of the degradation of PU and more precisely that of toxical agents from smoke.

27.1 Ecological Issue of Isocyanates and Pyrolysis of Polyurethane Nanocomposite

Amines and isocyanates are used in the production of polyurethane. Exposure to isocyanates is associated with respiratory disorders and may occur during the production or processing of PU. During thermal degradation of PU, amines and aminoisocyanates are formed in addition to isocyanates. Some isocyanates and several aromatic amines are sensitizers and listed as carcinogens. Isocyanates are a group of reactive compounds that are widely used in industry, mainly in the production of polyurethane. The industrial introduction of PU started in the middle of the 20th century and, subsequently, reports on isocyanate-induced occupational diseases started to appear.[1,2] Today, many people are still affected by isocyanate exposure, despite legislation and safety precautions.[3,4]

The most common isocyanates are the aromatic difunctional isocyanates toluene diisocyanate (TDI) and methylenediphenyl diisocyanate (MDI) together with polymeric MDI (pMDI), which is a mixture of MDI oligomer analogs. They account for more than 90% of the total world production.[5] The major use of TDI is for the production of flexible PU foam, elastomers and coatings, while pMDI is used for the production of rigid PU foam and as binder resin for reconstituted wood products or foundry cores. MDI is mainly used for the manufacturing of PUR elastomers. The more expensive aliphatic diisocyanates are mainly employed for the production of color-stable PU for coatings and elastomers.

To reduce exposure levels for isocyanates with high vapor pressure, such as hexamethylene diisocyanate (HDI), they are often used as prepolymers or adducts such as biuret, allofanat and isocyanurate adducts.

27.2 Occupational Exposure

Occupational exposure to isocyanates occurs mainly during the production or processing of PU. Numerous isocyanates are commercially available and technical grade qualities, which contain complex mixtures of isocyanate isomers and analogous, are often used in the production of PU. The different compounds released when processing PU (*e.g.* thermal degradation) are even more complex, and many unknown isocyanates and other compounds, such as amines and aminoisocyanates, are formed (Scheme 1).

During the production of PU-foam, volatile isocyanates may be emitted to the atmosphere and the exothermic reaction will increase the emission. Exposure to TDI is mainly described during the manufacturing of flexible PU-foam.[6,7] Exposure to MDI during production of rigid PUR foam has also

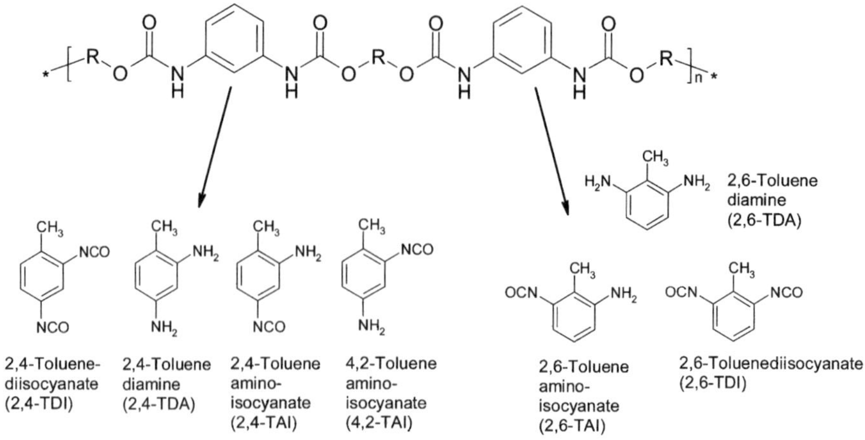

Scheme 1

been described Kaaria *et al.*[7] Due to the low volatility of MDI, the air concentrations when compared with TDI are much lower, during production of flexible foam. However, when isocyanates with low vapor pressure, such as MDI or prepolymerised HDI, are used, *e.g.* in spraying applications, high concentrations of isocyanates can be found in the working atmosphere.[8,9]

The urethane bonding in a PU polymer will start to dissociate at temperatures above 150–200°C.[10] During thermal degradation of PU, exposure to diisocyanates may occur in both the gas phase and the particle phase.[11] Exposure to thermal degradation products of PUR has been reported during work operations such as processing of PU-coated metal sheets in car repair shops[12] and flame lamination of PU with textiles.[6]

27.3 Health Effects

Exposure to isocyanates is irritating to the respiratory tract and some symptoms from occupationally exposed workers are cough, rhinitis and chest tightness.[4] Exposure to isocyanates is the most common cause of occupational asthma.[13] Even if the risk of developing asthma is greater at high exposure to isocyanates, asthma can occur at relatively low concentrations.[14]

Methyl isocyanate (MIC) has been reported to be irritating on the respiratory tract as well as to the skin and eyes. MIC has been shown to be a reproductive toxicant, neurotoxicant and to give rise to systemic effects.[15] Most studies have been performed on the victims of the tragic disaster in Bhopal in India. Information regarding occupational exposure to MIC is insufficient. TDI and MDI are suspected carcinogens based on animal studies, but only TDI is classified as a possible carcinogen to humans. Aliphatic amines, such as hexamethylene diamine (HDA) are classified as moderately toxic.[16]

27.3.1 GC-MS Pyrolysis

The additional aim of this study was to investigate the formation of the suspected carcinogen toluenediamine (TDA), toluene diisocyanate (TDI) and toluene isocyanate (TI) as a degradation product of PU and PU-OM nanocomposite.

Figures 1 and 2 show GC-MS analyses PU at 250 and 500°C in air. Peaks were assigned to the products of degradation by comparison of the mass spectrum with data from the Wiley275 spectral databases (Table 1). No detectable amounts of TDA and TDI could be found in extracts of both polyurethane compositions (PU and PU-OM 10% nanocomposite). By contrast, two isomers of toluene isocyanate TI: 1-isocyanato-2-methylbenzene (TI2), 1-isocyanato-4-methylbenzene (TI4) were found in extracts from both polyurethane compositions pyrolysis at 500°C.

Figure 3 shows mass spectra of TI2 (a) and TI4 (b). These findings confirm that the formation of TI2 and TI4 in, and GC-MS analysis of, polyurethane samples is 40% higher for PU pyrolysis than for PU-OM 10% (Table 1).

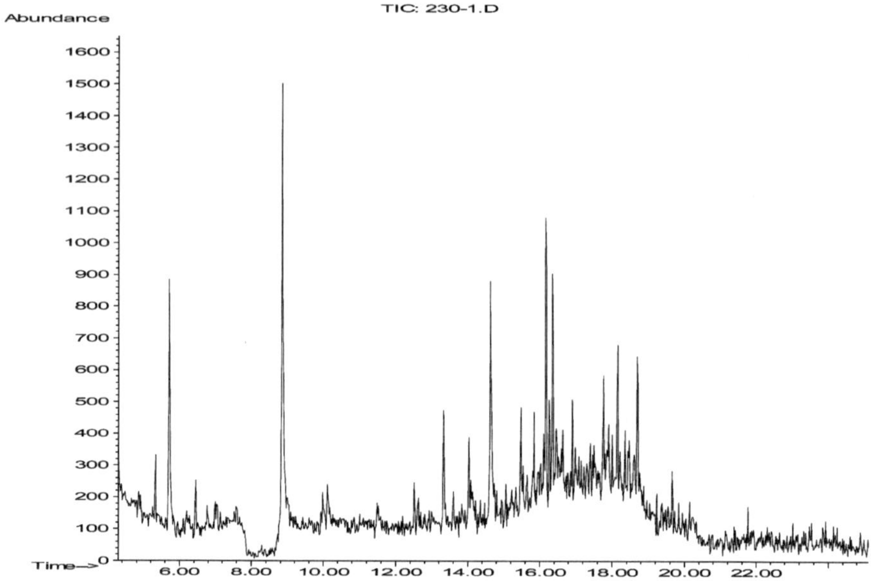

Figure 1 *CG analysis of PU degradation products at 250°C*

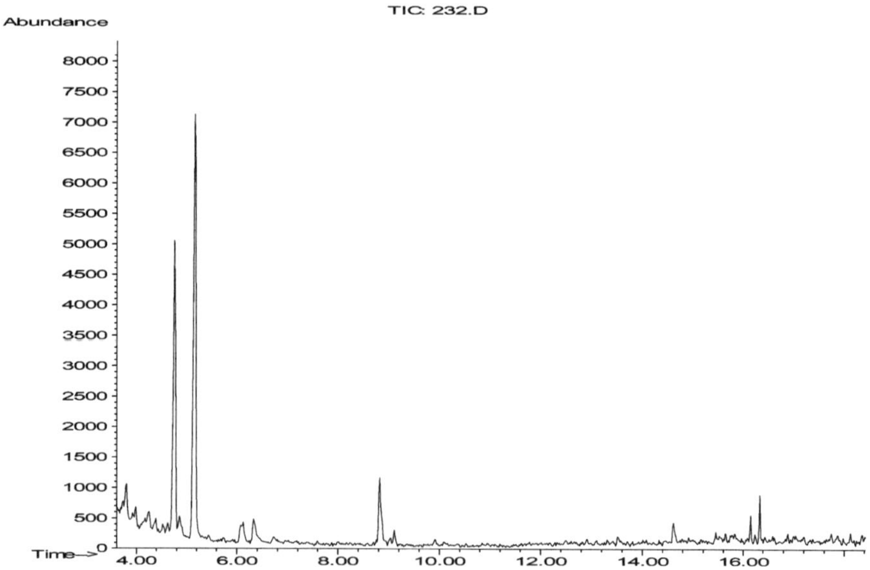

Figure 2 *CG analysis of PU degradation products at 500°C*

Table 1 *Products of thermal degradation of PU compositions*

Products	*Retention time* (min)	*Pyrolysis at* 250°C PPU (%)	*Pyrolysis at* 250°C PPU-OM 10% (%)	*Pyrolysis at* 500°C PPU (%)	*Pyrolysis at* 500°C PPU-OM 10% (%)
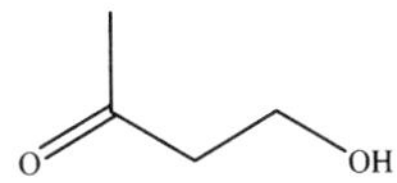 4-Hydroxybutan-2-one	4:15	2.3	4.9	1.2	3.2
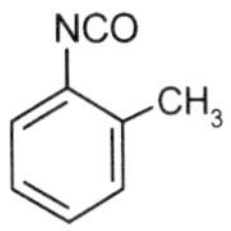 **1-Isocyanato-2-methylbenzene (γI_2)**	**4:44**	**0.0**	**0.0**	**27.0**	**18.2**
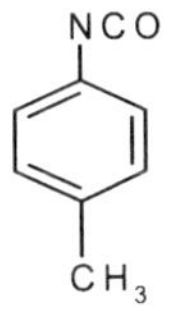 **1-Isocyanato-4-methylbenzene (γI_4)**	**5:09**	**0.0**	**0.0**	**44.0**	**27.2**
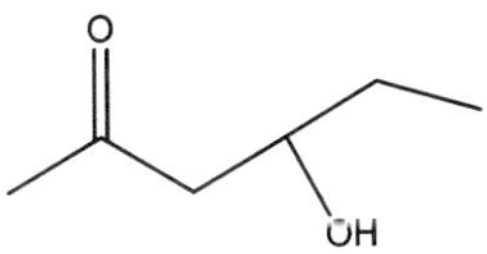 4-Hydroxyhexan-2-one	5:19	1.4	0.0	0.0	0.0
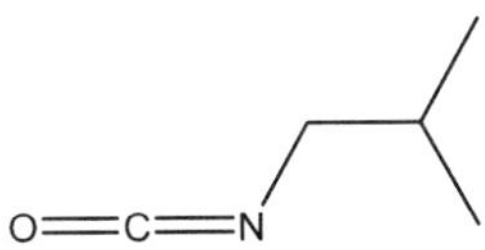 Isobutyl isocyanate	**5:68**	**5.3**	**1.9**	**0.0**	**0.0**
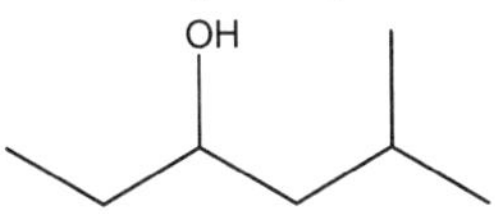 5-Methyl-3-hexanol	6:10	0.8	1.8	2.7	2.7
meta-Xylene Dioxolan	6:14	0.3	0.7	0.0	0.0
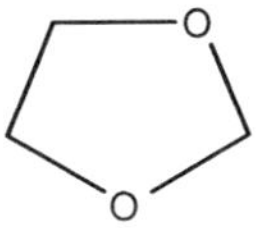	6:21	0.5	12.4	3.5	0.0
Styrene	7:02	0.8	0.0	0.0	0.0
2-Acetoxy-1-propanol	8:52	11.4	31.6	7.2	3.5

Table 1 *Continued*

Products	*Retention time* (min)	*Pyrolysis at 250°C* PPU (%)	*Pyrolysis at 250°C* PPU-OM 10% (%)	*Pyrolysis at 500°C* PPU (%)	*Pyrolysis at 500°C* PPU-OM 10% (%)
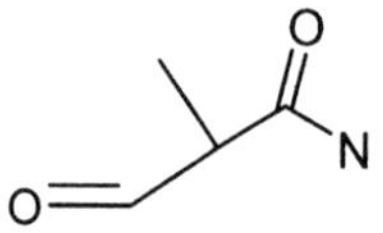					
2-Methyl-3-oxo-propionamide	9:06	0.0	0.0	1.4	1.7
Ethyl, methyl-benzene	9:56	1.6	0.0	0.6	1.7
Indene	11:28	1.1	0.0	0.0	0.0
Hexane-1-propoxy	12:37	1.2	1.2	0.0	0.0
2-Phenylbutadiene	14:00	2.5	0.0	0.0	0.0
Benzene, cyclopropyledene methyl	14:06	2.2	0.0	0.0	0.0
Naphthalene	14:37	9.1	4.3	1.7	5.3
	15:28	1.7	1.2	0.6	2.6
Aliphatics	15:32	0.7	0.0	0.4	1.0
7-Octen-2-one	15:50	0.9	1.7	0.0	0.7
Naphthalene-1-methyl	16:07	4.7	0.9	0.0	1.0
Naphthalene-2-methyl	16:10	6.4	3.5	1.4	1.1
Aliphatics	16:16	1.8	0.6	0.7	3.2
Naphthalene-methyl, ethyl	16:22	5.6	4.9	2.1	1.0
Aliphatics	16:26	3.0	2.4	0.3	3.3
	16:31	2.4	1.0	0.0	4.0
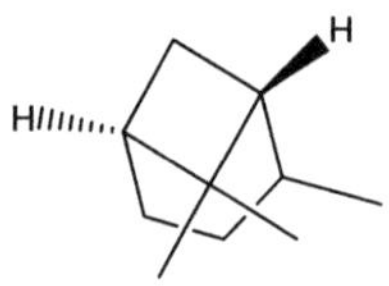					
trans-Pinane	17:01	0.9	0.0	0.0	0.0
Biphenyl	17:46	2.6	2.0	0.0	0.2
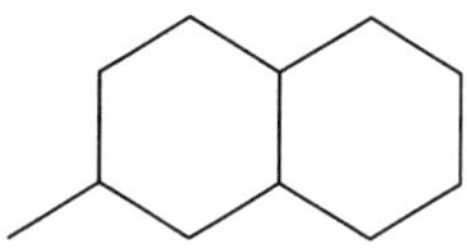					
Decahydro-2-	18:09	3.3	1.9	0.4	0.6
methylnaphthalene	18:21	3.3	3.0	0.0	1.0
Aliphatics	18:42	2.6	1.4	0.0	1.7
	19:40	2.0	0.0	0.0	0.0
Non-identified products		17.6	16.7	5.5	15.1
Total Σ (prod.)		100.0	100.0	100.0	100.0

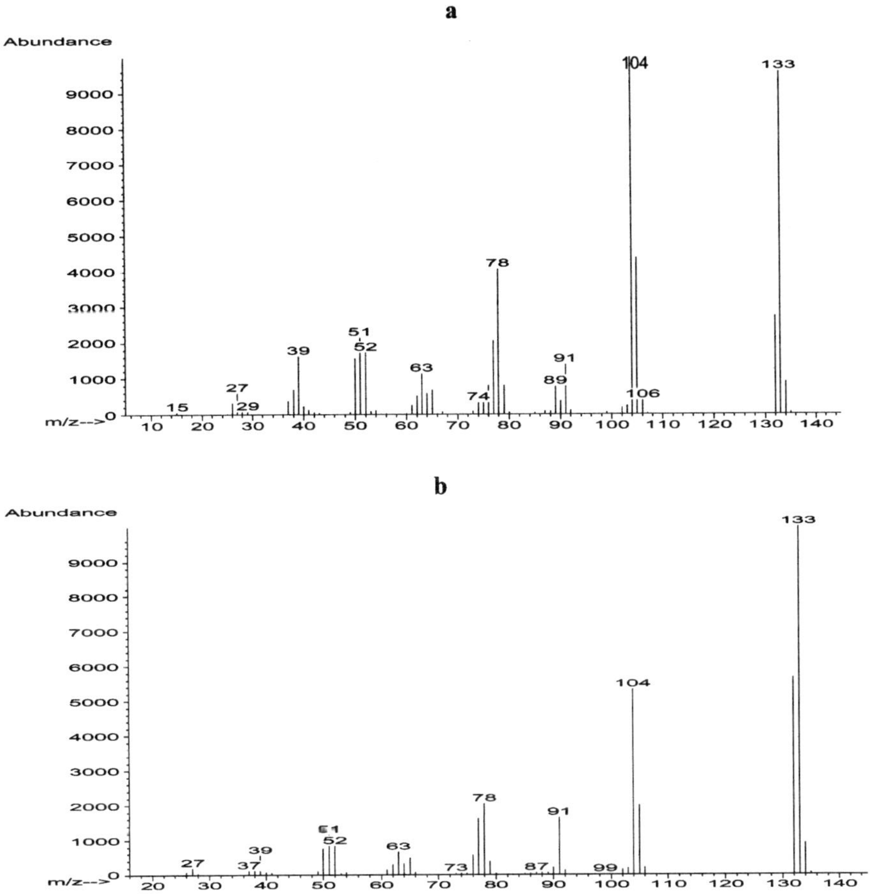

Figure 3 *Mass spectra of TI2* (a) *and TI4* (b)

All degradation products could be attributed to 7 specific groups and 1 unidentified group (Figure 4):

Group 1 4-Hydroxybutan-2-one, 4-hydroxyhexan-2-one, 5-methyl-3-hexanol, 2-acetoxy-1-propanol, hexane-1-propoxy, 7-octen-2-one, dioxolan.

Group 2 Aliphatics.

Group 3 1-Isocyanato-2-methylbenzene (TI2), 1-isocyanato-4-methylbenzene (TI4).

Group 4 Isobutyl isocyanate, 2-methyl-3-oxo-propionamide.

Group 5 *meta*-Xylene, styrene, ethyl, methyl-benzene, 2-phenyl butadiene, benzene, cyclopropyledene methyl.

Group 6 Indene, biphenyl, naphthalene, naphthalene-1-methyl, naphthalene-2-methyl, naphthalene-methyl, ethyl.

Group 7 *trans*-Pinane, decahydro-2-methylnaphthalene.

Group 8 Non-identified products.

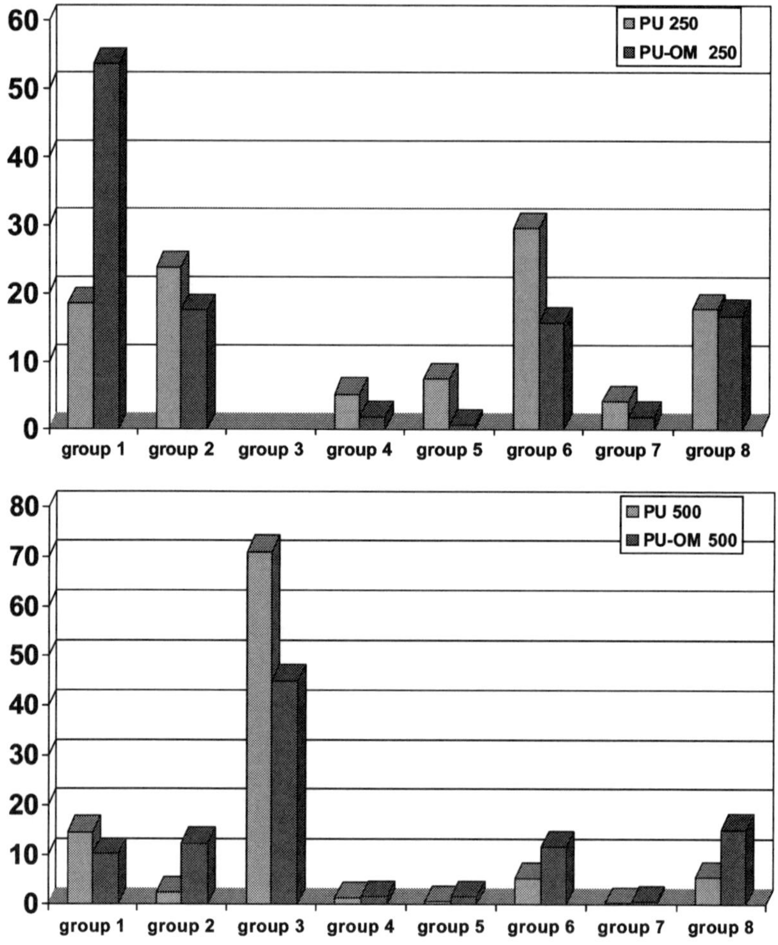

Figure 4 *PU and PU-OM degradation products*

27.4 Conclusion

Analysis of the degradation products indicates that depolycondensation and reactions involving the urethane functional group are not as favorable for PU-OM nanocomposite as for plain PU, since there is a reorientation in the mechanism of thermal degradation. The nano-structure of PU-OM interferes with formation of the toxic toluene isocyanates TI2 and TI4.

27.5 References

1. S. Fuchs and P. Valade, *Arch. Maladies Prof.*, 1951, **12**, 191–196.
2. A. Swensson, C.E. Holmquist and K.D. Lundgren., *Brit. J. Ind. Med.*, 1955, **12**, 50.

3. S.M. Tarlo, G.M. Liss, C. Dias and D.E. Banks, *Am. J. Ind. Med.*, 1997, **32**, 517–521.
4. M.G. Ott, J.E. Klees and S.L. Poche, *Occup. Environ. Med.*, 2000, **57**, 43–52.
5. H. Ulrich, in *Chemistry and Technology of Isocyanates*, John Wiley & Sons Ltd, Chichester, U.K., 1996, p. 344.
6. H. Tinnerberg, M. Spanne, M. Dalene and G. Skarping, *Analyst*, 1996, **121**, 1101–1106.
7. K. Kaaria, A. Hirvonen, H. Norppa, P. Piirila, H. Vainio and C. Rosenberg, *Analyst*, 2001, **126**, 1025–1031.
8. J. Crespo and J. Galan, *Ann. Occup. Hyg.*, 1999, **43(6)**, 415–419.
9. E. England, R. Key-Schwartz, J. Lesage, G. Carlton, R. Streicher and R. Song, *Appl. Occup. Env. Hyg.*, 2000, **15(6)**, 472–478.
10. D.C. Gupta, D.V. Wast, M.A. Tapaswi and B.N. Nigade, *Macromol. Rep.*, 1994, **A31**, 613–625.
11. R.P. Streicher, E.R. Kennedy and C.D. Loiberau, *Analyst*, 1994, **119**, 89–97.
12. G. Skarping, M. Dalene and L. Mathiasson, *J. Chromatogr.*, 1988, **435**, 453–468.
13. K.D. Rosenman, M.J. Reilly and D.J. Kalinowski, *J. Occup. Environ. Med.,* 1997, **39(5)**, 415.
14. S.K. Meredith, J. Bugler and R.L. Clark, *J. Occup. Environ. Med.*, 2000, **57**, 830–836.
15. D.R. Varma and I. Guest, *J. Toxicol. Environ. Health.*, 1993, **40**, 513.
16. O.L. Dashiell and G.L. Kennedy, *J. Appl. Toxical.*, 1984, **4**, 320–325.

CHAPTER 28

Mechanisms of Smoke and CO Suppression from EVA Composites

T. RICHARD HULL, CLAIRE L.WILLS, TANYA ARTINGSTALL, DENNIS PRICE AND G. JOHN MILNES

Fire Materials Laboratory, Bolton Institute, BL3 5AB, U.K.
(t.r.hull@bolton.ac.uk)

28.1 Introduction

Modern performance-based building and transport codes require evidence that designers have allowed sufficient time for escape in the event of an emergency. Underestimating the hazard could endanger life, while overestimation would constrain design options and escalate costs. Thus, as part of a design analysis, it is important to accurately predict the production and distribution of fire effluents resulting from the combustion of construction products and contents for different building layouts and the effects these may have on the occupants. In response to these needs, international standard criteria have recently been agreed for the assessment and measurement of smoke toxicity (ISO 13344, Estimation of lethal toxic potency of fire effluents) and how the presence of toxic fire products may impact life safety (ISO 13571 Life threat from fires – Guidance on the estimation of time available for escape using fire data). These will oblige designers to predict the fire toxicity associated with their buildings. The focus of this project is to develop a practical engineering tool capable of bridging the gap between this guidance and building design.

Smoke and toxic gas inhalation is the cause of most fire fatalities,[1] and the majority of these deaths[2] are attributed to CO poisoning. Most UK fire fatalities result from inhalation of toxic gas and smoke, and the number of casualties overcome by gas or smoke has doubled from around 3000 to 6000 per year over the last decade. In addition, while CO inhalation may be the actual cause of death, the presence of smoke and irritant gases within escape routes is an unquantified causative factor. Recent events, such as the Mont Blanc tunnel and Paddington rail disasters, clearly illustrate the effects of low ventilation on the number of toxic gas fatalities. This work sets out to understand the production of toxic gases in fires, so that this knowledge may be applied to prediction of the hazards associated with unwanted fire.

Fire gas toxicity is dependent both on material composition, and on fire conditions. Combustion toxicity is generally underestimated in small-scale tests,[2] because it is highly dependent on the fire conditions. The production of CO, the main killer, which is more prevalent in full-scale fires, has been shown to be highly dependant both on type of fire[3] and on the mode of action of flame retardants present.[4] The formation of CO in fires occurs at low temperatures in the early stages of fire development. As the fire develops, the higher temperature favours the formation of CO_2, but within any enclosure the fire spreads rapidly, until it is limited by the availability of oxygen. The intense heat drives the reaction on, despite the oxygen depletion, resulting in incomplete combustion. The gaseous products may contain CO_2, CO and a number of toxic and irritant species such as HCN and acrolein. These stages of flaming combustion have been characterised into fire types,[5] each typified by its own toxic product yield. Fully developed fires with low ventilation are the most difficult to replicate on a small scale, and are the most lethal, with CO_2/CO ratios as low as 5 or 10.

Extensive research, reviewed by Pitts,[6] on prediction of carbon monoxide evolution from flames of simple hydrocarbons, has shown the importance of the equivalence ratio ϕ (Table 1).

In a fully developed fire, with low ventilation, ϕ can be as large as 5. For many hydrocarbon polymers, CO yield increases rapidly with increase in ϕ almost independent of polymer. In addition, a close correlation between CO formation and HCN formation has been established in full-scale fire studies,[7] as the formation of both species appears to be favourable under the same poorly ventilated fire conditions. Few reliable measurements of the yields of toxic fire gas species from solid materials have been undertaken, because of the expense of large scale tests, and the lack of suitability of most small-scale tests to replicate the conditions occurring in real fires. Four standard tests exist for small-scale measurement of fire gas toxicity, the DIN 53436, the FM Fire Propagation Apparatus (ASTM E2058), the French NF-X test and the NBS smoke chamber

Table 1

$$\phi = \frac{actual\ fuel/air\ ratio}{stoichiometric\ fuel/air\ ratio}$$

with :		Typical CO yield (g g^{-1})
$\phi < 1$	Fuel lean flames	0.01
$\phi = 1$	Stoichiometric flames	0.05
$\phi > 1$	Fuel rich flames	0.2

and finally the Purser furnace (BS 7990 and ISO draft CD 19700). Only the Purser furnace can force a steady state under the most toxic oxygen-depleted conditions. It does so by feeding the sample and air into a tube furnace at fixed rates, so that the flame front is held stationary relative to the furnace. This makes it the only small-scale apparatus capable of giving reliable data on the product yields over the full range of fire conditions.

Ethylene-vinyl acetate copolymer (EVA) is widely used as a zero-halogen specification electric cable material. It is highly elastomeric, and tolerates high filler loadings while retaining its flexible properties. EVA decomposes by a two-step mechanism, with the loss of acetic acid during the first step (300–350°C), resulting in the formation of unsaturated polyenes.[8] The second decomposition step involves random chain scission of the remaining material, forming unsaturated vapour species (~430°C), such as butene and ethylene.[9] Deacetylation proceeds through β-elimination of the vinyl acetate groups present in the EVA molecules, with up to 100% conversion into polyethylene macromolecules, containing polyene sequences having up to four conjugated double bonds.[10] During thermal degradation the polymer cross-links rapidly, and appears to be autocatalytic. These cross-linking reactions lead to the formation of a protective layer that limits the access of oxygen to the remaining material, and impedes the flow of fuel to the gas phase. In earlier work on EVA[8] a protective layer surrounded the residue formed by loss of acetic acid. When this layer was physically broken, rapid decomposition occurred in a tube furnace in air at 400°C, when left unbroken decomposition was much slower.

EVA is frequently used in combination with metal hydroxide flame retardants, such as aluminium hydroxide (generally referred to as ATH) or magnesium hydroxide (MH), which release water endothermically:[11]

$$2Al(OH)_3(s) \xrightarrow{180\text{–}200°C} Al_2O_3(s) + 3H_2O(g) \qquad \Delta H = 1.3 \text{ kJ g}^{-1}$$

and

$$Mg(OH)_2(s) \xrightarrow{300\text{–}350°C} MgO(s) + H_2O(g) \qquad \Delta H = 1.45 \text{ kJ g}^{-1}$$

Zinc borate (ZB) has been extensively used as a synergist[12–16] with metal hydroxides in EVA formulations. It is thought that the zinc borate with ATH at high temperatures creates a fused ceramic-like residue; this, combined with degraded polymeric material, leads to a more protective surface layer during combustion.

$$2H_3BO_3 \xrightarrow{180\text{–}200°C} 2HBO_2 + H_2O \xrightarrow{260\text{–}270°C} B_2O_3 + H_2O$$

As a synergist it is suggested[17,18] that zinc borate slows down the degradation of the polymeric material. It is reported to be very effective with oxygen-containing polymers. Shen[14] studied the pyrolysis of EVA/ATH at 500°C, and found that ATH catalysed the combustion of EVA, while replacement of ATH with equal portions of ATH and ZB increased the char yield and changed the combustion mode from glowing to smouldering. The char yield in both cases was significantly reduced at 550°C, which is closer to the conditions of the current work. However, the exothermicity of the oxidative pyrolysis of EVA/ATH/ZB over the range 362–562°C measured using differential scanning calorimetry was significantly less than for EVA/ATH, also indicating solid-phase oxidation.

Magnesium tetraborate (MB) also decomposes to yield water:

$$MgB_4O_7{\cdot}9H_2O \longrightarrow MgB_4O_7 + 9H_2O$$

There are two possible reactions of the magnesium borate under fire condition, like ATH it could release water causing endothermic loss and dilution of the fuel present in the gas phase, and boric acid may be generated, which should aid char formation.

Zinc hydroxystannate (ZHS) has been used with ATH and MH to give better physical and flame retardant properties, and reduced smoke and toxic gas emissions at lower filler loadings. ZHS will act as a char promoter, complementing the water release and endothermic effect of ATH/MH. At 180°C ZHS loses 3 moles of water endothermically:

$$ZnSn(OH)_6 \xrightarrow{180°C} ZnSnO_3 + 3H_2O$$

ZHS is reported to promote a thermally stable cross-linked char instead of volatile and flammable products.[19] This char in turn restricts the supply of fuel to the flame and also reduces smoke and toxic emissions both for halogen and halogen-free systems.

Nanocomposites are particle-filled polymers for which at least one dimension of the dispersed particles is in the nanometre range. Polymer–clay nanocomposites have demonstrated flame retardant properties, such as reduction of peak heat release rate, formation of protective char, and decrease in the mass loss rate during combustion.[20–22] In decomposition and flaming combustion, these essentially ceramic clay layers should reinforce the protective layer formed by the charring polyene. In EVA-clay nanocomposites, acceleration of EVA deacetylation has been observed, and attributed to the strongly acidic sites on the clay exerting a catalytic effect on the β-elimination. This is followed by slower thermal degradation and protection against thermo-oxidation and delayed weight loss in air.[23] This slowing down of the volatilisation of the deacetylated polymer in nitrogen may be due to the labyrinth effect of the silicate layers in the polymer matrix. Despite the apparently disruptive influence of water evolution on the formation of a barrier layer, the fire retardancy of EVA/ATH nanocomposite clay material has been the subject of a patent.[24]

The aim of this work is to understand the formation of carbon monoxide and smoke of ethylene-vinyl acetate copolymer (EVA) materials containing fire retardants, in order to predict the fire toxicity, particularly of fully developed fires.

28.2 Experimental

28.2.1 Materials

This study correlates the results obtained from EVA materials from 3 different sources:

EVA1	Scapa Polymerics' commercially optimised formulation containing 27% vinyl acetate
EVA2	Escorene UL 00328 EVA with 28% vinyl acetate (Exxon Mobil Corp.)
EVA3	Escorene Ultra FL00328 (Exxon Chemicals, with vinyl acetate content of 27%)
EVA1/ATH1	70% ATH1 (Baco SF11E, Alcan Chemicals Europe)
EVA2/ATH2	37.5% ATH2 [of median particle size (d 50) 1.3–2.3 μm]
EVA3/ATH3	60% ATH3 (Superfine SF4 Alcan Chemicals Europe)
EVA1/ATH1/ZHS1	65% ATH1, 5% ZHS1 (Storflam ZHS, Joseph Storey & Co Ltd)
EVA3/ATH3/ZHS2	57% ATH3, 3% ZHS2 (Flamtard H, Alcan Chemicals Europe)
EVA3/ATH-ZHS	58% ATH coated with 2% ZHS (International Tin Research Institute[25])
EVA1/ATH1/ZB	65% ATH1, 5% ZB (Storflam ZB, Joseph Storey & Co Ltd)
EVA1/ATH1/MB	65% ATH1, 5% MB (Storflam MGB, Joseph Storey & Co Ltd)
EVA2/ATH2/Nano	36.4% ATH and 3% organoclay (of median particle size (d 50) 25 μm),
EVA2/Nano	4.8% organoclay (of median particle size (d 50) 25 μm)
EVA3/MH	60% MH (Magnifin H5, Martinswerk)
EVA3/MH/ZHS2	57% MH, 3% ZHS2 (Flamtard H, Alcan Chemicals Europe)
EVA3/MH-ZHS2	58% MH coated with 2% ZHS (International Tin Research Institute[25])

Materials containing EVA1 were prepared using a Banbury internal mixer and were all kindly supplied by Scapa Polymerics Ltd. The materials containing EVA2 were kindly supplied by Dr G Beyer, Kabelwerk Eupen AG. The materials containing EVA3 were kindly supplied by the International Tin Research Institute. All materials were granulated into pieces 2–8 mm long, except the EVA samples which were used as 5 mm diameter pellets.

28.2.2 Burning Behaviour

The cone calorimeter (ISO 5660) is a standard test method for studying the flammability of plastics. A Fire Testing Technology cone calorimeter was used predominantly in this study. The tube furnace method for assessing combustion

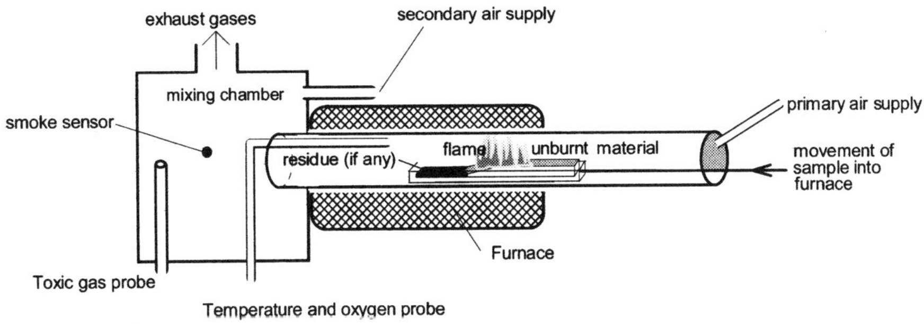

Figure 1 *Purser furnace*

toxicity (BS7990) – the Purser furnace[26] is a tube furnace with a moving sample and controlled temperature and air flow rate (Figure 1). The technique was developed to enable the study of smoke and toxic combustion product evolution from polymers under the different stages and types of fire. The technique provides for steady combustion conditions to be established, since the fuel feed rate, air flow rates, and hence the rate of burning are constant. The Purser furnace controls the rate of burning through the sample feed rate. Once steady state conditions have been established, the statistical fluctuations of burning are significantly reduced.

The apparatus was designed to compare the fire toxicity of different fire conditions. This was achieved, for a fixed fuel feed rate, by altering the ratio of primary to secondary air, while keeping the total air flow into the mixing chamber constant.

Fuel in this work refers to the copolymer and not the inorganic additives. To estimate the fuel component of the effluent gases, the diluted fire gases from the mixing chamber were passed through a secondary oxidiser containing quartz wool at 900°C. This converts all the fuel carbon into CO_2, giving a measure of the fuel component of the effluent gases.

Thermogravimetric analysis (TGA) was used to study the decomposition of the EVA and EVA/nanocomposite samples isothermally in air and dynamically in air (oxidative decomposition) and under nitrogen (non-oxidative decomposition). A Mettler M3 TGA with a Mettler TG50 Thermobalance and a Mettler TC10A Processor was the instrument used for all TGA in this study. Approximately 5 mg of sample was used for each determination and a heating rate of 10°C min^{-1} was used throughout. Differential scanning calorimetry (DSC) was carried out using a heating rate of 15°C min^{-1} under flowing nitrogen using a Polymer Laboratories DSC, and run from 20–700°C using copper pans.

28.3 Results

28.3.1 Correlation of Physical Fire Models

Carbon monoxide/carbon dioxide yield ratios from both cone tests and BS7990 tube furnace experiments are presented in Figure 2.

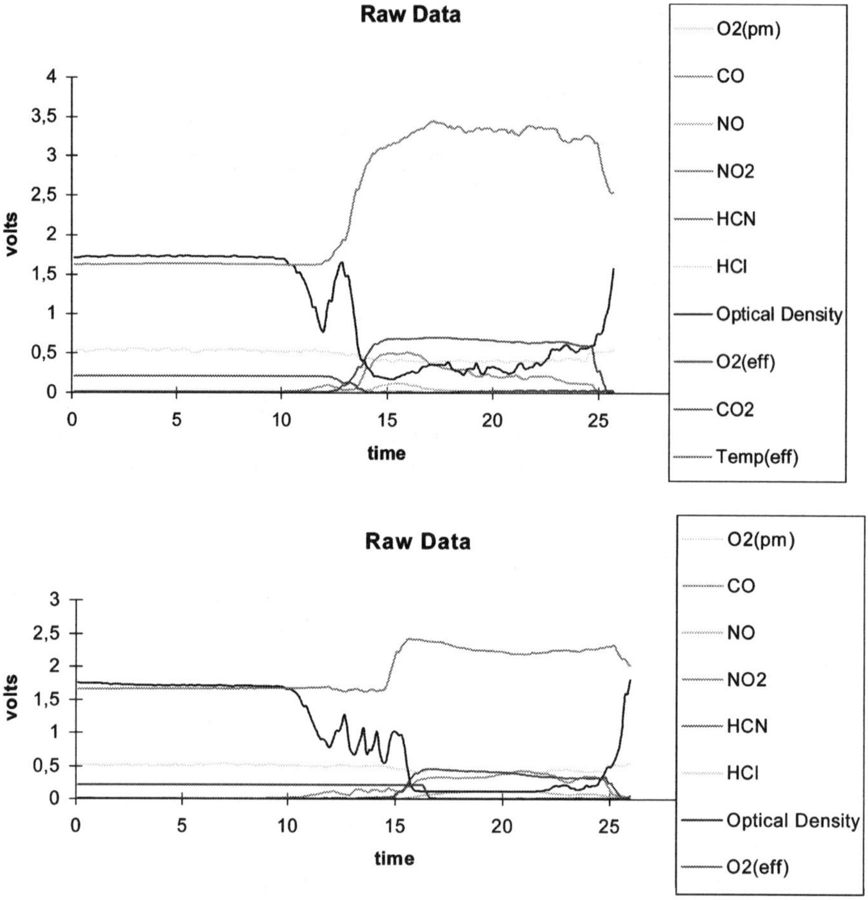

Figure 2 *Comparison between raw data for evolved gases amounts from both cone tests and BS7990 tube furnace*

Cone data were measured using a heat flux of 50 kW m^{-2} for the samples containing EVA3, and 75 kW m^{-2} for the samples containing EVA1. In the tube furnace, the fire condition is precisely known, and may be considered uniform across the area of flaming. In the cone, the outer areas are clearly well-ventilated, while the centre of the sample will be oxygen deficient.

The rapid quenching of the fire gases in the duct will prevent the formation of a hot layer for oxidation to go to completion. To estimate the average fire condition in the cone, the CO_2/CO ratio has been plotted for each sample (normally this is greater than 100 for fully developed flaming at relatively high ventilation, and less than 10 for fully developed flaming at relatively low ventilation and 100–200 for a developing fire[27]) (Figure 3). Insertion of the cone data between $\phi = 0.7$ and $\phi = 1.0$ has been based on the best fit with the tube furnace data. This shows that, for most of the samples, the fire condition in the cone lies between $\phi = 0.7$, well-ventilated, and $\phi = 1.0$, fully developed, high ventilation.

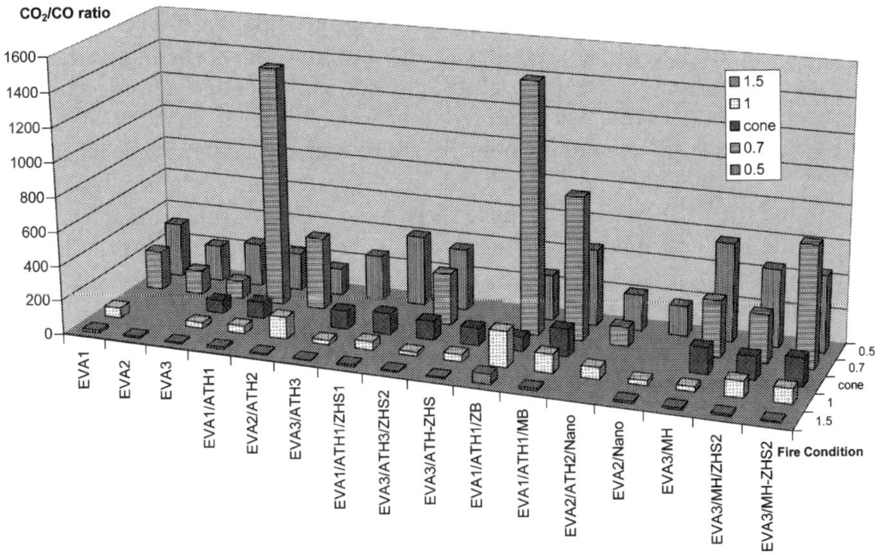

Figure 3 *Carbon dioxide/carbon monoxide yield ratios for filled and unfilled EVA materials*

The carbon monoxide yields shown in Figure 4 for the three samples of EVA show significant differences under conditions of reduced ventilation. This data is plotted alongside cone data, with a slight difference in the definition of the yield.

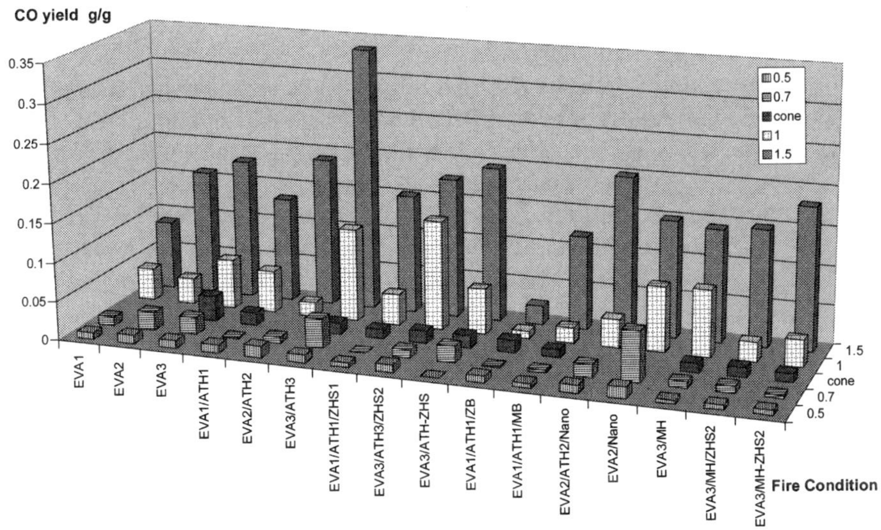

Figure 4 *Carbon monoxide yield for all materials in grams per gram of polymer, or gram per gram of mass lost for cone data (note the reversal of the fire condition axis for clarity of view)*

The tube furnace CO yields are grams CO per gram of polymer, where the cone yields are grams CO per gram of mass lost. With pure EVA there is no significant difference between the two definitions, but with ATH-containing samples almost half the mass loss will result from the loss of water from ATH. Thus the CO yield data from the cone would be about half the value expressed as g/g polymer. Thermal analysis of EVA1 shows a greater char yield at 500°C than was found for EVA2 or EVA3. Since there was no significant residual mass for any of the EVA samples, it is assumed that a greater proportion of the EVA1 remains as carbon (soot) particles, reducing the CO yield. All the EVA/ATH materials show higher CO yields than their base polymers under conditions of low ventilation, though with sample EVA3/ATH3 it proved impossible to impose steady state burning conditions under low ventilation. In this case the majority of the CO was evolved in the later 7 min of a 23 min run. This appears to be caused by incandescent afterglow, which forces the residual carbon to be at least partially oxidised. The samples containing ATH and ZHS showed some levelling out of the still evident afterglow phenomena, but there was little evidence to suggest that the observed reduction in CO yield under high ventilation was replicated under conditions of low ventilation.

The sample containing EVA1/ATH1/ZB showed the most surprising behaviour. Under conditions of low ventilation a very low CO yield was observed, and a significant amount of partially burnt carbon was detected by the secondary oxidiser. Conversely, the cone calorimeter showed the same EVA1/ATH1/ZB sample to give the highest yield of CO of any of the whole range of filled samples. The incomplete combustion resulting in higher CO yields in the cone probably results from quenching of the fire gases, rather than through oxygen depletion. Incandescence and afterglow are phenomena that occur in real fires, which would be expected to give significantly higher yields of CO for EVA/ATH samples that did not contain ~5% zinc borate. Rapid quenching of fire gases, coupled with the fact that only a single area is burnt, means that suppression of afterglow does not affect the CO yield in the cone calorimeter.

Several processes are occurring to the filled copolymer during the passage of the sample through the furnace, which are similar to the processes occurring during flame spread in a fully developed fire. These are illustrated in Figure 5.

Under fuel-rich conditions, the increase in CO yield for EVA/ATH, EVA/ATH/ZHS, and EVA/ATH/MB appears to result from solid-phase oxidation of the trapped organic material, while the low CO yield for EVA/ATH/ZB suggests that pyrolysis of the residual organic material occurred in preference to oxidation. This may be attributed to the formation of a zinc borate-alumina adduct that did not promote the oxidation reactions. The specific nature of this interaction is suggested by the failure of either the sample containing magnesium borate or zinc hydroxystannate to show a similar effect. When ATH and ZB were ground together in various ratios and studied by TGA, a slight difference was observed between the proportional sum of the individual thermograms and that of the mixtures showing the water loss of both substances merging to give a single slightly sharper peak. DSC analysis of the water loss process showed an increase in endothermicity of around 15% for the ATH/ZB in a 9:1 ratio.

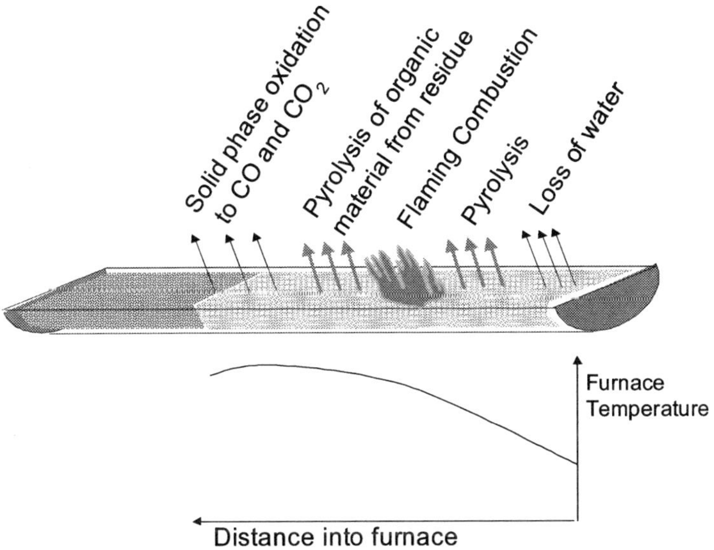

Figure 5 *Simultaneous processes occurring during steady state combustion of EVA/ATH in the Purser furnace*

Figure 6 shows the TG curves of each component in the EVA/ATH/ZB formulation as a percentage of the total original mass. These components are added together to give a calculated curve, which is compared to the actual residual mass of the composite. The TG of the EVA1/ATH1/ZB, Figure 6, also showed

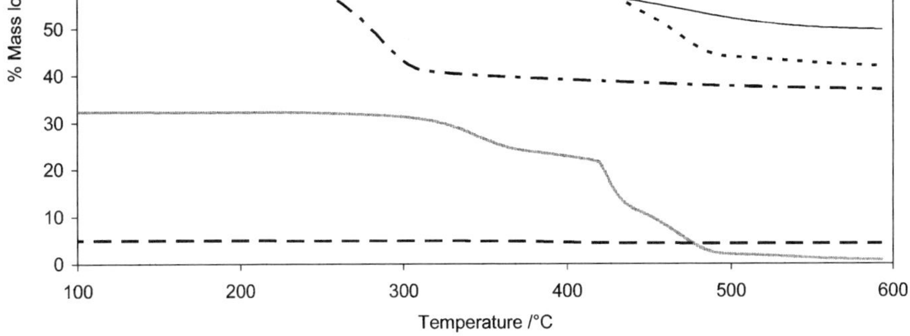

Figure 6 *TGA of EVA1/ATH1/ZB in air showing char enhancement by zinc borate*

a significant increase in the char yield, from 450–600°C, showing the poisoning effect of ZB on alumina towards char oxidation.

The EVA magnesium hydroxide material showed a modest reduction in CO yield compared with ATH, as did the MH ZHS containing samples. The advantages of coating ATH and MH with ZHS were not apparent when compared with slightly higher loadings of ZHS used with uncoated filler.

For EVA2/nano under intermediate ventilation conditions, there is just enough oxygen ($\phi = 0.7$ to 1.0), provided it can get access to the unburnt parts of the polymer; the EVA-clay shows higher yields of CO but, under fuel-rich conditions, the EVA-clay shows a similar yield (of around 0.15 g g^{-1}) of CO to EVA, and both the EVA/ATH and EVA/ATH/clay samples show a significantly higher yield of CO (of around 0.20 g g^{-1}). At $\phi = 1$, for the EVA-clay, the variation of CO concentration during the experiment showed a large initial peak of CO (of three times the steady state value) before the steady state was reached. In contrast, the EVA sample showed a smaller initial peak, which was about twice its (much lower) steady state value. Samples containing ATH again showed the apparently catalytic effect of alumina in increasing CO yields at low ventilation, as reported in earlier work on EVA.[28]

28.3.2 Smoke

The yields of all products of incomplete combustion are generally believed to increase with decrease in ventilation. Moreover, a number of flame retardant additives show smoke and CO suppression under well ventilated conditions. However, it is not clear the extent to which these additives perform, and whether good CO suppression coincides with good smoke suppression. The smoke yield from the tube furnace, calculated from the optical density, and the smoke data from the cone for comparison taken from Cross *et al.*,[29] are both expressed as average specific extinction area (SEA in m^2 kg^{-1}) and shown in Figure 7. The data differ slightly in the definition of the mass from which the smoke came, for the tube furnace work the kg^{-1} refers to kg of polymer, in the cone it is per kg of mass lost. This is only noticeable when comparing materials containing metal hydrate fillers, which lose some mass as water, which would not cause smoke.

The results show similar trends for smoke production for both the cone and the tube furnace. MH acts as a smoke suppressant under well-ventilated conditions, and is moderately effective under fuel-rich conditions. The enhanced smoke production for ATH containing materials, particularly at higher fuel to air ratios, suggests that pyrolysis of aromatics must compete with catalytic conversion of fuel carbon into CO or CO_2.

28.4 Conclusions

This work shows that the tube furnace method provides an important source of data on the fire toxicity of fire retarded polymer samples. It shows the inadequacy of cone data in assessing CO yields and smoke production in developed fires.

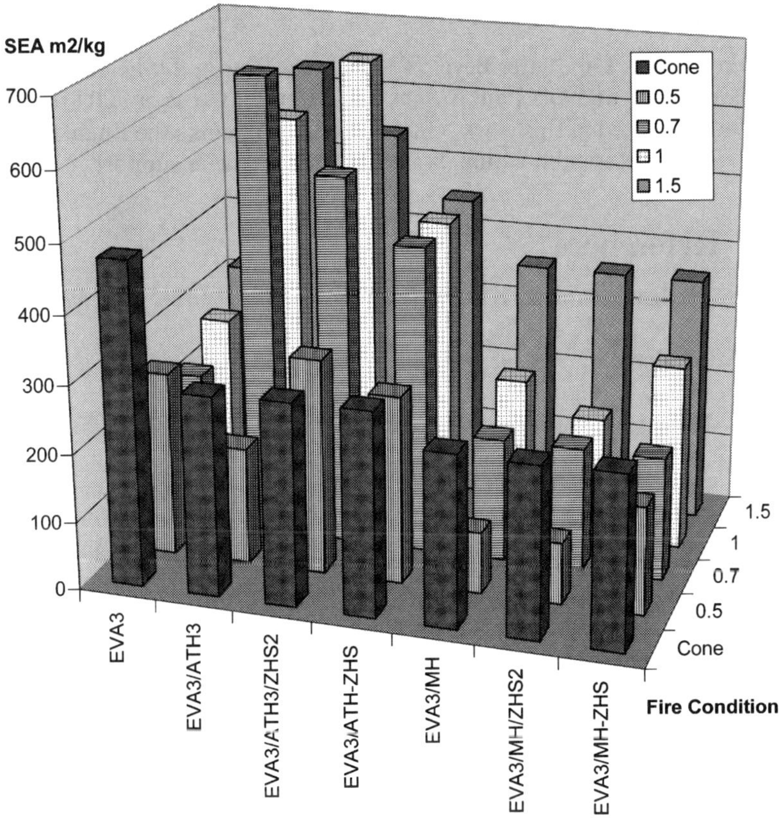

Figure 7 *Smoke production in tube furnace SEA (expressed per g polymer) and from the cone (average SEA expressed per g mass lost)*

The mechanisms of CO production vary with fire condition. Under poorly ventilated conditions with high concentrations of ATH, higher toxicity may occur. The synergy of zinc borate as an afterglow suppressant when used with aluminium hydroxide seems to depend on an interaction of the decomposition products of the two materials that poisons the catalytic activity.

Magnesium hydroxide shows a greater tendency to suppress smoke and a smaller tendency to lower CO yields under poorly ventilated conditions. This needs further investigation, and may result from more efficient catalysis, leading to less smoke and CO, or a subtly different type of behaviour.

Clay, which forms a protective layer, seems to have a negligible effect on the yield of carbon monoxide, both under fuel-lean and fuel-rich conditions. The influence of the residual alumina from ATH, in catalysing the conversion of more organic material into CO, also appears to be unaffected by the presence of the clay material. The increased yield of CO under stoichiometric conditions (when $\phi = 1$) probably arises from a reduction in access of oxygen caused by the presence of a protective layer.

28.5 Acknowledgements

The authors thank Dr Gunter Beyer of Kabelwerk, Eupen, Dr Peter Salthouse of Scapa Polymerics, and Drs Paul Cusack and Matthew Cross of ITRI for providing the samples used in this work. One of us (CLW) thanks the Engineering and Physical Science Research Council (EPSRC) for financial support.

28.6 References

1. "Fire Statistics United Kingdom 1997", *Home Office Statistical Bulletin*, 1998, Issue 25/98, p. 13.
2. M.M. Hirschler, *Recent Advances in Flame Retardancy of Polymeric Materials*, M. Begin (ed.), BCC Inc, Ct., U.S.A., 1999.
3. T.R. Hull, J.M. Carman and D.A. Purser, *Polym. Inter.*, 2000, **49**, 1259–1265.
4. T.R. Hull, R.E. Quinn, I.G. Areri and D.A. Purser, *Polym. Degrad. Stab.*, 2002, **77**, 235–242.
5. *Toxicity testing of fire effluents – Part 1*, ISO TR 9122-1 1989 (E).
6. W.M. Pitts, *Progress in Energy and Combustion Sci.*, 1995, **21**, 197–237.
7. D.A. Purser, *Polym. Int.*, 2000, **49**, 1232–1255.
8. K. McGarry, *Polym. Int.*, 2000, **49(10)**, 1193–1198.
9. M.B. Maurin, L.W. Dittert and M.A. Hussain, *Thermochim. Acta*, 1991, **186**, 97–102.
10. J.T. Yeh, M.J. Yang and S.H. Hsieh, *Polym. Degrad. Stab.*, 1998, **61**, 465–472.
11. R.N. Rothon, in *Particulate-Filled Polymer Composites*, Chapter 6: Effects of particulate fillers on flame-retardant properties of composites, R.N. Rothon (ed.), Harlow, Longman, UK, 1995.
12. *US Borax Technical Bulletin*, HF596.
13. M. Le Bras, N. Pecoul, S. Bourbigot and R. Delobel, in *Extended Abstracts of Eurofillers '97*, R.N. Rothon (ed.), British Plastics federation Filplas Committee & MOFFIS Committee Pub., London, 1997.
14. K. Shen, *Plastics Compound.*, Nov./Dec. 1988.
15. M. Le Bras, S. Bourbigot, F. Carpentier, R. Leeuwendal and D. Schubert, *GAK Gummi Fasern Kunstoffe*, 1998, **12**, 972–982.
16. K. Shen and D.F. Ferm, in *Proceedings of Recent Advances in Flame Retardancy of Polymeric Materials*, M. Lewin (ed.), B.C.C Pub., Stamford, U.S.A., 1997.
17. F. Carpentier, S. Bourbigot, M. Le Bras and R. Delobel, *Polym. Int.*, 2000, **49**, 1216.
18. A. Marchal, M. Le Bras, R. Delobel and J.-M. Leroy, *Polym. Degrad. Stab.*, 1994, **44**, 263–272.
19. M.S. Cross, P.A. Cusack and P.R. Hornsby, *Polym. Degrad. Stab.*, 2003, **79**, 309–318.
20. J.W. Gilman, *Appl. Clay Sci*, 1999, **15**, 31–49.
21. G. Beyer, *Fire Mater.*, 2001, **25**, 193–197.

22. G. Beyer, *Polym. News,* 2001, **26**, 334–342.
23. M. Zanetti, T. Kashiwagi, L. Falqui and G. Camino, *Chem. Mater.*, 2002, **14(2)**, 881–887.
24. "Flameproof Polymer Composition" *World Intellectual Property Organisation Patent: WO0068312A1* (2000).
25. P.A. Cusack, B. Patel, M.S. Heer and R.G. Baggaley, *European Patent 833,862* (1999).
26. D.A. Purser, P.J. Fardell, J. Rowley, S. Vollam and B. Bridgeman, in *Proceedings of the 6th International Conference Flame Retardants '94*, Interscience Communications, London, 26–27 Jan. 1994.
27. *Toxicity testing of fire effluents Part I*, Technical Report 9122-1, International Standards Organisation, 1989 (E).
28. T.R. Hull, *Polym. Degrad. Stab.*, 2002, **77(2)**, 235–242.
29. M.S. Cross, P.A. Cusack and P.R. Hornsby, *Polym. Degrad. Stab.*, 2003, **79**, 309–318.

CHAPTER 29

Products of Incomplete Combustion from Fire Studies in the Purser Furnace

CLAIRE L. WILLS,[1] JUDAH AROTSKY,[1] T. RICHARD HULL,[2] DENNIS PRICE,[2] DAVID A. PURSER[3] AND JENNY PURSER[3]

[1]Institute of Materials Research (Chemistry), University of Salford, Salford, M5 4WT, UK
[2]Centre for Materials Research and Innovation, Bolton Institute, Deane Road, Bolton, BL3 5AB, UK (t.r.hull@bolton.ac.uk)
[3]Fire and Risk Sciences, Building Research Establishment, Garston, Watford, WD25 9XX, UK

29.1 Introduction

Ethylene-vinyl acetate copolymer (EVA) is widely used to obtain a zero-halogen specification electric cable material. It is highly elastomeric, and tolerates high filler loadings while retaining its flexible properties. Work has been carried out by Tin Technologies, UK to develop material suitable for a fire retarded sheath for electric cables while maintaining zero halogen specification. EVA is frequently used in combination with metal hydroxide flame retardants, such as aluminium hydroxide (generally referred to as ATH) or magnesium hydroxide (MH), to overcome flammability problems by releasing water endothermically:[1]

$$2Al(OH)_3\,(S) \xrightarrow{180-200°C} Al_2O_3\,(s) + 3H_2O\,(g) \qquad \Delta H = 1.3\ kJ\ g^{-1}$$

and

$$Mg(OH)_2\,(S) \xrightarrow{300-350°C} MgO\,(s) + H_2O\,(g) \qquad \Delta H = 1.45\ kJ\ g^{-1}$$

Zinc hydroxystannate (ZHS) has been investigated as a potential synergist. The mechanism of ZHS in halogen systems involves formation in the vapour

22. G. Beyer, *Polym. News,* 2001, **26**, 334–342.
23. M. Zanetti, T. Kashiwagi, L. Falqui and G. Camino, *Chem. Mater.*, 2002, **14(2)**, 881–887.
24. "Flameproof Polymer Composition" *World Intellectual Property Organisation Patent: WO0068312A1* (2000).
25. P.A. Cusack, B. Patel, M.S. Heer and R.G. Baggaley, *European Patent 833,862* (1999).
26. D.A. Purser, P.J. Fardell, J. Rowley, S. Vollam and B. Bridgeman, in *Proceedings of the 6th International Conference Flame Retardants '94*, Interscience Communications, London, 26–27 Jan. 1994.
27. *Toxicity testing of fire effluents Part I*, Technical Report 9122-1, International Standards Organisation, 1989 (E).
28. T.R. Hull, *Polym. Degrad. Stab.*, 2002, **77(2)**, 235–242.
29. M.S. Cross, P.A. Cusack and P.R. Hornsby, *Polym. Degrad. Stab.*, 2003, **79**, 309–318.

CHAPTER 29

Products of Incomplete Combustion from Fire Studies in the Purser Furnace

CLAIRE L. WILLS,[1] JUDAH AROTSKY,[1] T. RICHARD HULL,[2] DENNIS PRICE,[2] DAVID A. PURSER[3] AND JENNY PURSER[3]

[1]Institute of Materials Research (Chemistry), University of Salford, Salford, M5 4WT, UK
[2]Centre for Materials Research and Innovation, Bolton Institute, Deane Road, Bolton, BL3 5AB, UK (t.r.hull@bolton.ac.uk)
[3]Fire and Risk Sciences, Building Research Establishment, Garston, Watford, WD25 9XX, UK

29.1 Introduction

Ethylene-vinyl acetate copolymer (EVA) is widely used to obtain a zero-halogen specification electric cable material. It is highly elastomeric, and tolerates high filler loadings while retaining its flexible properties. Work has been carried out by Tin Technologies, UK to develop material suitable for a fire retarded sheath for electric cables while maintaining zero halogen specification. EVA is frequently used in combination with metal hydroxide flame retardants, such as aluminium hydroxide (generally referred to as ATH) or magnesium hydroxide (MH), to overcome flammability problems by releasing water endothermically:[1]

$$2Al(OH)_3\,(S) \xrightarrow{180\text{–}200^\circ C} Al_2O_3\,(s) + 3H_2O\,(g) \qquad \Delta H = 1.3\ \text{kJ g}^{-1}$$

and

$$Mg(OH)_2\,(S) \xrightarrow{300\text{–}350^\circ C} MgO\,(s) + H_2O\,(g) \qquad \Delta H = 1.45\ \text{kJ g}^{-1}$$

Zinc hydroxystannate (ZHS) has been investigated as a potential synergist. The mechanism of ZHS in halogen systems involves formation in the vapour phase of SnX_4,[2] resulting in flame inhibition, but no work has been reported on its mechanism in zero halogen systems.

This work concentrates on the burning behaviour of EVA/ATH/ZHS and EVA/MH/ZHS combinations. Cusack *et al.* have indicated that ZHS and zinc stannate (ZS) could be used as a highly effective flame retardant.[3] Work on the samples has indicated that ZHS is an effective synergist with ATH and MH under well-ventilated conditions in the cone calorimeter, particularly with regard to suppression of CO and smoke. There is, hence, a need to investigate these materials under less well-ventilated conditions. The aim of this work is to investigate the effect of smoke and CO suppressants in EVA-based materials under different fire conditions.

29.2 Experimental

29.2.1 Materials

The seven samples of formulation shown in Table 1 were obtained for study from Tin Technologies, UK. The preparation of these formulations is documented[3] and involved compounding on a Prism 16 mm twin extruder, operating in the temperature range 155–165°C. Materials used to produce these samples were a cable grade EVA, Escorene Ultra FL00328 (Exxon Chemicals), with a vinyl acetate content of 27%; alumina trihydrate Superfine SF4 (Alcan Chemicals, Europe); magnesium hydroxide Magnifin H5 (Martinswerk), and ZHS used in the form of Flamtard H, the trade name (Alcan Chemicals, Europe) of commercial zinc hydroxystannate $ZnSn(OH)_6$ with *ca.* 40 wt% tin, was included at 5 wt% of total filler loading as a 'physical mixture'. The samples studied were not available in large quantities and were tested in granulated form.

29.2.2 Apparatus

To conduct the toxic yield assessments of highly fire retarded materials on a small scale, while forcing burning under conditions of reduced ventilation the best test available is the Purser furnace BS 7990. This allows small-scale replication of large-scale fire toxicity.

The Purser furnace is a tube furnace with a moving sample and controlled temperature and air flow rate (Figure 1). A more detailed description of the instrument is given by Purser *et al.*[4]

Table 1 *Samples studied and abbreviations used*

EVA	Unfilled Processed polymer EVA 27%VA
EVA/ATH	150phr ATH/EVA
EVA/MH	150phr MH
EVA/ATH/ZHS	142.5phr ATH +7.5phr Flamtard H
EVA/ATH-ZHS	150phr 5% ZHS coated ATH
EVA/MH/ZHS	142.5phr MH +7.5phr Flamtard H
EVA/MH-ZHS	150phr 5% ZHS coated MH

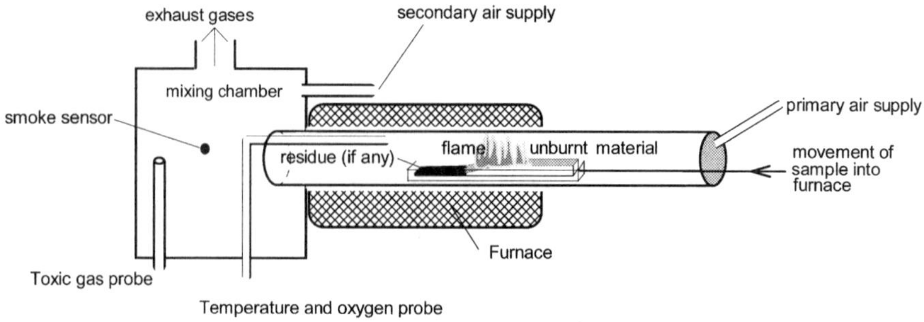

Figure 1 *Purser furnace*

The technique was developed to enable the study of smoke and toxic combustion product evolution from polymers under the different stages and types of fire, *e.g.*

- Non-flaming (oxidative) decomposition
- Developing fire (flaming)
- Fully developed (flaming) with (i) relatively high ventilation or (ii) relatively low ventilation

(taken from the ISO classification of fire stages in accordance with ISO/TR 9122-1.)

In this work, the fire conditions are defined in terms of the equivalence ratio, ϕ [Equation (1)].

$$\phi = \frac{\text{actual fuel/air ratio}}{\text{stiochiometric fuel/air ratio}} \tag{1}$$

The materials were tested at 750°C under four fire conditions of $\phi = 0.5$, 0.7, 1.0 and 1.5 (Table 3). At $\phi = 0.5$ the conditions correlate with a developing fire, at $\phi = 0.7$ and 1.0, the conditions are those of a fully developed, high ventilation fire, and at $\phi = 1.5$ conditions of low ventilation are produced. The fire type and air flow rates used for the four conditions are shown in Table 2.

Table 2 *Primary air flow rates used to obtain different ϕ conditions*

Equivalence ratio, ϕ at furnace temperature 750°C	*Primary air flow used in this work ($l\ min^{-1}$)*
0.5	19.1
0.7	13.6
1.0	9.5
1.5	6.4

The Purser furnace controls the rate of burning through the sample feed rate and fixed air flow rate. The apparatus was designed to compare the fire toxicity of different fire conditions. This was achieved, for a fixed fuel feed rate, by altering the ratio of primary to secondary air while keeping the total air flow into the mixing chamber constant. Fuel in this case refers to the copolymer and does not include the inorganic additives.

Temperature profiles of the furnace show a temperature maximum in the middle of the furnace, decreasing to just above ambient temperatures at the ends. The point at which the sample ignites occurs when the sample enters a part of the furnace that is hotter than its ignition temperature. Thus the ignition temperature is generally lower than the furnace maximum temperature, and the maximum furnace temperature corresponds to the upper layer temperature of the work reviewed by Pitts.[5]

29.2.3 Secondary Oxidiser

The products of incomplete combustion, having been mixed with secondary air in the chamber, may now be fully oxidised. A sample train leaves the mixing chamber and is drawn through the oxidiser at a flow rate of 2 l min^{-1}, operated at 900°C over quartz wool (obtained from BDH). The oxidiser consists of a quartz tube (internal diameter 34 mm) inserted into a tube furnace 34 cm long. After passing through the furnace the oxidised gases flow over a trap at 0°C to remove water. Species such as CO, hydrocarbons, and other products of incomplete combustion, including smoke, are oxidised to CO_2. CO_2 levels are monitored by a 0–3% non-dispersive infrared analyser calibrated with a 2% CO_2 cylinder from BOC Speciality Gases. Results are then used to calculate the % carbon recovery from the Purser furnace experiment.

The secondary oxidiser works along the same principle as a phi meter, a total hydrocarbon analyser. Complete oxidation of products of incomplete combustion in a large-scale fire is achieved in a similar manner using a phi ϕ meter.[6] This passes the fire gases over a heated oxidiser and measures the oxygen consumption. This is similar to the technique reported here except that in the present work the incomplete combustion is determined from the final yield of CO_2.

Steady state flaming combustion has been studied by driving samples through the furnace set to 750°C. In each case the same mass of EVA (19.7 g) was chosen for the determination. Thus, 49.25 g of each 150 phr filled sample was used.

29.3 Results

The results for mass loss, O_2, CO_2, CO_2/CO and secondary CO_2 are discussed to validate the physical fire model described. The yields of CO and smoke for the samples are then reported.

29.3.1 Mass Loss

The mass loss was studied in the Purser furnace, a small residue was left at the rear end of the boat. The mass loss was 59.3% for the ATH-containing samples and 57.6% for the MH-containing samples. The theoretical mass loss for the ATH-containing samples is 60.6% and for the MH-containing samples is 58.2%. The slightly greater mass loss falls within the limits of experimental error, and does not indicate a change in the dehydration chemistry of ATH, MH or ZHS.

29.3.2 Effluent Oxygen

Figure 2 shows the variation of O_2 concentration in the undiluted effluent. This shows that the oxygen concentration falls for all polymers from 7–11.5% at $\phi = 0.5$ to 0.7–3% at $\phi = 1.5$ (Figure 2). The point for EVA at $\phi = 1.5$ is not shown because a leak in the sample to the oxygen analyser prevented a valid measurement. The low oxygen concentrations at $\phi = 1.5$ for the other samples show that the conditions allow for adequate mixing.

29.3.3 Carbon Dioxide

As shown in Figure 3, there is a general decrease in CO_2 yield as the equivalence ratio increases, as expected. The size of the tube can account for lower values at $\phi = 0.5$ than expected, this has been documented in our previous work.[7] Omissions in the results are due to instrument logging problems and lack of further quantities of sample.

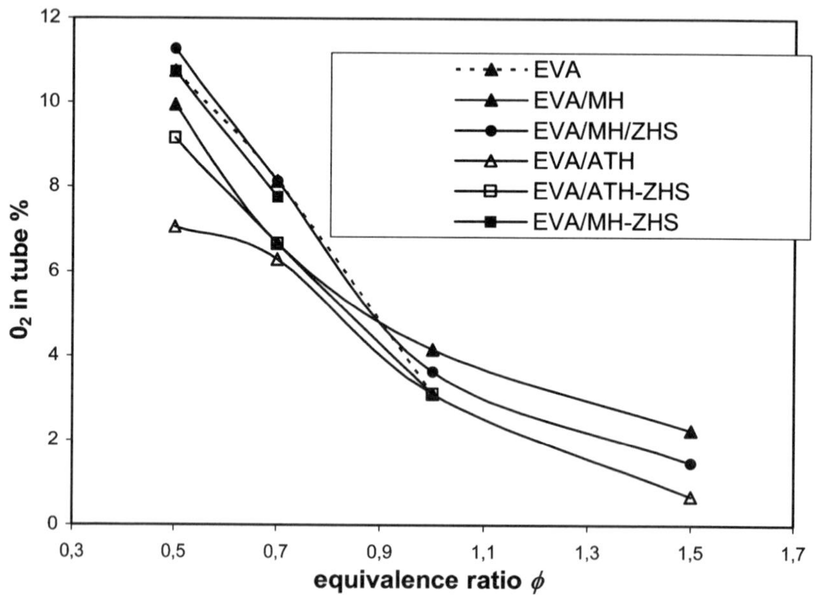

Figure 2 *Effluent oxygen concentrations at various equivalence ratios (ϕ)*

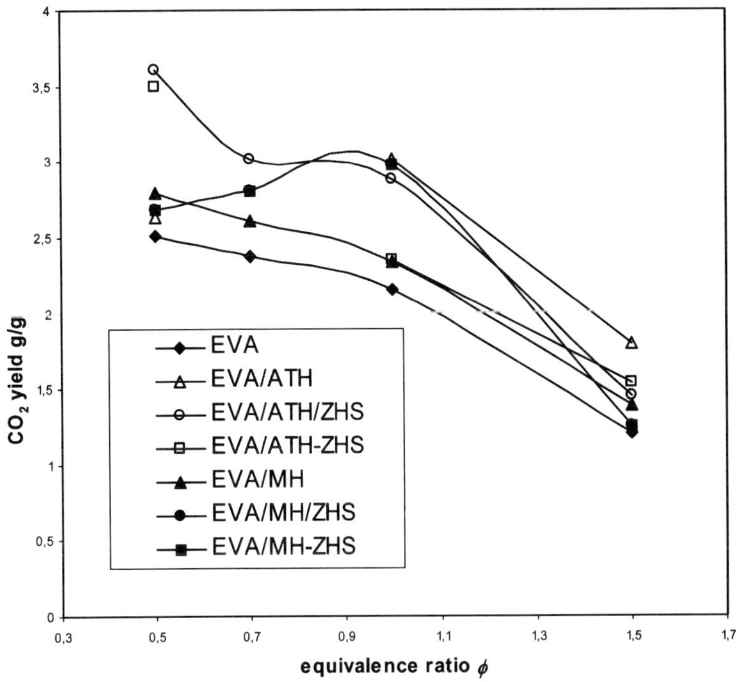

Figure 3 *CO_2 yield for all samples at various equivalence ratios (ϕ)*

29.3.4 CO_2/CO Ratio

Generally, at the beginning of a flaming fire, the CO_2/CO ratio in the fire atmosphere is relatively high, with high oxygen concentrations. As the fire develops, the ratio tends to fall, together with oxygen concentration. The relationship between CO_2/CO ratio and oxygen demonstrates the varying and complex processes during a fire, but affords one method of comparison with other fire atmospheres. The CO_2/CO ratio in Table 3 gives another indication of the

Table 3 *CO_2/CO ratios at various ϕ calculated by the authors; except cone calorimeter data calculated by* Cross,[3] *given for comparison*

Samples	*Purser CO_2/CO $\phi=0.5$*	*Purser CO_2/CO $\phi=0.7$*	*Cone calorimeter CO_2/CO*	*Purser CO_2/CO $\phi=1.0$*	*Purser CO_2/CO $\phi=1.5$*
EVA	114	74	74	22	7
EVA/ATH			106	20	
EVA/ATH/ZHS	263	111	109	16	10
EVA/ATH-ZHS	361		101	69	8
EVA/MH	279	372	151	28	10
EVA/MH/ZHS	257	403	160	94	10
EVA/MH-ZHS	298	685	136	68	7

fire condition created. It also shows an apparent correlation between the fire condition established in the cone calorimeter and how this corresponds to the Purser furnace study. However, while the CO from the Purser furnace results from vitiated combustion, it is more likely that the CO from the cone arises through rapid quenching of the fire gases.

29.3.5 Secondary Oxidiser

There is a wide spread of 72–157% in the calculated amount of carbon recovered (Table 4 and Figure 4). This data was calculated assuming a steady fuel feed and no accumulation of carbon in the sample boat. The phenomena of afterglow, observed elsewhere for samples containing ATH,[7] where the residual

Table 4 *Carbon recovered (relative %)*

ϕ equivalence ratio	*0.5*	*0.7*	*1.0*	*1.5*
EVA	72	94	85	84
EVA/ATH	91	133	157	
EVA/ATH/ZHS	129	96	115	90
EVA/ATH-ZHS	115	95	79	
EVA/MH	84	89	95	82
EVA/MH/ZHS	70	83	73	87
EVA/MH-ZHS	81	95	91	88

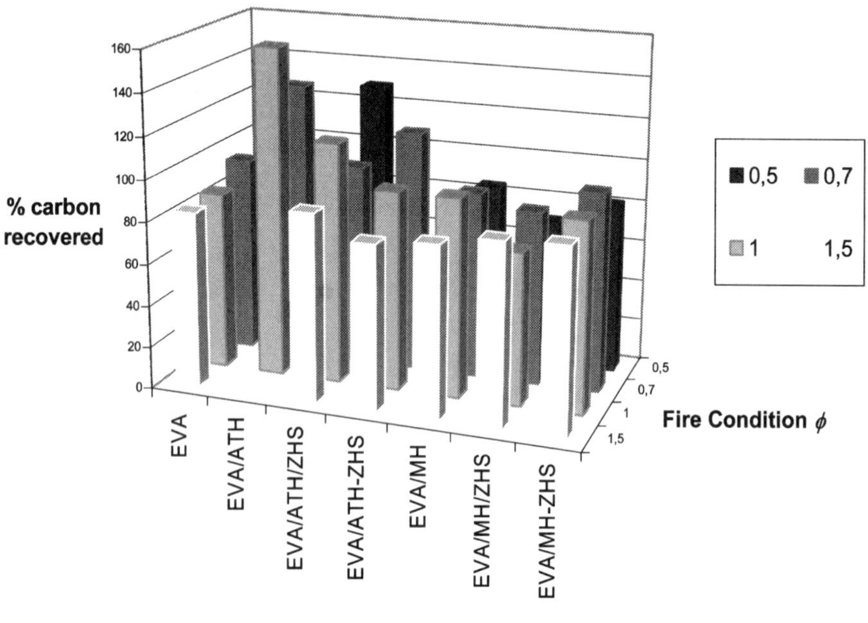

Figure 4 *Carbon recovered from all samples at various equivalence ratios (ϕ)*

alumina catalyses the decomposition of residual char trapped in the solid matrix, may account for the higher values of carbon recovered, particularly for the samples containing ATH. In these cases, after initial combustion, the residue is passed to a hotter part of the furnace where it is pyrolysed, combining with oxygen either in the tube or in the secondary oxidiser. Thus, during the apparent steady state, there are in fact two processes occurring. Smaller errors may also arise from smoke particles deposited as soot (the mass of which was not recorded), which is not all transferred to the secondary oxidiser. It could, however, be carbon evolved at an unsteady rate – a disproportionate amount of carbon may be lost outside the main steady state part of the burning process.

29.3.6 CO Yield

The results for the unfilled processed polymer show that the CO yield, measured by an electrochemical cell, increases as ventilation conditions become more vitiated. Carbon monoxide formation is a consequence of incomplete combustion, which can arise from a shortage of oxygen or insufficient temperature or time in the flame zone. Passing through the tube furnace at 750°C should be sufficient to ensure near equilibrium product yields are reached and the high CO yields at high ϕ arose from insufficient oxygen (Figure 5).

The introduction of ATH with EVA leads to an increase in the level of CO under all the ventilation conditions studied compared to that of the pure polymer. The incorporation of ZHS with EVA/ATH was seen to lower the CO yields under all conditions except $\phi = 1.0$. The CO yield for the EVA with the

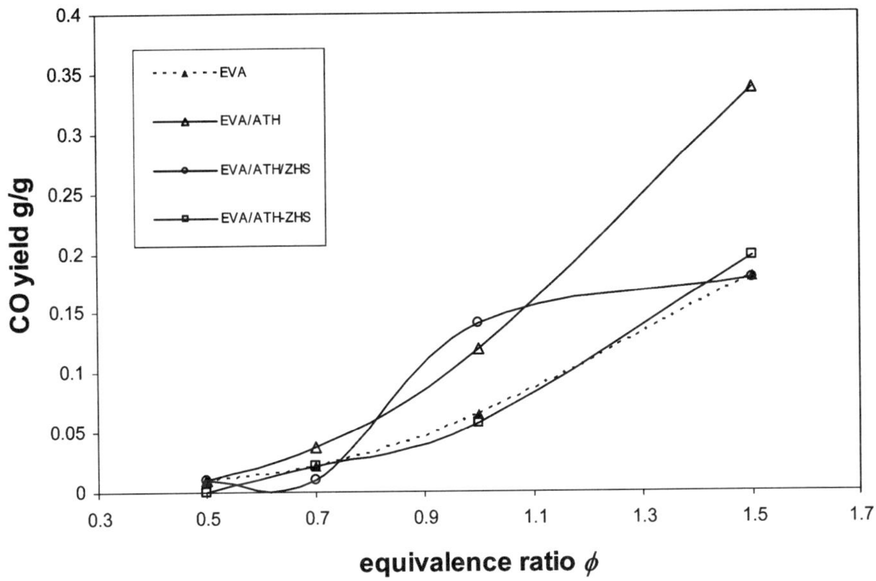

Figure 5 *CO yield for ATH samples at various equivalence ratios (ϕ)*

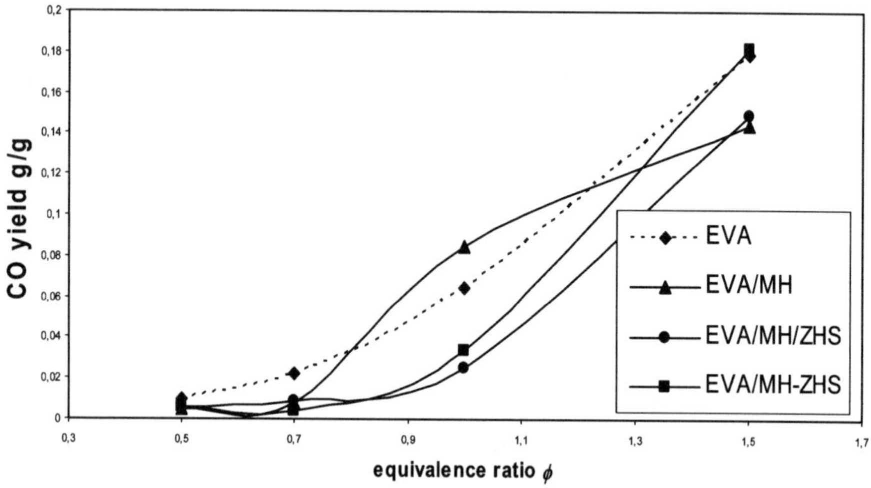

Figure 6 *CO yield for MH samples at various equivalence ratios* (ϕ)

ATH coated with ZHS (EVA/ATH-ZHS) lowers the CO yield to that of the pure polymer. The coated ATH-ZHS appears advantageous in CO yield compared to the ATH/ZHS sample especially at $\phi = 1.0$.

The effect of EVA with MH alone was to significantly reduce the CO yield at high ϕ compared to the corresponding EVA/ATH. In combination with ZHS, no further reduction was observed (Figure 6).

29.3.7 Smoke

The amount of smoke produced has been measured in terms of the optical density in the dilution chamber, obtained *via* obscuration of a horizontal laser beam through the mixing chamber. The optical system was calibrated with optical density filters. The results are expressed in terms of specific extinction area (SEA) based on the optical density, volume flow rate and fuel feed rate of the polymer (as distinct from the polymer plus filler). The smoke measurements show a general upward trend in results with decreasing ventilation ratio.

A plateau appears to be reached at $\phi = 0.7$, much lower than the plateau for CO at around $\phi = 1.5$ to 2 (Figure 7). The lower values at $\phi = 0.5$, especially for the pure EVA, can be attributed to the wind effect. The MH formulations lower the smoke production in comparison to that of the pure polymer under all conditions except $\phi = 1.5$. The ATH samples all produced greater amounts of smoke than did the pure polymer. For the EVA/MH-ZHS sample there is a definite increase in smoke production as vitiated conditions prevail, although for the other samples this is not observed.

29.4 Discussion

These results show that the parameters have been met to indicate that the appropriate fire conditions have been replicated. Specifically, oxygen

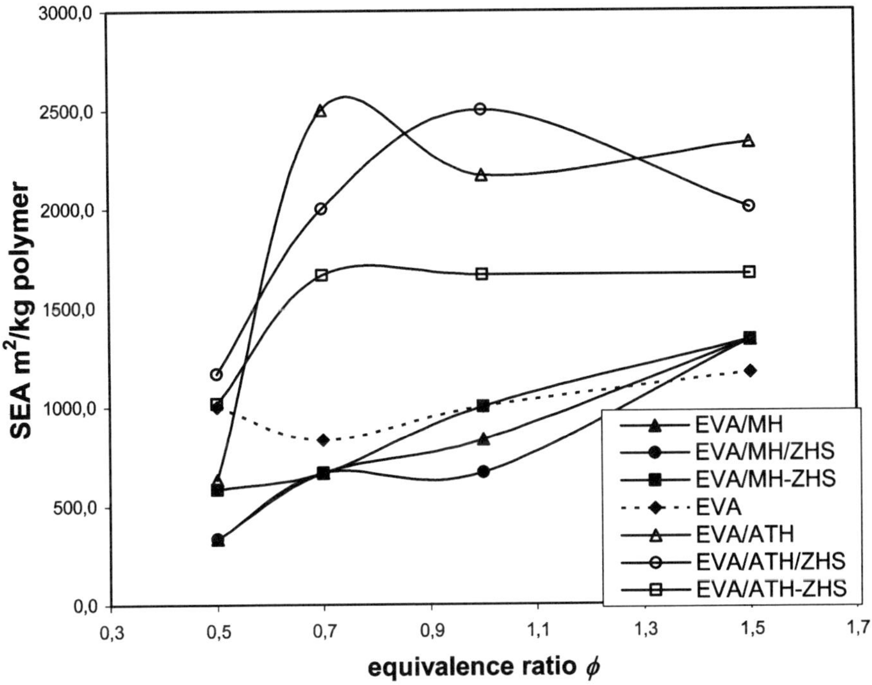

Figure 7 *Smoke results from all samples at various equivalence ratios (ϕ)*

concentrations and CO_2/CO ratios indicate the appropriate fire conditions have been established. The mass loss corresponds with complete loss of polymer material and agrees with results obtained from the samples in the cone calorimeter.[8]

Previous work[7] has also found that, at $\phi = 0.5$, the CO_2 levels are lower than expected. This is attributed to high air velocity through the tube, causing incomplete combustion through quenching of the fire plume.

In the present work a tube with internal diameter 55 mm is used with a primary air flow rate of 19.1 l min^{-1} under $\phi = 0.5$ conditions. This corresponds to an average air flow velocity of 13.3 cm s^{-1}. In another reported work[4] the tube internal diameter was 47.5 mm and the airflow rate was 22.6 l min^{-1} at $\phi = 0.5$, giving an air flow velocity of 21.2 cm s^{-1} under well ventilated conditions. Even lower levels of CO_2 were detected in this work. This shows that, under the conditions used here, the well-ventilated scenario has such a high level of ventilation that it is forcing incomplete combustion through excessive ventilation. The tube inside diameter (55 mm) is just above the middle of the range implied by the standard (36 to 66 mm).[9] The air flow velocity obtained is lower than the midpoint in the specified range of furnace diameters.

These samples had been previously investigated by Cusack *et al.* using the cone calorimeter[3] and it is interesting to compare results. The Purser furnace results have been compared with cone calorimeter results at a heat flux of

50 kW m^{-2} operated under normal ventilated conditions. The average CO_2 yield from the cone calorimeter is 2.36 g g^{-1} of polymer which is very close to our value of 2.38 g g^{-1} polymer found using the Purser furnace technique under $\phi = 0.7$ conditions (Figure 3). The theoretical yield of CO_2 corresponding to complete oxidation of all carbon to CO_2 is 2.84 g g^{-1} polymer. Interestingly, the degree of completeness of combustion in the cone calorimeter (as measured by conversion into CO_2) is 83.1% and 83.8% in the Purser furnace.

The results for smoke yields from the cone calorimeter obtained by Cross *et al.*[3,8] have been compared to those obtained in this work. However, the basis used in the cone calorimeter software for determining the SEA uses the mass loss rate to calculate the smoke yield per kg of sample, where the tube furnace method uses the fuel feed rate. These differ when studying hydrated metal oxide fillers in that some of the mass is lost as water (~20% for ATH and ~18% for MH), and therefore the SEA values will be proportionately lower. The cone calorimeter data showed that the EVA/ATH/ZHS combination resulted in lower smoke levels, the opposite of those found in the Purser furnace, Figure 8. This illustrates a significant difference between these two physical fire models. It can be concluded from this that prediction of smoke yield for a particular series of materials will not be reliable if measurement is made under a single condition, because smoke yields are highly dependent on fire type.

For pure EVA in the cone calorimeter the average CO yield was 0.032 g g^{-1} of polymer. The $\phi = 0.7$ condition in the Purser furnace also produces 0.032 g g^{-1} of polymer (Figure 5). The CO yield under well ventilated burning is around 0.005–0.01 g g^{-1} of polymer rising to 0.14–0.18 g g^{-1} of polymer for samples containing MH (Figure 6). For the samples containing ATH the CO yield rises from 0.005–0.01 g g^{-1} of polymer under well ventilated conditions to

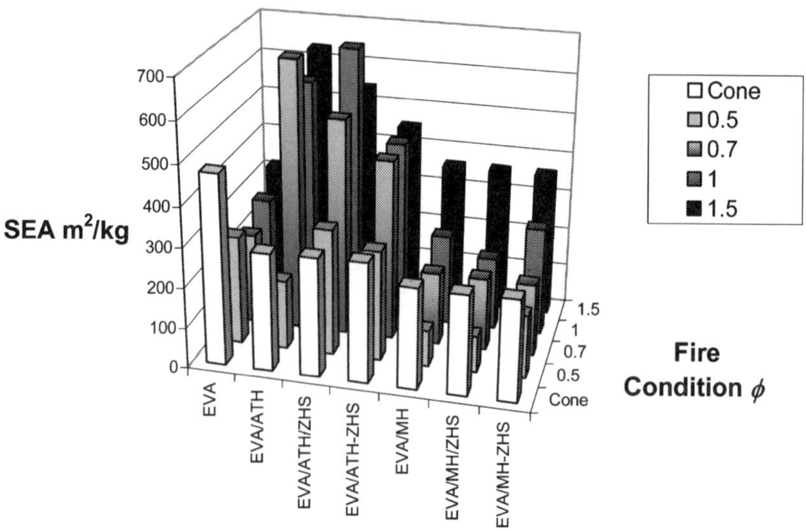

Figure 8 *Comparison of smoke data from cone calorimeter and Purser furnace*

0.18–0.33 g g^{-1} of polymer under fuel-rich conditions (Figure 5). The high CO yield observed for EVA/ATH alone is believed to arise from the freshly formed alumina catalysing afterglow of the polymer residue, resulting in incomplete oxidation.[7] This effect is suppressed by the presence of ZHS, particularly under fuel-rich conditions, where the CO yield falls from 0.33 for EVA/ATH to 0.18 g g^{-1} of polymer for EVA/ATH with ZHS.

The secondary oxidiser provides a useful means of identifying other types of combustion products. The error associated with the % carbon recovered from the dual reaction zones and the evenness of the steady state is less significant when compared to the orders of magnitude change in the CO yield on moving from well to poorly ventilated conditions.

29.5 Conclusions

Cusack *et al.* have indicated that ZHS and ZS could be used as a highly effective flame retardant.[3] Other work on the samples has indicated that ZHS is an effective synergist with ATH and MH under well-ventilated conditions in the cone calorimeter,[3] particularly with regard to suppression of CO and smoke. This work, investigating these materials under less well-ventilated conditions, has not shown a clear advantage of EVA/ATH/ZHS or EVA/MH/ZHS in terms of CO yield.

Different physical fire models give different smoke and CO yields. Care must be taken to choose the appropriate model when using small-scale tests to predict these yields. No clear advantage has been found in encapsulating ZHS, particularly in terms of smoke and toxicity, though there may, however, be advantages in physical properties.

The MH additives lowered the smoke levels measured by the Purser furnace more than the ATH additives did; however, at high ϕ ratios these samples did not exhibit superior smoke suppressant behaviour. The cone calorimeter ventilation conditions appear to correlate to a Purser ϕ ratio of between 0.7 and 1.0. The Purser furnace used in this work, with a tube diameter at the higher end of the range specified in the tube furnace standard, BS 7990:2003, still showed signs of quenching occurring at high ventilation.

29.6 Acknowledgements

The authors thank Drs Paul Cusack and Matthew Cross of Tin Technology for supplying the samples used in this study. One of us (CLW) thanks the Engineering and Physical Science Research Council (EPSRC) for financial support.

References

1. R.N. Rothon, in *Particulate-Filled Polymer Composites*, R.N. Rothon (ed.), Longman, Harlow, U.K., 1995, Chapter 6.

2. G.T. Linteris, V. Knyazev and V. Babushok (NIST), presented at the Halon Technical Working Conference, 2001.
3. M.S. Cross, P.A. Cusack and P.R. Hornsby, *Polym. Degrad. Stab.*, 2003, **79**, 309–318.
4. D.A. Purser, P.J. Fardell, J. Rowley, S. Vollam and B. Bridgeman, in *Proceedings of the 6th International Conference Flame Retardants '94'*, InterScience Communications, London, Jan. 1994, pp. 26–27.
5. W.M. Pitts, *Prog. Energy Combust. Sci.*, 1995, **21**, 197–237.
6. V. Babrauskas, W. Parker, G. Mullholland and W. Twilley, *Rev. Sci. Instrum.*, 1994, **65(7)**.
7. T.R. Hull, R.E. Quinn, I.G. Areri and D.A. Purser, *Polym. Degrad. Stab.*, 2002, **77**, 235–242.
8. M. Cross, Personal communication of unpublished data, 2003.
9. BS 7990, 2003, Tube furnace method for the determination of toxic products in fire effluents.

CHAPTER 30

Improved and Cost-Efficient Brominated Fire Retardant Systems for Plastics and Textiles by Reducing or Eliminating Antimony Trioxide

RUDI BORMS,[1] RONALD WILMER,[1a] MICHAEL PELED,[2] NURIT KORNBERG,[2] ROYI MAZOR,[2] YOAV BAR YAAKOV,[2] JACOB SCHEINERT[2] AND PIERRE GEORLETTE[2]

[1]Eurobrom B.V., Verrijn Stuartlaan 1, 2288 EK Rijswijk, The Netherlands
[2]DSBG, P.O. Box 180, 84101 Beer Sheva, Israel (georlettep@dsbg.com)
[a]Current address: NOVA Chemicals, Brede, The Netherlands 2288 EK Rijswijk

30.1 Introduction

Brominated fire retardants (FRs) are well known for their superiority[1] in fire safety and, in contrast to other FR types, they are not limited to specific materials. Among the synergistic agents, antimony trioxide, by far the most efficient,[2–4] further increases brominated FR effectiveness by enabling bromine to stay in the flame zone for longer periods.[5] However, for special applications, the use of antimony trioxide may be accompanied by certain detrimental effects.[6–9] Moreover, antimony trioxide prices have increased significantly since 2002. To address these needs, recent developments of new fire retardant systems are being offered to reduce or even eliminate the use of antimony trioxide. This chapter reviews some particular applications.

30.2 Polypropylene (PP)

The total worldwide consumption of PP was *ca.* 32 million mT (2002) with an estimated growth of 7.2% yr^{-1} during the following three years, but its flame retardant applications are proportionately much less important than for other resins such as styrenics and engineering thermoplastics. High standards of flame retardancy required in the electronics and building industries with PP compounds are indeed difficult to achieve at reasonable cost and with satisfactory properties. The cause of this difficulty is its high crystallinity and flammability.

SaFRon-5371, a new aliphatic bromine flame retardant system that has been introduced recently by DSBG,[10] can be formulated as an antimony trioxide free option for V-2 PP (homo- and block copolymers) and V-0 dripping homopolymer. This flame retardant system is recommended for indoor and outdoor applications where long-term UV/light stabilities are required. It is also suggested for use in the production of fibers for carpet applications where it is preferred to avoid use of antimony trioxide.

To obtain non-dripping UL 94 class V-0 PP, the high FR loading required results in loss of mechanical and physical properties.[11–13] The use of intumescent P/N technology limits thermal processing stability and is accompanied by the production of acidic species during processing often causing mold plate out, blooming and poor recycling due to water absorption.[11]

Most flame retarded PP compounds produced today are flame retarded by brominated FRs, in particular decabromodiphenyl oxide in combination with antimony trioxide.[14] The disadvantages of antimony trioxide based systems are an increase in smoke density[15] and loss in impact properties of the base resin. This is particularly true with PP block copolymers that require higher loadings of flame retardants compared with homopolymer systems.[14] Several publications describe ways to reduce antimony trioxide content in flame retarded PP compounds,[16,17] but even with the addition of 20 wt% talc, complete elimination of antimony trioxide is not achieved.

To address these limitations, DSBG has developed a hybrid filler type flame retardant system designed to produce UL 94 non-dripping class V-0 PP block copolymers without use of antimony trioxide. Typical properties of this new FR system, designated SaFRon-5202, are given in Table 1. This FR is surface treated to ensure its good dispersion during the processing steps. Unlike intumescent FR systems, it has very good thermal stability, allowing high processing temperatures during compounding and molding.

Table 1 *Properties of SaFRon-5202*

Appearance	*White to off-white micronised powder*
Active FR content (%)	98.6
Melting start, °C	>300
Specific gravity, g cm^{-3}	2.58
Thermogravimetric Analysis (TGA–10°C min^{-1} in air)	
Weight loss (%)	*Temperature* (°C)
2	355
5	376
10	395

Table 2 *Non-dripping V-0 PP block copolymer flame retarded by SaFRon-5202*

FR type	*Ref. no FR*	*SaFRon-5202 (Sb_2O_3 free)*	*Decabromo DPO (FR-1210 DSBG)*
Composition (wt%)			
PP Capilene SL50	100	46	61.5
Flame retardant	–	54	27
Antimony trioxide	–	–	11.5
Properties			
Flame retardancy			
UL 94 (1.6 mm) class	NR	V-0	V-0
MFI (230°C–2.16 kg), g (10 min)$^{-1}$	9	8	9
Tensile properties			
Strength at yield (MPa)	25	20	20
Elongation at break (%)	74	55	45
Modulus (MPa)	1200	2800	1600
Notched IZOD, J m^{-1}	120	73	47
HDT (1820 kPa), °C	50	57	55
NBS smoke density (ASTM E662-flaming mode)	100	400	>900

The main advantages offered by SafRon-5202 in PP applications are UV stability, low smoke generation, good impact properties and cost efficiency. Table 2 gives comparative formulations and properties for SaFRon-5202 *versus* the classical decabromodiphenyl oxide/antimony trioxide system.

30.3 High Impact Polystyrene (HIPS)

Flame retarded grades of HIPS are widely used for the production of housings of electric and electronic equipments (television, PC monitors, audio and video). Decabromodiphenyl oxide, by far the most popular FR for TV housing applications, is cost efficient but has poor UV stability and is not melt blendable during injection molding. It is thus not suitable for the production of large size TV housings with light grey colour recently introduced in the market.

For this case, tris(tribromophenyl) cyanurate (FR-245 DSBG), a joint development between the Japanese Company Dai-Ichi Kogyo Seiyaku (DKS) and DSBG, offers an optimal balance of properties: high melt flow during injection molding, excellent light stability, good impact properties and high heat distortion temperature (HDT). Its main properties and chemical structure are given in Figure 1.

The combination of 67% aromatic bromine and a cyanurate segment provides good flame retardant efficiency and UV/light stability. Use of FR-245 also enhances flow during injection molding as it melts during the process. In addition to its good thermal stability, FR-245 is designed and developed to be environmentally friendly.

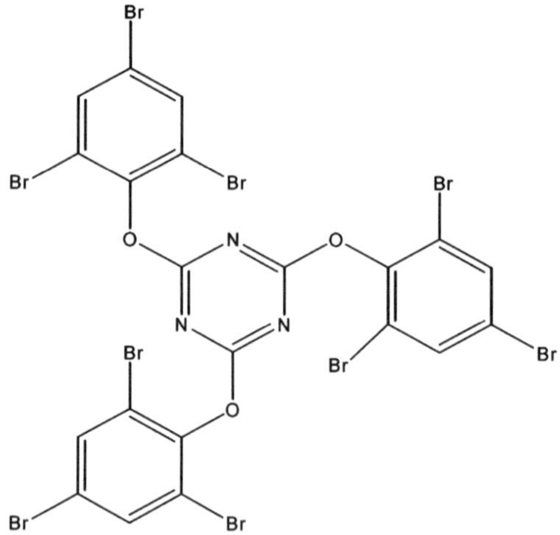

Bromine content, wt.-%	67
Nitrogen content, wt.-%	4
Melting point, °C	230
Specific gravity	2.44
Molecular weight	1067

Thermogravimetric analysis (TGA-10°C under air):

Weight loss, %	Temperature, °C
2	360
5	385
10	400

Figure 1 *Chemical structure and properties of FR-245 (Tris(tribromophenyl) cyanurate)*

However, for thin-wall molded parts, higher loadings of antimony trioxide would be needed to eliminate the dripping with this melt-blendable FR and this might be a cause for excessive after-glow.[18]

A proprietary modification of tris(tribromophenyl) cyanurate developed by DSBG allows processors to produce flame retarded HIPS UL 94 class V-0 (1.6 mm) with less antimony trioxide, shorter after-glow and no dripping as can be seen from results shown in Table 3.

In a previous publication,[10] a method to produce antimony trioxide free HIPS able to reach class V-2 according to the standard UL 94 has been described by the use of a heat-stabilized grade of hexabromocyclododecane with improved thermal stability and better corrosion resistance.

Table 3 *Properties of HIPS flame retarded by tris(tribromophenyl) cyanurate*

Type	*tris(Tribromophenyl) cyanurate (FR-245-DSBG)*	*Modified tris(tribromophenyl) cyanurate (Proprietary DSBG)*	
Composition (wt%)			
HIPS (Styron 472 – Dow)	79.9	79.9	81.6
Flame retardant	16.4	17.4	15.9
Antimony trioxide	3.7	2.7	2.5
Bromine content (%)	11	11.5	10.5
Flame retardancy (UL 94–1.6 mm)			
Total flaming time (s)	10	9	25
Total after-glow (s)	34	0	0
Non flaming dripping	1	0	0
Class	V-0	V-0	V-0

30.4 Styrenic Copolymers

Tribromophenol end-capped brominated epoxies have been designed to ensure optimal properties in styrenic copolymers (ABS and HIPS) and their alloys. F-3020 (properties in Figure 2) is a melt blendable flame retardant combining good IZOD notched impact properties with UV and light stability.

It is suitable for UL-94 V-0 ratings. A particular advantage of F-3020 is freedom from metal adhesion problems during lengthy injection molding operations.

However, ABS formulations containing tribromophenol end-capped brominated epoxies have very low Gardner impact properties. DSBG is presently

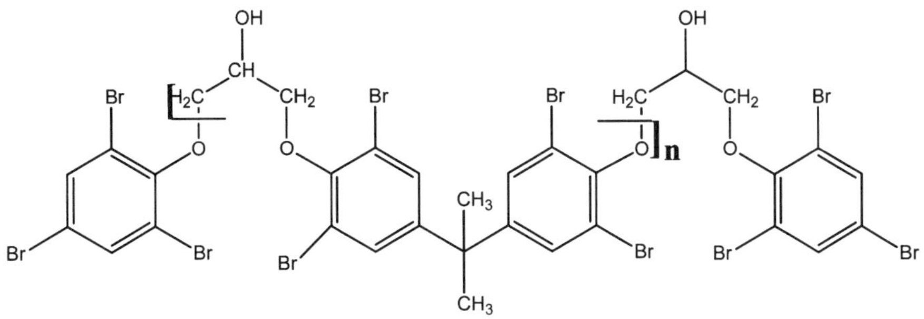

Bromine content, wt %	56
Softening range, ^{0}C	105-120
Specific gravity	1.9
Molecular weight	2000

Figure 2 *Chemical structure and properties of F-3020. tribromophenol end-capped brominated epoxy*

Table 4 *Properties of ABS flame retarded by tribromophenol end-capped brominated epoxies*

Type	*Tribromophenol end-capped BEO (F-3020-DSBG)*	*Modified tribromophenol end-capped BEO*
Composition (wt%)		
ABS medium impact	75.9	79.6
Flame retardant	17.8	16.2
Antimony trioxide	6.0	3.9
UV absorber	0.3	0.3
Bromine content (%)	10	9
Properties		
Flame retardancy		
UL-94, class (1.6 mm)	V-0	V-0
MFI (220°C-10 kg), g $(10\ \text{min})^{-1}$	28	25
Tensile		
Maximum strength (MPa)	41	42
Elongation at break (%)	3.5	4
Modulus (MPa)	2300	2600
IZOD notched impact (J m^{-1})	92	105
Gardner impact (kg cm)	16	85
HDT (1.81 MPa annealed) (°C)	91	90
UV stability, Delta E (Xenotest 300h-ASTM D4459-93)	8	7

developing proprietary modified grades of tribromophenol end-capped brominated epoxies that have enhanced fire retardant properties, providing class V-0 with less antimony trioxide. This reduction is accompanied by a significant improvement in Gardner impact properties (Table 4).

30.5 Polyamide

Data in a recent paper,[19] show that brominated flame retardants are superior to halogen-free systems for engineering resins in combining the highest standards of fire-safety (V-0) with demanding thermomechanical performance.

Moreover, according to incineration studies conducted by the GSF – Research Center for Environment and Health and the Technishe University of Munich, the toxicity of the combustion products from an engineering resin fire retarded by a brominated fire retardant is several times lower than for a resin fire retarded by a halogen-free system based on phosphinic acid salt and than beech wood used as a reference.[20] Non-FR engineering resins also emit highly toxic smoke during incineration.

Brominated trimethylphenyl indan (see properties in Figure 3), is a proprietary flame retardant, designated FR-1808, offered by DSBG and is particularly suitable for use with polyamide 6 and 6,6 with or without fiber

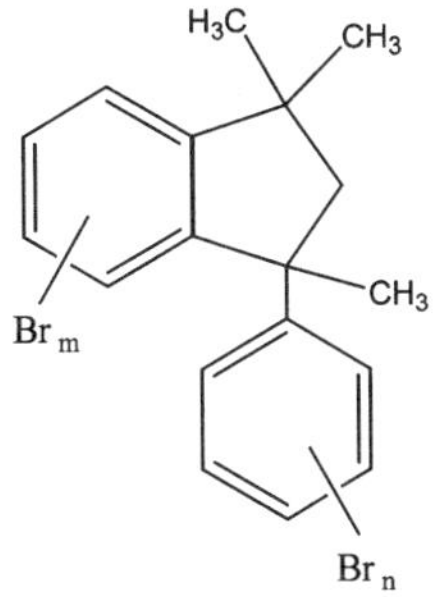

m+n=7;8

Bromine content, wt.-%	73
Melting range, °C	240-255
Specific gravity	2.6
Molecular weight	857

Figure 3 *Chemical structure and properties of FR-1808, brominated trimethylphenyl indan*

reinforcement. FR-1808 exhibits inherent advantages over other halogenated FR additives currently used for the same applications, as a result of its chemical structure, high bromine content and good thermal stability. In addition, the processability of polymers containing FR-1808 is very good.

The use of FR-1808 is advantageous when the following properties are required: cost/effective flame retardancy, good temperature stability, easy processability and high melt flow properties (for production of parts with thin walls and/or large dimensions with short injection molding cycles and high precision), and good impact properties and electrical properties

For special applications in nylon, the market is looking for cost efficient antimony trioxide free FR systems in order to get better electrical properties such as a high tracking index while maintaining the highest level of fire safety. SaFRon-5201 is a proprietary fire retardant sytem using the FR efficiency of brominated trimethylphenyl indane to address this need. Typical properties of this FR system, designated SaFRon-5201, are given in Table 5. This FR is

Table 5 *Properties of SaFRon-5201*

Appearance	*White to off-white powder*
Active FR content (%)	98.9
Melting start (°C)	240
Specific gravity (g cm^{-3})	2.5
Thermogravimetric analysis (TGA – 10°C min^{-1} in air)	
Weight loss (%)	*Temperature* (°C)
2	301
5	333
10	350

Table 6 *Glass-reinforced polyamide 6,6 flame retarded by SaFRon-5201*

FR type	*Ref. no FR*	*SaFRon-5201 (Sb_2O_3 free)*	*Brominated trimethylphenyl indane (FR-1808 DSBG)*
Composition (wt%)			
Polyamide	85	48.5	60.9
Glass fiber	15	15	15
Flame retardant	–	36.5	16.5
Antimony trioxide	–	–	7.6
Properties			
Flame retardancy:			
UL 94 (0.8 mm) class	NR	V-0	V-0
Limited oxygen index (LOI) (%)	22	57	51
Spiral flow (inch at 310°C)	43	39	58
Tensile properties			
Strength at break (MPa)	115	125	117
Elongation at break (%)	2.1	2.4	2.4
Modulus (MPa)	6600	6400	7300
Flexural properties			
Strength (MPa)	195	197	182
Modulus (MPa)	5300	8200	6100
Notched IZOD (J m^{-1})	37	53	37
Unnotched IZOD (J m^{-1})	302	350	320
HDT (1820 kPa) (°C)	242	236	240
Comparative tracking index (CTI) (V)	575	350	275

surface treated to ensure its good dispersion during the processing steps and good compatibility with the resin.

The main advantages offered by SaFRon-5201 in glass reinforced polyamide applications are high fire retardancy and tracking index, good thermomechanical properties and cost efficiency. Table 6 gives comparative formulations and properties for SaFRon-5201 *versus* the reference based on brominated triphenyl indane/antimony trioxide system.

30.6 Polycarbonate (PC) and its Alloys with ABS

PC-ABS alloys flame retarded by phosphate esters are often used in the production of halogen-free housings of E & E equipments. In this case a fire retardant of choice is resorcinol diphosphate (RDP).

Halogen-free PC ABS are not easy to compound as the fire retardant system is in a liquid state. Moreover, their heat distortion temperatures (HDT) are rather low *versus* the non-flame-retarded alloys.

DSBG found that modified grades of high molecular weight brominated epoxy enable production of fire retarded PC ABS alloys with the following advantages: class V-0 achievable without antimony trioxide, HDT very close to that of the non fire retarded alloys and high IZOD impact.

Figures 4 and 5 show the excellent performance achieved by PC ABS fire retarded by such a modified brominated epoxy in comparison with a similar one fire retarded by a RDP based system.

Similar results have also been obtained in polycarbonate (Figures 6 and 7). In this case the use of a modified high molecular weight brominated epoxy even significantly improves the impact properties of the polycarbonate while

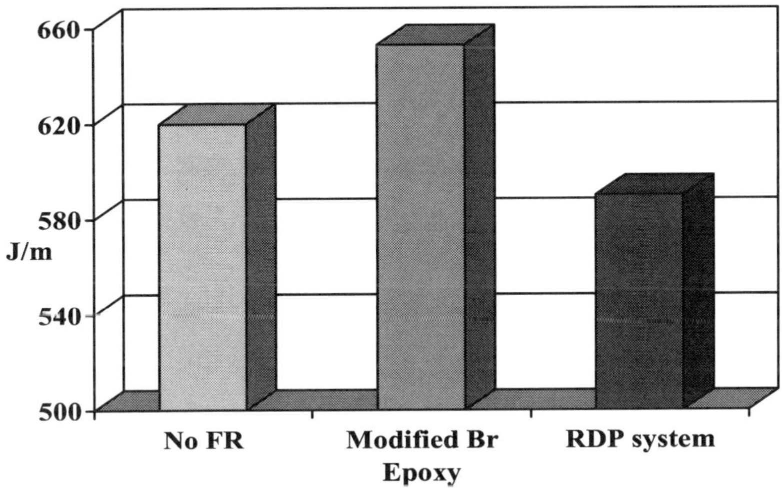

Figure 4 *IZOD notched in PC ABS (class V-0; 1.6 mm)*

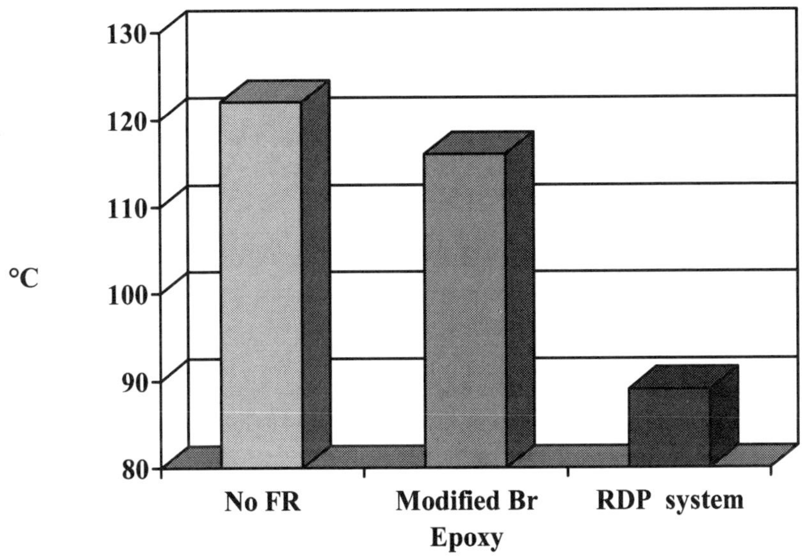

Figure 5 *HDT in PC ABS (1820 kPa-class V-0; 1.6 mm)*

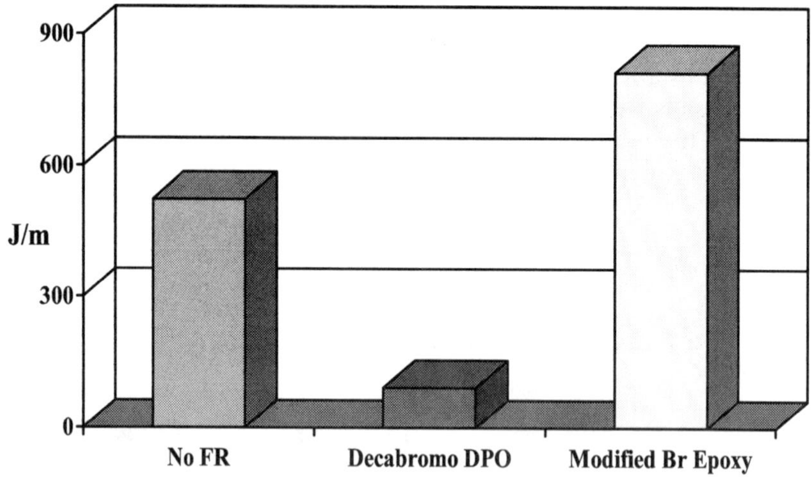

Figure 6 *IZOD notched in PC (V-0; 1.6 mm)*

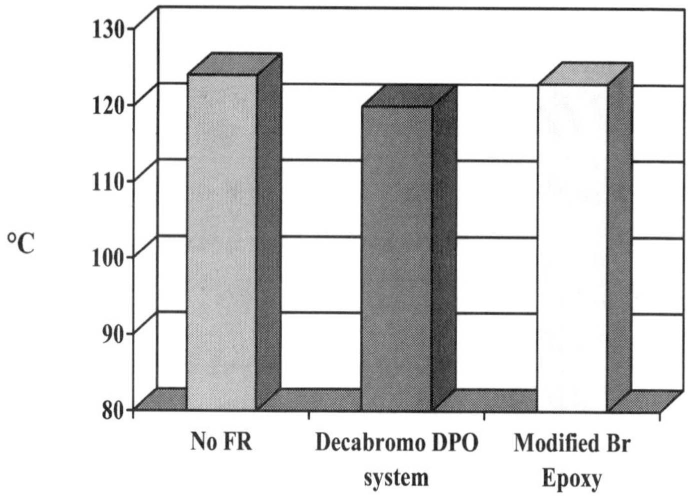

Figure 7 *HDT in PC ABS (1820 kPa-class V-0; 1.6 mm)*

maintaining the HDT. Interestingly, a similar decabromodiphenyl oxide (Decabromo DPO) based system causes a severe drop in impact.

30.7 Textile Back-Coating

Back-coating by brominated fire retardant systems is the major method for controlling efficiently the flammability of textile products.[21] This technique is used for applications such as upholstery, wall coverings and seats for the automotive industry.

Table 7 *Properties of SaFRon-5700 series*

Appearance	*White powder*
Bromine content (%)	67
Melting start (°C)	100
Specific gravity (g cm^{-3})	2.3–2.4
(TGA – 10°C min^{-1} in air)	
Weight loss (%)	*Temperature* (°C)
2	294
5	316
10	324

Antimony trioxide is used as a synergist, but problems, including price volatility, pigmentation and stiffening, have led to a search for antimony trioxide free systems. The SaFRon-5700 series are now being tested successfully in antimony trioxide free back-coating systems for application on cotton. Properties of the SaFRon-5700 series are given in Table 7.

Cotton fabric treated by FR latices containing 40% of these fire retardants with a pick-up of 50 g m^{-2} pass the severe standard BS-5852/1-2 (cigarette and match). The handle of the treated fabrics is very soft and treatment by the SaFRon-5700 series is semi-clear.

30.8 Conclusion

Antimony trioxide is considered as the most efficient synergist for brominated fire retardants. Its use has been driven by its effectiveness in reducing the amount of bromine-containing compound needed to meet a high level of fire-safety. But, for some applications, antimony trioxide has several drawbacks and there is a need to limit or avoid its use.

Some factors limiting its use in fire retardant systems are a significant price increase during 2002, a high density, melt flow reduction and detrimental effect on impact and electrical properties, loss of transparency, smoke density and after-glow and, finally, good handling.

This chapter has introduced recent developments of fire-retardant systems enabling elimination or a sharp reduction in the use of antimony trioxide. Application areas are in styrenics and their alloys, polypropylene, engineering thermoplastics and textiles.

30.9 Aknowledgement

The authors thank Drs Joe Simons, A. Staimetz, M. Manor and M. Ben Simon for their contributions, which rendered this publication possible.

30.10 References

1. R. C. Kidder, J. H. Troitzsch, E. Naumann and H. J. Roux, from Course Work Materials in New Developments and Future Trends in Europe and the United States for Fire Retardant Polymer Products, (1989).

2. E. D. Weil in: A. F. Grand & C. A. Wilkie (Eds)., Fire Retardancy of Polymeric Materials, Marcel Dekker Publisher, New-York, 2000, pp. 120–122.
3. A. R. Horrocks in: A. R. Horrocks and D. Price (Eds)., Fire retardant materials, Woodhead Publishing limited, Cambridge England, 2001, pp. 148–149.
4. H. Zweifel in: Plastics Additives Handbook, Hanser 5th Edition, 2000, pp. 682–684.
5. J. W. Hastie in: High Temperature Vapors, Academic Press, Inc., New York, 1975, p. 353.
6. Wern-Shiarng Jou *et al.* in: The influence of red phosphorus upon the flame properties and dielectric properties of glass fiber reinforced nylon-66, SPE-ANTEC, 1520 (2001), Society of Plastics Engineers, Brookfield, CT, USA.
7. P. S. Murfitt *et al.* in: Some alternative Synergists for Halogen-Containing flame retardants polymers, Flame Retardants 92, The Plastics and Rubber Institute, Elsevier Applied Science 1992, pp. 176–186.
8. R. Schmidt, R. Herbit, M. Amberg in: Smoke Gets in Your Eyes, Conference Proceedings ANTEC 98, Society of Plastic Engineers, CD Rom.
9. C. Wild: Spin Finishes for Polypropylene Staple Fibres used in the Spunlace Process, Proceedings of the Polypropylene in Textiles World Congress; Huddersfield 5–6th July 2000, Nonwoven.co.uk, Eathorpe, Warwickshire, UK.
10. T. Geran, I. Finberg, G. Reznick, S. Hini, D. Plewinsky and Y. Bar Yaakov: Development for fire retarded plastics with reduced or no presence of antimony trioxide, the 14th annual BCC Conference on Flame retardancy, June 2–4, 2003, Business Communications Company, Inc., Norwalk, Connecticut, USA.
11. J. F. Day in: Improved non-halogen fire retardant technology for polyolefins, Proceedings of the Fire Retardant Chemicals Association (March 12–15, 2000), pp129–137, Fire Retardant Chemicals Association, Lancaster, Pennsylvania, USA.
12. M. J. Keough in: Past, Present and Future Developments in Flame Retarded Polyolefins, Proceedings of Business Communications Co., Inc. Second Symposium, Stamford CT, (May 1991), Business Communications Company, Inc., Norwalk, Connecticut, USA.
13. M. Tono and M. Ogasa in: Fire-resistant Polyolefin Resin Compositions containing Ammonium Polyphosphate and Metal Oxides, Japanese patent 7330968 (1994).
14. S. Munro and R. Farner in: High performance N-P Flame retardants for PP, Proceedings of Maack Polypropylene 2001, (11th–13th September 2001), Maack Business Services, CH-8804 AU, Switzerland.
15. Baljinder K. Kandola and A. Richard Horrocks in: A. R. Horrocks and D. Price (Eds)., Fire retardant materials, Woodhead Publishing Limited, Cambridge England, 2001 p. 198.
16. R. L. Markezich and R. F. Mundhenke in: Review of synergist used with halogen flame retardants, Proceedings of the Fire Retardant Chemicals

Association, October 26–29, 1997, pp. 1–11, Fire Retardant Chemicals Association, Lancaster, Pennsylvania, USA.
17. D. M. Schubert in: The use of borates as fire retardant synergists in talc-filled polypropylene, Proceedings of the Fire Retardant Chemicals Association, March 22–25, 1998, pp 185–194, Fire Retardant Chemicals Association, Lancaster, Pennsylvania, USA.
18. R. C. Nametz in: Bromine compounds for flame retarding polymer compositions, Proceedings of the Fire Retardant Chemicals Association, March 28–30, 1984, pp 55–131, Fire Retardant Chemicals Association, Lancaster, Pennsylvania, USA.
19. M. Wagner in: Modern Plastics International February 2003, pp 70–71.
20. N. Milanov, K. Doods, K. W. Schramm, G. Matuschek, D. Lenoir and A. Kettrup in: Comparison between halogenated and phosphorus-containing flam retardants in polybutyleneterephthalate: toxicological and ecotoxicological evaluation of the combustion products, Organohalogen Compounds, Vol. 55 (2002).
21. L. Costa, P. Georlette and J. Simons, in: Arthur F. Grand and Charles A. Wilkic (Eds)., Fire Retardancy of Polymeric Materials, Marcel Dekker, New-York, 2000, pp. 279–280.

Subject Index